GUIDED NOTEBOOK WITH INTEGRATED REVIEW

BEGINNING AND INTERMEDIATE ALGEBRA

EIGHTH EDITION

Margaret L. Lial
American River College

John Hornsby
University of New Orleans

Terry McGinnis

Callie J. Daniels
St. Charles Community College

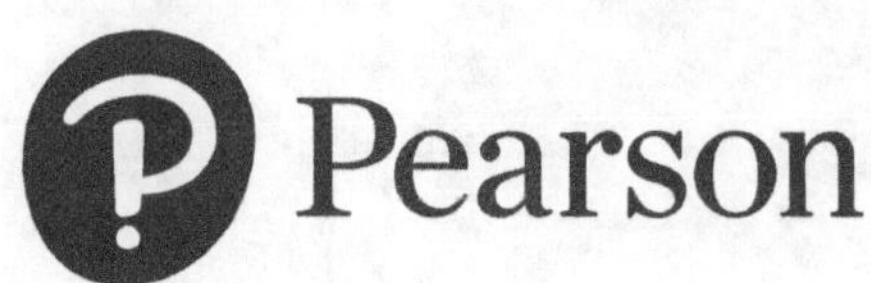

ISBN-13: 978-0-13-823388-4
ISBN-10: 0-13-823388-8

Contents

Name: Date:
Instructor: Section:

Chapter R PREALGEBRA REVIEW

R.1   Fractions

Learning Objectives	
1	Write numbers in factored form.
2	Write fractions in lowest terms.
3	Convert between improper fractions and mixed numbers.
4	Multiply and divide fractions.
5	Add and subtract fractions.
6	Solve applied problems that involve fractions.
7	Interpret data in a circle graph.

Key Terms

Use the vocabulary terms listed below to complete each statement in exercises 1−9.

numerator	denominator	proper fraction
improper fraction	**equivalent fractions**	**lowest terms**
prime number	**composite number**	**prime factorization**

1. Two fractions are _________________________________ when they represent the same portion of a whole.

2. A fraction whose numerator is larger than its denominator is called an _____________________________.

3. In the fraction $\frac{2}{9}$, the 2 is the _________________________________.

4. A fraction whose denominator is larger than its numerator is called a _____________________________.

5. The _________________________________ of a fraction shows the number of equal parts in a whole.

6. A _________________________________ has at least one factor other than itself and 1.

7. In a _________________________________ every factor is a prime number.

8. The factors of a _________________________________ are itself and 1.

9. A fraction is written in _________________________________ when its numerator and denominator have no common factor other than 1.

Name: Date:
Instructor: Section:

Objective 1   Write numbers in factored form.

Review these examples for Objective 1:   **Now Try:**
1. Identify each number as *prime*, *composite*, or   1. Identify each number as *prime*,
    neither. If the number is composite, write it as a   *composite*, or *neither*. If the
    product of prime factors.   number is composite, write it as
                                                               a product of prime factors.

    a. 41   **a.** 37

    There are no natural numbers other than 1 and 41
    itself that divide *evenly* into 41, so the number
    41 is prime.
                                                          ____
    c. 54   **c.** 210

    The number 48 is composite. We show a factor
    tree on the right, with prime factors boxed.
                                                          ____
    Divide by the   54
    least prime factor   /   \
    of 54, which is 2. $54 = 2 \cdot 27$ 2 · 27
                                         / \
    Divide 27 by 3   3 · 9
    to find two $54 = 2 \cdot 3 \cdot 9$
    factors of 27.   / \
                                          3 · 3
    Now factor $54 = 2 \cdot 3 \cdot 3 \cdot 3$
    9 as $3 \cdot 3$.

Objective 1 Practice Exercises

For extra help, see Example 1 on page 2 of your text.

Identify each number as prime, composite, *or* neither. *If the number is composite, write it
as a product of prime factors.*

   1. 98   1. _____________________

   2. 101   2. _____________________

   3. 546   3. _____________________

Objective 2   Write fractions in lowest terms.

Review this example for Objective 2:   **Now Try:**
2. Write the fraction in lowest terms.   2. Write the fraction in lowest
                                                               terms.

    $$\frac{15}{25}$$ $$\frac{9}{15}$$

    $$\frac{15}{25} = \frac{3 \cdot 5}{5 \cdot 5} = \frac{3}{5} \cdot \frac{5}{5} = \frac{3}{5} \cdot 1 = \frac{3}{5}$$

                                                          ____

Name: Date:

Instructor: Section:

Objective 2 Practice Exercises

For extra help, see Example 2 on pages 3–4 of your text.

Write each fraction in lowest terms.

4. $\dfrac{42}{150}$ 4. _______________

5. $\dfrac{180}{216}$ 5. _______________

6. $\dfrac{132}{292}$ 6. _______________

Objective 3   Convert between improper fractions and mixed numbers.

Review these examples for Objective 3:

Now Try:

3. Write $\dfrac{53}{6}$ as a mixed number.

We divide the numerator of the improper fraction by the denominator.

$$6\overline{)53} \quad \dfrac{53}{6} = 8\dfrac{5}{6}$$

$$\begin{array}{r} 8 \\ 6\overline{)53} \\ \underline{48} \\ 5 \end{array}$$

3. Write $\dfrac{74}{5}$ as a mixed number.

____

4. Write $5\dfrac{3}{8}$ as an improper fraction.

We multiply the denominator of the fraction by the whole number and add the numerator to get the numerator of the improper fraction.

$$8 \cdot 5 + 3 = 40 + 3 = 43$$

The denominator of the improper fraction is the same as the denominator in the mixed number, which is 8 here. Thus, $5\dfrac{3}{8} = \dfrac{43}{8}$

4. Write $12\dfrac{2}{7}$ as an improper fraction.

____

Name: Date:

Instructor: Section:

Objective 3 Practice Exercises

For extra help, see Examples 3–4 on page 4 of your text.

Write the improper fraction as a mixed number.

7. $\dfrac{321}{15}$ 7. _________________

Write each mixed number as an improper fraction.

8. $13\dfrac{5}{9}$ 8. _________________

9. $22\dfrac{2}{11}$ 9. _________________

Objective 4   Multiply and divide fractions.

Review these examples for Objective 4:

5. Find the product, and write it in lowest terms.

$$\frac{5}{12}\cdot\frac{3}{10}$$

$$\frac{5}{12}\cdot\frac{3}{10}=\frac{5\cdot 3}{12\cdot 10} \qquad \text{Multiply numerators.}$$
$$\text{Multiply denominators.}$$

$$=\frac{5\cdot 3}{4\cdot 3\cdot 2\cdot 5} \qquad \text{Factor the denominator.}$$

$$=\frac{1}{4\cdot 2} \qquad \frac{3}{3}=1 \text{ and } \frac{5}{5}=1$$

$$=\frac{1}{8} \qquad \text{Write in lowest terms.}$$

6. Find the quotient, and write it in lowest terms.

$$\frac{2}{5}\div\frac{8}{7}$$

$$\frac{2}{5}\div\frac{8}{7}=\frac{2}{5}\cdot\frac{7}{8} \qquad \text{Multiply by the reciprocal.}$$

$$=\frac{2\cdot 7}{5\cdot 4\cdot 2} \qquad \text{Multiply and factor.}$$

$$=\frac{7}{20}$$

Now Try:

5. Find the product, and write it in lowest terms.

$$\frac{7}{15}\cdot\frac{3}{14}$$

____

6. Find the quotient, and write it in lowest terms.

$$\frac{6}{7}\div\frac{9}{8}$$

____

Objective 4 Practice Exercises

For extra help, see Examples 5–6 on pages 5–7 of your text.

Find each product or quotient, and write it in lowest terms.

10. $\dfrac{25}{11} \cdot \dfrac{33}{10}$

10. _______________________

11. $\dfrac{5}{4} \div \dfrac{25}{28}$

11. _______________________

12. $4\dfrac{3}{8} \cdot 2\dfrac{4}{7}$

12. _______________________

Objective 5   Add and subtract fractions.

Review these examples for Objective 5:

7. Find the sum, and write it in lowest terms as needed.

$$\frac{5}{24} + \frac{7}{24}$$

Add numerators. Keep the same denominator.
$$\frac{5}{24} + \frac{7}{24} = \frac{5+7}{24} = \frac{12}{24}, \text{ or } \frac{1}{2}$$
Write in lowest terms.

8. Find the sum, and write it in lowest terms as needed.

$$\frac{5}{21} + \frac{3}{14}$$

Step 1 To find the LCD, factor the denominators to prime factored form.
$$21 = 3 \cdot 7 \quad \text{and} \quad 14 = 2 \cdot 7$$
7 is a factor of both denominators.

$$\begin{array}{cc} 21 & 14 \\ \wedge & \wedge \end{array}$$

Step 2 LCD $= 3 \cdot 7 \cdot 2 = 42$
In this example, the LCD needs one factor of 3, one factor of 7 and one factor of 2.

Now Try:

7. Find the sum, and write it in lowest terms as needed.

$$\frac{5}{16} + \frac{7}{16}$$

____

8. Find the sum, and write it in lowest terms as needed.

$$\frac{7}{12} + \frac{3}{8}$$

____

Step 3 Now we can use the second property of 1 to write each fraction with 42 as the denominator.

$$\frac{5}{21} = \frac{5}{21} \cdot \frac{2}{2} = \frac{10}{42} \quad \text{and} \quad \frac{3}{14} = \frac{3}{14} \cdot \frac{3}{3} = \frac{9}{42}$$

Now add the two equivalent fractions to get the sum.

$$\frac{5}{21} + \frac{3}{14} = \frac{10}{42} + \frac{9}{42}$$
$$= \frac{19}{42}$$

9. Find the difference, and write it in lowest terms as needed.

$$\frac{26}{9} - \frac{5}{9}$$

Subtract numerators. Keep the same denominator.

$$\frac{26}{9} - \frac{5}{9} = \frac{26-5}{9}$$
$$= \frac{21}{9}$$
$$= \frac{7}{3}, \text{ or } 2\frac{1}{3}$$

9. Find the difference, and write it in lowest terms as needed.

$$\frac{11}{18} - \frac{7}{18}$$

____

Objective 5 Practice Exercises

For extra help, see Example 7–9 on pages 7–10 of your text.

Find each sum or difference, and write it in lowest terms.

13. $\dfrac{23}{45} + \dfrac{47}{75}$

13. _______________

14. $2\dfrac{3}{4} + 7\dfrac{2}{3}$

14. _______________

15. $12\dfrac{5}{6} - 7\dfrac{7}{8}$

15. _______________

Objective 6   Solve applied problems that involve fractions.

| **Review this example for Objective 6:** | **Now Try:** |

Review this example for Objective 6:

10. Pauline and her two children picked cherries. Pauline picked $2\frac{3}{4}$ quarts, Jennie picked $1\frac{2}{3}$ quarts, and Dan picked $1\frac{1}{2}$ quarts. How many quarts of cherries did they pick?

Use Method 2.

$$2\frac{3}{4} = 2\frac{9}{12}$$
$$1\frac{2}{3} = 1\frac{8}{12}$$
$$+1\frac{1}{2} = 1\frac{6}{12}$$
$$4\frac{23}{12}$$

Since $\frac{23}{12} = 1\frac{11}{12}$, $4\frac{23}{12} = 4 + 1\frac{11}{12} = 5\frac{11}{12}$ quarts.

Now Try:

10. A punch is made with $3\frac{1}{3}$ cups of ginger ale, $1\frac{1}{2}$ cups of orange juice, $1\frac{2}{3}$ cups of lemonade, and $2\frac{1}{4}$ cups of pineapple juice. Find the total number of cups in the punch.

____

Objective 6 Practice Exercises

For extra help, see Example 10 on page 10 of your text.

Solve each applied problem. Write each answer in lowest terms.

16.   Arnette worked $24\frac{1}{2}$ hours and earned \$9 per hour. How much did she earn?

16. _____________________

17.   Debbie made a shirt with $3\frac{1}{8}$ yards of material, a dress with $4\frac{7}{8}$ yards, and a jacket with $3\frac{3}{4}$ yards. How many yards of material did she use?

17. _____________________

18.   Three sides of a parking lot are $35\frac{1}{4}$ yards, $42\frac{7}{8}$ yards, and $32\frac{3}{4}$ yards. If the total distance around the lot is $145\frac{1}{2}$ yards, find the length of the fourth side.

18. _____________________

Objective 7   Interpret data in a circle graph.

In August, 1300 workers were surveyed on where they eat during their lunch time. The circle graph shows the approximate fractions of locations.

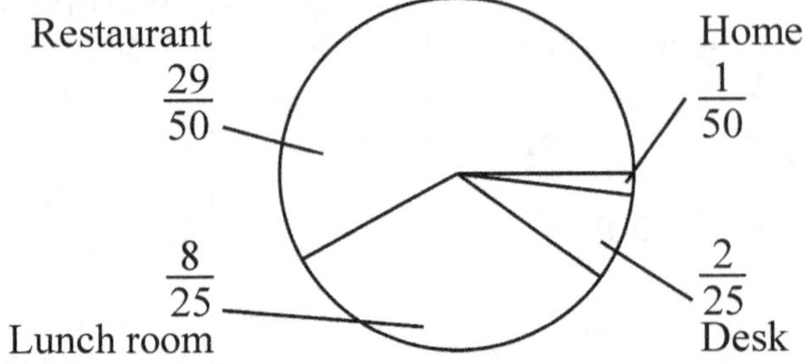

Review these examples for Objective 7:

11.

a. Which location has the largest share of workers? What was the share?

In the circle graph, the sector for Restaurant is the largest, so Restaurant had the largest share of workers, $\frac{29}{50}$.

b. How many actual workers ate in the Lunch Room?

Multiply the actual fraction from the graph of the Lunch Room by the number of workers surveyed.

$$\frac{8}{25}\cdot 1300 = \frac{8}{25}\cdot\frac{1300}{1}$$
$$= \frac{10,400}{25}$$
$$= 416$$

Thus, 416 workers ate in the Lunch Room.

Now Try:

11.

a. Which location had the smallest share of workers? What was the share?

____

b. How many actual workers ate at their Desk?

____

Objective 7 Practice Exercises

For extra help, see Example 11 on page 11 of your text.

In August, 1300 workers were surveyed on where they eat during their lunch time. The circle graph shows the approximate fractions of locations.

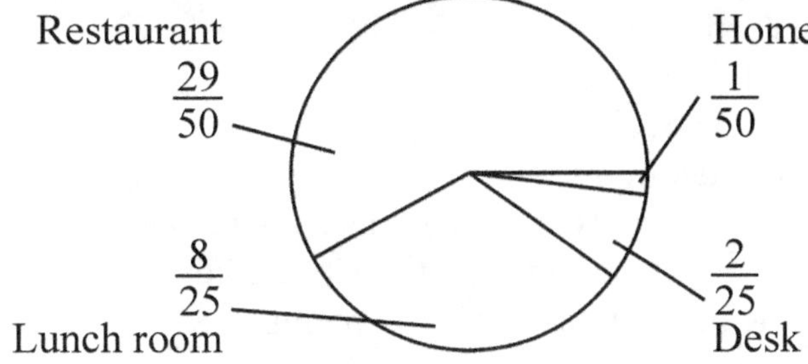

19. What location had the second-largest number of workers?

19. _______________

20. Estimate the number of workers who eat at a Restaurant.

20. _______________

21. How many actual workers ate at a Restaurant?

21. _______________

Chapter R PREALGEBRA REVIEW

R.2   Decimals and Percents

Learning Objectives
1.   Write decimals as fractions.
2.   Add and subtract decimals.
3.   Multiply and divide decimals.
4.   Write fractions as decimals.
5.   Write percents as decimals and decimals as percents.
6.   Write percents as fractions and fractions as percents.
7.   Solve applied problems that involve percents.

Key Terms

Use the vocabulary terms listed below to complete each statement in exercises 1−3.

     decimals   **place value**   **percent**

1.   We use _______________________ to show parts of a whole.

2.   A _______________________________ is assigned to each place to the left or right of the decimal point.

3.   _______________________ means per one hundred.

Objective 1   Write decimals as fractions.

Review these examples for Objective 1:

1. Write each decimal as a fraction. (Do not write in lowest terms.)

   a. 0.87

   We read 0.87 as "eighty-seven hundredths," so the fraction form is $\dfrac{87}{100}$. Using the shortcut method, since there are two places to the right of the decimal point, there will be two zeros in the denominator.

$$0.87 = \frac{87}{100}$$

   2 places   2 zeros

   b. 0.043

   We read 0.043 as "forty-three thousandths."

$$0.043 = \frac{43}{1000}$$

   3 places   3 zeros

Now Try:

1. Write each decimal as a fraction. (Do not write in lowest terms.)

   a. 0.72

   ____

   b. 0.053

   ____

c. 5.3084

Here we have 4 places.
$$5.3084 = 5 + 0.3084$$

$$= \frac{50,000}{10,000} + \frac{3084}{10,000} \quad \text{The LCD is 10,000.}$$

$$= \frac{53,084}{10,000}$$

4 zeros

c. 3.7058

____

Objective 1 Practice Exercises

For extra help, see Example 1 on page 17 of your text.

Write each decimal as a fraction. Do not write in lowest terms.

1.   0.007

1. _______________

2.   18.03

2. _______________

3.   30.0005

3. _______________

Objective 2   Add and subtract decimals.

Review these examples for Objective 2:

2.   Add or subtract as indicated.

a. $7.34 + 15.6 + 2.419$

Place the digits of the numbers in columns, so that tenths are in one column, hundredths in another column, and so on.

$$
\begin{array}{r}
7.34 \\
15.6 \\
+ \; 2.419 \\
\hline
25.359
\end{array}
\quad \text{Align decimal points.}
$$

To avoid errors, attach zeros to make all the numbers the same length.

$$
\begin{array}{r}
7.34 \\
15.6 \\
+ \; 2.419 \\
\hline
\end{array}
\quad \text{becomes} \quad
\begin{array}{r}
7.340 \\
15.600 \\
+ \; 2.419 \\
\hline
25.359
\end{array}
$$

Now Try:

2.   Add or subtract as indicated.

a. $5.23 + 28.9 + 3.741$

____

b. $56.8 - 42.307$ **b.** $64.5 - 37.218$

$$\begin{array}{r} 56.8 \\ -\ 42.307 \\ \hline \end{array} \quad \text{becomes} \quad \begin{array}{r} 56.800 \\ -\ 42.307 \\ \hline 14.493 \end{array}$$

Objective 2 Practice Exercises

For extra help, see Example 2 on pages 17–18 of your text.

Add or subtract as indicated.

4. $45.83 + 20.923 + 5.7$ **4.** _______________

5. $768.5 - 13.402$ **5.** _______________

6. $689 - 79.832$ **6.** _______________

Objective 3   Multiply and divide decimals.

Review these examples for Objective 3:   **Now Try:**

3. Multiply.   **3.** Multiply.

    a. 37.4×5.26 **a.** 26.8×9.37

There is 1 decimal place in the first number and 2 decimal places in the second number. Therefore, there are $1 + 2 = 3$ decimal places in the answer.

$$\begin{array}{r} 37.4 \\ \times\ 5.26 \\ \hline 2244 \\ 748 \\ 1870 \\ \hline 196.724 \end{array}$$

    b. 0.08×0.6 **b.** 0.04×0.7

Here $8 \times 6 = 48$. There are 2 decimal places in the first number and 1 decimal place in the second number. Therefore, there are $2 + 1 = 3$ decimal places in the answer.

    $0.08 \times 0.6 = 0.048$

4. Divide.

   a. $1191.45 \div 23.5$

Write the problem as follows.

$$23.5\overline{)1191.45}$$

To change 23.5 into a whole number, move the decimal point one place to the right. Move the decimal point in 1191.45 the same number of places to the right, to get 11,914.5.

$$235\overline{)11,914.5}$$

Move the decimal point straight up and divide as with whole numbers.

$$
\begin{array}{r}
50.7 \\
235\overline{)11,914.5} \\
\underline{1175} \\
1645 \\
\underline{1645} \\
0
\end{array}
$$

b. $9.581 \div 5.78$ (Round the answer to two decimal places.)

Move the decimal point two places to the right in 5.78, to get 578. Do the same thing with 9.581 to get 958.1.

$$578\overline{)958.1}$$

Move the decimal point straight up and divide as with whole numbers.

$$
\begin{array}{r}
1.657 \\
578\overline{)958.100} \\
\underline{578} \\
3801 \\
\underline{3468} \\
3330 \\
\underline{2890} \\
4400 \\
\underline{4046} \\
354
\end{array}
$$

We carried out the division to three decimal places so that we could round to two decimal places, obtaining the quotient 1.66.

4. Divide.

   a. $2472.12 \div 76.3$

____

b. $7.648 \div 5.36$ (Round the answer to two decimal places.)

____

Objective 3 Practice Exercises

For extra help, see Examples 3–5 on pages 18–20 of your text.

Multiply or divide as indicated.

   7. 14.64×0.16 **7.** _______________

   8. $498.624 \div 21.2$ **8.** _______________

   9. $429.2 \div 1000$ **9.** _______________

Objective 4   Write fractions as decimals.

Review these examples for Objective 4:	**Now Try:**
6. Write each fraction as a decimal.	**6.** Write each fraction as a decimal.
a. $\dfrac{27}{8}$	**a.** $\dfrac{7}{20}$

Divide 27 by 8. Add a decimal point and as many 0s as necessary.

$$
\begin{array}{r}
3.375 \\
8\overline{)27.000} \\
\underline{24} \\
30 \\
\underline{24} \\
60 \\
\underline{56} \\
40 \\
\underline{40} \\
0
\end{array}
$$

$$\frac{27}{8} = 3.375$$

b. $\dfrac{26}{9}$ **b.** $\dfrac{32}{9}$

$$
\begin{array}{r}
2.888\ldots \\
9\overline{)26.000\ldots} \\
\underline{18} \\
80 \\
\underline{72} \\
80 \\
\underline{72} \\
80 \\
\underline{72} \\
8
\end{array}
$$

____

$\dfrac{26}{9} = 2.888\ldots$

The remainder is never 0. Because 8 is always left after the subtraction, this quotient is a repeating decimal. A convenient notation for a repeating decimal is a bar over the digit (or digits) that repeats.

$$\frac{26}{9} = 2.888\ldots \ \text{ or } \ 2.\overline{8}$$

Objective 4 Practice Exercises

For extra help, see Example 6 on page 20 of your text.

Write each fraction as a decimal. For repeating decimals, write the answer two ways: using the bar notation and rounding to the nearest thousandth.

10. $\dfrac{3}{7}$ **10.** ________________

11. $\dfrac{4}{9}$ **11.** ________________

12. $\dfrac{151}{200}$ **12.** ________________

                      15

Objective 5   Write percents as decimals and decimals as percents.

Review these examples for Objective 5:	**Now Try:**
9. Convert each percent to a decimal and each decimal to a percent.	9. Convert each percent to a decimal and each decimal to a percent.

9. Convert each percent to a decimal and each decimal to a percent.

 a. 54%

 $54\% = 0.54$

 b. 3%

 $3\% = 0.03$

 c. 0.29

 $0.29 = 29\%$

 d. 4.6

 $4.6 = 460\%$

Now Try:

9. Convert each percent to a decimal and each decimal to a percent.

 a. 91%

 ____

 b. 6%

 ____

 c. 0.43

 ____

 d. 5.2

 ____

Objective 5 Practice Exercises

For extra help, see Examples 7–9 on pages 21–22 of your text.

Convert each percent to a decimal and each decimal to a percent.

13.   362%   13. _______________

14.   0.4%   14. _______________

15.   0.084   15. _______________

Objective 6   Write percents as fractions and fractions as percents.

Review these examples for Objective 6:

10. Write each percent as a fraction. Give answers in lowest terms.

 a. 12%

 Recall writing 12% as a decimal.
 $12\% = 12 \div 100 = 0.12$
 Because 0.12 means 12 hundredths,

 $$0.12 = \frac{12}{100} = \frac{12 \div 4}{100 \div 4} = \frac{3}{25}$$

Now Try:

10. Write each percent as a fraction. Give answers in lowest terms.

 a. 30%

 ____

b. 350%

$$350\% = \frac{350}{100} = \frac{350 \div 50}{100 \div 50} = \frac{7}{2} = 3\frac{1}{2}$$

11. Write each fraction as a percent. Round to the nearest tenth as necessary.

a. $\frac{3}{5}$

$$\frac{3}{5} = \left(\frac{3}{5}\right)(100\%) = \left(\frac{3}{5}\right)\left(\frac{100}{1}\%\right) = \left(\frac{3}{5}\right)\left(\frac{20 \cdot 5}{1}\%\right)$$

$$= \frac{60}{1}\% = 60\%$$

b. $\frac{1}{15}$

$$\frac{1}{15} = \left(\frac{1}{15}\right)(100\%) = \left(\frac{1}{15}\right)\left(\frac{100}{1}\%\right)$$

$$= \left(\frac{1}{3 \cdot 5}\right)\left(\frac{5 \cdot 20}{1}\%\right)$$

$$= \frac{20}{3}\% = 6\frac{2}{3}\%$$

$6\frac{2}{3}\%$ rounds to 6.7%.

b. 125%

____

11. Write each fraction as a percent. Round to the nearest tenth as necessary.

a. $\frac{3}{20}$

____

b. $\frac{1}{18}$

____

Objective 6 Practice Exercises

For extra help, see Examples 10–11 on pages 22–23 of your text.

Convert each percent to a fraction and each fraction to a percent.

16.   140%

16. ________________

17.   55.6%

17. ________________

18. $\frac{11}{40}$

18. ________________

Objective 7   Solve applied problems that involve percents.

Review this example for Objective 7:

12. A calendar with a regular price of $12 is on sale at 46% off. Find the amount of the discount and the sale price of the calendar.

The discount is 46% of 12.

46%   of   12

$\downarrow$ $\downarrow$ $\downarrow$

0.46   ·   12

$= 5.52$

The discount is $5.52. The sale price is found by subtracting.

Original price – discount = sale price

$12 – $5.52 = $6.48

Now Try:

12. Olympic t-shirts are on sale at 56% off. The regular price is $19. Find the amount of the discount and the sale price?

Objective 7 Practice Exercises

For extra help, see Example 12 on page 24 of your text.

Solve each problem.

19.   Ranee bought a pair of shoes with a regular price of $70, on sale at 20% off. Find the amount of the discount and the sale price?

19. ___________________

20.   Geishe's Shoes sells shoes at $33\frac{1}{3}\%$ off the regular price. Find the price of a pair of shoes normally priced at $54, after the discount is given.

20. ___________________

21.   At the end of the season, a swim suit is on sale at 75% off. The regular price is $56. Find the amount of the discount and the sale price?

21. ___________________

Chapter 1 THE REAL NUMBER SYSTEM

1.1   Exponents, Order of Operations, and Inequality

Learning Objectives
Learning Objectives
1   Use exponents.
2   Use the rules for order of operations.
3   Use more than one grouping symbol.
4   Use inequality symbols in statements.
5   Translate word statements to symbols.
6   Write statements that change the direction of inequality symbols.

Key Terms

Use the vocabulary terms listed below to complete each statement in exercises 1−4.

exponent   base   exponential expression   inequality

1.   A number written with an exponent is an ________________________________.

2.   Examples of __________________ symbols are $\neq$, $<$, $>$, $\leq$, and $\geq$.

3.   The __________________ is the number that is a repeated factor when written
       with an exponent.

4.   An __________________ is a number that indicates how many times a factor is
       repeated.

Objective 1   Use exponents.

Review this example for Objective 1:

1.   Find the value of the exponential expression.

   6^2

   6^2 means $6 \cdot 6$, which equals 36.

Now Try:

1.   Find the value of the
     exponential expression.
     7^2

     ____

Objective 1 Practice Exercises

For extra help, see Example 1 on page 28 of your text.

Find the value of each exponential expression.

1. 3^3

     1. ________________

2. $\left(\dfrac{2}{3}\right)^4$

     2. ________________

3. $(0.4)^2$

     3. ________________

Objective 2   Use the rules for order of operations.

Review this example for Objective 2:

2.   Find the value of the expression.

$$6+7\cdot 3$$

Multiply, then add.
$$6+7\cdot 3 = 6+21$$
$$= 27$$

Now Try:

2.   Find the value of the expression.

$$5+2\cdot 9$$

____

Objective 2 Practice Exercises

For extra help, see Example 2 on pages 29–30 of your text.

Find the value of each expression.

4. $20\div 5-3\cdot 1$

4.   _______________

5. $3\cdot 5^2-3\cdot 7-9$

5.   _______________

6. $6^2\div 3^2-4\cdot 3-2\cdot 5$

6.   _______________

Objective 3   Use more than one grouping symbol.

Review this example for Objective 3:

3.   Find the value of the expression.

$$3[9+4(7+8)]$$

Start by adding inside the parentheses.
$$3[9+4(7+8)] = 3[9+4(15)] \quad \text{Add.}$$
$$= 3[9+60] \quad \text{Multiply.}$$
$$= 3[69] \quad \text{Add.}$$
$$= 207 \quad \text{Multiply.}$$

Now Try:

3.   Find the value of the expression.

$$5[3+4(8+2)]$$

____

                Copyright © 2025 Pearson Education, Inc.

Objective 3 Practice Exercises

For extra help, see Example 3 on pages 30–31 of your text.

Find the value of each expression.

7. $\dfrac{10(5-3)-9(6-2)}{2(4-1)-2^2}$

7. _______________________

8. $19-3\big[8(5-2)+6\big]$

8. _______________________

9. $4\big[5+2(8-6)\big]+12$

9. _______________________

Objective 4   Use inequality symbols in statements.

Review this example for Objective 4:

4. Determine whether the statement is true or false.

$$5\cdot6-12\le22$$

$$5\cdot6-12\le22$$
$$30-12\le22$$
$$18\le22$$

The statement is true.

Now Try:

4. Determine whether the statement is true or false.

$$7\cdot14-15\le13$$

____

Objective 4 Practice Exercises

For extra help, see Example 4 on page 32 of your text.

Tell whether each statement is true *or* false.

10. $3\cdot4\div2^2\ne3$

10. _______________________

11. $3.25>3.52$

11. _______________________

12. $2\big[7(4)-3(5)\big]\le45$

12. _______________________

Objective 5   Translate word statements to symbols.

Review this example for Objective 5:	**Now Try:**
5. Write each word statement in symbols. Thirteen is greater than or equal to nine plus four. $13 \geq 9 + 4$	5. Write each word statement in symbols. Nineteen is less than or equal to eleven plus 8. ___________

Objective 5 Practice Exercises

For extra help, see Example 5 on page 32 of your text.

Write each word statement in symbols.

13.   Seven equals thirteen minus six.   13. ___________

14.   Five times the sum of two and nine is less than one   14. ___________
       hundred six.

15.   Twenty is greater than or equal to the product of two   15. ___________
       and seven.

Objective 6   Write statements that change the direction of inequality symbols.

Review this example for Objective 6:	**Now Try:**
6. Write each statement as another true statement with the inequality symbol reversed. **a.** $9 > 7$ $9 > 7$ is equivalent to $7 < 9$	6. Write each statement as another true statement with the inequality symbol reversed. **a.** $15 > 11$ ___________

Objective 6 Practice Exercises

For extra help, see Example 6 on page 32 of your text.

Write each statement with the inequality symbol reversed.

16. $\dfrac{3}{4} > \dfrac{2}{3}$ 16. ___________

17. $12 \geq 8$ 17. ___________

18. $0.002 > 0.0002$ 18. ___________

Chapter 1 THE REAL NUMBER SYSTEM

1.2   Variables, Expressions, and Equations

Learning Objectives
1   Evaluate algebraic expressions, given values for the variables.
2   Translate word phrases to algebraic expressions.
3   Identify solutions of equations.
4   Identify solutions of equations from a set of numbers.
5   Distinguish between *equations* and *expressions*.

Key Terms

Use the vocabulary terms listed below to complete each statement in exercises 1−7.

variable	constant	algebraic expression
equation	solution	set elements

1.   A(n) ____________________ is a statement that says two expressions are equal.

2.   The objects that belong to a set are its ____________________ .

3.   A ____________________ is a symbol, usually a letter, used to represent an unknown number.

4.   A collection of numbers, variables, operation symbols, and grouping symbols is an____________________.

5.   A ____________________ is collection of objects.

6.   Any value of a variable that makes an equation true is a(n) ____________________ of the equation.

7.   A ____________________ is a fixed, unchanging number.

Objective 1   Evaluate algebraic expressions, given values for the variables.

Review these examples for Objective 1:	**Now Try:**
1. Evaluate each expression for $x = 4$.	1. Evaluate each expression for $x = 6$.
b. $3x - 5$	**b.** $4x - 7$
$\begin{aligned} 3x - 5 &= 3(4) - 5 \quad \text{Let } x = 4. \\ &= 12 - 5 \quad \text{Multiply.} \\ &= 7 \quad \text{Subtract.} \end{aligned}$	_______________
c. $5x^2$	**c.** $7x^2$
$\begin{aligned} 5x^2 &= 5 \cdot 4^2 \quad \text{Let } x = 4. \\ &= 5 \cdot 16 \quad \text{Square 4.} \\ &= 80 \quad \text{Multiply.} \end{aligned}$	_______________

2. Find the value of each expression for $x = 7$ and $y = 6$.

$$3x + 4y + 2$$

Replace x with 7 and y with 6.

$$3x + 4y + 2 = 3 \cdot 7 + 4 \cdot 6 + 2$$
$$= 21 + 24 + 2 \qquad \text{Multiply.}$$
$$= 47 \qquad \text{Add.}$$

2. Find the value of each expression for $x = 8$ and $y = 4$.

$$5x + 6y + 1$$

____

Objective 1 Practice Exercises

For extra help, see Examples 1–2 on pages 36–37 of your text.

Find the value of each expression if $x = 2$ and $y = 4$.

1. $9x - 3y + 2$

1. _______________

2. $\dfrac{2x + 3y}{3x - y + 2}$

2. _______________

3. $\dfrac{3y^2 + 2x^2}{5x + y^2}$

3. _______________

Objective 2 Translate word phrases to algebraic expressions.

Review this example for Objective 2: | **Now Try:**

3. Write the word phrase as an algebraic expression, using x as the variable.

The product of 15 and a number

$15 \cdot x$, or $15x$

3. Write the word phrase as an algebraic expression, using x as the variable.
The product of 20 and a number

____

Objective 2 Practice Exercises

For extra help, see Example 3 on pages 37–38 of your text.

Write each word phrase as an algebraic expression. Use x as the variable.

4. Ten times a number, added to 21

4. __________________

5. 11 fewer than eight times a number

5. __________________

6. Half a number subtracted from two-thirds of the number

6. __________________

Objective 3 Identify solutions of equations.

Review this example for Objective 3: | **Now Try:**

4. Decide whether the given number is a solution of the equation.

$$8n - 7(n - 4) = 41; \quad 9$$

$$8n - 7(n - 4) = 41$$
$$8 \cdot 9 - 7(9 - 4) \overset{?}{=} 41$$
$$8 \cdot 9 - 7 \cdot 5 \overset{?}{=} 41$$
$$72 - 35 \overset{?}{=} 41$$
$$37 = 41 \quad \text{False – the left side does not equal the right side.}$$

The number 9 is not a solution of the equation.

4. Decide whether the given number is a solution of the equation.

$$9m - 4(m - 3) = 41; \quad 5$$

____

Objective 3 Practice Exercises

For extra help, see Example 4 on page 38 of your text.

Decide whether the given number is a solution of the equation.

7. $5 + 3x^2 = 19;\ 2$

7. ___________________

8. $\dfrac{m+2}{3m-10} = 1;\ 8$

8. ___________________

9. $3y + 5(y-5) = 7;\ 4$

9. ___________________

Objective 4   Identify solutions of equations from a set of numbers.

Review this example for Objective 4:

5. Write the sentence as an equation. Use x as the variable. Then find the solution of the equation from the following set.
    $\{0, 2, 4, 6, 8, 10\}$

    The sum of a number and five is eleven.

    the sum of a
    number and five   is   eleven.
    $$x + 5 = 11$$
    Because $6 + 5 = 11$ is true, 6 is the only solution.

Now Try:

5. Write the sentence as an equation. Use x as the variable. Then find the solution of the equation from the following set.
    $\{0, 2, 4, 6, 8, 10\}$
    The sum of a number and seven is eleven.

    ____

Objective 4 Practice Exercises

For extra help, see Example 5 on page 39 of your text.

Write each sentence as an equation. Use x as the variable. Then find the solution of the equation from the following set. $\{1, 3, 5, 7, 9, 11\}$

10. Ten divided by a number is nine more than the number.

10. ___________________

11. Five more than a number is fourteen.

11. ___________________

12. Five times a number is 12 plus the number.

12. ___________________

          Copyright © 2025 Pearson Education, Inc.

Objective 5   Distinguish between *equations* and *expressions*.

Review this example for Objective 5:

6. Decide whether each is an equation or an expression.

$$5x - 4(x - 6)$$

Ask, "Is there an equality symbol?" The answer is no, so this is an expression.

Now Try:

6. Decide whether each is an equation or an expression.

$$2(x - 4) - 4x$$

____

Objective 5 Practice Exercises

For extra help, see Example 6 on page 40 of your text.

Identify each as an **expression** *or an* **equation**.

13. $y^2 - 4y - 3$

13. _______________________

14. $\dfrac{x + 4}{5}$

14. _______________________

15. $8x = 2y$

15. _______________________

Chapter 1 THE REAL NUMBER SYSTEM

1.3   Real Numbers and the Number Line

Learning Objectives
1   Classify numbers and graph them on number lines.
2   Use inequality symbols with real numbers.
3   Find the additive inverse of a real number.
4   Find the absolute value of a real number.
5   Interpret meanings of real numbers from a table of data.

Key Terms

Use the vocabulary terms listed below to complete each statement in exercises 1−15.

natural numbers	**whole numbers**	**number line**	**additive inverse**
integers	**negative number**	**positive number**	**signed numbers**
rational number	**set-builder notation**		**coordinate**
irrational number	**real numbers**	**absolute value**	**graph**

1.   The set {0, 1, 2, 3, …} is called the set of ________________________________.

2.   The __________________ of a number is the same distance from 0 on the number line as the original number, but located on the opposite side of 0.

3.   The whole numbers together with their opposites and 0 are called ____________________.

4.   The set { 1, 2, 3, …} is called the set of ________________________________.

5.   The ________________________________ of a number is the distance between 0 and the number on the number line.

6.   A ________________________________ shows the ordering of the real numbers on a line.

7.   A real number that is not a rational number is called a(n) __________________.

8.   The number that corresponds to a point on the number line is the ____________________ of that point.

9.   A number located to the left of 0 on a number line is a __________________.

10.   A number located to the right of 0 on a number line is a __________________.

11.   Numbers that can be represented by points on the number line are ____________________.

          Copyright © 2025 Pearson Education, Inc.

12. To ______________________ a number, place a dot on the number line at the point that corresponds to the number.

13. ______________________ uses a variable and a description to describe a set.

14. A number that can be written as the quotient of two integers is a ______________________.

15. Positive numbers and negative numbers are ______________________.

Objective 1   Classify numbers and graph them on number lines.

Review these examples for Objective 1:

1. Use an integer to express the boldface italic number in the application.

In August, 2012, the National Debt was approximately $*16* trillion.

Use –$16 trillion because "debt" indicates a negative number.

2. Graph each number on a number line.
$$-3\frac{1}{2}, \ -\frac{3}{2}, \ 0, \ \frac{7}{2}, \ 1$$

To locate the improper fractions on the number line, write them as mixed numbers or decimals.

3. List the numbers in the following set that belong to each set of numbers.
$$\left\{-6, \ -\frac{5}{6}, \ 0, \ 0.\overline{3}, \ \sqrt{3}, \ 4\frac{1}{5}, \ 6, \ 6.7\right\}$$

a. Whole numbers

Answer: 0 and 6

b. Integers

Answer: –6, 0, and 6

c. Rational numbers

Answer: $-6, \ -\frac{5}{6}, \ 0, \ 0.\overline{3}, \ 4\frac{1}{5}, \ 6, \ 6.7$

d. Irrational numbers

Answer: $\sqrt{3}$

Now Try:

1. Use an integer to express the boldface italic number in the application.
Death Valley is *282* feet below sea level.

____

2. Graph each number on a number line.
$$\frac{1}{2}, \ 0, \ -3, \ -\frac{5}{2}$$

3. List the numbers in the following set that belong to each set of numbers.
$$\left\{-10, -\frac{5}{8}, \ 0, \ 0.\overline{4}, \ \sqrt{5}, \ 5\frac{1}{2}, \ 7, \ 9.9\right\}$$

a. Whole numbers

____

b. Integers

____

c. Rational numbers

____

d. Irrational numbers

____

Name: _______________________ Date: _______________________

Instructor: _______________________ Section: _______________________

Objective 1 Practice Exercises

For extra help, see Examples 1–3 on pages 43–45 of your text.

Use a real number to express each number in the following applications.

1. Last year Nina lost 75 pounds.

1. _______________________

2. Between 1970 and 1982, the population of Norway increased by 279,867.

2. _______________________

Graph the group of rational numbers on a number line.

3. –4.5, –2.3, 1.7, 4.2

3. 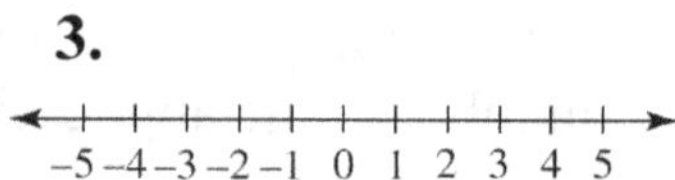

Objective 2 Use inequality symbols with real numbers.

Review this example for Objective 2:

4. Is the statement $-4 < -2$ true or false?

Because –4 is to the left of –2 on the number line, –4 is less than –2. The statement $-4 < -2$ is true.

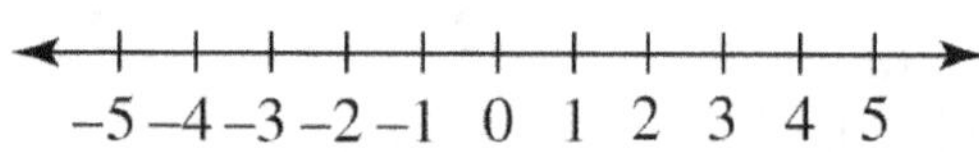

Now Try:

4. Is the statement $-10 < -8$ true or false.

____

Objective 2 Practice Exercises

For extra help, see Example 4 on page 46 of your text.

*Decide whether each statement is **true** or **false**.*

4. $-76 < 45$

4. _______________________

5. $-5 > -5$

5. _______________________

6. $-12 > -10$

6. _______________________

Objective 3 Find the additive inverse of a real number.

Review this example for Objective 3:

5. Find the additive inverse each number: 12, –6, 0.75, and 0.

Number	Additive Inverse
12	–12
–6	6
0.75	–0.75
0	0

Now Try:

5. Find the additive inverse each number: 8, –4, 0.26, and 0.

____

Objective 3 Practice Exercises

For extra help, see pages 46–47 of your text.

Find the additive inverse of each number.

  7. -25 **7.** ________________

  8. $\dfrac{3}{8}$ **8.** ________________

  9. 4.5 **9.** ________________

Objective 4   Find the absolute value of a real number.

Review these examples for Objective 4:

6.   Simplify by finding the absolute value.

    a. $|16|$

    $|16| = 16$

    b. $|-16|$

    $|-16| = -(-16) = 16$

    c. $-|-16|$

    $-|-16| = -(16) = -16$

Now Try:

6.   Simplify by finding the absolute value.

    a. $|-10|$

    ____

    b. $-|10|$

    ____

    c. $|10-7|$

    ____

Objective 4 Practice Exercises

For extra help, see Example 6 on page 48 of your text.

Simplify.

  10. $-|49-39|$ **10.** ________________

  11. $|-7.52+6.3|$ **11.** ________________

  12. $|16-14|$ **12.** ________________

Objective 5   Interpret meanings of real numbers from a table of data.

Review this example for Objective 5:

7. In the table, which category represents a decrease for both years?

Category	Change from 2012 to 2013	Change from 2013 to 2014
Eggs	−0.3	3.9
Milk	1.6	0.7
Orange Juice	−8.8	−3.9
Electricity	0.8	3.9

Source: U.S. Bureau of Labor and Statistics

Since a decrease implies a negative number, the category Orange Juice has a negative number for both years. So the answer is Orange Juice.

Now Try:

7. In the table to the left, which category represents an increase for both years?

____

Objective 5 Practice Exercises

For extra help, see Example 7 on page 48 of your text.

The Consumer Price Index (CPI) measures the average change in prices of goods and services purchased by urban consumers in the United States. The table shows the percent change in CPI for selected categories of goods and services from 2012 to 2013 and from 2013 to 2014. Use the table to answer each question.

Category	Change from 2012 to 2013	Change from 2013 to 2014
Gasoline	−1.2	−0.9
Eggs	−0.3	3.9
Milk	1.6	0.7
Electricity	0.8	3.9

13. Which category represents a decrease for both years?

13. ________________

14. Which category in which year represents the greatest percent decrease?

14. ________________

15. Which category in which year represents the least change?

15. ________________

                    Copyright © 2025 Pearson Education, Inc.

Chapter 1 THE REAL NUMBER SYSTEM

1.4   Adding and Subtracting Real Numbers

Learning Objectives
1   Add two numbers with the same sign.
2   Add numbers with different signs.
3   Use the definition of subtraction.
4   Use the rules for order of operations when adding and subtracting signed numbers.
5   Translate words and phrases involving addition and subtraction.
6   Use signed numbers to interpret data.

Key Terms

Use the vocabulary terms listed below to complete each statement in exercises 1−5.

sum   **addends**   **minuend**

subtrahend   **difference**

1.   The number from which another number is being subtracted is called the

_____________________.

2.   The _____________________ is the number being subtracted.

3.   The answer to a subtraction problem is called the _____________________.

4.   The answer to an addition problem is called the _____________________.

5.   In an addition problem, the numbers being added are the _____________________.

Objective 1   Add two numbers with the same sign.

Review these examples for Objective 1:	**Now Try:**
1. Use a number line to find the sum. $-3 + (-5)$. *Step 1* Start at 0 and draw an arrow 3 units to the left. *Step 2* From the left end of that arrow, draw another arrow 5 units to the left. The number below the end of this second arrow is -5, so $-3 + (-5) = -8$. 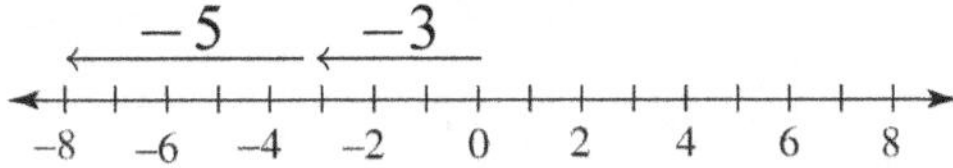	1. Use a number line to find the sum. $-4 + (-1)$ 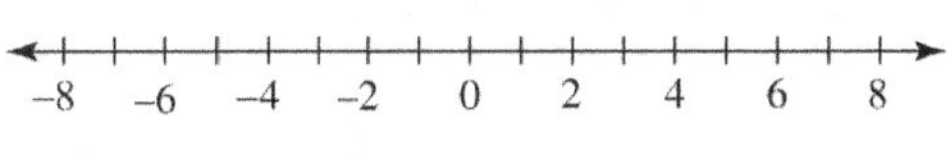 _____________________

2. Find the sum.

$-3 + (-7)$

$-3 + (-7) = -10$

2. Find the sum.

$-8 + (-4)$

____

Objective 1 Practice Exercises

For extra help, see Examples 1–2 on pages 51–52 of your text.

Find each sum.

1. $-7 + (-11)$

1. ______________

2. $-9 + (-9)$

2. ______________

3. $-2\frac{3}{8} + \left(-3\frac{1}{4}\right)$

3. ______________

Objective 2   Add numbers with different signs.

Review these examples for Objective 2:

3. Use the number line to find the sum $-3 + 4$.

Step 1 Start at 0 and draw an arrow 3 units to the left.

Step 2 From the left end of that arrow, draw a second arrow 4 units to the right.

The number below the end of this second arrow is 1, so $-3 + 4 = 1$.

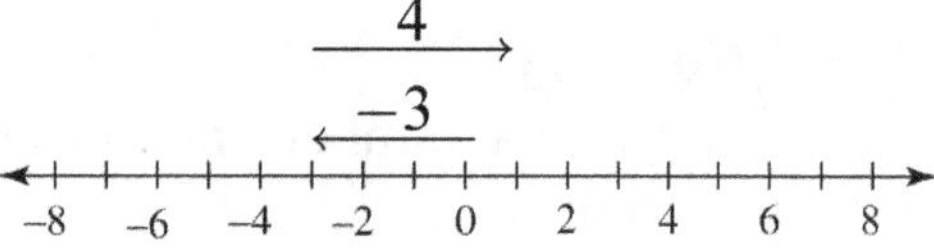

Now Try:

3. Use the number line to find the sum $7 + (-4)$.

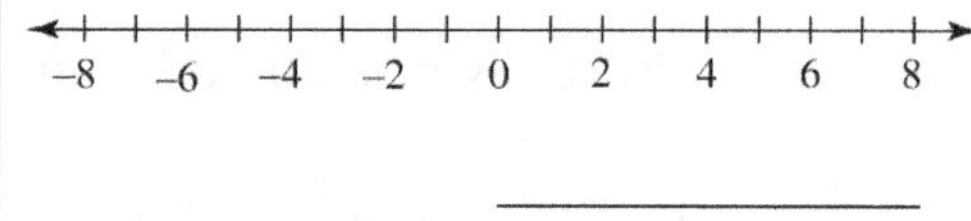

____

4. Find the sum.

$\frac{5}{8} + \left(-1\frac{1}{4}\right)$

$\frac{5}{8} + \left(-1\frac{1}{4}\right) = \frac{5}{8} + \left(-\frac{5}{4}\right)$

$= \frac{5}{8} + \left(-\frac{10}{8}\right) = +\left(\frac{5}{8} - \frac{10}{8}\right)$

$= -\frac{5}{8}$

4. Find the sum.

$\frac{3}{5} + \left(-1\frac{3}{10}\right)$

____

Objective 2 Practice Exercises

For extra help, see Examples 3–4 on pages 53–54 of your text.

Use a number line to find the sum.

4. $-8 + 5$ 4. _______________

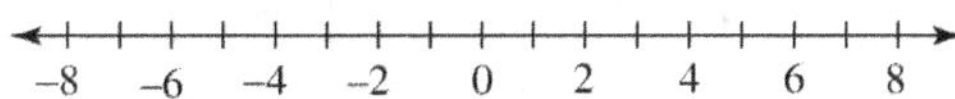

Find each sum.

5. $\dfrac{7}{12} + \left(-\dfrac{3}{4}\right)$ 5. _______________

6. $-\dfrac{4}{7} + \dfrac{3}{5}$ 6. _______________

Objective 3   Use the definition of subtraction.

Review these examples for Objective 3:   Now Try:

6.   Subtract.   6.   Subtract.

    a. $2 - 8$ **a.** $3 - 6$

    $2 - 8 = 2 + (-8) = -6$

                                                ____

    b. $-6 - (-9)$ **b.** $-5 - (-7)$

    $-6 - (-9) = -6 + (9) = 3$

                                                ____

Objective 3 Practice Exercises

For extra help, see Examples 5–6 on pages 54–56 of your text.

Subtract.

7. $-14 - 11$ 7. _______________

8. $15 - (-2)$ 8. _______________

9. $-\dfrac{3}{10} - \left(-\dfrac{3}{10}\right)$ 9. _______________

Objective 4   Use the rules for order of operations when adding and subtracting signed numbers.

Review this example for Objective 4:

7.   Perform each operation.

$$-4-[7-(9+6)]$$

$$-4-[7-(9+6)]=-4-[7-15]$$
$$=-4-[-8]$$
$$=-4+8$$
$$=4$$

Now Try:

7.   Perform each operation.

$$-5-[10-(7+4)]$$

____

Objective 4 Practice Exercises

For extra help, see Example 7 on pages 56–57 of your text.

Find each sum.

10. $-2+\left[4+(-18+13)\right]$ 10. _______________

11. $\left[(-7)+14\right]+\left[(-16)+3\right]$ 11. _______________

12. $-8.9+\left[6.8+(-4.7)\right]$ 12. _______________

Objective 5   Translate words and phrases involving addition and subtraction.

Review these examples for Objective 5:

8.   Write a numerical expression for the phrase, and simplify the expression.

The sum of –9 and 5 and 3

–9 + 5 + 3 simplifies to –4 + 3, which equals –1.

9.   Write a numerical expression for the phrase, and simplify the expression.

The difference between –10 and 7

$-10-7$ simplifies to $-10+(-7)$, which equals –17

Now Try:

8.   Write a numerical expression for the phrase, and simplify the expression.
The sum of –10 and 11 and 2

____

9.   Write a numerical expression for the phrase, and simplify the expression.
The difference between –17 and 9

____

10. The early morning temperature on a mountain in California was –8°F. At noon the temperature was 38°F. What was the rise in temperature?

We must subtract the lowest temperature from the highest temperature.
$$38 - (-8) = 38 + 8 = 46$$
The rise was 46°F.

10. The floor of Death Valley is 282 ft below sea level. A nearby mountain has an elevation of 5182 ft above sea level. Find the difference between the highest and lowest elevations.

____

Objective 5 Practice Exercises

For extra help, see Examples 8–10 on pages 57–59 of your text.

Write a numerical expression for each phrase, and then simplify the expression.

13. 4 less than −4

13. ______________

14. The sum of −4 and 12, decreased by 9

14. ______________

Solve the problem.

15. Dr. Somers runs an experiment at −43.3°C. He then lowers the temperature by 7.9°C. What is the new temperature for the experiment?

15. ______________

Objective 6   Use signed numbers to interpret data.

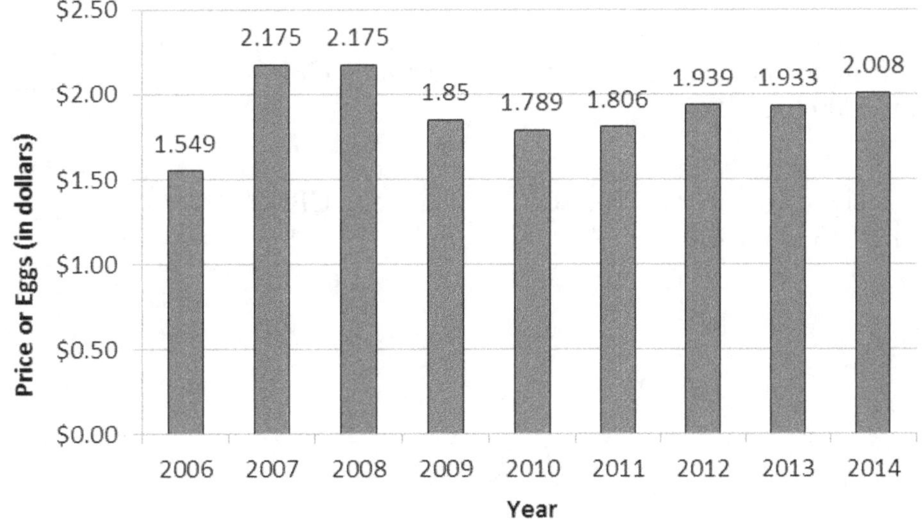

The bar graph above shows the Consumer Price Index (CPI) for a dozen of grade A large Eggs between 2006 and 2014.   Source: U.S. Bureau of Labor and Statistics

Review this example for Objective 6:

11. Use a signed number to represent the change in CPI from 2008 to 2009.

$$\$1.85 - \$2.175 = -\$0.325$$

Now Try:

11. Use a signed number to represent the change in CPI from 2012 to 2013.

____

Objective 6 Practice Exercises

For extra help, see Example 11 on page 59 of your text.

The bar graph below shows the Consumer Price Index (CPI) for a dozen of grade A large Eggs between 2006 and 2014. Source: U.S. Bureau of Labor and Statistics

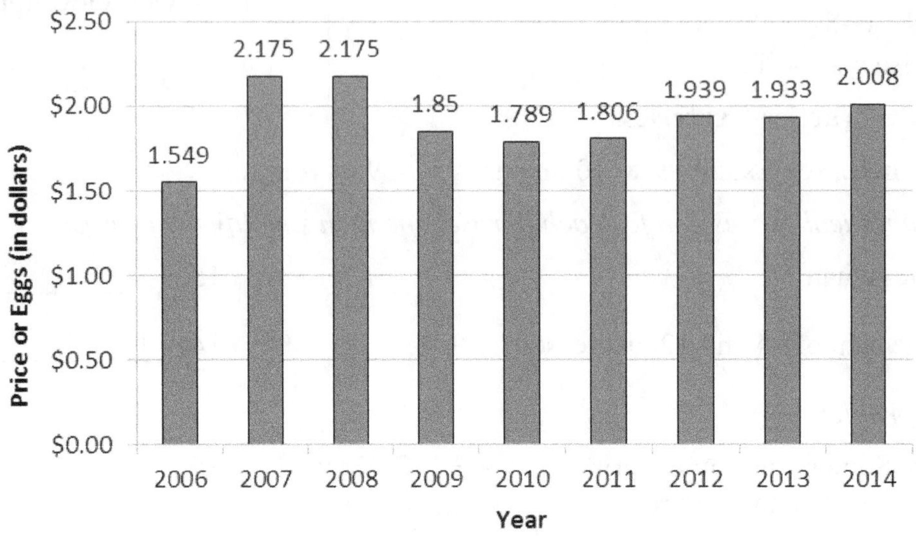

16. Use a signed number to represent the change in CPI from 2006 to 2007.

16. ______________________

17. Use a signed number to represent the change in CPI from 2009 to 2010.

17. ______________________

18. Use a signed number to represent the change in CPI from 2013 to 2014.

18. ______________________

Chapter 1 THE REAL NUMBER SYSTEM

1.5   Multiplying and Dividing Real Numbers

Learning Objectives	
1	Find the product of a positive number and a negative number.
2	Find the product of two negative numbers.
3	Identify factors of integers.
4	Use the reciprocal of a number to apply the definition of division.
5	Use the rules for order of operations when multiplying and dividing signed numbers.
6	Evaluate algebraic expressions given values for the variables.
7	Translate words and phrases involving multiplication and division.
8	Translate simple sentences into equations.

Key Terms

Use the vocabulary terms listed below to complete each statement in exercises 1−3.

> **product**   **quotient**   **reciprocals**

1. The answer to a division problem is called the ___________________.

2. Pairs of numbers whose product is 1 are called ___________________.

3. The answer to a multiplication problem is called the ___________________.

Objective 1   Find the product of a positive number and a negative number.

Review this example for Objective 1:

1. Find the product using the multiplication rule.

$$9(-6)$$

$$9(-6) = -(9 \cdot 6) = -54$$

Now Try:

1. Find the product using the multiplication rule.

$$8(-7)$$

____

Objective 1 Practice Exercises

For extra help, see Example 1 on page 66 of your text.

Find each product.

1. $7(-4)$ 1. ___________________

2. $\left(\frac{1}{5}\right)\left(-\frac{2}{3}\right)$ 2. ___________________

3. $(-3.2)(4.1)$ 3. ___________________

Objective 2   Find the product of two negative numbers.

Review this example for Objective 2:

2.   Find the product using the multiplication rule.

a. $-7(-3)$

$-7(-3) = 21$

Now Try:

2.   Find the product using the multiplication rule.

a. $-5(-6)$

____

Objective 2 Practice Exercises

For extra help, see Example 2 on page 67 of your text.

Find each product.

4. $(-4)(-10)$

4. _______________

5. $\left(-\dfrac{2}{7}\right)\left(-\dfrac{14}{5}\right)$

5. _______________

6. $(-0.4)(-3.4)$

6. _______________

Objective 3   Identify factors of integers.

Review this example for Objective 3:

3.   Find all integer factors of the number 52.

Positive integer factors are
    1 and 52, 2 and 26,   and 4 and 13.
Negative integer factors are
    -1 and -52, -2 and -26, and -4 and -13.

The integer factors of 52 are $-52, -26, -13,$
$-4, -2, -1, 1, 2, 4, 13, 26,$ and 52.

Now Try:

3.   Find all integer factors of the number 12.

____

Objective 3 Practice Exercises

For extra help, see Example 3 on page 67 of your text.

Find all integer factors of each number.

7.   8

7. _______________

                Copyright © 2025 Pearson Education, Inc.

8.   38

8. _______________

9.   42

9. _______________

Objective 4   Use the reciprocal of a number to apply the definition of division.

Review these examples for Objective 4:

4.   Find each quotient.

a. $\dfrac{30}{-6}$

$\dfrac{30}{-6} = -5$

b. $\dfrac{-15}{-3}$

$\dfrac{-15}{-3} = 5$

Now Try:

4.   Find each quotient.

a. $\dfrac{16}{-8}$

____

b. $\dfrac{-18}{-6}$

____

Objective 4 Practice Exercises

For extra help, see Example 4 on page 68 of your text.

Find each quotient.

10. $\dfrac{-120}{-20}$

10. _______________

11. $\dfrac{0}{-2}$

11. _______________

12. $\dfrac{10}{0}$

12. _______________

Objective 5   Use the rules for order of operations when multiplying and dividing signed numbers.

Review these examples for Objective 5:

5. Simplify.

    a. $-6(-1-4)$

$$-6(-1-4)=-6(-5)$$
$$=30$$

    d. $\dfrac{6(-4)-5(3)}{3(2-7)}$

$$\dfrac{6(-4)-5(3)}{3(2-7)}=\dfrac{-24-15}{3(-5)}$$
$$=\dfrac{-39}{-15}$$
$$=\dfrac{13}{5}$$

Now Try:

5. Simplify.

    a. $-4(-5-2)$

    d. $\dfrac{-9(-3)+4(-8)}{-4(5-6)}$

Objective 5 Practice Exercises

For extra help, see Example 5 on page 70 of your text.

Perform the indicated operations.

13. $-4\big[(-2)(7)-2\big]$

13. ______________________

14. $\dfrac{-7(2)-(-3)}{5+(-3)}$

14. ______________________

15. $\dfrac{-4\big[8-(-3+7)\big]}{-6\big[3-(-2)\big]-3(-3)}$

15. ______________________

Objective 6   Evaluate algebraic expressions given values for the variables.

Review this example for Objective 6:	**Now Try:**
6. Evaluate the expression for $x = -2$, $y = -4$, and $m = -5$.	6. Evaluate the expression for $x = -5$, $y = -3$, and $p = -4$.

Review example:

6. Evaluate the expression for $x = -2$, $y = -4$, and $m = -5$.

$$(5x + 6y)(-3m)$$

Substitute the given values for the variables. Then simplify.

$$(5x + 6y)(-3m)$$
$$= [5(-2) + 6(-4)][-3(-5)]$$
$$= [-10 + (-24)][15]$$
$$= [-34]15$$
$$= -510$$

Now Try:

6. Evaluate the expression for $x = -5$, $y = -3$, and $p = -4$.

$$(6x + 2y)(-3p)$$

____

Objective 6 Practice Exercises

For extra help, see Example 6 on pages 70–71 of your text.

Evaluate the following expressions if $x = -3$, $y = 2$, and $a = 4$.

16. $-x + [(-a + y) - 2x]$

16. ____________

17. $(-4 + x)(-a) - |x|$

17. ____________

18. $\dfrac{4a - x}{y^2}$

18. ____________

Objective 7   Translate words and phrases involving multiplication and division.

Review this example for Objective 7:	**Now Try:**
7. Write a numerical expression for the phrase, and simplify the expression.	7. Write a numerical expression for the phrase, and simplify the expression.

Review example:

7. Write a numerical expression for the phrase, and simplify the expression.

Three fifths of the sum of –6 and –7

$\dfrac{3}{5}[-6 + (-7)]$ simplifies to $\dfrac{3}{5}[-13]$,

which equals $-\dfrac{39}{5}$.

Now Try:

7. Write a numerical expression for the phrase, and simplify the expression.

Five-sixths of the sum of –8 and –4

____

Objective 7 Practice Exercises

For extra help, see Examples 7–8 on pages 71–72 of your text.

Write a numerical expression for each phrase and simplify.

19.   The product of –7 and 3, added to –7   **19.** __________________

20.   Three-tenths of the difference between 50 and –10,   **20.** __________________
        subtracted from 85

21.   The sum of –12 and the quotient of 49 and –7   **21.** __________________

Objective 8   **Translate simple sentences into equations.**

Review this example for Objective 8:

9.   Write the sentence in symbols, using x to represent the number.

The quotient of 27 and a number is –3.

$$\frac{27}{x} = -3$$

Now Try:

9.   Write the sentence in symbols, using x to represent the number.

The quotient of 36 and a number is –4

____

Objective 8 Practice Exercises

For extra help, see Example 9 on pages 72–73 of your text.

Write each statement in symbols, using x as the variable.

22.   Two-thirds of a number is –7.   **22.** __________________

23.   –8 times a number is 72.   **23.** __________________

24.   When a number is divided by –4, the result is 1.   **24.** __________________

Chapter 1 THE REAL NUMBER SYSTEM

1.6   Properties of Real Numbers

<table>
<tr><td colspan="2">Learning Objectives</td></tr>
<tr><td>1</td><td>Use the commutative properties.</td></tr>
<tr><td>2</td><td>Use the associative properties.</td></tr>
<tr><td>3</td><td>Use the identity properties.</td></tr>
<tr><td>4</td><td>Use the inverse properties.</td></tr>
<tr><td>5</td><td>Use the distributive property.</td></tr>
</table>

Key Terms

Use the vocabulary terms listed below to complete each statement in exercises 1−2.

> **identity element for addition**

> **identity element for multiplication**

1.   When the _________________________________, which is 0, is added to a number, the number is unchanged.

2.   When a number is multiplied by the ___________________________________, which is 1, the number is unchanged.

Objective 1   Use the commutative properties.

Review these examples for Objective 1:

1.   Use a commutative property to complete each statement.

   a. $-7+6=6+$ ______

   Using the commutative property of addition,
   $-7+6=6+(-7)$

   b. $(-3)5=$ ____ (-3)

   Using the commutative property of multiplication,
   $(-3)5=5(-3)$

Now Try:

1.   Use a commutative property to complete each statement.

   a. $-12+8=8+$ ______

   ____

   b. $(-4)2=$ ____ (-4)

   ____

Objective 1 Practice Exercises

For extra help, see Example 1 on page 78 of your text.

Complete each statement. Use a commutative property.

   1. $y+4=$ ______ $+y$

   1. ______________

2. $5(2) = \underline{\hspace{1cm}}(5)$ **2.** _______________

3. $-4(4+z) = \underline{\hspace{1cm}}(-4)$ **3.** _______________

Objective 2   Use the associative properties.

Review these examples for Objective 2:

2. Use an associative property to complete each statement.

   a. $-5+(3+7)=(-5+\underline{\hspace{1cm}})+7$

Using the associative property of addition,
$-5+(3+7)=(-5+3)+7$

   b. $[4\cdot(-9)]\cdot 2 = 4\cdot\underline{\hspace{1cm}}$

Using the associative property of multiplication,
$[4\cdot(-9)]\cdot 2 = 4\cdot[(-9)\cdot 2]$

3. Decide whether each statement is an example of a commutative property, an associative property, or both.

   a. $(5+9)+11=5+(9+11)$

The order of the three numbers is the same, but the change is in grouping. This is an example of the associative property.

   b. $7\cdot(9\cdot 11)=7\cdot(11\cdot 9)$

The only change involves the order of the number, so this is an example of the commutative property.

   c. $(12+3)+6=12+(6+3)$

Both the order and the grouping are changed. This is an example of both the associative and commutative properties.

Now Try:

2. Use an associative property to complete each statement.

   a. $-8+(4+6)=(-8+\underline{\hspace{0.8cm}})+6$

____

   b. $[8\cdot(-3)]\cdot 4 = 8\cdot\underline{\hspace{1cm}}$

____

3. Decide whether each statement is an example of a commutative property, an associative property, or both.

   a. $(13+8)+25=13+(8+25)$

____

   b. $4\cdot(15\cdot 30)=4\cdot(30\cdot 15)$

____

   c. $(21+19)+4=21+(4+19)$

____

4. Find the sum.

$$21x + 3 + 17x + 29$$

$$21x + 3 + 17x + 29$$
$$= 21x + (3 + 17x) + 29$$
$$= (21x + 17x) + (3 + 29)$$
$$= 38x + 32$$

4. Find the sum

$$15x + 12 + 24x + 8$$

____

Objective 2 Practice Exercises

For extra help, see Examples 2–4 on pages 79–80 of your text.

Complete each statement. Use an associative property.

4. $4(ab) = \underline{\quad} \cdot b$

4. ________________

5. $\left[x + (-4)\right] + 3y = x + \underline{\quad}$

5. ________________

6. $4r + (3s + 14t) = \underline{\quad} + 14t$

6. ________________

Objective 3   Use the identity properties.

Review these examples for Objective 3:

5. Use an identity property to complete each statement.

a. $-6 + \underline{\quad} = -6$

Use the identity property for addition.
$$-6 + 0 = -6$$

b. $\underline{\quad} \cdot \dfrac{1}{6} = \dfrac{1}{6}$

Use the identity property for multiplication.
$$1 \cdot \dfrac{1}{6} = \dfrac{1}{6}$$

Now Try:

5. Use an identity property to complete each statement.

a. $8 + \underline{\quad} = 8$

____

b. $-9 \cdot \underline{\quad} = -9$

____

6.

  a. Write $\dfrac{56}{35}$ in lowest terms.

$$\dfrac{56}{35} = \dfrac{8 \cdot 7}{5 \cdot 7}$$

$$= \dfrac{8}{5} \cdot \dfrac{7}{7}$$

$$= \dfrac{8}{5} \cdot 1$$

$$= \dfrac{8}{5}$$

  b. Perform the operation: $\dfrac{5}{6} - \dfrac{7}{18}$

$$\dfrac{5}{6} - \dfrac{7}{18} = \dfrac{5}{6} \cdot 1 - \dfrac{7}{18}$$

$$= \dfrac{5}{6} \cdot \dfrac{3}{3} - \dfrac{7}{18}$$

$$= \dfrac{15}{18} - \dfrac{7}{18}$$

$$= \dfrac{8}{18}$$

$$= \dfrac{4}{9}$$

6.

  a. Write $\dfrac{49}{63}$ in lowest terms.

            ____

  b. Perform the operation: $\dfrac{3}{7} + \dfrac{5}{21}$

            ____

Objective 3 Practice Exercises

For extra help, see Examples 5–6 on pages 80–81 of your text.

Use an identity property to complete each statement.

  7. $4 + 0 = $ _____

  8.   _____ $\cdot 1 = 12$

Use an identity property to simplify the expression.

  9. $\dfrac{30}{35}$

  7. _________________

  8. _________________

  9. _________________

Objective 4   Use the inverse properties.

Review these examples for Objective 4:

Now Try:

7. Use an inverse property to complete each statement.

7. Use an inverse property to complete each statement.

a. $\underline{\quad} + \dfrac{1}{4} = 0$

a. $-11 + \underline{\quad} = 0$

Use the inverse property of addition.

$$-\dfrac{1}{4} + \dfrac{1}{4} = 0$$

____

b. $5 + \underline{\quad} = 0$

b. $8 + \underline{\quad} = 0$

Use the inverse property of addition.

$$5 + (-5) = 0$$

____

d. $\underline{\quad} \cdot \dfrac{6}{7} = 1$

d. $\dfrac{8}{5} \cdot \underline{\quad} = 1$

Use the inverse property of multiplication.

$$\dfrac{7}{6} \cdot \dfrac{6}{7} = 1$$

____

e. $-9(\underline{\quad}) = 1$

e. $-\dfrac{1}{10}(\underline{\quad}) = 1$

Use the inverse property of multiplication.

$$-9\left(-\dfrac{1}{9}\right) = 1$$

____

8. Simplify $-4x + 1 + 4x$.

8. Simplify $-\dfrac{1}{4}x + 6 + \dfrac{1}{4}x$.

$-4x + 1 + 4x$

$$
\begin{aligned}
&= (-4x + 1) + 4x && \text{Order of operations}\\
&= [1 + (-4x)] + 4x && \text{Commutative property}\\
&= 1 + [(-4x) + 4x] && \text{Associative property}\\
&= 1 + 0 && \text{Inverse property}\\
&= 1 && \text{Identity property}
\end{aligned}
$$

____

Objective 4 Practice Exercises

For extra help, see Examples 7–8 on pages 81–82 of your text.

Complete the statements so that they are examples of either an identity property or an inverse property. Identify which property is used.

10. $-4 + \underline{\quad} = 0$

10. _______________

11. $-9 + \underline{\quad} = -9$

11. _______________

12. $-\dfrac{3}{5}\cdot \underline{\quad\quad} = 1$ **12.** ________________

Objective 5   Use the distributive property.

Review these examples for Objective 5:

9. Use the distributive property to rewrite each expression.

    a. $4(8+7)$

$$4(8+7) = 4\cdot 8 + 4\cdot 7$$
$$= 32 + 28$$
$$= 60$$

    b. $5(3+x+m)$

$$5\cdot 3 + 5x + 5m = 5(3+x+m)$$

    e. $7(p-6)$

$$7(p-6) = 7[p+(-6)]$$
$$= 7p + 7(-6)$$
$$= 7p - 42$$

    f. $-3(5x-2)$

$$-3(5x-2) = -3[5x+(-2)]$$
$$= -3(5x) + (-3)(-2)$$
$$= (-3\cdot 5)x + (-3)(-2)$$
$$= -15x + 6$$

10. Rewrite each expression.

    a. $-(5x+7)$

$$-(5x+7) = -1\cdot(5x+7)$$
$$= -1\cdot 5x + (-1)\cdot 7$$
$$= -5x - 7$$

    b. $-(-6c-7)$

$$-(-6c-7) = -1(-6c-7)$$
$$= -1(-6c) - 1(-7)$$
$$= 6c + 7$$

Now Try:

9. Use the distributive property to rewrite each expression.

    a. $3(11+7)$

____

    b. $12(y+6+x)$

____

    e. $17(x-6)$

____

    f. $-4(2x-5)$

____

10. Rewrite each expression.

    a. $-(3x+4)$

____

    b. $-(-8k-9)$

____

c. $-(-p-5r+9x)$ **c.** $-(-4x-5y+z)$

$-(-p-5r+9x)$

$=-1\cdot(-1p-5r+9x)$

$=-1\cdot(-1p)-1\cdot(-5r)-1\cdot(9x)$

$=p+5r-9x$

Objective 5 Practice Exercises

For extra help, see Examples 9–10 on pages 83–84 of your text.

Use the distributive property to rewrite each expression. Simplify if possible.

13. $n(2a-4b+6c)$ **13.** ______________

14. $-2(5y-9z)$ **14.** ______________

15. $-(-2k+7)$ **15.** ______________

Chapter 1 THE REAL NUMBER SYSTEM

1.7   Simplifying Expressions

Learning Objectives
1   Simplify expressions.
2   Identify terms and numerical coefficients.
3   Identify like terms.
4   Combine like terms.
5   Simplify expressions from word phrases.

Key Terms

Use the vocabulary terms listed below to complete each statement in exercises 1–3.

term   **numerical coefficient**   **like terms**

1.   In the term $4x^2$, "4" is the____.

2.   A number, a variable, or a product or quotient of a number and one or more variables raised to powers is called a ____.

3.   Terms with exactly the same variables, including the same exponents, are called _________________________________.

Objective 1   Simplify expressions.

Review these examples for Objective 1:

1.   Simplify each expression.

   a. $8(4m-6n)$

   Use the distributive property.
   $$8(4m-6n)=8(4m)+8(-6n)$$
   $$=32m-48n$$

   d. $9-(4y-6)$

   $$9-(4y-6)=9-1(4y-6)$$
   $$=9-4y+6$$
   $$=15-4y$$

Now Try:

1.   Simplify each expression.

   a. $7(5x-3y)$

   ____

   d. $8-(7x-3)$

   ____

Objective 1 Practice Exercises

For extra help, see Example 1 on page 88 of your text.

Simplify each expression.

    1. $4(2x+5)+7$ **1.** ________________

    2. $-4+s-(12-21)$ **2.** ________________

    3. $-2(-5x+2)+7$ **3.** ________________

Objective 2   Identify terms and numerical coefficients.

Objective 2 Practice Exercises

For extra help, see pages 88–89 of your text.

Give the numerical coefficient of each term.

    4. $-2y^2$ **4.** ________________

    5. $\dfrac{7x}{9}$ **5.** ________________

    6. $5.6r^5$ **6.** ________________

Objective 3   Identify like terms.

Objective 3 Practice Exercises

For extra help, see page 89 of your text.

*Identify each group of terms as **like** or **unlike**.*

    7. $4x^2, -7x^2$ **7.** ________________

    8. $-8m, -8m^2$ **8.** ________________

    9. $7xy, -6xy^2$ **9.** ________________

Objective 4   Combine like terms.

Review these examples for Objective 4: | **Now Try:**

2. Combine like terms in each expression.

2. Combine like terms in each expression.

b. $8r + 5r + 4r$

b. $14r + 7r + 2r$

$$8r + 5r + 4r = (8 + 5 + 4)r$$
$$= 17r$$

____

c. $9x + x$

c. $18x + x$

$$9x + x = 9x + 1x$$
$$= (9 + 1)x$$
$$= 10x$$

____

d. $15x^2 - 8x^2$

d. $17x^2 - 9x^2$

$$15x^2 - 8x^2 = (15 - 8)x^2$$
$$= 7x^2$$

____

3. Simplify each expression.

3. Simplify each expression.

b. $8k - 5 - 4(7 - 3k)$

b. $7k - 9 - 5(3 - 6k)$

$$8k - 5 - 4(7 - 3k) = 8k - 5 - 4(7) - 4(-3k)$$
$$= 8k - 5 - 28 + 12k$$
$$= 20k - 33$$

____

f. $-\dfrac{3}{5}(x - 10) - \dfrac{1}{10}x$

f. $-\dfrac{3}{4}(x - 8) - \dfrac{1}{2}x$

$$-\frac{3}{5}(x - 10) - \frac{1}{10}x = -\frac{3}{5}x - \frac{3}{5}(-10) - \frac{1}{10}x$$
$$= -\frac{3}{5}x + 6 - \frac{1}{10}x$$
$$= -\frac{6}{10}x + 6 - \frac{1}{10}x$$
$$= -\frac{7}{10}x + 6$$

____

Objective 4 Practice Exercises

For extra help, see Examples 2–3 on pages 89–91 of your text.

Simplify.

10. $12y - 7y^2 + 4y - 3y^2$

10. _______________

11. $-4(x+4)+2(3x+1)$ **11.** ______________________

12. $2.5(3y+1)-4.5(2y-3)$ **12.** ______________________

Objective 5   Simplify expressions from word phrases.

Review this example for Objective 5:

4.   Translate the phrase into a mathematical expression and simplify.

The sum of 8, three times a number, nine times a number, and seven times a number.

Use x for the number.

$8+3x+9x+7x$ simplifies to $8+19x$.

Now Try:

4.   Translate the phrase into a mathematical expression and simplify.

The sum of 11, ten times a number, eight times a number, and four times a number

____

Objective 5 Practice Exercises

For extra help, see Example 4 on page 91 of your text.

Write each phrase as a mathematical expression and simplify by combining like terms. Use x as the variable.

13.   The sum of six times a number and 12, added to four   **13.** ______________________
times the number.

14.   The sum of seven times a number and 2, subtracted   **14.** ______________________
from three times the number.

15.   Four times the difference between twice a number   **15.** ______________________
and six times the number, added to six times the sum
of the number and 9.

Chapter 2   LINEAR EQUATIONS AND INEQUALITIES IN ONE VARIABLE

2.1   The Addition Property of Equality

Learning Objectives
1   Identify linear equations.
2   Use the addition property of equality.
3   Simplify, and then use the addition property of equality.

Key Terms

Use the vocabulary terms listed below to complete each statement in exercises 1−3.

linear equation   solution set   equivalent equations

1.   Equations that have exactly the same solutions sets are called

_________________________________.

2.   An equation that can be written in the form $Ax + B = C$, where A, B, and C are real numbers and $A \neq 0$, is called a _________________________________.

3.   The set of all numbers that satisfy an equation is called its _________________________.

Objective 1   Identify linear equations.

Objective 1 Practice Exercises

For extra help, see page 104 of your text.

Tell whether each of the following is a linear equation.

1. $3x^2 + 4x + 3 = 0$ 1. _________________

2. $\dfrac{5}{x} - \dfrac{3}{2} = 0$ 2. _________________

3. $4x - 2 = 12x + 9$ 3. _________________

Name: Date:
Instructor: Section:

Objective 2   Use the addition property of equality.

Review these examples for Objective 2:

1. Solve $x - 15 = 8$.

$$x - 15 = 8$$
$$x - 15 + 15 = 8 + 15$$
$$x = 23$$

Check $x - 15 = 8$
$$23 - 15 \overset{?}{=} 8$$
$$8 = 8 \quad \text{True}$$

The solution is 23, and the solution set is $\{23\}$.

3. Solve $y - \dfrac{3}{4} = -\dfrac{1}{6}$.

$$y - \frac{3}{4} = -\frac{1}{6}$$
$$y - \frac{3}{4} + \frac{3}{4} = -\frac{1}{6} + \frac{3}{4}$$
$$y = -\frac{2}{12} + \frac{9}{12}$$
$$y = \frac{7}{12}$$

Check $y - \dfrac{3}{4} = -\dfrac{1}{6}$
$$\frac{7}{12} - \frac{3}{4} \overset{?}{=} -\frac{1}{6}$$
$$-\frac{1}{6} = -\frac{1}{6} \quad \text{True}$$

The solution set is $\left\{\dfrac{7}{12}\right\}$.

4. Solve $-5 = x + 17$.

$$-5 = x + 17$$
$$-5 - 17 = x + 17 - 17$$
$$-22 = x$$

Check $-5 = x + 17$
$$-5 \overset{?}{=} -22 + 17$$
$$-5 = -5 \quad \text{True}$$

The solution set is $\{-22\}$.

Now Try:

1. Solve $x - 12 = 9$.

____

3. Solve $w - \dfrac{3}{5} = -\dfrac{1}{3}$.

____

4. Solve $-10 = x + 9$.

____

6. Solve $\frac{5}{6}k+9=\frac{11}{6}k$.

$$\frac{5}{6}k+9=\frac{11}{6}k$$

$$\frac{5}{6}k+9-\frac{5}{6}k=\frac{11}{6}k-\frac{5}{6}k$$

$$9=1k$$

$$9=k$$

Check by substituting 9 in the original equation. The solution set is {9}.

7. Solve $9-5p=-6p+3$.

$$9-5p=-6p+3$$

$$9-5p+6p=-6p+3+6p$$

$$9+p-9=3-9$$

$$p=-6$$

Check by substituting –6 in the original equation. The solution set is {–6}.

6. Solve $\frac{4}{7}k+13=\frac{11}{7}k$.

____

7. Solve $10-8p=-9p+7$.

____

Objective 2 Practice Exercises

For extra help, see Examples 1–7 on pages 105–108 of your text.

Solve each equation by using the addition property of equality. Check each solution.

4. $y-4=16$

4. ______________

5. $\frac{9}{8}p-\frac{1}{2}=\frac{1}{8}p$

5. ______________

6. $9.5y-2.4=10.5y$

6. ______________

Objective 3   Simplify, and then use the addition property of equality.

Review these examples for Objective 3:

8. Solve $5t-16+t+4=9+5t+6$.

$$5t-16+t+4=9+5t+6$$
$$6t-12=15+5t$$
$$6t-12-5t=15+5t-5t$$
$$t-12=15$$
$$t-12+12=15+12$$
$$t=27$$

Check by substituting 27 in the original equation. The solution set is {27}.

9. Solve $4(3+6x)-(5+23x)=19$.

$$4(3+6x)-(5+23x)=19$$
$$4(3)+4(6x)-1(5)-1(23x)=19$$
$$12+24x-5-23x=19$$
$$x+7=19$$
$$x+7-7=19-7$$
$$x=12$$

Check by substituting 12 in the original equation. The solution set is {12}.

Now Try:

8. Solve
$$8t-9+t+7=12+8t+15.$$

____

9. Solve
$$5(7+8x)-(29+39x)=14.$$

____

Objective 3 Practice Exercises

For extra help, see Examples 8–9 on page 109 of your text.

Solve each equation. First simplify each side of the equation as much as possible. Check each solution.

7. $3(t+3)-(2t+7)=9$

7. _______________

8. $-4(5g-7)+3(8g-3)=15-4+3g$

8. _______________

9. $3.6p+4.8+4.0p=8.6p-3.1+0.7$

9. _______________

Chapter 2   LINEAR EQUATIONS AND INEQUALITIES IN ONE VARIABLE

2.2   The Multiplication Property of Equality

> **Learning Objectives**
> 1   Use the multiplication property of equality.
> 2   Simplify, and then use the multiplication property of equality.

Key Terms

Use the vocabulary terms listed below to complete each statement in exercises 1−2.

multiplication property of equality   **addition property of equality**

1.   The _______________________________ states that multiplying both sides of an equation by the same nonzero number will not change the solution.

2.   When the same quantity is added to both sides of an equation, the _______________________________ is being applied.

Objective 1   Use the multiplication property of equality.

Review these examples for Objective 1:

1.   Solve $6x = 78$.

$$6x = 78$$

$$\frac{6x}{6} = \frac{78}{6}$$

$$x = 13$$

Check $6x = 78$

$$6(13) \overset{?}{=} 78$$

$$78 = 78 \quad \text{True}$$

The solution set is $\{13\}$.

3.   Solve $4.3x = 10.32$.

$$4.3x = 10.32$$

$$\frac{4.3x}{4.3} = \frac{10.32}{4.3}$$

$$x = 2.4$$

Check by substituting 2.4 in the original equation. The solution set is $\{2.4\}$.

Now Try:

1.   Solve $4x = 56$.

____

3.   Solve $3.6x = 20.52$.

____

4. Solve $\dfrac{x}{7} = 5$.

$$\dfrac{x}{7} = 5$$

$$\dfrac{1}{7}x = 5$$

$$7 \cdot \dfrac{1}{7}x = 7 \cdot 5$$

$$x = 35$$

Check by substituting 35 in the original equation. The solution set is {35}.

5. Solve $-\dfrac{5}{6}x = -15$.

$$-\dfrac{5}{6}x = -15$$

$$-\dfrac{6}{5} \cdot \left(-\dfrac{5}{6}x\right) = -\dfrac{6}{5} \cdot (-15)$$

$$1 \cdot x = -\dfrac{6}{5} \cdot \left(-\dfrac{15}{1}\right)$$

$$x = 18$$

Check by substituting 18 in the original equation. The solution set is {18}.

6. Solve $-x = 5$.

$$-x = 5$$

$$-1 \cdot x = 5$$

$$-1(-1 \cdot x) = -1(5)$$

$$[-1(-1)] \cdot x = -5$$

$$1 \cdot x = -5$$

$$x = -5$$

Check by substituting –5 in the original equation. The solution set is {–5}.

4. Solve $\dfrac{x}{8} = 3$.

____

5. Solve $-\dfrac{7}{9}h = -28$.

____

6. Solve $-x = 3$.

____

Objective 1 Practice Exercises

For extra help, see Examples 1–6 on pages 113–115 of your text.

Solve each equation and check your solution.

1. $-3w = 51$

1. _______________

2. $\dfrac{3p}{7} = -6$ **2.** _______________

3. $-2.7v = -17.28$ **3.** _______________

Objective 2   Simplify, and then use the multiplication property of equality.

Review these examples for Objective 2:

Now Try:

7. Solve $9m + 4m = 39$.

7. Solve $12m + 8m = 80$.

$$9m + 4m = 39$$
$$13m = 39$$
$$\frac{13m}{13} = \frac{39}{13}$$
$$m = 3$$

Check by substituting 3 in the original equation. The solution set is $\{3\}$.

____

8. Solve $3(2x - 7) + 21 = -18$.

8. Solve $4(2x - 5) + 20 = -32$.

$$3(2x - 7) + 21 = -18$$
$$3(2x) + 3(-7) + 21 = -18$$
$$6x - 21 + 21 = -18$$
$$6x = -18$$
$$\frac{6x}{6} = \frac{-18}{6}$$
$$x = -3$$

Check by substituting –3 in the original equation. The solution set is $\{-3\}$.

____

Objective 2 Practice Exercises

For extra help, see Examples 7–8 on pages 115–116 of your text.

Solve each equation and check your solution.

4. $-7b + 12b = 125$ **4.** _______________

5. $3w - 7w = 20$ **5.** _______________

6. $-11h - 6h + 14h = -21$ **6.** _______________

Chapter 2   LINEAR EQUATIONS AND INEQUALITIES IN ONE VARIABLE

2.3   Solving Linear Equations Using Both Properties of Equality

Learning Objectives
1   Use the four steps for solving a linear equation.
2   Solve equations with no solution or infinitely many solutions.
3   Write expressions for two related unknown quantities.

Key Terms

Use the vocabulary terms listed below to complete each statement in exercises 1–3.

conditional equation   **identity**   **contradiction**   **empty set**

1.   An equation that is true for all values of the variable is called a(n) _____________.

2.   A(n) _______________________________ is an equation that is true for some values of the variable and false for other values.

3.   An equation with no solution is called a(n) _____________________________; its solution set is the ___________________.

Objective 1   Use the four steps for solving a linear equation.

Review these examples for Objective 1:

1.   Solve $-5x + 8 = 23$.

Step 1 There are no parentheses, fractions, or decimals in this equation, so this step is not necessary.

$$-5x + 8 = 23$$

Step 2 $-5x + 8 - 8 = 23 - 8$

$$-5x = 15$$

Step 3 $$\frac{-5x}{-5} = \frac{15}{-5}$$

$$x = -3$$

Step 4 Check by substituting –3 for x in the original equation.

$$-5x + 8 = 23$$
$$-5(-3) + 8 \overset{?}{=} 23$$
$$15 + 8 \overset{?}{=} 23$$
$$23 = 23 \quad \text{True}$$

The solution, –3, checks, so the solution set is $\{-3\}$.

Now Try:

1.   Solve $-8x + 11 = 59$.

____

2. Solve $4x + 3 = 6x - 11$.

Step 1 There are no parentheses, fractions, or decimals in this equation, so begin with Step 2.
$$4x + 3 = 6x - 11$$

Step 2 $\quad 4x + 3 - 4x = 6x - 11 - 4x$
$$3 = 2x - 11$$
$$3 + 11 = 2x - 11 + 11$$
$$14 = 2x$$

Step 3 $\qquad \dfrac{14}{2} = \dfrac{2x}{2}$
$$7 = x$$

Step 4 Check by substituting 7 for x in the original equation.
$$4x + 3 = 6x - 11$$
$$4(7) + 3 \stackrel{?}{=} 6(7) - 11$$
$$28 + 3 \stackrel{?}{=} 42 - 11$$
$$31 = 31 \quad \text{True}$$
The solution, 7, checks, so the solution set is $\{7\}$.

4. Solve $9a - (4 + 3a) = 2a + 5$.

$$9a - (4 + 3a) = 2a + 5$$

Step 1 $\quad 9a - 4 - 3a = 2a + 5$
$$6a - 4 = 2a + 5$$

Step 2 $\quad 6a - 4 - 2a = 2a + 5 - 2a$
$$4a - 4 = 5$$
$$4a - 4 + 4 = 5 + 4$$
$$4a = 9$$

Step 3 $\qquad \dfrac{4a}{4} = \dfrac{9}{4}$
$$a = \dfrac{9}{4}$$

Step 4 Check that the solution set is $\left\{\dfrac{9}{4}\right\}$.

2. Solve $5x + 4 = 8x - 20$.

____

4. Solve $10a - (11 + 3a) = 5a + 4$.

____

Objective 1 Practice Exercises

For extra help, see Examples 1–5 on pages 118–121 of your text.

Solve each equation and check your solution.

1. $\quad 7t + 6 = 11t - 4$

1. _______________________

2. $3a - 6a + 4(a-4) = -2(a+2)$ 2. _________________

3. $3(t+5) = 6 - 2(t-4)$ 3. _________________

Objective 2   Solve equations with no solution or infinitely many solutions.

Review these examples for Objective 2: | **Now Try:**

6.   Solve $6x - 18 = 6(x-3)$.

$$6x - 18 = 6(x-3)$$
$$6x - 18 = 6x - 18$$
$$6x - 18 - 6x = 6x - 18 - 6x$$
$$-18 = -18$$
$$-18 + 18 = -18 + 18$$
$$0 = 0$$

The solution set is {all real numbers}.

6.   Solve $3x + 4(x-5) = 7x - 20$.

____

7.   Solve $3x + 4(x-5) = 7x + 5$.

$$3x + 4(x-5) = 7x + 5$$
$$3x + 4x - 20 = 7x + 5$$
$$7x - 20 = 7x + 5$$
$$7x - 20 - 7x = 7x + 5 - 7x$$
$$-20 = 5 \quad \text{False}$$

There is no solution. The solution set is $\varnothing$.

7.   Solve $-5x + 17 = x - 6(x+3)$.

____

Objective 2 Practice Exercises

For extra help, see Examples 6–7 on pages 121–122 of your text.

Solve each equation and check your solution.

4. $3(6x-7) = 2(9x-6)$ 4. _________________

5. $6y - 3(y+2) = 3(y-2)$ 5. _________________

6. $3(r-2)-r+4=2r+6$ 6. ___________________

Objective 3   Write expressions for two related unknown quantities.

Review this example for Objective 3:

8.   Two numbers have a sum of 51. If one of the numbers is represented by x, find an expression for the other number.

If one number is x, then the other number is obtained by subtracting x from 51.
$51-x$.
To check, we find the sum of the two numbers.
$x+(51-x)=51$

Now Try:

8.   Two numbers have a sum of 67. If one of the numbers is represented by t, find an expression for the other number.

____

Objective 3 Practice Exercises

For extra help, see Example 8 on page 123 of your text.

Write an expression for the two related unknown quantities.

7.   Two numbers have a sum of 36. One is m. Find the other number.   7. ___________________

8.   The product of two numbers is 17. One number is p. What is the other number?   8. ___________________

9.   Admission to the circus costs x dollars for an adult and y dollars for a child. Find the total cost of 6 adults and 4 children.   9. ___________________

Chapter 2   LINEAR EQUATIONS AND INEQUALITIES IN ONE VARIABLE

2.4   Clearing Fractions and Decimals When Solving Linear Equations

Learning Objectives
1   Solve equations with fractions as coefficients.
2   Solve equations with decimals as coefficients.

Objective 1   Solve equations with fractions as coefficients.

Review these examples for Objective 1: | **Now Try:**

1. Solve $\dfrac{x}{3} - 8 = -\dfrac{x}{5}$.   1. Solve $\dfrac{x}{2} - 5 = -\dfrac{x}{3}$.

$$\frac{x}{3} - 8 = -\frac{x}{5}$$

$$\textit{Step 1} \qquad 15\left(\frac{x}{3} - 8\right) = 15\left(-\frac{x}{5}\right)$$

$$15\left(\frac{x}{3}\right) + 15(-8) = 15\left(-\frac{x}{5}\right)$$

$$5x - 120 = -3x$$

$$\textit{Step 2} \qquad 5x - 120 - 5x = -3x - 5x$$

$$-120 = -8x$$

$$\textit{Step 3} \qquad \frac{-120}{-8} = \frac{-8x}{-8}$$

$$x = 15$$

Step 4 Check by substituting 15 for x in the original equation.

$$\frac{x}{3} - 8 = -\frac{x}{5}$$

$$\frac{15}{3} - 8 \stackrel{?}{=} -\frac{15}{5}$$

$$5 - 8 \stackrel{?}{=} -3$$

$$-3 = -3 \quad \text{True}$$

The check confirms that the solution set is $\{15\}$.

2.   Solve $\dfrac{3}{4}x - \dfrac{1}{2}x = -\dfrac{1}{8}x - 6$.

Multiply each side by **8**, the LCD.

$$\frac{3}{4}x - \frac{1}{2}x = -\frac{1}{8}x - 6$$

Step 1 $8\left(\dfrac{3}{4}x - \dfrac{1}{2}x\right) = 8\left(-\dfrac{1}{8}x - 6\right)$

$$8\left(\frac{3}{4}x\right) + 8\left(-\frac{1}{2}x\right) = 8\left(-\frac{1}{8}x\right) + 8(-6)$$

$$6x - 4x = -x - 48$$

$$2x = -x - 48$$

Step 2 $2x + x = -x - 48 + x$

$$3x = -48$$

Step 3 $\dfrac{3x}{3} = -\dfrac{48}{3}$

$$x = -16$$

Step 4 $\dfrac{3}{4}x - \dfrac{1}{2}x = -\dfrac{1}{8}x - 6$

$$\frac{3}{4}(-16) - \frac{1}{2}(-16) \overset{?}{=} -\frac{1}{8}(-16) - 6$$

$$-12 + 8 \overset{?}{=} 2 - 6$$

$$-4 = -4 \quad \text{True}$$

The solution set is $\{-16\}$.

3.   Solve $\dfrac{1}{4}(x+3) - \dfrac{2}{5}(x+1) = 2$.

To clear fractions, multiply by **20**, the LCD.

$$\frac{1}{4}(x+3) - \frac{2}{5}(x+1) = 2$$

Step 1 $20\left[\dfrac{1}{4}(x+3) - \dfrac{2}{5}(x+1)\right] = 20(2)$

$$20\left[\frac{1}{4}(x+3)\right] + 20\left[-\frac{2}{5}(x+1)\right] = 20(2)$$

$$5(x+3) - 8(x+1) = 40$$

$$5x + 15 - 8x - 8 = 40$$

$$-3x + 7 = 40$$

Step 2 $-3x + 7 - 7 = 40 - 7$

$$-3x = 33$$

Step 3 $\dfrac{-3x}{-3} = \dfrac{33}{-3}$

$$x = -11$$

Step 4 Check to confirm that $\{-11\}$ is the solution set.

2.   Solve $\dfrac{2}{9}x - \dfrac{1}{6}x = \dfrac{2}{3}x + 11$.

———————

3.   Solve

$$\frac{1}{7}(x+5) - \frac{1}{2}(x+4) = -2.$$

———————

Objective 1 Practice Exercises

For extra help, see Examples 1–3 on pages 126–127 of your text.

Solve each equation and check your solution.

1. $\dfrac{2w}{3} + \dfrac{w}{4} = \dfrac{11}{2}$

1. _________________

2. $\dfrac{3}{8}x - \dfrac{1}{3}x = \dfrac{1}{12}$

2. _________________

3. $\dfrac{1}{3}(2m-1) - \dfrac{3}{4}m = \dfrac{5}{6}$

3. _________________

Objective 2   Solve equations with decimals as coefficients.

Review these examples for Objective 2:

4. Solve $1.9x - 5 = 2.3x + 9$.

$$1.9x - 5 = 2.3x + 9$$

Step 1 $10(1.9x - 5) = 10(2.3x + 9)$

$$10(1.9x) + 10(-5) = 10(2.3x) + 10(9)$$

$$19x - 50 = 23x + 90$$

Step 2 $19x - 50 - 19x = 23x + 90 - 19x$

$$-50 = 4x + 90$$

$$-50 - 90 = 4x + 90 - 90$$

$$-140 = 4x$$

Step 3 $\dfrac{-140}{4} = \dfrac{4x}{4}$

$$-35 = x$$

Step 4 Check by substituting -35 for x in the original equation.

$$1.9x - 5 = 2.3x + 9$$

$$1.9(-35) - 5 \stackrel{?}{=} 2.3(-35) + 9$$

$$-66.5 - 5 \stackrel{?}{=} -80.5 + 9$$

$$-71.5 = -71.5 \quad \text{True}$$

The check confirms that the solution set is $\{-35\}$.

Now Try:

4. Solve $2.2x - 5 = 3.5x + 8$.

____

5. Solve $0.2x + 0.04(10 - x) = 0.06(4)$.

To clear decimals, multiply by 100.

$$0.2x + 0.04(10 - x) = 0.06(4)$$

Step 1 $100[0.2x + 0.04(10 - x)] = 100[0.06(4)]$

$$100(0.2x) + 100[0.04(10 - x)] = 100[0.06(4)]$$

$$20x + 4(10) + 4(-x) = 24$$

$$20x + 40 - 4x = 24$$

$$16x + 40 = 24$$

Step 2 $16x + 40 - 40 = 24 - 40$

$$16x = -16$$

Step 3 $\dfrac{16x}{16} = \dfrac{-16}{16}$

$$x = -1$$

Step 4 Check to confirm that $\{-1\}$ is the solution set.

5. Solve
$$0.5x + 0.04(5 - 8x) = 0.07(8).$$

____

Objective 2 Practice Exercises

For extra help, see Examples 4–5 on pages 128–129 of your text.

Solve each equation and check your solution.

4. $0.2x - 3.4 = -9$

4. ____________________

5. $0.45a - 0.35(20 - a) = 0.02(50)$

5. ____________________

6. $0.6(z - 30) - 1.85z = 7$

6. ____________________

Chapter 2   LINEAR EQUATIONS AND INEQUALITIES IN ONE VARIABLE

2.5   Applications of Linear Equations

Learning Objectives
1   Learn the six steps for solving applied problems.
2   Solve problems involving unknown numbers.
3   Solve problems involving sums of quantities.
4   Solve problems involving consecutive integers.
5   Solve problems involving complementary and supplementary angles.

Key Terms

Use the vocabulary terms listed below to complete each statement in exercises 1–5.

complementary angles   right angle   supplementary angles

straight angle   consecutive integers

1.   Two angles whose measures sum to 180° are ___________________________________.

2.   Two angles whose measures sum to 90° are ___________________________________.

3.   An angle whose measure is exactly 90° is a ___________________________________.

4.   An angle whose measure is exactly 180° is a ___________________________________.

5.   Two integers that differ by 1 are ___________________________________.

Objective 1   Learn the six steps for solving applied problems.

Objective 1 Practice Exercises

For extra help, see page 132 of your text.

1.   Write the six problem-solving steps.   1.   _______________

____

____

____

____

____

Name: Date:

Instructor: Section:

Objective 2   Solve problems involving unknown numbers.

Review this example for Objective 2:

2. The product of 5, and a number decreased by 8, is 150. What is the number?

Step 1 Read the problem carefully. We are asked to find a number.

Step 2 Assign a variable to represent the unknown quantity.

   Let x = the number.

Step 3 Write an equation.

The product   a   decreased
    of 5,   and number   by   8, is 150.
     ↓   ↓   ↓   ↓ ↓ ↓
     $5 \cdot$ $(x$ $-$ $8) = 150$

Step 4 Solve the equation.

$$5(x-8)=150$$
$$5x-40=150$$
$$5x-40+40=150+40$$
$$5x=190$$
$$\frac{5x}{5}=\frac{190}{5}$$
$$x=38$$

Step 5 State the answer. The number is 38.

Step 6 Check. The number 38 decreased by 8 is 30. The product of 5 and 30 is 150. The answer, 38, is correct.

Now Try:

2. The product of 8, and a number decreased by 11, is 40. What is the number?

____

Objective 2 Practice Exercises

For extra help, see Examples 1–2 on pages 132–133 of your text.

Write an equation for each of the following and then solve the problem. Use x as the variable.

2. If 4 is added to 3 times a number, the result is 7. Find the number.

2. _________________

   ____

3. If -2 is multiplied by the difference between 4 and a number, the result is 24. Find the number.

3. ______________________

____

4. If four times a number is added to 7, the result is five less than six times the number. Find the number.

4. ______________________

____

Objective 3   Solve problems involving sums of quantities.

Review these examples for Objective 3:

3. George and Al were opposing candidates in the school board election. George received 21 more votes than Al, with 439 votes cast. How many votes did Al receive?

Step 1 Read the problem carefully. We are given total votes and asked to find the number of votes Al received.

Step 2 Assign a variable.
Let x = the number of votes Al received.
Then $x + 21$ = the number of votes George received.

Step 3 Write an equation.

The total	is	votes for Al	plus	votes for George
↓	↓	↓	↓	↓
439	=	x	+	$(x+21)$

Step 4 Solve the equation.
$$439 = x + (x+21)$$
$$439 = 2x + 21$$
$$439 - 21 = 2x + 21 - 21$$
$$418 = 2x$$
$$\frac{418}{2} = \frac{2x}{2}$$
$$209 = x \quad \text{or} \quad x = 209$$

Step 5 State the answer. Al received 209 votes.

Now Try:

3. On a psychology test, the highest grade was 38 points more than the lowest grade. The sum of the two grades was 142. Find the lowest grade.

____

Step 6 Check. George won $209 + 21 = 230$ votes. The total number of votes is $209 + 230 = 439$. The answer checks.

6. You are making punch for a party. The recipe requires twice as much orange juice as cranberry juice and 8 times as much ginger ale as cranberry juice. If you plan to make 176 ounces of punch, how much of each ingredient should you use?

Step 1 Read the problem. The three amounts of ingredients must be found.

Step 2 Assign a variable.
   Let $x =$ the number of ounces of cranberry juice.
Then $2x =$ the number of ounces of orange juice, and $8x =$ the number of ounces of ginger ale.

Step 3 Write an equation.
Cranberry plus orange plus ginger ale is total
$$x \quad + \quad 2x \quad + \quad 8x \quad = 176$$
Step 4 Solve the equation.
$$x + 2x + 8x = 176$$
$$11x = 176$$
$$\frac{11x}{11} = \frac{176}{176}$$
$$x = 16$$

Step 5 State the answer. There are 16 ounces of cranberry juice, $2(16) = 32$ ounces of orange juice, and $8(16) = 128$ ounces of ginger ale.

Step 6 Check. The sum is 176. All conditions of the problem are satisfied.

6. You wish to build a rectangular dog pen using 52 feet of fence and the back of your house, which is 36 feet long to enclose the pen. How wide will the dog pen be if the pen is 36 feet long?

____

Objective 3 Practice Exercises

For extra help, see Examples 3–6 on pages 134–137 of your text.

Write an equation for each of the following and then solve the problem. Use x as the variable.

5. Denali in Alaska is 5910 feet higher than Mount Rainier in Washington. Together, their heights total 34,730 feet. How high is each mountain?

5. ________________

Mt. Rainier __________

Denali _____________

6. Charles bought five general admission tickets and four student tickets for a movie. He paid $35.25. If each student ticket cost $3.50, how much did each general admission ticket cost?

6. _________________

____

7. Pablo, Faustino, and Mark swim at a public pool each day for exercise. One day Pablo swam five more than three times as many laps as Mark, and Faustino swam four times as many laps as Mark. If the men swam 29 laps altogether, how many laps did each one swim?

7. _________________

Mark _________________

Pablo _________________

Faustino _________________

Objective 4   Solve problems involving consecutive integers.

Review this example for Objective 4:

8. Find two consecutive odd integers such that if three times the smaller is added to twice the larger, the sum is 69.

Step 1 Read the problem. We must find two consecutive odd integers.

Step 2 Assign a variable.
   Let x = the lesser consecutive odd integer.
Then $x + 2$ = the greater consecutive odd integer.

Step 3 Write an equation.
 Three times is added   twice the
  the smaller   to   larger   is   69.
       $\downarrow$ $\downarrow$ $\downarrow$ $\downarrow$ $\downarrow$
      $3x$ $+$ $2(x+2)$ $=$ 69

Step 4 Solve the equation.
$$3x + 2x + 4 = 69$$
$$5x + 4 = 69$$
$$5x = 65$$
$$x = 13$$

Step 5 State the answer. The lesser integer is 13.
The greater is $13 + 2 = 15$.

Now Try:

8. The sum of four consecutive even integers is 4. Find the integers.

____

Step 6 Check. Three times the smaller is 39, added to twice the larger, 30, is a sum of 69. The answers check.

Objective 4 Practice Exercises

For extra help, see Examples 7–8 on pages 137–139 of your text.

Solve each problem.

8. Find two consecutive even integers such that the smaller, added to twice the larger, is 292.

8. _______________

9. Find two consecutive integers such that the larger, added to three times the smaller, is 109.

9. _______________

10. Find three consecutive odd integers whose sum is 363.

10. _______________

Objective 5   Solve problems involving complementary and supplementary angles.

Review this example for Objective 5:

10. Find the measure of an angle if its supplement measures 4° less than three times its complement.

Step 1 Read the problem. We must find the measure of an angle.

Step 2 Assign a variable.
    Let x = the degree measure of the angle
Then $90 - x$ = the degree measure of its complement,
and $180 - x$ = the degree measure of its supplement.

Step 3 Write an equation.

The supplement	is	Three times the complement	minus	4
↓	↓	↓	↓	↓
$180 - x$	$=$	$3(90 - x)$	$-$	4

Now Try:

10. Find the measure of an angle such that the sum of the measures of its complement and its supplement is 138°.

____

Step 4 Solve the equation.
$$180 - x = 270 - 3x - 4$$
$$180 - x = 266 - 3x$$
$$180 - x + 3x = 266 - 3x + 3x$$
$$180 + 2x = 266$$
$$180 + 2x - 180 = 266 - 180$$
$$2x = 86$$
$$x = 43$$

Step 5 State the answer. The angle is 43°.

Step 6 Check. If the angle measures 43°, then its complement measures 90° – 43° = 47°, and the supplement measures 180° – 43° = 137°. Also, 137 is equal to 4 less than 3 times 47° (that is, 137° = 3(47) – 4). The answer is correct.

Objective 5 Practice Exercises

For extra help, see Examples 9–10 on pages 139–140 of your text.

Solve each problem.

11. Find the measure of an angle if the measure of the angle is 8° less than three times the measure of its supplement.

11. ______________________

12. Find the measure of an angle whose supplement measures 20° more than twice its complement.

12. ______________________

13. Find the measure of an angle whose complement is 9° more than twice its measure.

13. ______________________

                    Copyright © 2025 Pearson Education, Inc.

Chapter 2   LINEAR EQUATIONS AND INEQUALITIES IN ONE VARIABLE

2.6   Formulas and Additional Applications from Geometry

Learning Objectives
1   Solve a formula for one variable, given the values of the other variables.
2   Use a formula to solve an applied problem.
3   Solve problems involving vertical angles and straight angles.
4   Solve a formula for a specified variable.

Key Terms

Use the vocabulary terms listed below to complete each statement in exercises 1–4.

formula   area   perimeter   vertical angles

1.   The nonadjacent angles formed by two intersecting lines are called

_______________________________.

2.   An equation in which variables are used to describe a relationship is called a(n)

______________________.

3.   The distance around a figure is called its ______________________.

4.   A measure of the surface covered by a figure is called its _____________________.

Objective 1   Solve a formula for one variable, given the values of the other variables.

Review this example for Objective 1:

1.   Find the value of the remaining variable in each formula.

$A = LW; \;\; A = 54, L = 8$

Substitute the given values for A and L into the formula.

$$A = LW$$

$$54 = 8W$$

$$\frac{54}{8} = \frac{8W}{8}$$

$$6.75 = W$$

The width is 6.75. Since $8(6.75) = 54$, the answer checks.

Now Try:

1.   Find the value of the remaining variable in each formula.

$A = LW; \;\; A = 88, L = 16$

____

Objective 1 Practice Exercises

For extra help, see Example 1 on page 146 of your text.

In the following exercises, a formula is given, along with the values of all but one of the variables in the formula. Find the value of the variable that is not given.

1. $S = \dfrac{a}{1-r}$; $S = 60$, $r = 0.4$

    1. ______________________

2. $I = prt$; $I = 288$, $r = 0.04$, $t = 3$

    2. ______________________

3. $A = \frac{1}{2}(b + B)h$; $b = 6$, $B = 16$, $A = 132$

    3. ______________________

Objective 2   Use a formula to solve an applied problem.

Review these examples for Objective 2:

2. Find the dimensions of a rectangle. The length is 4 m less than three times the width. The perimeter is 96 m.

    Step 1 Read the problem. We must find the dimensions of the rectangle.

    Step 2 Assign a variable.
    Let W = the width of the rectangle, in meters.
    Then $L = 3W - 4$ is the length, in meters.

    Step 3 Write an equation. Use the formula for the perimeter of a rectangle. Substitute $3W - 4$ for the length.
    $$P = 2L + 2W$$
    $$96 = 2(3W - 4) + 2W$$

    Step 4 Solve.
    $$96 = 6W - 8 + 2W$$
    $$96 = 8W - 8$$
    $$96 + 8 = 8W - 8 + 8$$
    $$104 = 8W$$
    $$\frac{104}{8} = \frac{8W}{8}$$
    $$13 = W$$

Now Try:

2. Ruth has 42 feet of binding for a rectangular rug that she is weaving. If the rug is 9 feet wide, how long can she make the rug if she wishes to use all the binding on the perimeter of the rug?

    ____

                    Copyright © 2025 Pearson Education, Inc.

Step 5 State the answer. The width is 13 m. The length is $3(13) - 4 = 35$ m.

Step 6 Check. The perimeter is $2(13) + 2(35) = 96$ m. The answer checks.

3. The longest side of a triangle is 4 feet longer than the shortest side. The medium side is 2 feet longer than the shortest side. If the perimeter is 36 feet, what are the lengths of the three sides?

Step 1 Read the problem. We must find the lengths of the sides.

Step 2 Assign a variable.
Let s = the length of the shortest side, in feet.
Then $s + 2$ = the length of the medium side, in feet,
and $s + 4$ = the length of the longest side, in feet.

Step 3 Write an equation. Use the formula for the perimeter of a triangle.
$$P = a + b + c$$
$$36 = s + (s + 2) + (s + 4)$$

Step 4 Solve.
$$36 = 3s + 6$$
$$30 = 3s$$
$$10 = s$$

Step 5 State the answer. Since s represents the length of the shortest side, its measure is 10 ft.
$s + 2 = 10 + 2 = 12$ ft is the length of the medium side.
$s + 4 = 10 + 4 = 14$ ft is the length of the longest side.

Step 6 Check. The perimeter is $10 + 12 + 14 = 36$ ft, as required.

3. The longest side of a triangle is twice as long as the shortest side. The medium side is 5 feet longer than the shortest side. If the perimeter is 65 feet, what are the lengths of the three sides?

____

Objective 2 Practice Exercises

For extra help, see Examples 2–4 on pages 147–148 of your text.

Use a formula to write an equation for each of the following applications; then solve the application. (Use 3.14 as an approximation for π.)

4. Find the height of a triangular banner whose area is 48 square inches and base is 12 inches.

4. ______________________

5. Lucia invests \$5000 at 6% simple interest and earns   **5.** ___________________
\$450. How long did Lucia invest the money?

6. The circumference of a circular garden is 628 feet.   **6.** _________________
Find the area of the garden. (Hint: First find the
radius of the garden.)

Objective 3   Solve problems involving vertical angles and straight angles.

Review this example for Objective 3: | **Now Try:**

5. | **5.**

Find the measure of the marked angles in the figure below.

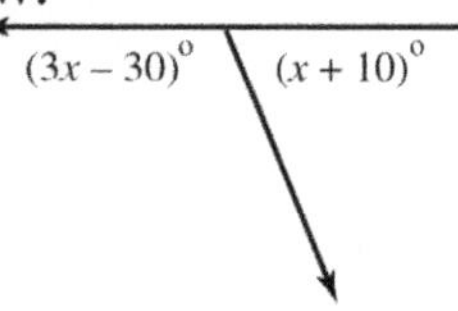

Find the measure of the marked angles in the figure below.

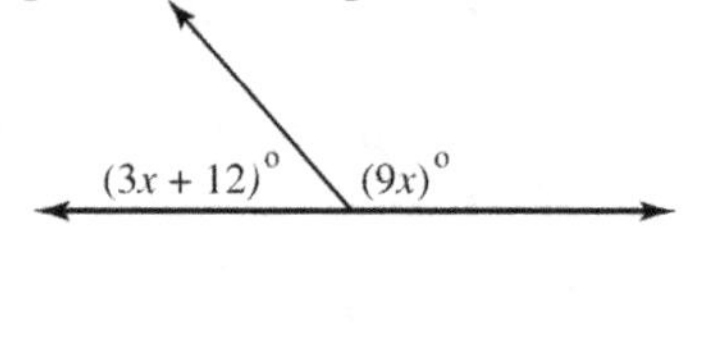

____

The measures of the marked angles must add to
180° because together they form a straight angle.
The angles are supplements of each other.

$$(3x - 30) + (x + 10) = 180$$
$$4x - 20 = 180$$
$$4x = 200$$
$$x = 50$$

Replace x with 50 in the measure of each marked
angle.

$$3x - 30 = 3(50) - 30 = 150 - 30 = 120$$
$$x + 10 = 50 + 10 = 60$$

The two angles measure 120° and 60°.

              Copyright © 2025 Pearson Education, Inc.

Objective 3 Practice Exercises

For extra help, see Example 5 on page 149 of your text.

Find the measure of each marked angle.

7.

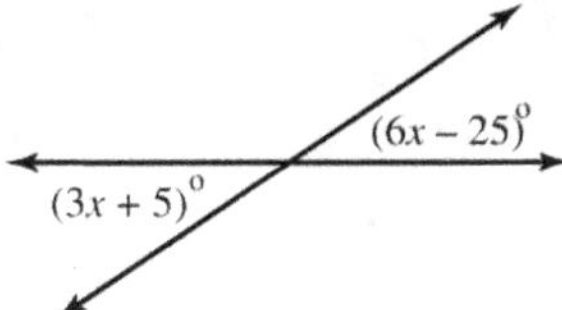

7. _________________

8.

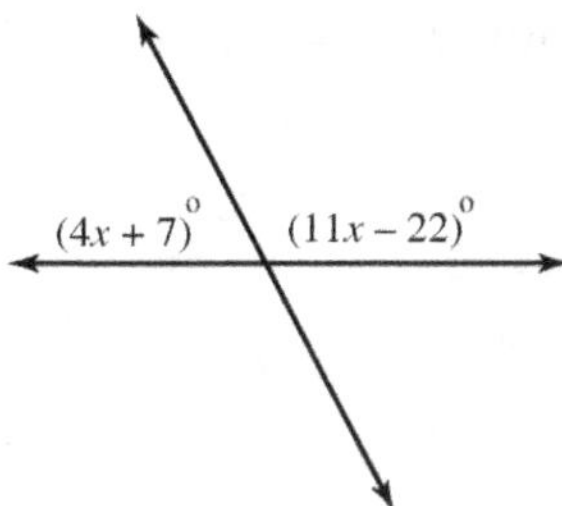

8. _________________

9.

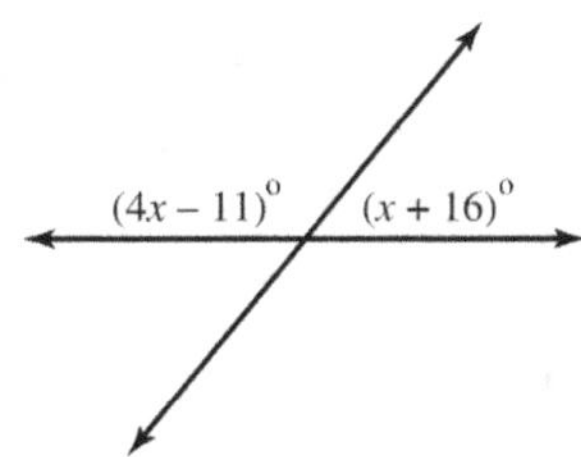

9. _________________

Objective 4   Solve a formula for a specified variable.

Review these examples for Objective 4:

6. Solve $A = \dfrac{1}{2}bh$ for h.

$$A = \frac{1}{2}bh$$

$$2A = bh$$

$$\frac{2A}{b} = \frac{bh}{b}$$

$$\frac{2A}{b} = h \quad \text{or} \quad h = \frac{2A}{b}$$

Now Try:

6. Solve $d = rt$ for r.

____

7. Solve $A = p + prt$ for r.

$$A = p + prt$$
$$A - p = p + prt - p$$
$$A - p = prt$$
$$\frac{A - p}{pt} = \frac{prt}{pt}$$
$$\frac{A - p}{pt} = r \quad \text{or} \quad r = \frac{A - p}{pt}$$

8. Solve $V = k + gt$ for t.

$$V = k + gt$$
$$V - k = k + gt - k$$
$$V - k = gt$$
$$\frac{V - k}{g} = \frac{gt}{g}$$
$$\frac{V - k}{g} = t$$
$$t = \frac{V - k}{g}$$

7. Solve $P = a + b + c$ for a.

____

8. Solve $A = \frac{1}{2}h(b + B)$ for h.

____

Objective 4 Practice Exercises

For extra help, see Examples 6–9 on pages 150–151 of your text.

Solve each formula for the specified variable.

10. $V = LWH$ for H

10. ______________

11. $S = (n - 2)180$ for n

11. ______________

12. $V = \frac{1}{3}\pi r^2 h$ for h

12. ______________

Chapter 2   LINEAR EQUATIONS AND INEQUALITIES IN ONE VARIABLE

2.7   Ratio, Proportion, and Percent

Learning Objectives
1   Write ratios.
2   Solve proportions.
3   Solve applied problems using proportions.
4   Solve percents problems.

Key Terms

Use the vocabulary terms listed below to complete each statement in exercises 1−4.

> ratio   proportion   extremes   means   cross products   terms

1.   A _____________________ is a statement that two ratios are equal.

2.   A _____________________ is a comparison of two quantities using a quotient.

3.   In the proportion, $\dfrac{a}{b} = \dfrac{c}{d}$, a, b, c, and d are called the _____________, while a and d are called the _____________, and b and c are called the _____________.

4.   To see whether a proportion is true, determine if the _____________________ are equal.

Objective 1   Write ratios.

Review these examples for Objective 1:

1.   Write a ratio for each word phrase.

a. 7 hr to 9 hr

$$\frac{7 \text{ hr}}{9 \text{ hr}} = \frac{7}{9}$$

b. 5 dollars to 25 dollars

$$\frac{5 \text{ dollars}}{25 \text{ dollars}} = \frac{5}{25} = \frac{1}{5}$$

c. 15 hr to 4 days

First convert 4 days to hours.

   4 days $= 4 \cdot 24 = 96$ hr

Now write the ratio using the common unit of measure, hours.

$$\frac{15 \text{ hr}}{4 \text{ days}} = \frac{15 \text{ hr}}{96 \text{ hr}} = \frac{15}{96}, \quad \text{or} \quad \frac{5}{32}$$

Now Try:

1.   Write a ratio for each word phrase.

a. 11 hr to 17 hr

____

b. 12 people to 32 people

____

c. 32 hr to 5 days

____

2. The local grocery store charges the following prices for a bottle of olive oil.
    16-ounce bottle: $6.99
    25.5-ounce bottle: $9.99
    32-ounce bottle: $12.99
    44-ounce bottle: $14.99
Which size is the best buy? That is, which size has the lowest unit price?

To find the best buy, write ratios comparing the price for each size bottle to the number of units (ounces) per bottle.

Size	Unit price (dollars per ounce)
16 oz	$\dfrac{\$6.99}{16} = \$0.437$
25.5 oz	$\dfrac{\$9.99}{25.5} = \$0.392$
32 oz	$\dfrac{\$12.99}{32} = \$0.406$
44 oz	$\dfrac{\$14.99}{44} = \$0.341$

Because the 44-oz size has the lowest unit price, $0.341, it is the best buy.

2. The local grocery store charges the following prices for a jar of applesauce.
    16-ounce jar: $1.19
    24-ounce jar: $1.29
    48-ounce jar: $2.69
    64-ounce jar: $3.49
Which size is the best buy? That is, which size has the lowest unit price?

____

Objective 1 Practice Exercises

For extra help, see Examples 1–2 on pages 157–158 of your text.

Write a ratio for each word phrase. Write fractions in lowest terms.

1. 8 men to 3 men

1. _______________

2. 9 dollars to 48 quarters

2. _______________

A supermarket was surveyed and the following prices were charged for items in various sizes. Find the best buy (based on price per unit) for each of the following items.

3. Trash bags
    10-count box: $2.89
    20-count box: $5.29
    45-count box: $6.69
    85-count box: $13.99

3. _______________

Objective 2   Solve proportions.

Review these examples for Objective 2:	**Now Try:**
3. Decide whether the proportion is *true* or *false*.	**3.** Decide whether the proportion is *true* or *false*.

3. Decide whether the proportion is *true* or *false*.

$$\frac{2}{5} = \frac{12}{30}$$

Check to see whether the cross product are equal.

$$5 \cdot 12 = 60$$

$$\frac{2}{5} = \frac{12}{30}$$

$$2 \cdot 30 = 60$$

The cross products are equal, so the proportion is true.

Now Try:

3. Decide whether the proportion is *true* or *false*.

$$\frac{5}{6} = \frac{20}{24}$$

____

4. Solve the proportion $\frac{6}{11} = \frac{x}{88}$.

Solve for x.

$$\frac{6}{11} = \frac{x}{88}$$

$$6 \cdot 88 = 11 \cdot x$$

$$528 = 11x$$

$$48 = x$$

Check by substituting 48 for x in the proportion. The solution set is $\{48\}$.

4. Solve the proportion $\frac{x}{15} = \frac{42}{90}$.

____

5. Solve the equation $\frac{n-1}{3} = \frac{2n+1}{4}$.

$$\frac{n-1}{3} = \frac{2n+1}{4}$$

$$3(2n+1) = 4(n-1)$$

$$6n+3 = 4n-4$$

$$2n+3 = -4$$

$$2n = -7$$

$$n = -\frac{7}{2}$$

A check confirms that the solution is $-\frac{7}{2}$, so the solution set is $\left\{-\frac{7}{2}\right\}$.

5. Solve the equation

$$\frac{2x+1}{2} = \frac{7x+3}{9}.$$

____

                87

Objective 2 Practice Exercises

For extra help, see Examples 3–5 on pages 159–160 of your text.

Solve each equation.

4. $\dfrac{z}{20} = \dfrac{25}{125}$ 4. ______________

5. $\dfrac{m}{5} = \dfrac{m-2}{2}$ 5. ______________

6. $\dfrac{z+1}{4} = \dfrac{z+7}{2}$ 6. ______________

Objective 3   Solve applied problems using proportions.

Review these examples for Objective 3:

6. If four pounds of fertilizer will cover 50 square feet of garden, how many pounds would be needed for 125 square feet?

To solve this problem, set up a proportion, with pounds in the numerator and square feet in the denominator.

$$\frac{4}{50} = \frac{x}{125}$$
$$4(125) = 50x$$
$$500 = 50x$$
$$10 = x$$

10 lb of fertilizer are needed.

Now Try:

6. Mei earns \$423.28 in 26 hours. How much does Mei earn in 40 hours?

____

                                 Copyright © 2025 Pearson Education, Inc.

7. Suppose you are the 18th person in the express line to pay for items on Black Friday. You order pizza take out when you get in line.

a. In 160 seconds, 3 people have completed their purchases. Assuming this rate stays the same, how long (in minutes) will you be waiting in line?

Let x = the number of minutes you will be waiting in line.

$$\frac{160 \text{ sec}}{3 \text{ people}} = \frac{x \text{ sec}}{18 \text{ people}}$$

$$18(160) = 3x$$

$$2880 = 3x$$

$$960 = x$$

Because 60 sec = 1 min, the number of minutes waiting in line will be

$\dfrac{960}{60}$, or 16 minutes.

b. It takes 10 min to get to the pizza take out place. If the pizza will be ready 30 minutes after the order is placed, will you arrive at the pizza take out place in time?

Yes. Based on the answer to part (a), you will arrive $30 - 16 - 10 = 4$ minutes early.

7. Suppose you are the 8th person in line to pay for items on Black Friday. You order pizza for delivery when you get in line.
a. In 240 seconds, 2 people have completed their purchases. Assuming this rate stays the same, how long (in minutes) will you be waiting in line?

____

b. It takes 15 min to get home. If the pizza will be delivered 30 minutes after the order is placed, will you arrive at home in time?

____

Objective 3 Practice Exercises

For extra help, see Examples 6–7 on page 161 of your text.

Solve each problem.

7. On a road map, 6 inches represents 50 miles. How many inches would represent 125 miles?

      7. _______________

8. If 12 rolls of tape cost $4.60, how much will 15 rolls cost?

      8. _______________

9. A garden service charges \$30 to install 50 square feet of sod. Find the charge to install 225 square feet.

9. _______________

Objective 4   Solve percents problems.

Review these examples for Objective 4:

8. Solve each problem.

a. What is 18% of 700?

Let n = the number. The word of indicates multiplication.

$$\text{What} \quad \text{is} \quad 18\% \quad \text{of} \quad 700$$
$$\downarrow \qquad \downarrow \qquad \downarrow \qquad \downarrow \qquad \downarrow$$
$$n \quad = \quad 0.18 \quad \cdot \quad 700$$
$$n = 126$$

Thus, 126 is 18% of 700.

b. 54% of what number is 162?

$$54\% \quad \text{of} \quad \text{what number} \quad \text{is} \quad 162$$
$$\downarrow \qquad \downarrow \qquad \downarrow \qquad \downarrow \qquad \downarrow$$
$$0.54 \quad \cdot \qquad n \qquad = \quad 162$$
$$n = \frac{162}{0.54}$$
$$n = 300$$

54% of 300 is 162.

c. 75 is what percent of 500?

$$75 \quad \text{is} \quad \text{what percent} \quad \text{of} \quad 500$$
$$\downarrow \quad \downarrow \qquad \downarrow \qquad \downarrow \quad \downarrow$$
$$75 \quad = \qquad p \qquad \cdot \quad 500$$
$$\frac{75}{500} = p$$
$$0.15 = p$$

Thus, 75 is 15% of 500.

Now Try:

8. Solve each problem.

a. What is 35% of 400?

____

b. 42% of what number is 399?

____

c. 102 is what percent of 120?

____

9. An advertisement for a bicycle gives a sale price of $175.50. The regular price is $225. Find the percent discount on this bicycle.

The savings amounted to $225 − $175.50 = $49.50. We can now restate the problem: What percent of 225 is 49.50?

$$\text{What percent} \quad \text{of} \quad 225 \quad \text{is} \quad 49.50$$
$$p \qquad \cdot \quad 225 \;=\; 49.50$$
$$p = \frac{49.50}{225}$$
$$p = 0.22 \quad \text{or} \quad 22\%$$

The discount is 22%.

9. The number of students enrolled in a calculus course is 145. If 40% of these students are from in-state, how many are from out-of-state?

____

10. An Uber ride cost $28.58. You want to leave a 10% tip. Estimate how much you should leave to the nearest dollar?

Round $28.58 to $30.00 We can restate the problem: 10% of $30.00 *is what number?*

$$10\% \quad \text{of} \quad 30 \quad \text{is} \quad \text{what number?}$$
$$0.10 \quad \cdot \quad 30 \;=\; \qquad n$$
$$3 = n \qquad\qquad \text{or} \quad 22\%$$

You should leave a $3 tip.

10. A restaurant bill (before tax) was $51.35. You want to leave a 20% tip. Estimate how much you should leave to the nearest dollar.

____

Objective 4 Practice Exercises

For extra help, see Examples 8–10 on pages 162–163 of your text.

Answer each question about percent.

10. What is 2.5% of 3500?

10. ______________

11. What percent of 5200 is 104?

11. ______________

Solve the problem.

12. Rodrigo recently bought a duplex for $144,000. Rodrigo expects to earn $6120 per year on this investment. What percent of the purchase price will Rodrigo earn?

12. ______________

Chapter 2   LINEAR EQUATIONS AND INEQUALITIES IN ONE VARIABLE

2.8   Further Applications of Linear Equations

Learning Objectives
1   Solve percent problems involving rates.
2   Solve problems involving mixtures
3   Solve problems involving simple interest.
4   Solve problems involving denominations of money.
5   Solve problems involving distance, rate, and time.

Objective 1   Solve percent problems involving rates.

Review this example for Objective 1:

1. The purchase price of a car is $12,500. In order to finance the car a purchaser is required to make a minimum down payment of 20% of the purchase price. What is the minimum down payment required?

$$12,500 \cdot 0.20 = 2500$$

The down payment is $2500

Now Try:

1. A certificate of deposit pays 2.6% simple interest in one year on a principal of $4500. What interest is being paid on this deposit?

____

Objective 1 Practice Exercises

For extra help, see Example 1 on page 170 of your text.

Solve each problem.

1. Twelve percent of a college student body has a grade point average of 3.0 or better. If there are 1250 students enrolled in the college, how many have a grade point average of less than 3.0?

1. ______________

2. In a class of first year and second year students only, there are 85 students. If 60% of the class is second year students, how many students are second year students?

2. ______________

3. At a large university, 40% of the student body is   **3.** _______________
from out-of-state. If the total enrollment at the
university is 25,000 students, how many are from
in-state?

Objective 2   Solve problems involving mixtures.

Review this example for Objective 2:

2. How many gallons of a 20% alcohol solution must be mixed with 15 gallons of a 12% alcohol solution to obtain a 14% alcohol solution?

Step 1 Read the problem. Find the amount of the 20% alcohol solution.

Step 2 Assign a variable. Let x = the amount of the 20% alcohol solution.

Liters of Solution	Rate (as a decimal)	Liters of Pure Acid
x	0.20	$0.2x$
15	0.12	0.12(15)
$x+15$	0.14	$0.14(x+15)$

Step 3 Write an equation.
$$0.20x + 0.12(15) = 0.14(x+15)$$

Step 4 Solve.
$$0.20x + 1.8 = 0.14x + 2.1$$
$$0.06x = 0.3$$
$$x = 5$$

Step 5 State the answer. 5 gallons of the 20% alcohol solution must be added.

Step 6 Check.
$$0.20(5) + 0.12(15) \stackrel{?}{=} 0.14(5+15)$$
$$2.8 = 2.8$$
The answer checks.

Now Try:

2. How many ounces of a 35% solution of acid must be mixed with a 60% solution to get 20 ounces of a 50% solution?

____

Objective 2 Practice Exercises

For extra help, see Examples 2–3 on pages 170–172 of your text.

Solve each problem.

4. How many pounds of peanuts worth $3 per pound must be mixed with mixed nuts worth $5.50 per pound to make 40 pounds of a mixture worth $5 per pound?

4. ______________________

5. How many pounds of candy worth $7 per pound must be mixed with candy worth $4.50 per pound to make 100 pounds of candy worth $6 per pound?

5. ______________________

Objective 3   Solve problems involving simple interest.

Review this example for Objective 3:

4. Luan invested some money at 8% simple interest and $700 less than this amount at 7%. Luan's total annual income from the interest was $584. How much was invested at each rate?

Step 1 Read the problem. Find the two amounts.

Step 2 Assign a variable.
Let x = the amount at 8%.
Let $x - 700$ = the amount at 7%.

Amount invested (in dollars)	Rate (as a decimal)	Interest for One Year (in dollars)
x	0.08	$0.08x$
$x - 700$	0.07	$0.07(x - 700)$

Step 3 Write an equation.
$$0.08x + 0.07(x - 700) = 584$$

Now Try:

4. Taniel invested some money at 5% and $300 more than twice this amount at 7%. Taniel's total annual income from the two investments is $325. How much is invested at each rate?

____

Step 4 Solve.
$$0.08x + 0.07x - 49 = 584$$
$$0.15x - 49 = 584$$
$$0.15x = 633$$
$$x = 4220$$

Step 5 State the answer. Luan must invest $4220 at 8%, and 4220 − 700 = $3520 at 7%.

Step 6 Check. The sum of the two amounts should be $584.
$$0.08(4220) + 0.07(3520) = 584$$

Objective 3 Practice Exercises

For extra help, see Example 4 on page 173 of your text.

Solve each problem.

6. Louisa invested $16,000 in bonds paying 7% simple interest. How much additional money should she invest at 4% simple interest so that the average return on the two investments is 6%?

6. ______________________

7. Desiree invested some money at 9% and $100 less than three times that amount at 7%. Her total annual interest was $83. How much did she invest at each rate?

7. 7%: ______________

    9%: ______________

8. Jacob has $48,000 invested in stocks paying 6%. How much additional money should he invest in certificates of deposit paying 2.5% so that the average return on the two investments is 4%?

8. ______________________

Objective 4   Solve problems involving denominations of money.

Review this example for Objective 4:

5. A collection of dimes and nickels has a total value of $2.70. The number of nickels is 2 more than twice the number of dimes. How many of each type of coin are in the collection?

Step 1 Read the problem. Find the number of dimes and the number of nickels.

Step 2 Assign a variable.
Let d = the number of dimes.
Then $2d + 2$ = the number of nickels.

Number of coins	Denomination (in dollars)	Total Value (in dollars)
d	0.10	$0.10d$
$2d + 2$	0.05	$0.05(2d + 2)$

Step 3 Write an equation.
$$0.10d + 0.05(2d + 2) = 2.70$$

Step 4 Solve.
$$10d + 5(2d + 2) = 270$$
$$10d + 10d + 10 = 270$$
$$20d + 10 = 270$$
$$20d = 260$$
$$d = 13$$

Step 5 State the answer. There are 13 dimes and $2(13) + 2 = 28$ nickels.

Step 6 Check.
$$0.10(13) + 0.05(28) = 1.30 + 1.40 = 2.70$$
The answer is correct.

Now Try:

5. Erika's piggy bank has quarters and nickels in it. The total number of coins in the piggy bank is 50. Their total value is $8.90. How many of each type are in the piggy bank?

____

____

Objective 4 Practice Exercises

For extra help, see Example 5 on page 174 of your text.

Solve each problem.

9. Anthony sells two different size jars of peanut butter. The large size sells for $2.60 and the small size sells for $1.80. He has 80 jars worth $164. How many of each size jar does he have?

9.
large ____________________

small ____________________

10. Stan has 14 bills in his wallet worth $95 altogether. If the wallet contains only $5 and $10 bills, how many bills of each denomination does he have?

10.
$5 bills ______________

$10 bills ______________

11. Twice as many general admission tickets to a basketball game were sold as reserved seat tickets. General admission tickets cost $10 and reserved seat tickets cost $15. If the total value of both kinds of tickets was $26,250, how many tickets of each kind were sold?

11.
general admission______

reserved seats ________

Objective 5   Solve problems involving distance, rate, and time.

| **Review these examples for Objective 5:** | **Now Try:** |

6. A driver averaged 58 mph and took 10 hours to drive from Little Rock to Indianapolis. What is the distance between Little Rock and Indianapolis?

We must find the distance, given the rate and time using $rt = d$.

$$58 \cdot 10 = 580 \text{ miles}$$

The distance is 580 miles.

6. A driver averaged 54 mph and took 5 hours to drive from Los Angeles to Las Vegas. What is the distance between Los Angeles and Las Vegas?

____

7. A car and a truck leave Oklahoma City at the same time and travel west on the same route. The car travels at a constant rate of 62 mph. The truck travels at a constant rate of 68 mph. In how many hours will the distance between them be 30 miles?

7. A car and a truck leave Dallas at the same time and travel north on the same route. The car travels at a constant rate of 67 mph. The truck travels at a constant rate of 72 mph. In how many hours will the distance between them be 25 miles?

____

Step 1 Read the problem.

Step 2 Assign a variable. We are looking for time.
Let t = the number of hours until the distance between them is 30 miles.

	Rate	Time	Distance
Truck	68	t	$68t$
Car	62	t	$62t$

Step 3 Write an equation.
$$68t - 62t = 30$$

Step 4 Solve.
$$6t = 30$$
$$t = 5$$

Step 5 State the answer. It will take 5 hours for the truck and car to be 30 miles apart.

Step 6 Check. After 5 hours, the truck travels $68 \cdot 5 = 340$ miles and the car travels $62 \cdot 5 = 310$ miles. The difference is $340 - 310 = 30$, as required.

Objective 5 Practice Exercises

For extra help, see Examples 6–8 on pages 175–176 of your text.

Solve each problem.

12. A driver averages 52 mph and took 10 hours to drive   **12.** _______________
     from Charlotte, North Carolina to Orlando, Florida.
     What is the distance between Charlotte and Orlando?

13. A driver averages 50 mph and took 9 hours to drive   **13.** _______________
     from Denver, Colorado to Albuquerque, New
     Mexico. What is the distance between Denver and
     Albuquerque?

14. A car and a truck leave Billings, Montana at the   **14.** _______________
     same time and travel east on the same route. The
     truck travels at a constant rate of 78 mph. The car
     travels at a constant rate of 67 mph. In how many
     hours will the distance between them be 33 miles?

Name: _______________________   Date: _______________________
Instructor: _______________________   Section: _______________________

Chapter 2   LINEAR EQUATIONS AND INEQUALITIES IN ONE VARIABLE

2.9   Solving Linear Inequalities

Learning Objectives
1   Graph intervals on a number line.
2   Use the addition property of inequality.
3   Use the multiplication property of inequality.
4   Solve linear inequalities using both properties of inequality.
5   Solve applied problems using inequalities.
6   Solve linear inequalities with three parts.

Key Terms

Use the vocabulary terms listed below to complete each statement in exercises 1−5.

inequalities   **interval**   **interval notation**

linear inequality   **three-part inequality**

1.   An inequality that says that one number is between two other numbers is a(n)_______________________________________.

2.   A portion of a number line is called a(n) _______________________________________.

3.   A(n) _______________________________ can be written in the form $Ax + B < C$, $Ax + B \leq C$, $Ax + B > C$, or $Ax + B \geq C$, where A, B, and C are real numbers with $A \neq 0$.

4.   Algebraic expressions related by $<$, $\leq$, $>$, or $\geq$ are called _______________________.

5.   The _______________________________ for $a \leq x < b$ is $[a, b)$.

Objective 1   Graph intervals on a number line.

Review this example for Objective 1:

1.   Write the inequality in interval notation, and graph the interval.

$x > -3$

The statement $x > -3$ says that x can represent any value greater than -3, but cannot equal -3, written $(-3, \infty)$. We graph this interval by placing a parenthesis at -3 and drawing an arrow to the right. The parenthesis indicates that -3 is not part of the graph.

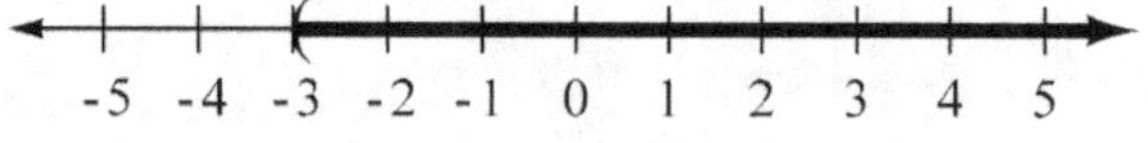

Now Try:

1.   Write the inequality in interval notation, and graph the interval.

$x > -1$

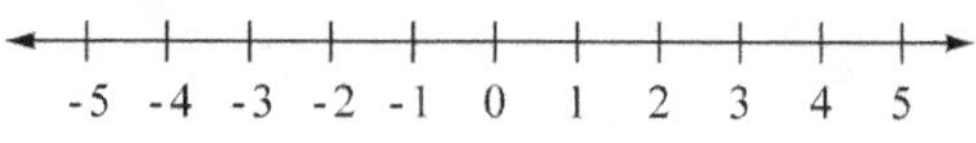

   Copyright © 2025 Pearson Education, Inc.

Objective 1 Practice Exercises

For extra help, see Example 1 on page 183 of your text.

Write each inequality in interval notation and graph the interval.

1. $3 < a$

1. _________________________
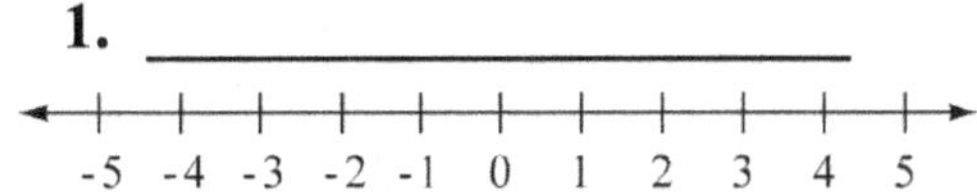

2. $y \geq -2$

2. _________________________
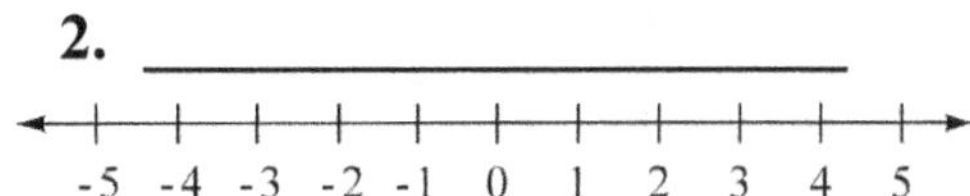

3. $x < -4$

3. _________________________
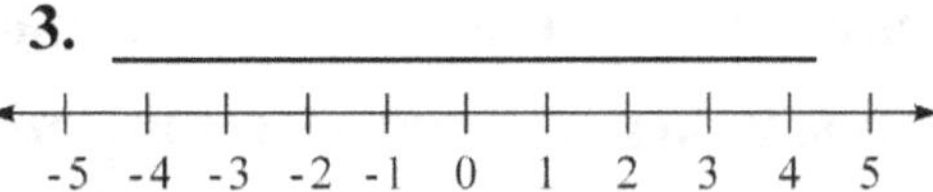

Objective 2   Use the addition property of inequality.

Review this example for Objective 2:

2.   Solve $8 + 4x \geq 3x + 3$ and graph the solution set.

$$8 + 4x \geq 3x + 3$$
$$8 + 4x - 3x \geq 3x + 3 - 3x$$
$$8 + x \geq 3$$
$$8 + x - 8 \geq 3 - 8$$
$$x \geq -5$$

The solution set, $[-5, \infty)$ is graphed below.

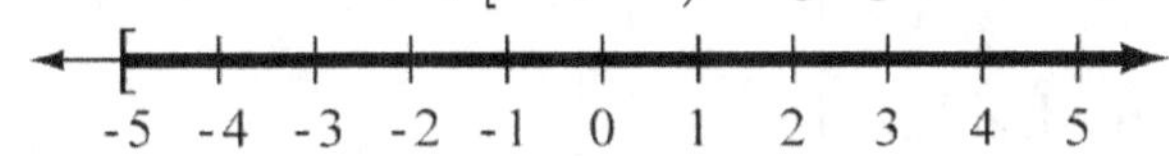

Now Try:

2.   Solve $5 + 9x \geq 8x + 2$ and graph the solution set.

____
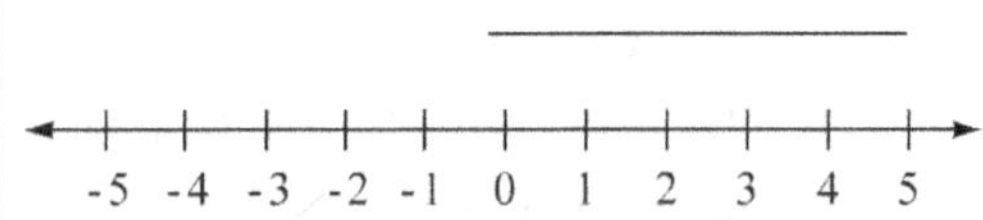

Objective 2 Practice Exercises

For extra help, see Example 2 on page 184 of your text.

Solve each inequality. Write the solution set in interval notation and then graph it.

4. $5a + 3 \leq 6a$

4. _________________________
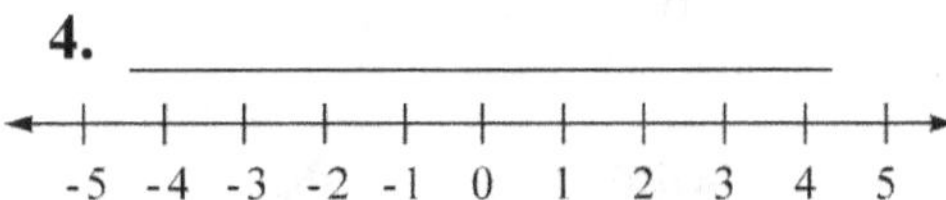

       101

5. $\quad 6+3x<4x+4$

5. _______________________

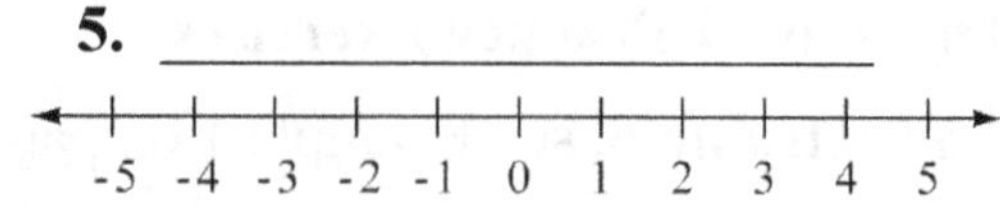

6. $\quad 3+5p\le 4p+3$

6. _______________________

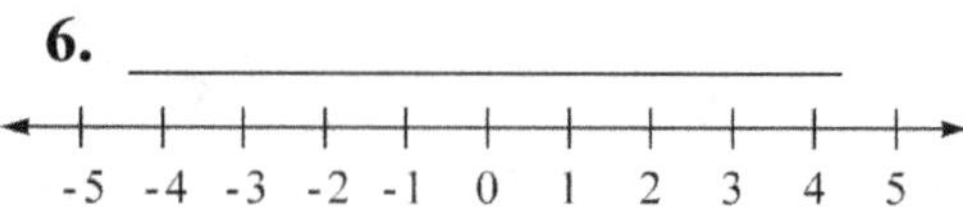

Objective 3   Use the multiplication property of inequality.

Review these examples for Objective 3:

3. Solve each inequality, and graph the solution set.

   a. $6x<-24$

We divide each side by 6.

$$6x<-24$$

$$\frac{6x}{6}<\frac{-24}{6}$$

$$x<-4$$

The graph of the solution set $(-\infty,-4)$, is shown below.

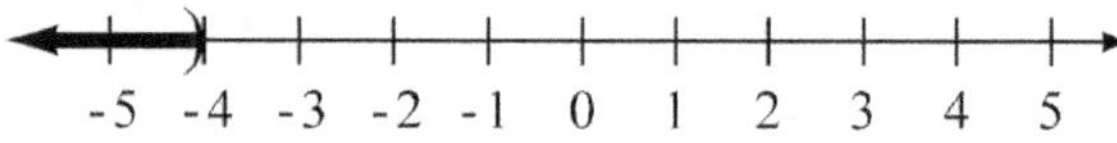

   b. $-6x\ge 30$

Here each side of the inequality must be divided by -6, a negative number, which does require changing the direction of the inequality symbol.

$$-6x\ge 30$$

$$\frac{-6x}{-6}\le\frac{30}{-6}$$

$$x\le -5$$

The solution set, $(-\infty,-5]$, is graphed below.

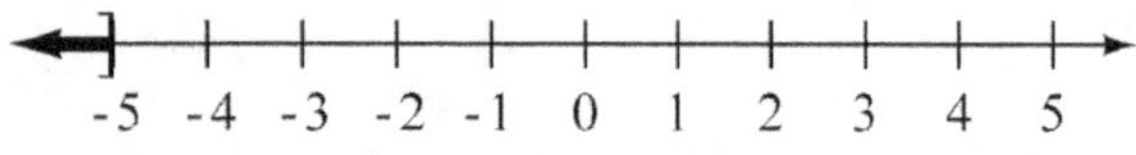

Now Try:

3. Solve each inequality, and graph the solution set.

   a. $8x\le -40$

____

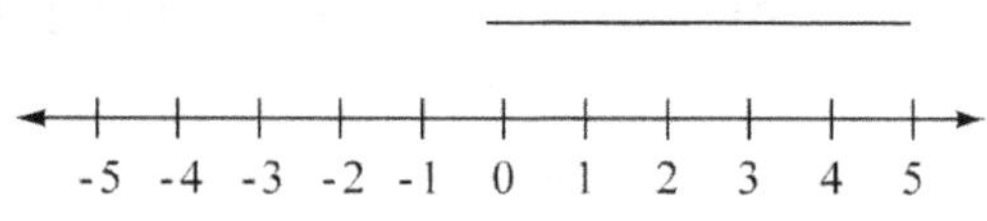

   b. $-9t>36$

____

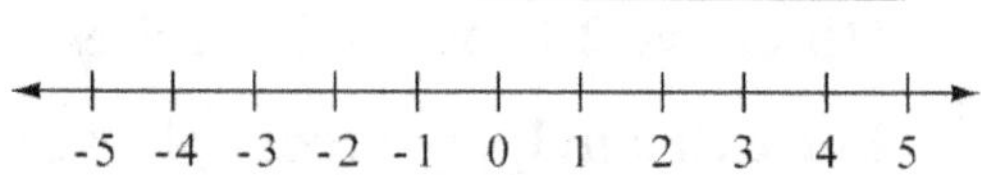

Objective 3 Practice Exercises

For extra help, see Example 3 on pages 186–187 of your text.

Solve each inequality. Write the solution set in interval notation and then graph it.

7. $-2s < 4$

7. ___________________________

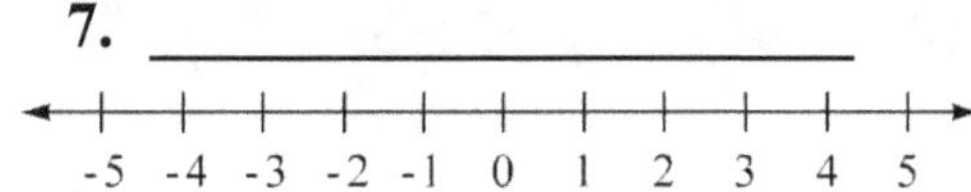

8. $4k \geq -16$

8. ___________________________

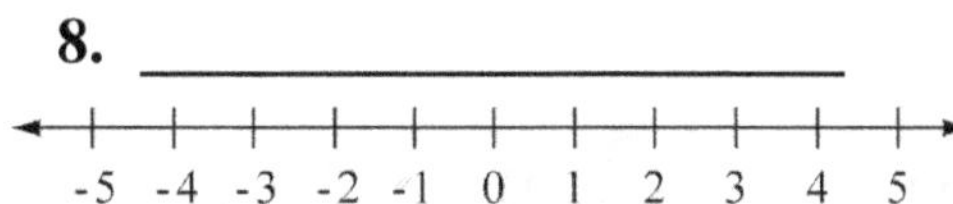

9. $-9m \geq -36$

9. ___________________________

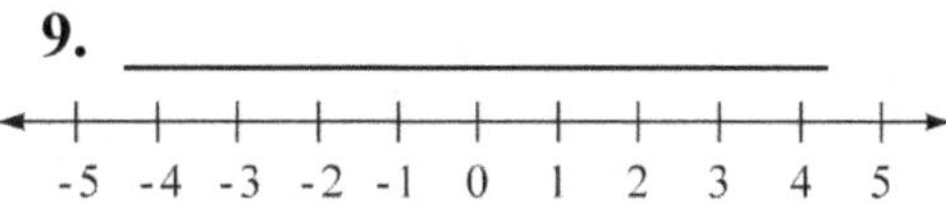

Objective 4   Solve linear inequalities using both properties of inequality.

Review this example for Objective 4:

4. Solve $4x + 3 - 7 > -2x + 8 + 3x$. Graph the solution set.

Step 1 Combine like terms and simplify.
$$4x + 3 - 7 > -2x + 8 + 3x$$
$$4x - 4 > x + 8$$

Step 2 Use the addition property of inequality.
$$4x - 4 - x > x + 8 - x$$
$$3x - 4 > 8$$
$$3x - 4 + 4 > 8 + 4$$
$$3x > 12$$

Step 3 Use the multiplication property of inequality.
$$\frac{3x}{3} > \frac{12}{3}$$
$$x > 4$$

The solution set is $(4, \infty)$. The graph is shown below.

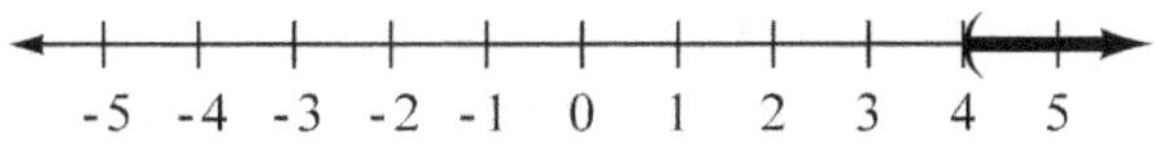

Now Try:

4. Solve $8x - 5 + 4 \geq 6x - 3x + 9$. Graph the solution set.

____

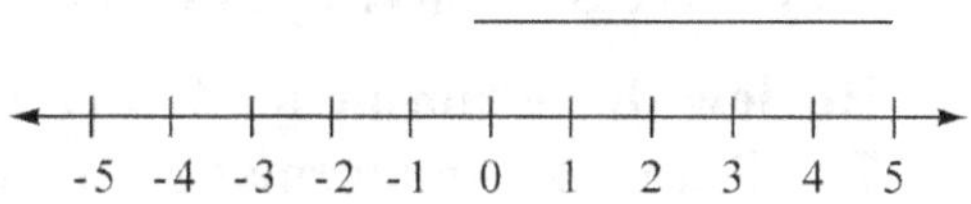

Objective 4 Practice Exercises

For extra help, see Example 4–6 on pages 187–188 of your text.

Solve each inequality. Write the solution set in interval notation and then graph it.

10. $4(y-3)+2>3(y-2)$

10. ______________________________

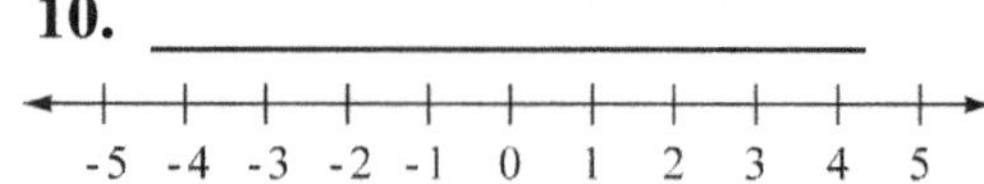

11. $-3(m+2)+3\le-4(m-2)-6$

11. ______________________________

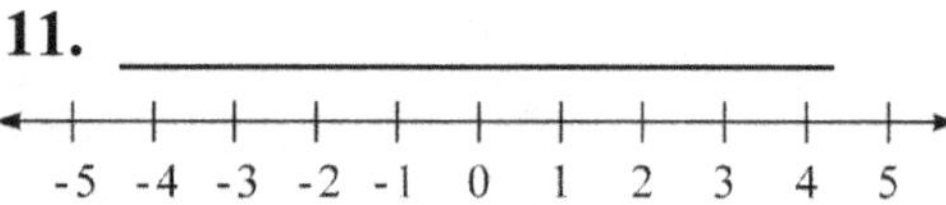

12. $7(2-x)\le-2(x-3)-x$

12. ______________________________

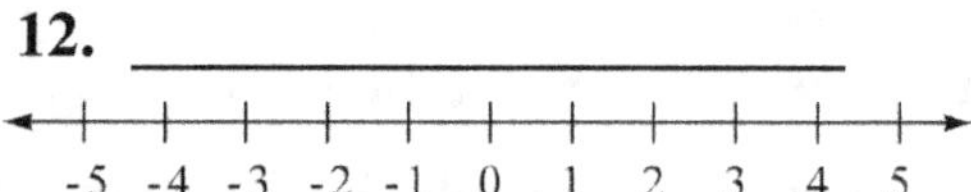

Objective 5   **Solve applied problems using inequalities.**

Review this example for Objective 5:

7.   Ruth tutors mathematics in the evenings in an office for which she pays \$600 per month rent. If rent is her only expense and she charges each student \$40 per month, how many students must she teach to make a profit of at least \$1600 per month?

Step 1 Read the problem again.

Step 2 Assign a variable.
   Let x = the number of students.

Step 3 Write an inequality.
   $40x-600\ge1600$

Step 4 Solve.
   $40x-600+600\ge1600+600$

          $40x\ge2200$

          $\dfrac{40x}{40}\ge\dfrac{2200}{40}$

            $x\ge55$

Now Try:

7.   Two sides of a triangle are equal in length, with the third side 8 feet longer than one of the equal sides. The perimeter of the triangle cannot be more than 38 feet. Find the largest possible value for the length of the equal sides.

____

Step 5 State the answer. Ruth must have 55 or more students to have at least $1600 profit.

Step 6 Check. $40(55) - 600 = 1600$ Also, any number greater than 55 makes the profit greater than $1600.

Objective 5 Practice Exercises

For extra help, see Example 7 on page 189 of your text.

Solve each problem.

13. Lauren has grades of 98 and 86 on her first two chemistry quizzes. What must she score on her third quiz to have an average of at least 91 on the three quizzes?

13. _________________

14. Nina has a budget of $230 for gifts for this year. So far she has bought gifts costing $47.52, $38.98, and $26.98. If she has three more gifts to buy, find the average amount she can spend on each gift and still stay within her budget.

14. _________________

15. If twice the sum of a number and 7 is subtracted from three times the number, the result is more than −9. Find all such numbers.

15. _________________

Objective 6   Solve linear inequalities with three parts.

Review this example for Objective 6:

9. Solve the inequality, and graph the solution set.

$$3 \le 4x - 5 < 7$$

$$3 \le 4x - 5 < 7$$
$$3 + 5 \le 4x - 5 + 5 < 7 + 5$$
$$8 \le 4x < 12$$
$$\frac{8}{4} \le \frac{4x}{4} < \frac{12}{4}$$
$$2 \le x < 3$$

The solution set is $[2,\ 3)$. The graph is shown below.

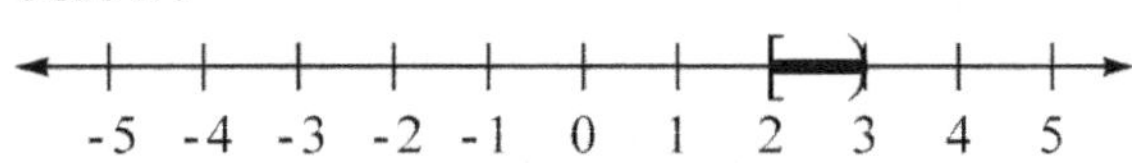

Now Try:

9. Solve the inequality, and graph the solution set.

$$8 \le 6x - 4 < 20$$

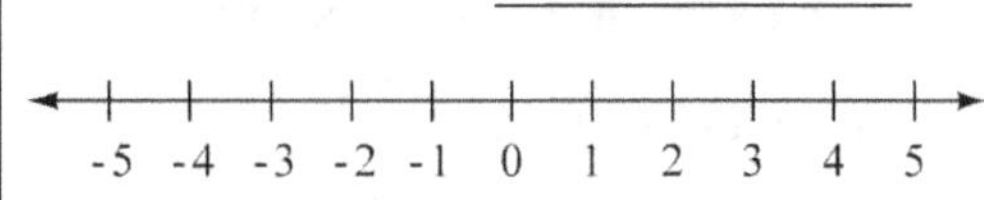

Objective 6 Practice Exercises

For extra help, see Examples 8–9 on pages 190–191 of your text.

Solve each inequality. Write the solution set in interval notation and then graph it.

16. $7 < 2x + 3 \le 13$

16. ___________________

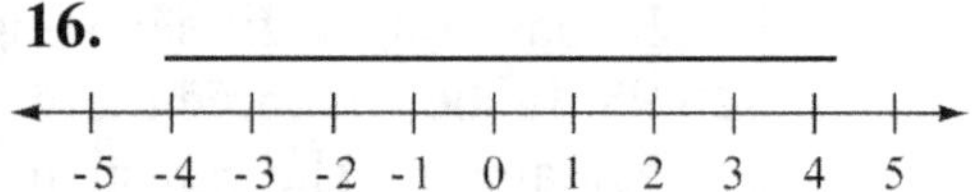

17. $-17 \le 3x - 2 < -11$

17. ___________________

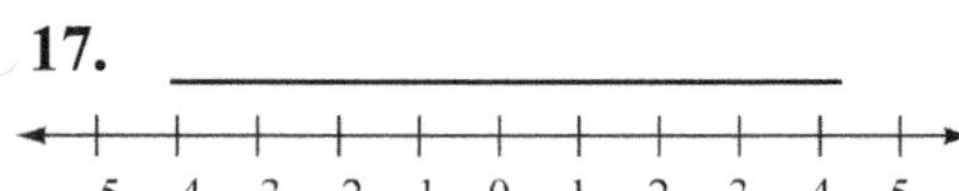

18. $1 < 3z + 4 < 19$

18. ___________________

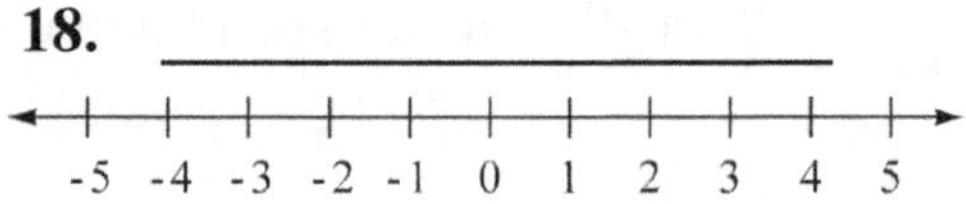

Chapter 3 LINEAR EQUATIONS IN TWO VARIABLES

3.1   Linear Equations and Rectangular Coordinates

Learning Objectives	
1	Interpret line graphs.
2	Write a solution as an ordered pair.
3	Decide whether a given ordered pair is a solution of a given equation.
4	Complete ordered pairs for a given equation.
5	Complete a table of values.
6	Plot ordered pairs.

Key Terms

Use the vocabulary terms listed below to complete each statement in exercises 1–13.

line graph	**linear equation in two variables**
ordered pair	**table of values** ***x*-axis**
***y*-axis**	**rectangular (Cartesian) coordinate system**
origin **quadrants**	**plane** **coordinates**
plot **scatter diagram**	

1. A ________________________________ uses dots connected by lines to show trends.

2. An equation that can be written in the form $Ax + By = C$, where A, B, and C are real numbers and $A, B \neq 0$, is called a ________________________________.

3. ________________________________ are the numbers in the ordered pair that specify the location of a point on a rectangular coordinate system.

4. In a coordinate system, the horizontal axis is called the ________________________.

5. In a coordinate system, the vertical axis is called the ________________________.

6. A pair of numbers written between parentheses in which order is important is called a(n) ________________________.

7. Together, the *x*-axis and the *y*-axis form a ________________________.

8. A coordinate system divides the plane into four regions called

    ________________________.

9. The axis lines in a coordinate system intersect at the ________________________.

10. To ________________________ an ordered pair is to find the corresponding point on a coordinate system.

11. A graph of ordered pairs is called a _________________________________.

12. A table showing selected ordered pairs of numbers that satisfy an equation is called a ______________________________.

13. A flat surface determined by two intersecting lines is a ____________________.

Objective 1   Interpret line graphs.

The line graph shows the number of degrees awarded by a university for the years 2020−2025.

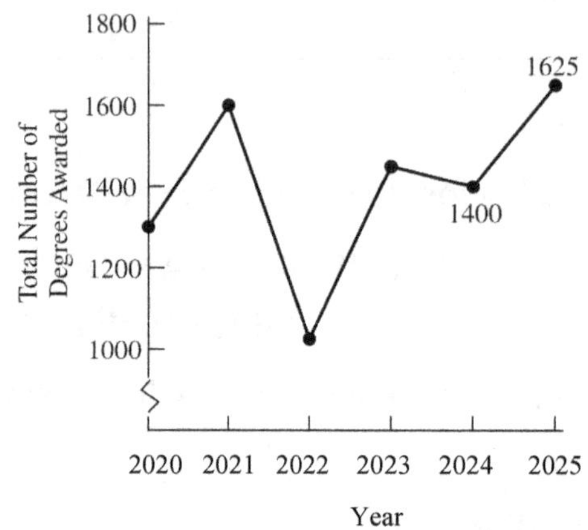

Review these examples for Objective 1:

1.

a. Between which years did the number of degrees awarded decrease?

The line between 2021 and 2022 and between 2023 and 2024 falls, so the number of degrees awarded decreased between 2021-2022 and 2023-2024.

b. Estimate the total number of degrees awarded in 2022 and in 2025. About how many more degrees were awarded in 2025?

Move up from 2022 on the horizontal scale to the point plotted for 2022. This point is about 1000. So about 1000 degrees were awarded in 2022.
Similarly, locate the point plotted for 2025. Moving across to the vertical scale, the graph indicates that the number of degrees awarded in 2025 was 1625.
Between 2022 and 2025, the increase was
$$1625 - 1000 = 625.$$

Now Try:

1.

a. Between which years did the number of degrees awarded increase?

____

b. Estimate the total number of degrees awarded in 2020 and in 2021. About how many more degrees were awarded in 2021?

____

Objective 1 Practice Exercises

For extra help, see Example 1 on page 208 of your text.
The line graph shows the number of degrees awarded by a university for the years 2020–2025. Use this graph to answer exercises 1–2.

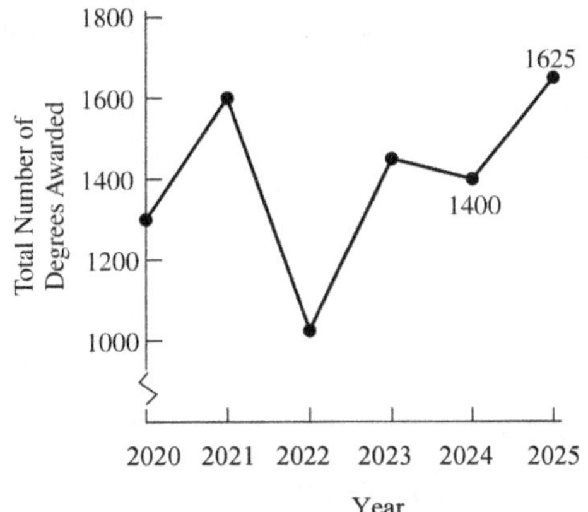

1. Between what pairs of consecutive years did the number of degrees decrease?

1. _________________

2. If 20% of the degrees awarded in 2025 were MBA degrees, how many MBAs were awarded in 2025?

2. _________________

Objective 2   Write a solution as an ordered pair.

Objective 2 Practice Exercises

For extra help, see page 209 of your text.

Write each solution as an ordered pair.

3. $x = 4$ and $y = 7$

3. _________________

4. $y = \frac{1}{3}$ and $x = 0$

4. _________________

5. $x = 0.2$ and $y = 0.3$

5. _________________

Objective 3   Decide whether a given ordered pair is a solution of a given equation.

Review these examples for Objective 3:

2. Decide whether each ordered pair is a solution of the equation $4x + 5y = 40$.

a. $(5,\ 4)$

Substitute 5 for x and 4 for y in the given equation.
$$4x + 5y = 40$$
$$4(5) + 5(4) \overset{?}{=} 40$$
$$20 + 20 \overset{?}{=} 40$$
$$40 = 40 \quad \text{True}$$

This result is true, so $(5,\ 4)$ is a solution of $4x + 5y = 40$.

b. $(-3,\ 6)$

Substitute -3 for x and 6 for y in the given equation.
$$4x + 5y = 40$$
$$4(-3) + 5(6) \overset{?}{=} 40$$
$$-12 + 30 \overset{?}{=} 40$$
$$18 = 40 \quad \text{False}$$

This result is false, so $(-3,\ 6)$ is not a solution of $4x + 5y = 40$.

Now Try:

2. Decide whether each ordered pair is a solution of the equation $3x - 4y = 12$.

a. $(8,\ 3)$

____

b. $(5, -4)$

____

Objective 3 Practice Exercises

For extra help, see Example 2 on page 210 of your text.

Decide whether the given ordered pair is a solution of the given equation.

6. $4x - 3y = 10;\ (1, 2)$

6. _________________

7. $2x - 3y = 1;\ \left(0, \frac{1}{3}\right)$

7. _________________

8. $x = -7;\ (-7, 9)$

8. _________________

Objective 4   Complete ordered pairs for a given equation.

Review this example for Objective 4:

3. Complete the ordered pair for the equation $y = 5x + 8$.

$$(3, \underline{\quad})$$

Replace x with 3.

$$y = 5x + 8$$

$$y = 5(3) + 8$$

$$y = 15 + 8$$

$$y = 23$$

The ordered pair is $(3, 23)$.

Now Try:

3. Complete the ordered pair for the equation $y = 4x - 7$.

$$(5, \underline{\quad})$$

____

Objective 4 Practice Exercises

For extra help, see Example 3 on pages 210–211 of your text.

For each of the given equations, complete the ordered pairs beneath it.

9. $y = 2x - 5$

     (a) $(2, \quad)$

     (b) $(0, \quad)$

     (c) $(\quad, 3)$

     (d) $(\quad, -7)$

     (e) $(\quad, 9)$

9.

     (a) _______________

     (b) _______________

     (c) _______________

     (d) _______________

     (e) _______________

10. $y = 3 + 2x$

     (a) $(-4, \quad)$

     (b) $(2, \quad)$

     (c) $(\quad, 0)$

     (d) $(-2, \quad)$

     (e) $(\quad, -7)$

10.

     (a) _______________

     (b) _______________

     (c) _______________

     (d) _______________

     (e) _______________

Objective 5   Complete a table of values.

Review these examples for Objective 5:

4. Complete the table of values for the equation. Then write the results as ordered pairs.

a. $2x - 3y = 6$

x	y
9	
6	
	-2
	8

From the table, we can write the ordered pairs: $(9, \underline{\quad}), (6, \underline{\quad}), (\underline{\quad}, -2), (\underline{\quad}, 8)$.

From the first row of the table, let $x = 9$ in the equation. From the second row of the table, let $x = 6$.

$$\begin{array}{l|l}
\text{If } x = 9, & \text{If } x = 6, \\
2x - 3y = 6 & 2x - 3y = 6 \\
2(9) - 3y = 6 & 2(6) - 3y = 6 \\
18 - 3y = 6 & 12 - 3y = 6 \\
-3y = -12 & -3y = -6 \\
y = 4 & y = 2
\end{array}$$

The first two ordered pairs are $(9, 4)$ and $(6, 2)$.

From the third and fourth rows of the table, let $y = -2$ and $y = 8$, respectively.

$$\begin{array}{l|l}
\text{If } y = -2, & \text{If } y = 8, \\
2x - 3y = 6 & 2x - 3y = 6 \\
2x - 3(-2) = 6 & 2x - 3(8) = 6 \\
2x + 6 = 6 & 2x - 24 = 6 \\
2x = 0 & 2x = 30 \\
x = 0 & x = 15
\end{array}$$

The last two ordered pairs are $(0, -2)$ and $(15, 8)$. The completed table and corresponding ordered pairs follow.

x	y	Ordered pairs
9	4	$\to (9, 4)$
6	2	$\to (6, 2)$
0	-2	$\to (0, -2)$
15	8	$\to (15, 8)$

Now Try:

4. Complete the table of values for the equation. Then write the results as ordered pairs.

a. $4x - y = 8$

x	y
1	
5	
	0
	4

____

c. $y = -5$

x	y
-4	
0	
3	

c. $y = -4$

x	y
-5	
0	
4	

____

The given equation is $y = -5$. No matter which value of x is chosen, the value of y is *always* -5.

x	y	Ordered pairs
-4	$-5 \rightarrow$	$(-4, -5)$
0	$-5 \rightarrow$	$(0, -5)$
3	$-5 \rightarrow$	$(3, -5)$

Objective 5 Practice Exercises

For extra help, see Example 4 on pages 211–212 of your text.

Complete each table of values. Write the results as ordered pairs.

11. $2x + 5 = 7$

x	y
	-3
	0
	5

11. _________________

12. $y - 4 = 0$

x	y
-4	
0	
6	

12. _________________

13. $4x + 3y = 12$

x	y
0	
	0
	-1

13. _________________

Name: Date:
Instructor: Section:

Objective 6   Plot ordered pairs.

Review this example for Objective 6:

5. Plot the given points in a coordinate system.
   (5, 4)   (−2,−1)   (−3, 5)   (4,−2)
   (1,−2.5)   (3, 0)   (0, 4)

Step 1 Move right or left the number of units
that correspond to the x-coordinate in the
ordered pair—right if the x-coordinate is
positive and left if it is negative.

Step 2 Then turn and move up or down
the number of units that corresponds to the
y-coordinate—up if the y-coordinate is
positive or down if it is negative.

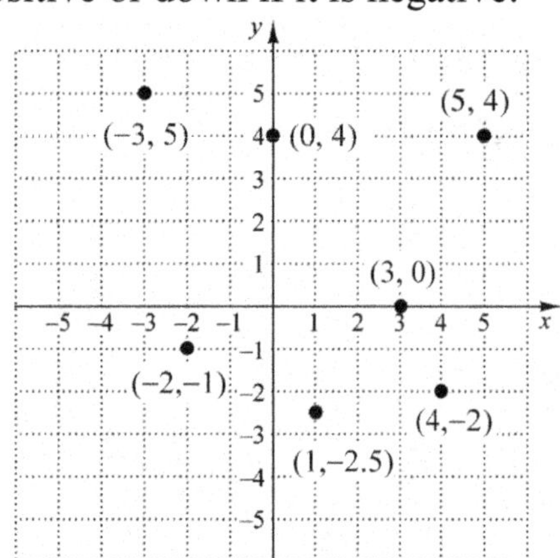

Now Try:

5. Plot the given points in a
   coordinate system.
   (2, 4)   (−5, 1)   (−4,−2)
   (3,−5)   (2,−1.5)   (6, 0)
   (0,−6)

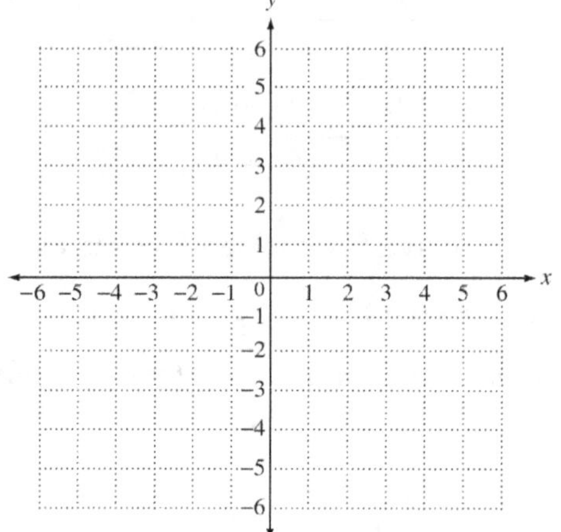

Objective 6 Practice Exercises

For extra help, see Examples 5–6 on pages 213–214 of your text.

Plot the each ordered pair on a coordinate system.

14. $(0,-2)$

14.

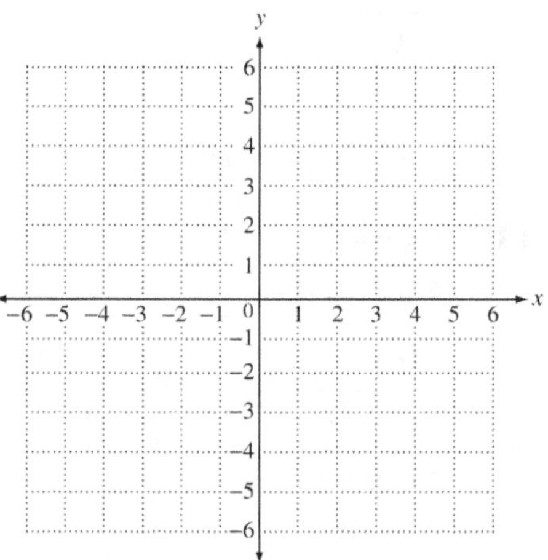

            Copyright © 2025 Pearson Education, Inc.

15. $(-3, 4)$

15.

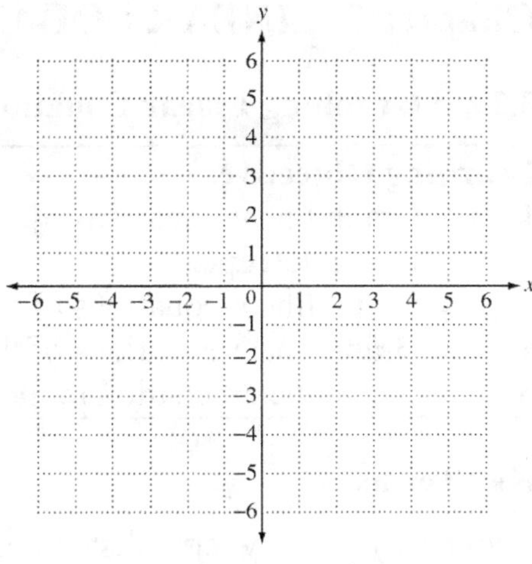

16. $(2, -5)$

16.

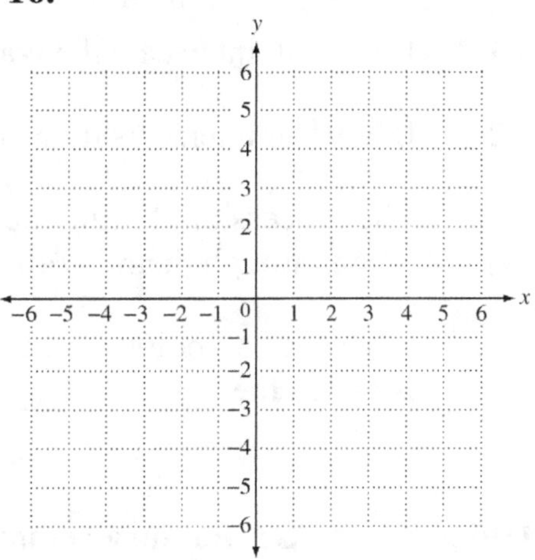

Chapter 3 LINEAR EQUATIONS IN TWO VARIABLES

3.2   Graphing Linear Equations in Two Variables

Learning Objectives
1   Graph linear equations by plotting ordered pairs.
2   Find intercepts.
3   Graph linear equations of the form $Ax + By = 0$.
4   Graph linear equations of the form $y = b$ or $x = a$.
5   Use a linear equation to model data.

Key Terms

Use the vocabulary terms listed below to complete each statement in exercises 1−4.

graph   graphing   *y*-intercept   *x*-intercept

1.   If a graph intersects the *y*-axis at *k*, then the _____________________ is $(0, k)$.

2.   If a graph intersects the *x*-axis at *k*, then the _____________________ is $(k, 0)$.

3.   The process of plotting the ordered pairs that satisfy a linear equation and drawing a line through them is called _____________________.

4.   The set of all points that correspond to the ordered pairs that satisfy the equation is called the _____________________ of the equation.

Objective 1   Graph linear equations by plotting ordered pairs.

Review this example for Objective 1:
2.   Graph $2x + 3y = 6$.

First let $x = 0$ and then let $y = 0$ to determine two ordered pairs.

$$\begin{array}{c|c} 2(0)+3y=6 & 2x+3(0)=6 \\ 0+3y=6 & 2x+0=6 \\ 3y=6 & 2x=6 \\ y=2 & x=3 \end{array}$$

The ordered pairs are $(0, 2)$ and $(3, 0)$. Find a third ordered pair by choosing a number other than 0 for x or y. We choose $y = 4$.

$$2x+3(4)=6$$
$$2x+12=6$$
$$2x=-6$$
$$x=-3$$

Now Try:
2.   Graph $x + y = 3$.

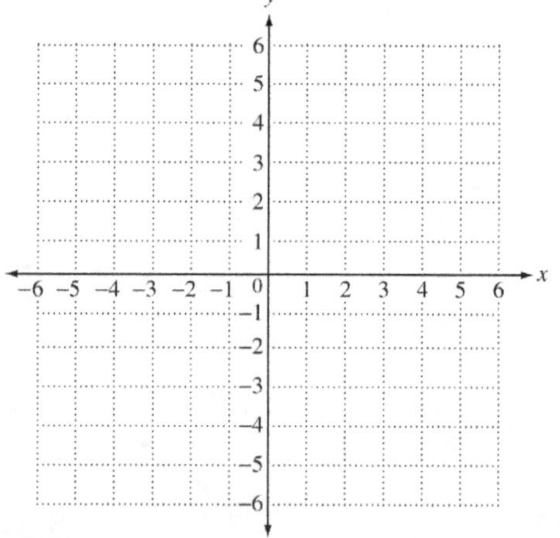

This gives the ordered pair (–3, 4). We plot the three ordered pairs (0, 2), (3, 0), and (–3, 4) and draw a line through them.

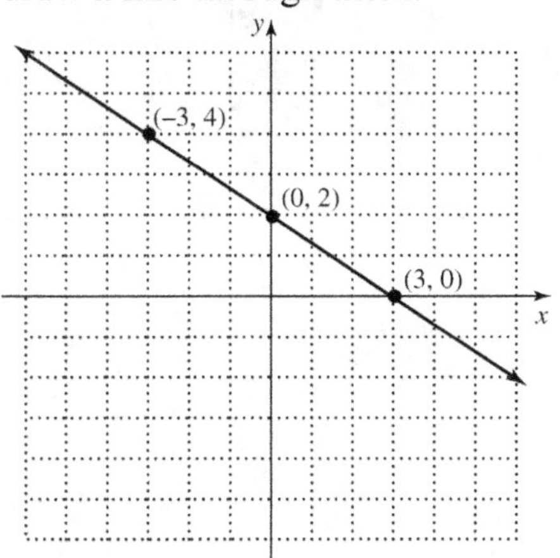

Objective 1 Practice Exercises

For extra help, see Examples 1–2 on pages 220–221 of your text.

Complete the ordered pairs for each equation. Then graph the equation by plotting the points and drawing a line through them.

1. $y = 3x - 2$

       (0,)

       (,0)

       (2,)

1.

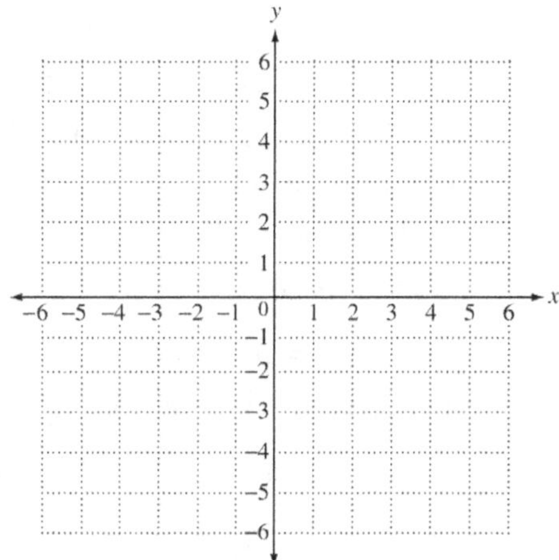

2. $x - y = 4$

       (0,)

       (,0)

       (–2,)

2.

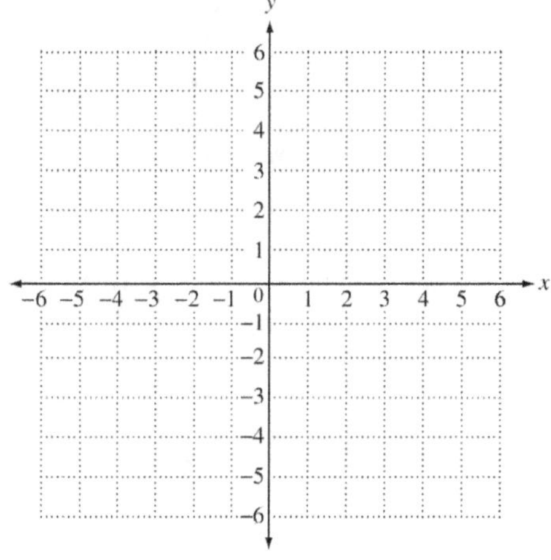

3. $x = 2y + 1$

$(0, \quad)$

$(\quad, 0)$

$(\quad, -2)$

3.

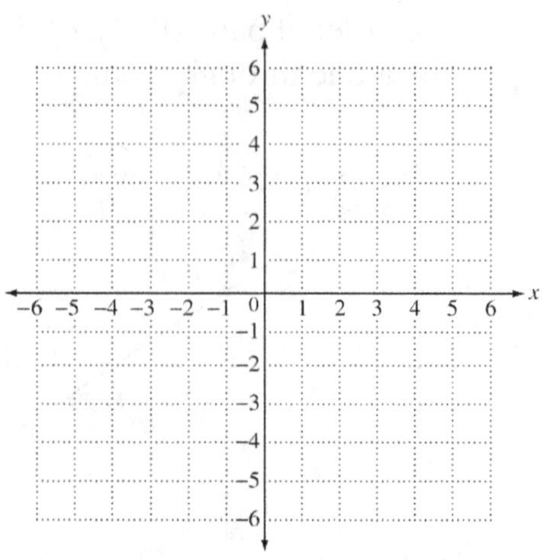

Objective 2   Find intercepts.

Review this example for Objective 2:

3. Graph $3x + y = 6$ using intercepts.

To find the y-intercept, let $x = 0$.
To find the x-intercept, let $y = 0$.

$$3(0) + y = 6 \quad \big| \quad 3x + 0 = 6$$
$$0 + y = 6 \quad \big| \qquad 3x = 6$$
$$y = 6 \quad \big| \qquad x = 2$$

The intercepts are (0, 6) and (2, 0). To find a third point, as a check, we let $x = 1$.

$$3(1) + y = 6$$
$$3 + y = 6$$
$$y = 3$$

This gives the ordered pair (1, 3).

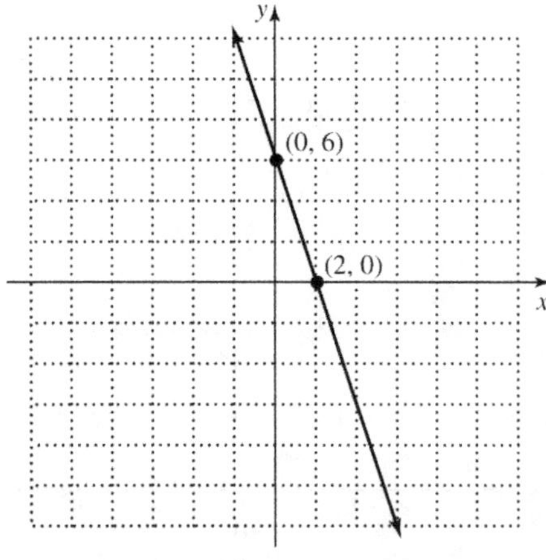

Now Try:

3. Graph $5x - 2y = -10$ using intercepts.

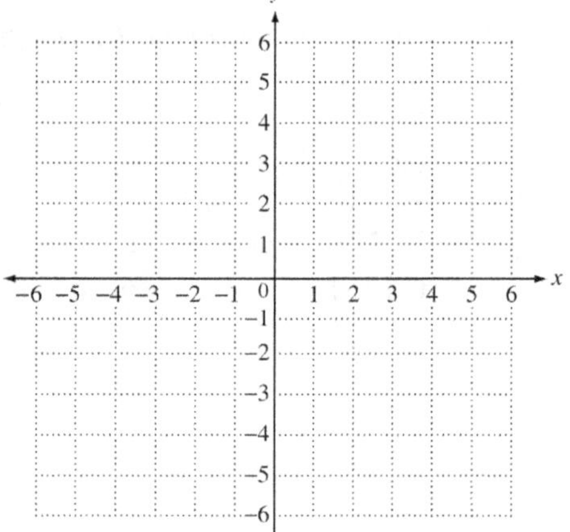

4. Graph $y = \dfrac{3}{2}x - 4$.

To find the y-intercept, let $x = 0$.
To find the x-intercept, let $y = 0$.

$$y = \frac{3}{2}(0) - 4 \quad\bigg|\quad 0 = \frac{3}{2}x - 4$$

$$y = 0 - 4 \quad\bigg|\quad 4 = \frac{3}{2}x$$

$$y = -4 \quad\bigg|\quad \frac{8}{3} = x$$

The intercepts are $(0, -4)$ and $\left(\dfrac{8}{3},\ 0\right)$. To find a third point, as a check, we let $x = 2$.

$$y = \frac{3}{2}(2) - 4$$

$$y = 3 - 4$$

$$y = -1$$

This gives the ordered pair $(2, -1)$.

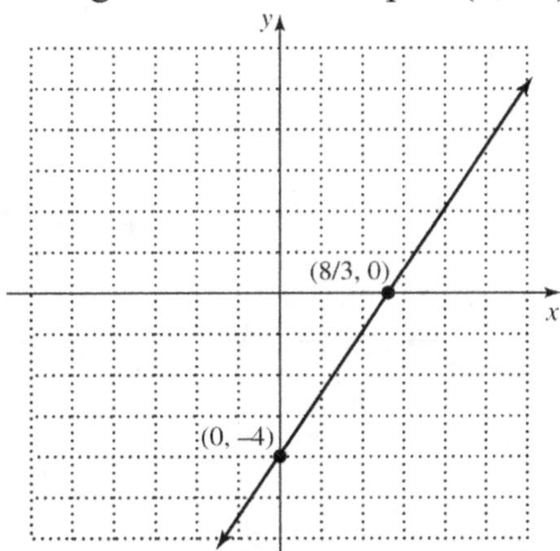

4. Graph $y = 4x - 4$.

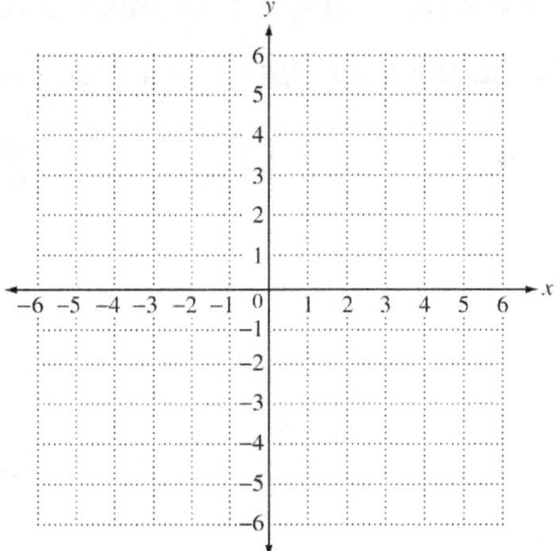

Name: Date:

Instructor: Section:

Objective 2 Practice Exercises

For extra help, see Examples 3–4 on pages 221–223 of your text.

Find the intercepts for each equation. Then graph the equation.

4. $y = \dfrac{2}{3}x - 2$

4.

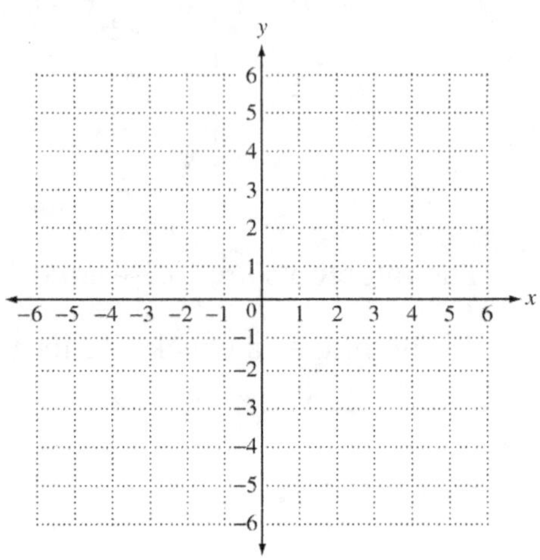

5. $4x - 7y = -8$

5.

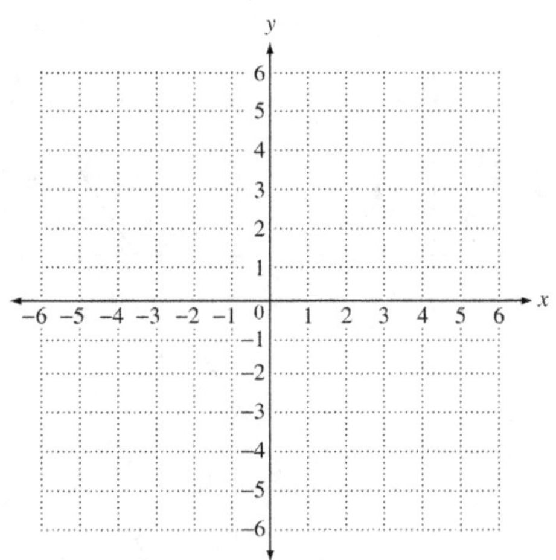

          Copyright © 2025 Pearson Education, Inc.

Objective 3   Graph linear equations of the form $Ax + By = 0$.

<table>
<tr><td valign="top" width="55%">

Review this example for Objective 3:

5.   Graph $x + 5y = 0$.

To find the y-intercept, let $x = 0$.
To find the x-intercept, let $y = 0$.

$$\begin{array}{c|c} 0 + 5y = 0 & x + 5(0) = 0 \\ 5y = 0 & x + 0 = 0 \\ y = 0 & x = 0 \end{array}$$

The x- and y-intercepts are the same point $(0, 0)$. We must select two other values for x or y to find two other points. We choose $y = 1$ and $y = -1$.

$$\begin{array}{c|c} x + 5(1) = 0 & x + 5(-1) = 0 \\ x + 5 = 0 & x - 5 = 0 \\ x = -5 & x = 5 \end{array}$$

We use $(-5, 1)$, $(0, 0)$, and $(5, -1)$ to draw the graph.

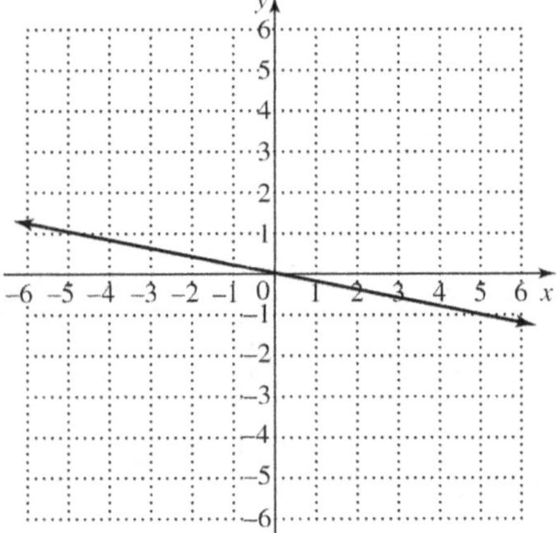

</td><td valign="top" width="45%">

Now Try:

5.   Graph $3x - y = 0$.

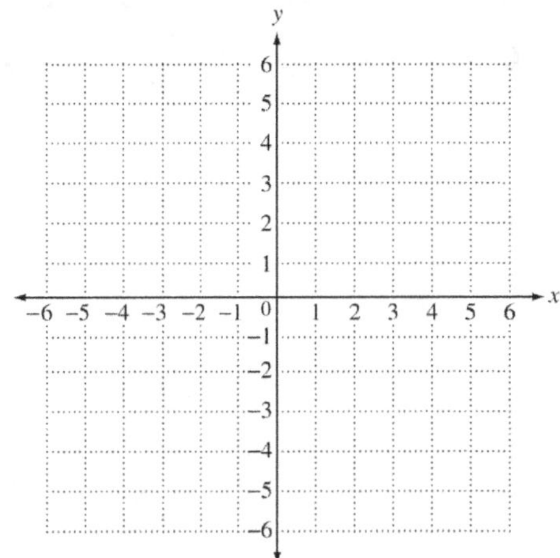

</td></tr>
</table>

Objective 3 Practice Exercises

For extra help, see Example 5 on page 223 of your text.

Graph each equation.

6. $-3x - 2y = 0$

6.

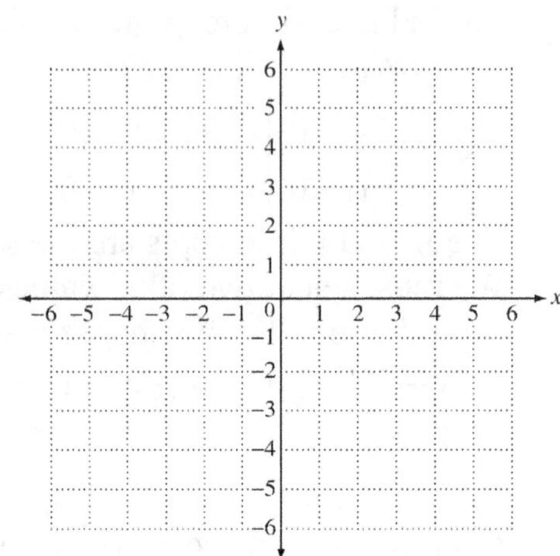

7. $x + y = 0$

7.

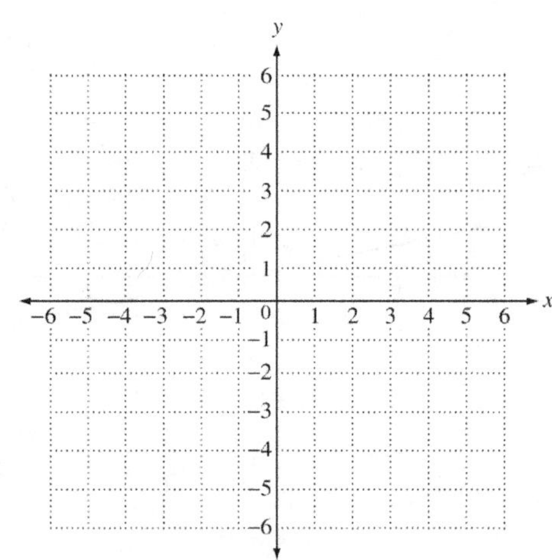

8. $y = 2x$

8.

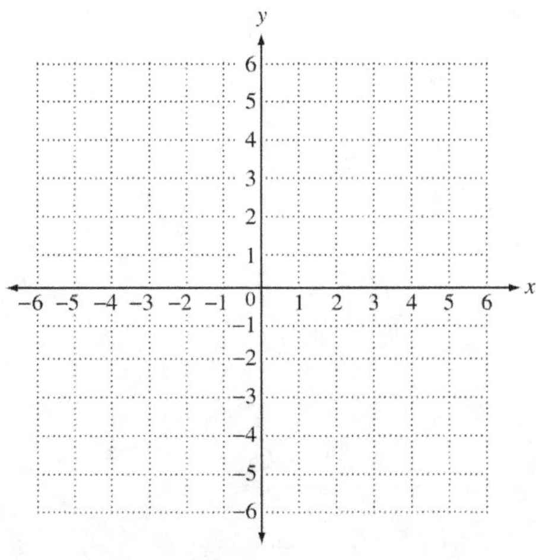

          Copyright © 2025 Pearson Education, Inc.

Objective 4   Graph linear equations of the form $y = b$ or $x = a$.

Review these examples for Objective 4: | **Now Try:**

6. Graph $y = -2$.

For any value of x, y is always –2. Three ordered pairs that satisfy the equation are (–4, –2), (0, –2) and (2, –2). Drawing a line through these points gives the horizontal line. The y-intercept is (0, –2). There is no x-intercept.

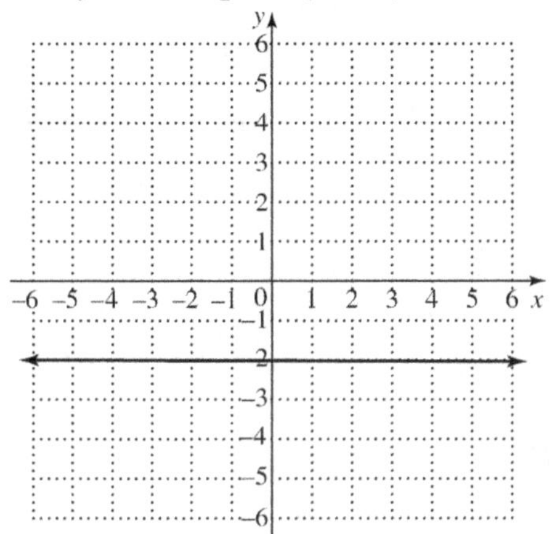

6. Graph $y = 4$.

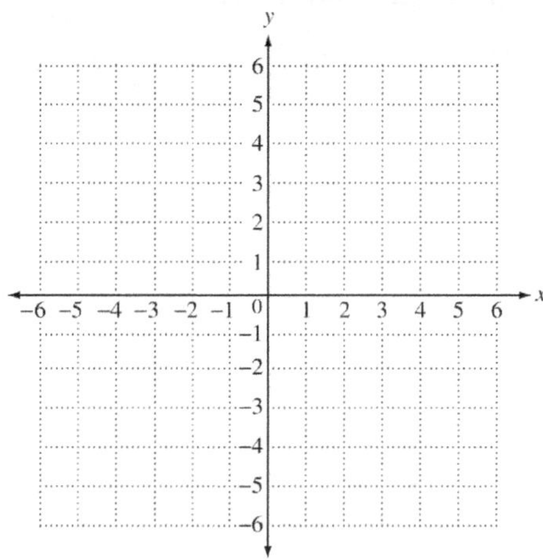

7. Graph $x + 4 = 0$.

First we subtract 4 from each side of the equation to get the equivalent equation $x = -4$. All ordered-pair solutions of this equation have x-coordinate –4.
Three ordered pairs that satisfy the equation are (–4, –1), (–4, 0), and (–4, 3). The graph is a vertical line. The x-intercept is (–4, 0). There is no y-intercept.

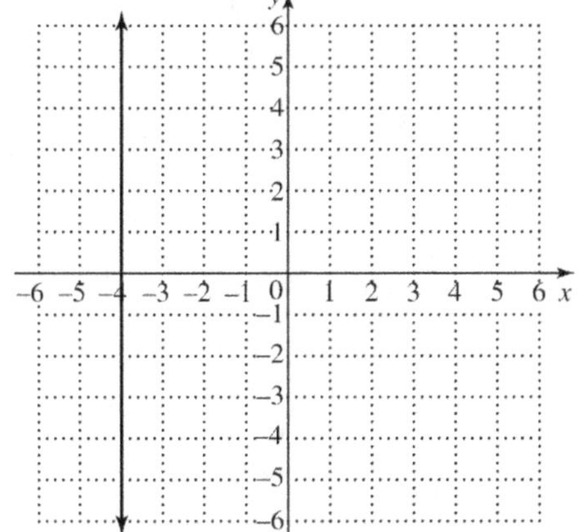

7. Graph $x = 0$.

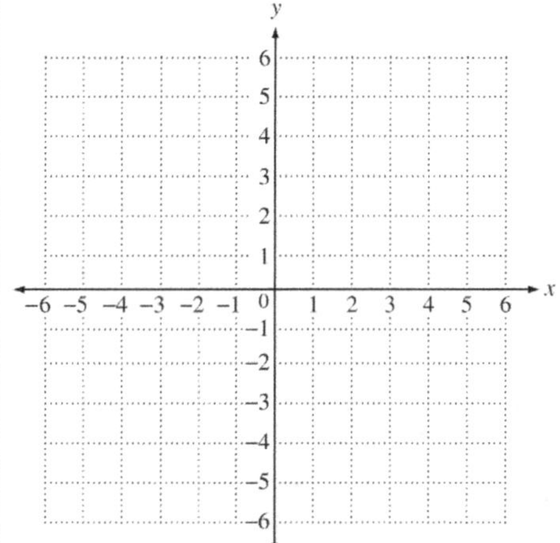

Objective 4 Practice Exercises

For extra help, see Examples 6–7 on page 224 of your text.

Graph each equation.

9. $x - 1 = 0$

9.

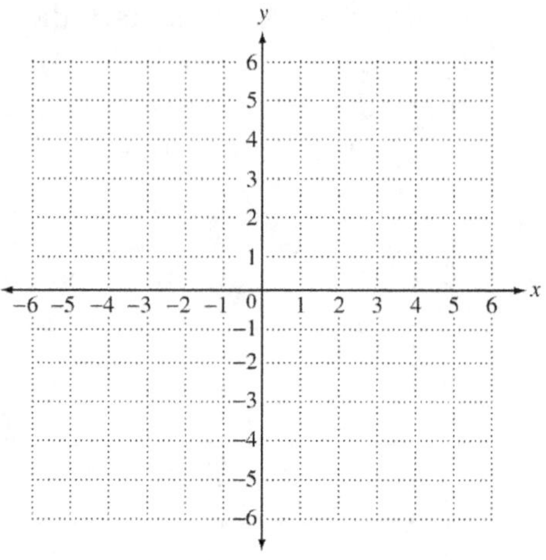

10. $y + 3 = 0$

10.

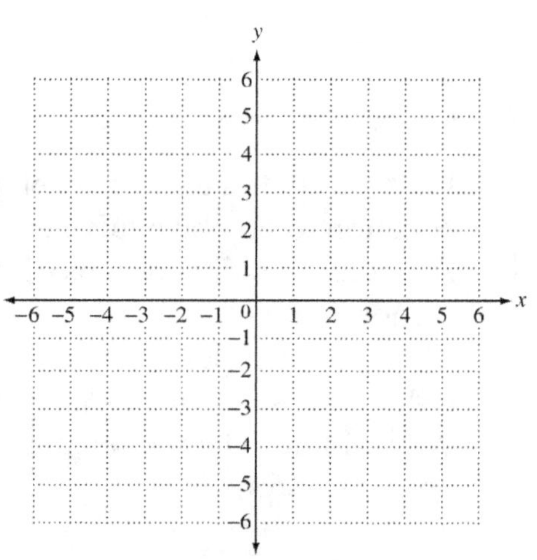

          Copyright © 2025 Pearson Education, Inc.

Objective 5   Use a linear equation to model data.

Review these examples for Objective 5:

Now Try:

8. Every year sea turtles return to a certain group of islands to lay eggs. The number of turtle eggs that hatch can be approximated by the equation $y = -70x + 3260,$ where y is the number of eggs that hatch and $x = 0$ representing 1990.

8. Suppose that the demand and price for a certain model of calculator are related by the equation $y = 45 - \dfrac{3}{5}x,$ where y is the price (in dollars) and x is the demand (in thousands of calculators).

a. Use this equation to find the number of eggs that hatched in 1995, 2000, and 2005, and 2015.

Substitute the appropriate value for each year x to find the number of eggs hatched in that year.

For 1995:
$$y = -70(5) + 3260 \qquad 1995 - 1990 = 5$$
$$y = 2910 \text{ eggs} \qquad \text{Replace } x \text{ with 5.}$$

For 2000:
$$y = -70(10) + 3260 \qquad 2000 - 1990 = 10$$
$$y = 2560 \text{ eggs} \qquad \text{Replace } x \text{ with 10.}$$

For 2005:
$$y = -70(15) + 3260 \qquad 2005 - 1990 = 15$$
$$y = 2210 \text{ eggs} \qquad \text{Replace } x \text{ with 15.}$$

For 2015:
$$y = -70(25) + 3260 \qquad 2015 - 1990 = 25$$
$$y = 1510 \text{ eggs} \qquad \text{Replace } x \text{ with 25.}$$

b. Write the information from part (a) as four ordered pairs, and use them to graph the given linear equation.

Since x represents the year and y represents the number of eggs, the ordered pairs are (5, 2910), (10, 2560), (15, 2210), and (25, 1510).

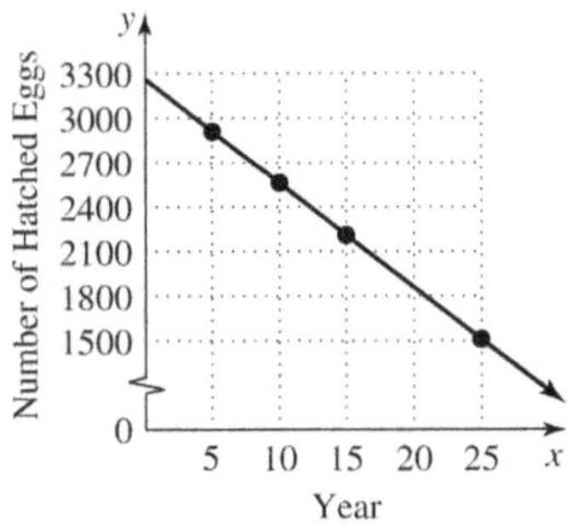

a. Assuming that this model is valid for a demand up to 50,000 calculators, use this equation to find the price of calculators at each level of demand.

0 calculators _____________

5000 calculators _____________

20,000 calculators _____________

45,000 calculators _____________

b. Write the information from part (a) as four ordered pairs, and use them to graph the given linear equation.

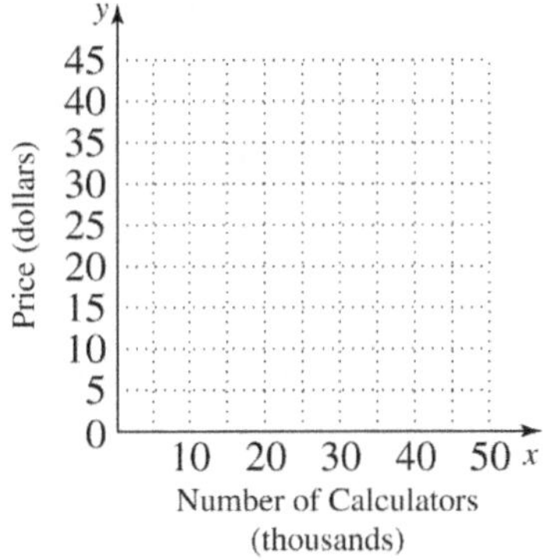

c. Use this graph and the equation to estimate the number of eggs that will hatch in 2010.

For 2010, $x = 20$. On the graph, find 20 on the horizontal axis, move up to the graphed line and then across to the vertical axis. It appears that in 2010, there were about 1900 eggs.

To use the equation, substitute 20 for x.

$$y = -70(20) + 3260$$

$$y = 1860 \text{ eggs}$$

This result for 2020 is close to our estimate of 1900 eggs from the graph.

c. Use this graph and the equation to estimate the price of 30,000 calculators.

____

Objective 5 Practice Exercises

For extra help, see Example 8 on pages 225–226 of your text.

Solve each problem. Then graph the equation.

11. The profit y in millions of dollars earned by a small computer company can be approximated by the linear equation $y = 0.63x + 4.9$, where $x = 0$ corresponds to 2020, $x = 1$ corresponds to 2021, and so on. Use this equation to approximate the profit in each year from 2020 through 2023.

11. 2020 ____________

2021 ____________

2022 ____________

2023 ____________

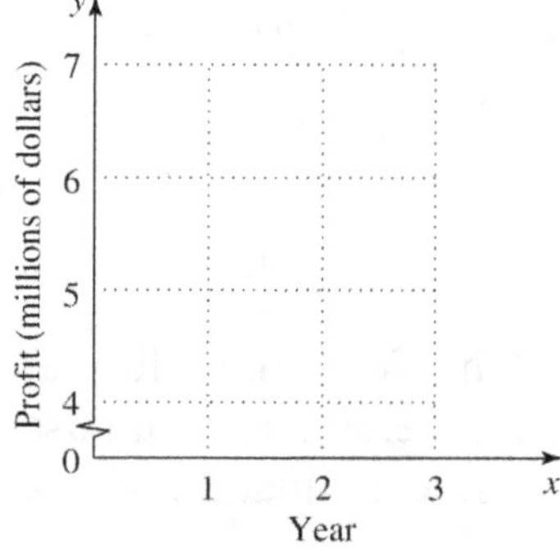

Copyright © 2025 Pearson Education, Inc.

12. The number of band instruments sold by Elmer's Music Shop can be approximated by the equation $y = 325 + 42x$, where y is the number of instruments sold and x is the time in years, with $x = 0$ representing 2019. Use this equation to approximate the number of instruments sold in each year from 2019 through 2022.

12. 2019 _____________

2020 _____________

2021 _____________

2022 _____________

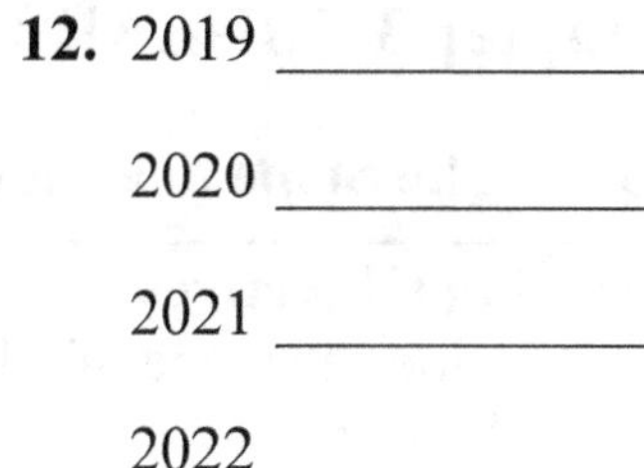
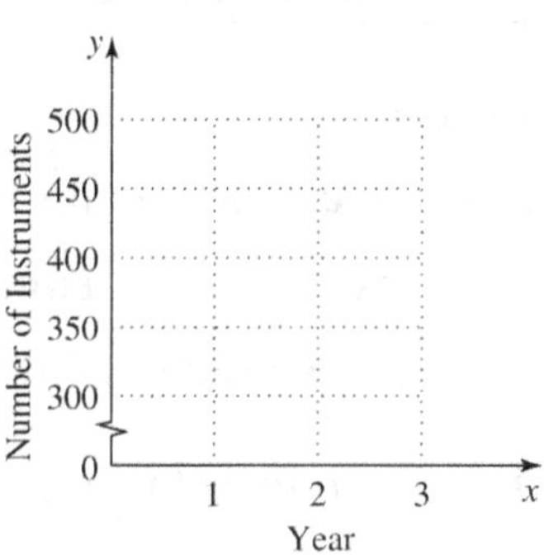

13. According to *The Old Farmer's Almanac*, the temperature in degrees Celsius can be determined by the equation $y = \frac{1}{3}x + 4$, where x is the number of cricket chirps in 25 seconds and y is the temperature in degrees Celsius. Use this equation to find the temperature when there are 48 chirps, 54 chirps, 60 chirps, and 66 chirps.

13. 48 _____________

54 _____________

60 _____________

66 _____________

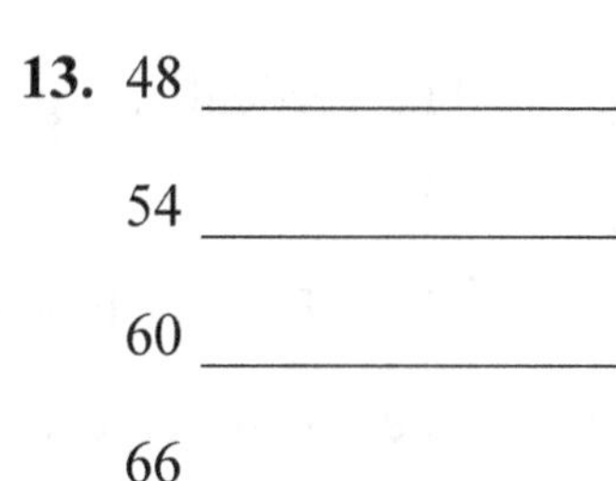
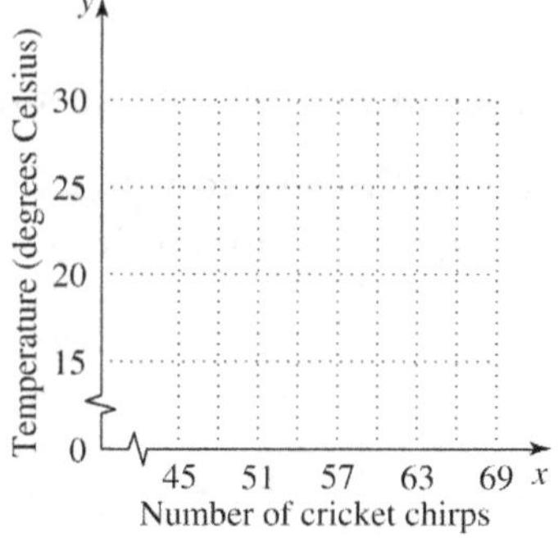

Chapter 3 LINEAR EQUATIONS IN TWO VARIABLES

3.3   The Slope of a Line

Learning Objectives
1   Find the slope of a line given two points.
2   Find the slope from the equation of a line.
3   Use slopes to determine whether two lines are parallel, perpendicular, or neither.

Key Terms

Use the vocabulary terms listed below to complete each statement in exercises 1−5.

rise   run   slope   parallel lines

perpendicular lines

1.   Two lines that intersect in a 90° angle are called ________________________________.

2.   The ________________ of a line is the ratio of the change in y compared to the change in x when moving along the line from one point to another.

3.   The vertical change between two different points on a line is called the ____________________.

4.   Two lines in a plane that never intersect are called ________________________________.

5.   The horizontal change between two different points on a line is called the ____________________.

Objective 1   Find the slope of a line given two points.

Review these examples for Objective 1:
1.   Find the slope of the line.

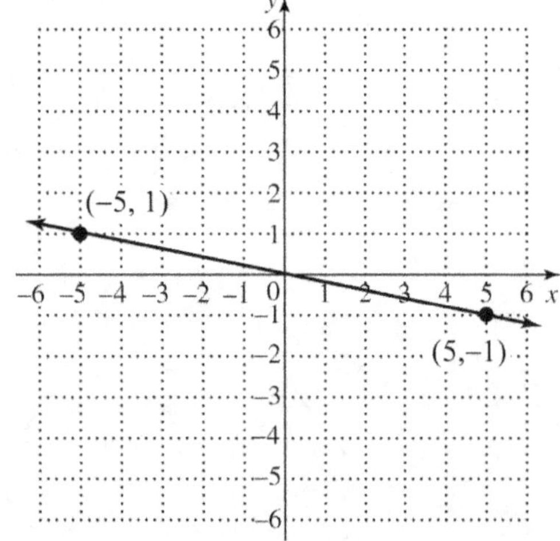

Now Try:
1.   Find the slope of the line.

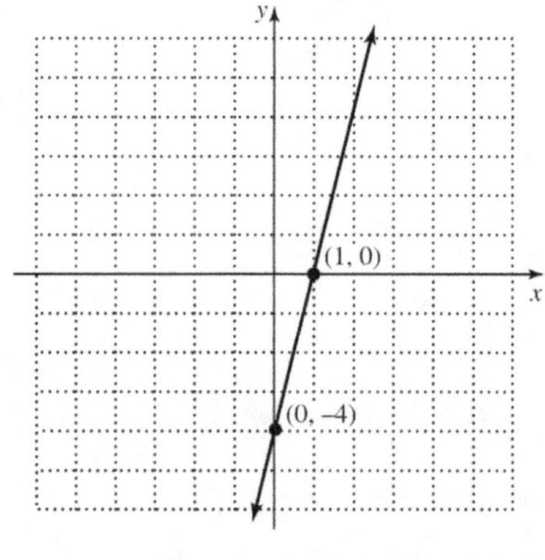

                        Copyright © 2025 Pearson Education, Inc.

We use the two points shown on the line.
The vertical change is the difference in the y-values, or $-1 - 1 = -2$, and the horizontal change is the difference in the x-values or $5 - (-5) = 10$.
Thus, the line has

$$\text{slope} = \frac{-2}{10}, \text{ or } -\frac{1}{5}.$$

2.   Find the slope of the line.

The line passing through $(-5, 4)$ and $(2, -6)$

Apply the slope formula.
$$(x_1, y_1) = (-5,\ 4) \text{ and } (x_2, y_2) = (2, -6)$$

$$\text{slope } m = \frac{y_2 - y_1}{x_2 - x_1} = \frac{-6 - 4}{2 - (-5)}$$

$$= \frac{-10}{7}, \text{ or } -\frac{10}{7}$$

3.   Find the slope of the line passing through $(-9, 3)$ and $(4, 3)$.

$$(x_1, y_1) = (-9,\ 3) \text{ and } (x_2, y_2) = (4,\ 3)$$

$$m = \frac{y_2 - y_1}{x_2 - x_1} = \frac{3 - 3}{4 - (-9)} = \frac{0}{13} = 0$$

4.   Find the slope of the line passing through $(-5, 3)$ and $(-5, 8)$.

$$(x_1, y_1) = (-5,\ 3) \text{ and } (x_2, y_2) = (-5,\ 8)$$

$$m = \frac{y_2 - y_1}{x_2 - x_1} = \frac{8 - 3}{-5 - (-5)} = \frac{5}{0} \text{ undefined slope}$$

2.   Find the slope of the line.

The line passing through $(-6, 7)$ and $(3, -9)$

____

3.   Find the slope of the line passing through $(8, -5)$ and $(-7, -5)$.

____

4.   Find the slope of the line passing through $(9, 11)$ and $(9, -7)$.

____

Objective 1 Practice Exercises

For extra help, see Examples 1–4 on pages 232–235 of your text.

Find the slope of the line through the given points.

1. $(4, 3)$ and $(3, 5)$

1. ______________

2. $(-4, 6)$ and $(-4, -1)$

2. ______________

3. $(-3, 3)$ and $(6, 3)$

3. ______________________

Objective 2   Find the slope from the equation of a line.

Review this example for Objective 2:

5. Find the slope of the line.

$$4x - 3y = 7$$

Step 1 Solve the equation for y.

$$4x - 3y = 7$$

$$-3y = -4x + 7$$

$$y = \frac{4}{3}x - \frac{7}{3}$$

Step 2 The slope is given by the coefficient of x, so the slope is $\frac{4}{3}$.

Now Try:

5. Find the slope of the line.

$$7x - 4y = 8$$

____

Objective 2 Practice Exercises

For extra help, see Example 5 on page 237 of your text.

Find the slope of each line.

4. $7y - 4x = 11$

4. ______________________

5. $3y = 2x - 1$

5. ______________________

6. $y = -\frac{2}{5}x - 4$

6. ______________________

Date:
Section:

Objective 3   Use slope to determine whether two lines are parallel, perpendicular, or neither.

Review these examples for Objective 3:

6. Decide whether each pair of lines is parallel, perpendicular, or neither.

   a. $5x - y = 3$

       $15x - 3y = 12$

Solve each equation for y.

   $y = 5x - 3$

   $y = 5x - 4$

Both lines have slope 5, so the lines are parallel.

b. $x + 3y = 8$ and $-3x + y = 5$

Find the slope of each line by first solving each equation for y.

   $3y = -x + 8$ $\Big|$ $y = 3x + 5$

   $y = -\dfrac{1}{3}x + \dfrac{8}{3}$

The slope is $-\dfrac{1}{3}$. $\Big|$ The slope is 3.

Because the slopes are not equal, the lines are not parallel.

Check the product of the slopes: $-\dfrac{1}{3}(3) = -1$.

The two lines are perpendicular because the product of their slopes is -1.

Now Try:

6. Decide whether each pair of lines is parallel, perpendicular, or neither.

  a. $2x - 4y = 7$

     $3x - 6y = 8$

____

  b. $9x - y = 7$ and $x + 9y = 11$

____

Objective 3 Practice Exercises

For extra help, see Example 6 on pages 239–240 of your text.

In each pair of equations, give the slope of each line, and then determine whether the two lines are **parallel, perpendicular,** *or* **neither**.

   7. $-x + y = -7$

       $x - y = -3$

                                   7. _______________

8. $4x + 2y = 8$

$x + 4y = -3$

8. _______________

9. $9x + 3y = 2$

$x - 3y = 5$

9. _______________

Chapter 3   LINEAR EQUATIONS IN TWO VARIABLES

3.4   Slope-Intercept Form of a Linear Equation

Learning Objectives

1   Use the slope-intercept form of the equation of a line.
2   Graph a line using its slope and a point on the line.
3   Write an equation of a line using its slope and any point on the line.
4   Graph and write equations of horizontal and vertical lines.

Key Terms

Use the vocabulary terms listed below to complete each statement in exercises 1–3.

        slope-intercept form   **point-slope form**   **standard form**

1.   A linear equation in the form $y - y_1 = m(x - x_1)$ is written in

     ________________________________ .

2.   A linear equation in the form $Ax + By = C$ is written in

     ________________________________ .

3.   A linear equation in the form $y = mx + b$ is written in

     ________________________________ .

Objective 1   Use the slope-intercept form of the equation of a line.

Review these examples for Objective 1:	Now Try:
1. Identify the slope and y-intercept of the line with each equation.	1. Identify the slope and y-intercept of the line with each equation.
a. $y = -8x + 7$	**a.** $y = -12x + 6$
The slope is -8, and the y-intercept is $(0, 7)$.	
	________________
b. $y = \dfrac{x}{9} - \dfrac{5}{4}$	**b.** $y = -\dfrac{x}{7} - \dfrac{7}{5}$
The equation can be written as $y = \dfrac{1}{9}x + \left(-\dfrac{5}{4}\right).$	
The slope is $\dfrac{1}{9}$, and the y-intercept is $\left(0, -\dfrac{5}{4}\right).$	________________

Objective 1 Practice Exercises

For extra help, see Example 1 on page 245 of your text.

Identify the slope and y-intercept of the line with each equation.

1. $y = \dfrac{3}{2}x - \dfrac{2}{3}$

1. ___________________

____

2. $y = -4x$

2. ___________________

____

Objective 2   Graph a line by using its slope and a point on the line.

| **Review this example for Objective 2:** | **Now Try:** |

Review this example for Objective 2:

2. Graph the equation by using the slope and *y*-intercept.

$$2x - 3y = 6$$

Step 1 Solve for *y* to write the equation in slope-intercept form.

$$2x - 3y = 6$$
$$-3y = -2x + 6$$
$$y = \frac{2}{3}x - 2$$

Step 2 The *y*-intercept is (0, –2). Graph this point.

Step 3 The slope is $\dfrac{2}{3}$. By definition,

$$\text{slope } m = \frac{\text{change in } y \text{ (rise)}}{\text{change in } x \text{ (run)}} = \frac{2}{3}$$

From the *y*-intercept, count up 2 units and to the right 3 units to obtain the point (3, 0).

Step 4 Draw the line through the points (0, –2) and (3, 0) to obtain the graph.

Now Try:

2. Graph the equation by using the slope and *y*-intercept.

$$y = \frac{2}{3}x$$

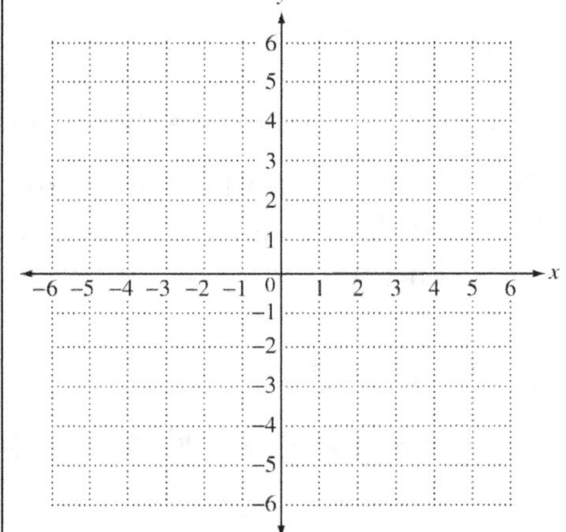

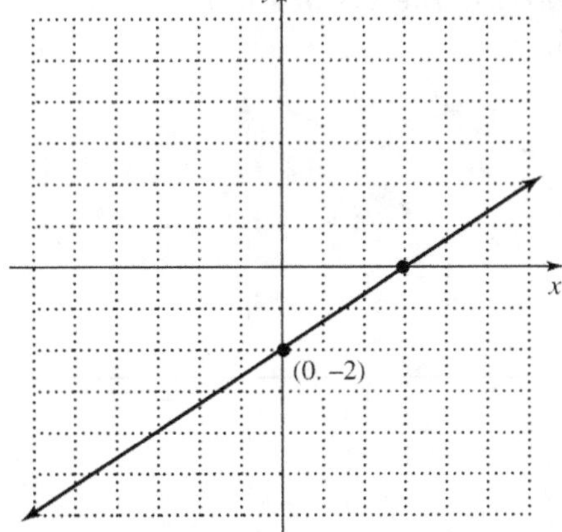

Objective 2 Practice Exercises

For extra help, see Examples 2–3 on pages 246–247 of your text.

Graph each equation by using the slope and y-intercept.

3. $4x - y = 4$

3.

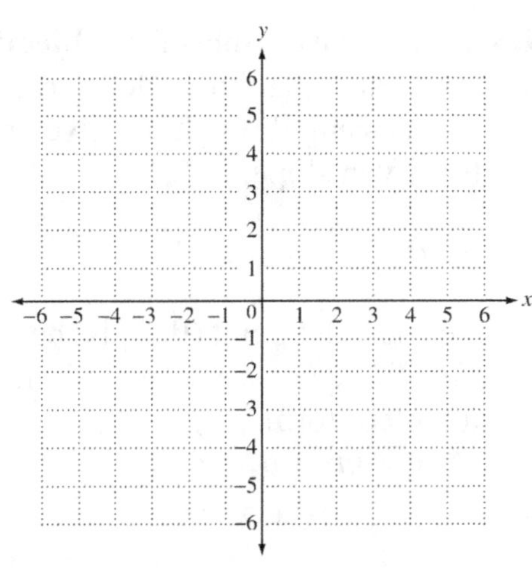

4. $y = -3x + 6$

4.

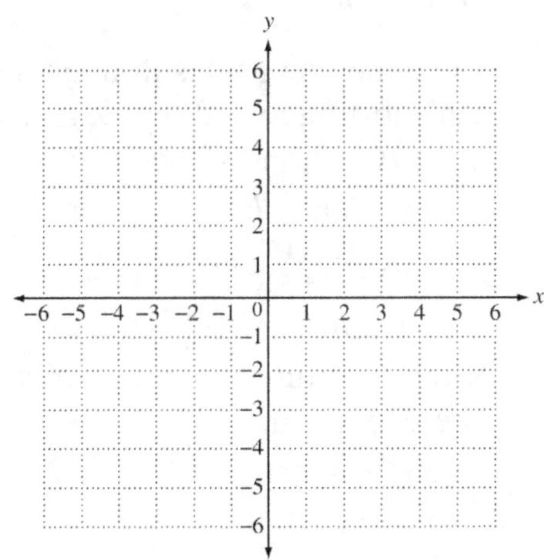

                    135

Graph the line passing through the given point and having the given slope.

5. $(4, -2)$; $m = -1$

5.

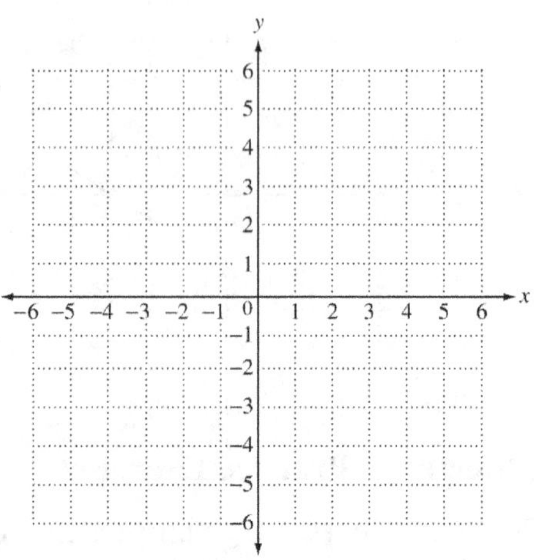

Objective 3 Write an equation of a line using its slope and any point on the line.

Review these examples for Objective 3:

4. Write an equation in slope-intercept form of the line passing through the given point and having the given slope.

a. $(0, 2)$, $m = -3$

Because the point $(0, 2)$ is the y-intercept, $b = 2$. Substitute $b = 2$ and $m = -3$ directly in the slope-intercept form.

$$y = mx + b$$
$$y = -3x + 2$$

b. $(2, 9)$, $m = 5$

Since the line passes through the point $(2, 9)$, we can substitute $x = 2$, $y = 9$, and slope $m = 5$ into $y = mx + b$ and solve for b.

$$y = mx + b$$
$$9 = 5(2) + b$$
$$-1 = b$$

Now substitute the values of m and b into slope-intercept form.

$$y = mx + b$$
$$y = 5x - 1$$

Now Try:

4. Write an equation in slope-intercept form of the line passing through the given point and having the given slope.

a. $(0, -5)$, $m = \dfrac{3}{4}$

____

b. $(-1, 4)$, $m = 6$

____

136

5. Write an equation in slope-intercept form of the line graphed.

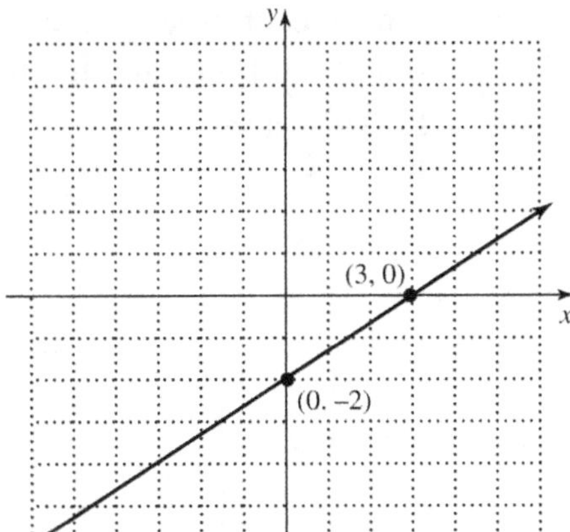

From the graph, the y-intercept is $(0,-2)$.
Thus $b = -2$. To find the slope, count grid squares up 2 units from the y-intercept and to the right 3 units to the x-intercept $(3, 0)$.

$$\text{slope } m = \frac{2}{3}$$

Thus, $y = mx + b$

$$y = \frac{2}{3}x - 2$$

5. Write an equation in slope-intercept form of the line graphed.

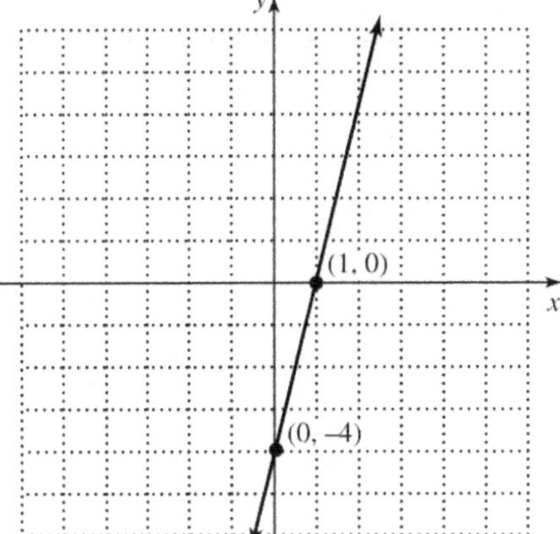

____

Objective 3 Practice Exercises

For extra help, see Examples 4–5 on pages 247–248 of your text.

Write an equation in slope-intercept form of the line passing through the given point and having the given slope.

6. $(0,-4)$, $m = \frac{2}{3}$

6. _______________

7. $(3, 6)$, $m = -2$

7. _______________

8. $(-2, 0)$, $m = 1$

8. _______________

Objective 4   Graph and write equations of horizontal and vertical lines.

Review these examples for Objective 4:	**Now Try:**

Review these examples for Objective 4:

6. Graph each line passing through the given point and having the given slope.

a. $(-5,-2)$, $m=0$

Horizontal lines have slope 0. Plot the point $(-5,-2)$ and draw a horizontal line through it.

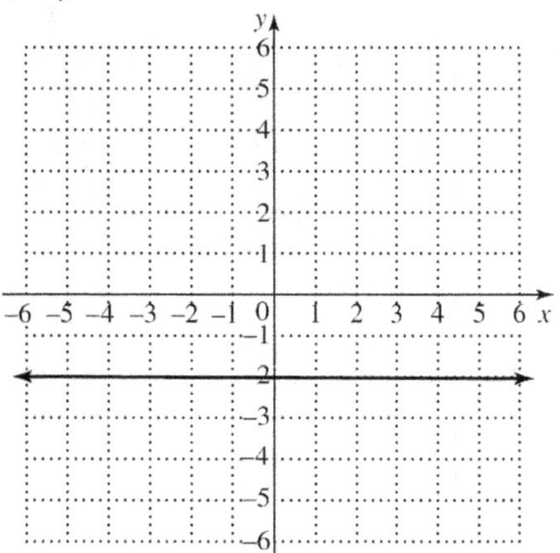

b. $(-4, 3)$, undefined slope

Vertical lines have undefined slope. Plot the point $(-4, 3)$ and draw a vertical line through it.

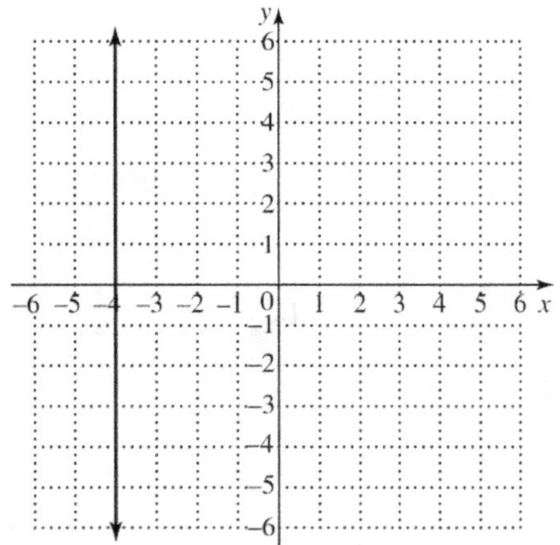

7. Write an equation of the line passing through the point $(2,-2)$ that satisfies the given condition.

a. The line has slope 0.

Since the slope is 0, this is a horizontal line.
$$y=-2.$$

Now Try:

6. Graph each line passing through the given point and having the given slope.

a. $(-2,-2)$, $m=0$

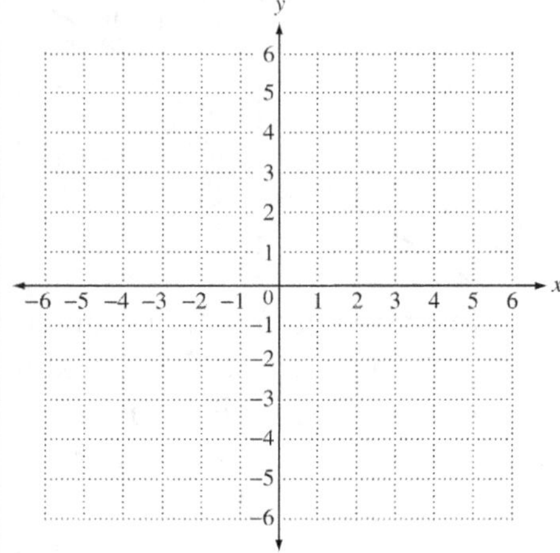

b. $(-3,-1)$, undefined slope

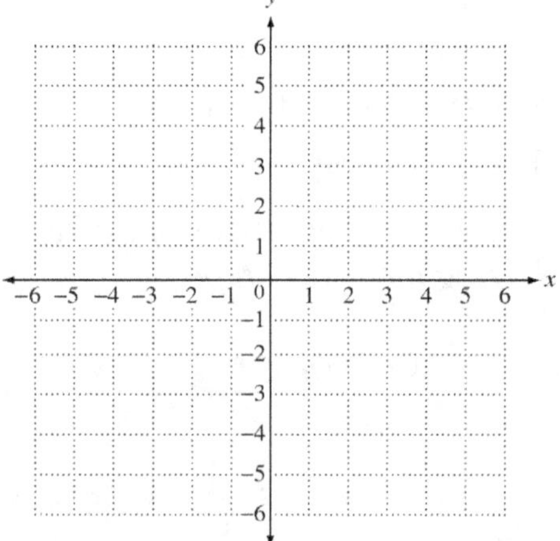

7. Write an equation of the line passing through the point $(-5, 5)$ that satisfies the given condition.

a. The line has slope 0.

____

b. The line has undefined slope.

This is a vertical line, since the slope is undefined.

$$x = 2$$

b. The line has undefined slope.

____

Objective 4 Practice Exercises

For extra help, see Examples 6–7 on pages 248–249 of your text.

Graph the line passing through the given point and having the given slope.

9. $(-1,\ 4);\ m = 0$

9. 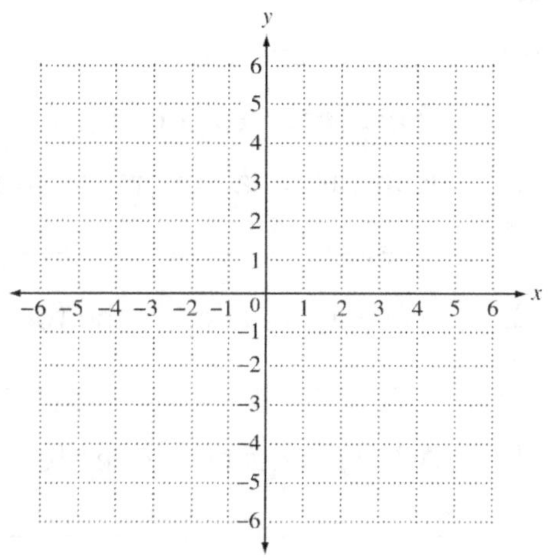

Write an equation of the line passing through (–3, 3) and having the given slope.

10.   Slope 0

10. _______________

11.   Undefined slope

11. _______________

Chapter 3 LINEAR EQUATIONS IN TWO VARIABLES

3.5   Point-Slope Form of a Linear Equation and Modeling

Learning Objectives
1   Use point-slope form to write an equation of a line.
2   Write an equation of a line using two points on the line.
3   Write an equation of a line that fits a data set.

Key Terms

Use the vocabulary terms listed below to complete each statement in exercises 1–3.

slope-intercept form   point-slope form   standard form

1.   A linear equation in the form $Ax + By = C$ is written in

_______________________________.

2.   A linear equation in the form $y = mx + b$ is written in

_______________________________.

3.   A linear equation in the form $y - y_1 = m(x - x_1)$ is written in

_______________________________.

Objective 1   Use point-slope form to write an equation of a line.

Review this example for Objective 1:

1.   Write an equation of each line. Give the final answer in slope-intercept form.

The line passing through $(5, -3)$ with slope $\dfrac{5}{4}$.

$$y - y_1 = m(x - x_1)$$
$$y - (-3) = \frac{5}{4}(x - 5)$$
$$y + 3 = \frac{5}{4}x - \frac{25}{4}$$
$$y = \frac{5}{4}x - \frac{37}{4}$$

Now Try:

1.   Write an equation of each line. Give the final answer in slope-intercept form.
The line passing through $(6, 11)$ with slope $-\dfrac{2}{3}$.

____

Objective 1 Practice Exercises

For extra help, see Example 1 on page 254 of your text.

Write an equation for the line passing through the given point and having the given slope. Write the equations in slope-intercept form, if possible.

1. $(-3,\ 4);\ \ m = -\dfrac{3}{5}$ 1. ________________

2. $(-4,-3);\ \ m = -2$ 2. ________________

3. $(2,\ 2);\ \ m = -\dfrac{3}{2}$ 3. ________________

Objective 2   Write an equation of a line using two points on the line.

Review these examples for Objective 2:

2. Write the equation of the line passing through the points (6, 8) and (–3, 5). Give the final answer in slope-intercept form and then in standard form.

   First, find the slope of the line.

   $(x_1, y_1) = (6,\ 8)$ and $(x_2, y_2) = (-3,\ 5)$

   $$\text{slope } m = \frac{y_2 - y_1}{x_2 - x_1} = \frac{5-8}{-3-6} = \frac{-3}{-9} = \frac{1}{3}$$

   Now use (x_1, y_1), here (6, 8) and point-slope form.

   $$y - y_1 = m(x - x_1)$$

   $$y - 8 = \frac{1}{3}(x - 6)$$

   $$y - 8 = \frac{1}{3}x - 2$$

   $$y = \frac{1}{3}x + 6 \quad \text{Slope-intercept form}$$

   $$3y = x + 18$$

   $$-x + 3y = 18$$

   $$x - 3y = -18 \qquad \text{Standard form}$$

Now Try:

2. Write the equation of the line passing through the points (7, 15) and (15, 9). Give the final answer in slope-intercept form and then in standard form.

   ____

3. Write an equation of the line graphed below. Give the final answer in slope-intercept form and then in standard form.

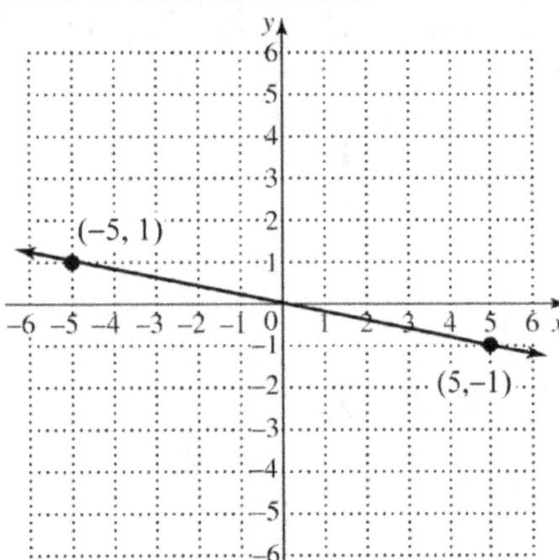

3. Write an equation of the line graphed below. Give the final answer in slope-intercept form and then in standard form.

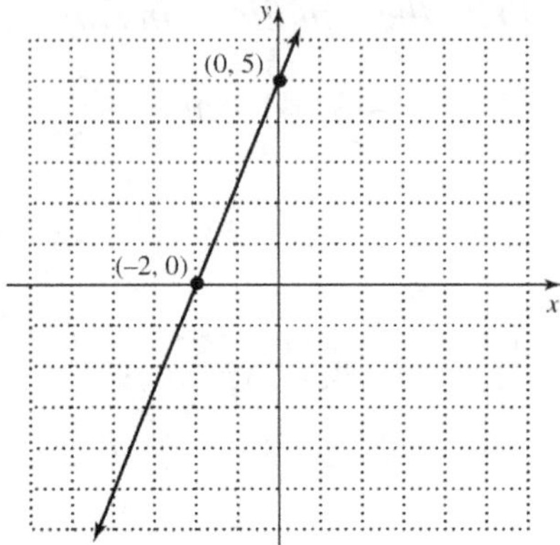

____

First, find the slope of the line. Count grid squares down 2 units from the point $(-5, 1)$, and to the right 10 units to the point $(5, -1)$.

$$\text{slope } m = \frac{-2}{10} = -\frac{1}{5}$$

Now use $m = -\frac{1}{5}$ and $(-5, 1)$ or $(5, -1)$ in point-slope form. We choose $(5, -1)$.

$$y - y_1 = m(x - x_1)$$

$$y - (-1) = -\frac{1}{5}(x - 5)$$

$$y + 1 = -\frac{1}{5}x + 1$$

$$y = -\frac{1}{5}x \quad \text{Slope-intercept form}$$

$$-5y = x$$

$$-x - 5y = 0$$

$$x + 5y = 0 \qquad \text{Standard form}$$

Objective 2 Practice Exercises

For extra help, see Example 2 on page 255 of your text.

Write an equation for the line passing through each pair of points. Write the equations in standard form.

4. $(-2, 1)$ and $(3, 11)$ 4. _________________

5. $(2, 3)$ and $(-2, -3)$ 5. _________________

6. $(3, -4)$ and $(2, 7)$.   6. _________________

Objective 3   Write an equation of a line that fits a data set.

Review this example for Objective 3:

4.   The table shows the number of internet users in the world from 1998 to 2005, where year 0 represents 1998.

Year	Number of Internet Users (millions)
0	147
2	361
4	587
6	817
8	1093

Plot the data and find an equation that approximates it.

Letting y represent the number of internet users in year x, we plot the data.

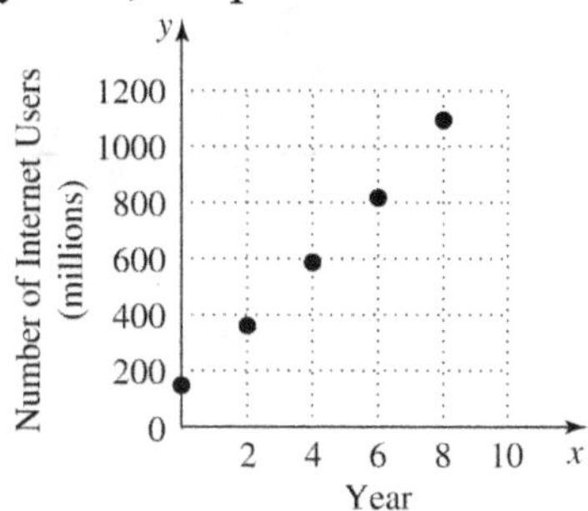

The points appear to lie approximately in a straight line. To find an equation of the line, we choose the ordered pairs (0, 147) and (8, 1093) from the table and find the slope of the line through these points.

$$(x_1, y_1) = (0, 147) \text{ and } (x_2, y_2) = (8, 1093)$$

$$\text{slope } m = \frac{y_2 - y_1}{x_2 - x_1} = \frac{1093 - 147}{8 - 0} = \frac{946}{8}$$
$$= 118.25$$

Use the slope, 118.25, and the point (0, 147) in slope-intercept form.

$$y = mx + b$$
$$147 = 118.25(0) + b$$
$$147 = b$$

Thus, $m = 118.25$ and $b = 147$, so the equation of the line is $y = 118.25x + 147$.

Now Try:

4.   The table shows the average annual telephone expenditures for residential and pay telephones from 2001 to 2006, where year 0 represents 2001.

Year	Annual Telephone Expenditures
0	$686
2	$620
3	$592
4	$570
5	$542

Plot the data and find an equation that approximates it.

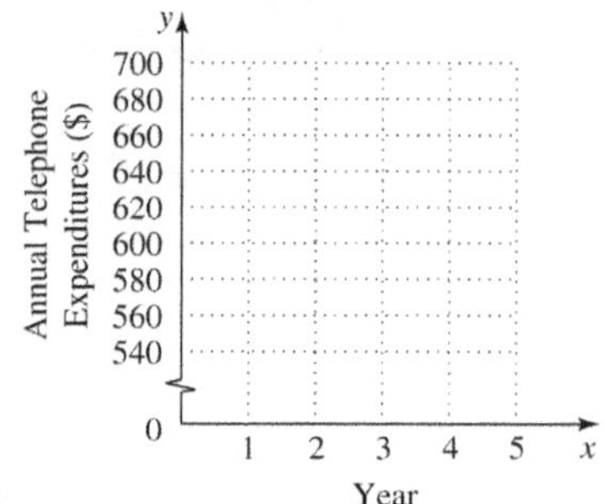

____

Objective 3 Practice Exercises

For extra help, see Example 4 on pages 256–257 of your text.

Plot the data and find an equation that approximates it.

7. The table shows the U.S. municipal solid waste recycling percents since 1985, where year 0 represents 1985.

Year	Recycling Percent
0	10.1
5	16.2
10	26.0
15	29.1
20	32.5

7.

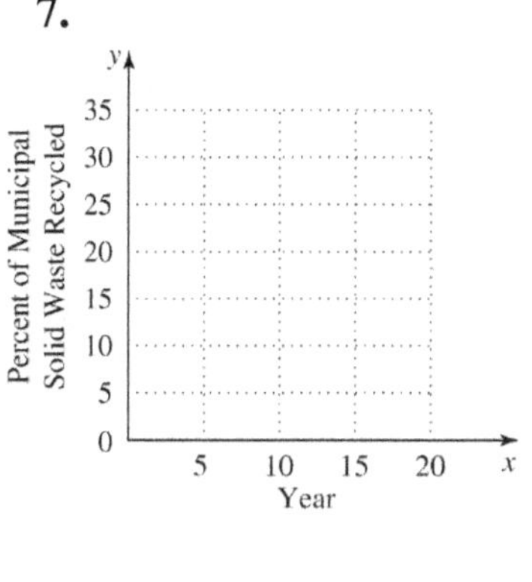

____

8. The table shows the approximate consumer expenditures for food in the U.S. in billions of dollars for selected years, where year 0 represents 1985.

Year	Food Expenditures (billions of dollars)
0	233
5	298
10	343
15	417
20	515

8.

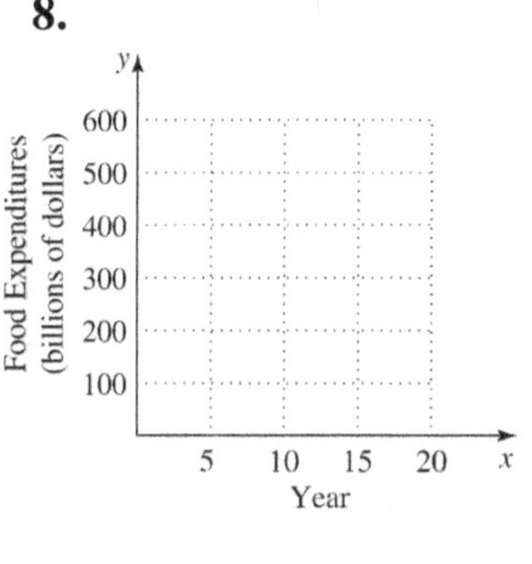

____

Chapter 4 EXPONENTS AND POLYNOMIALS

4.1   The Product Rule and Power Rules for Exponents

Learning Objectives
1   Use exponents.
2   Use the product rule for exponents.
3   Use the rule $(a^m)^n = a^{mn}$.
4   Use the rule $(ab)^m = a^m b^m$.
5   Use the rule $\left(\dfrac{a}{b}\right)^m = \dfrac{a^m}{b^m}$.
6   Use combinations of the rules for exponents.
7   Use the rules for exponents in a geometry application.

Key Terms

Use the vocabulary terms listed below to complete each statement in exercises 1−3.

| **exponential expression** | **base** | **power** |

1. 2^5 is read "2 to the fifth _____________________".

2.   A number written with an exponent is called a(n)
      _____________________.

3.   The _______________ is the number being multiplied repeatedly.

Objective 1   Use exponents.

Review these examples for Objective 1:

1.   Write $5 \cdot 5 \cdot 5$ in exponential form.

Since 5 occurs as a factor three times, the base is 5 and the exponent is 3.

$$5 \cdot 5 \cdot 5 = 5^3$$

2.   Name the base and exponent of each expression. Then evaluate.

a. 3^4

Base: 3
Exponent: 4
Value: $3^4 = 3 \cdot 3 \cdot 3 \cdot 3 = 81$

Now Try:

1.   Write $4 \cdot 4 \cdot 4 \cdot 4 \cdot 4$ in exponential form.

      ____

2.   Name the base and exponent of each expression. Then evaluate.

a. 2^6

      ____

b. $(-3)^4$

Base: -3
Exponent: 4
Value: $(-3)^4 = (-3)(-3)(-3)(-3) = 81$

b. $(-2)^6$

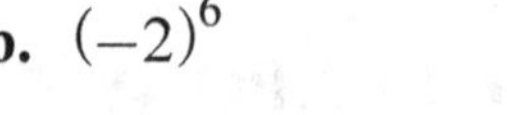

____

Objective 1 Practice Exercises

For extra help, see Examples 1–2 on page 272 of your text.

Write the expression in exponential form and evaluate, if possible.

1. $\left(\frac{1}{3}\right)\left(\frac{1}{3}\right)\left(\frac{1}{3}\right)\left(\frac{1}{3}\right)\left(\frac{1}{3}\right)$

1. _______________

Evaluate each exponential expression. Name the base and the exponent.

2. $(-4)^4$

2. _______________

base_______________

exponent___________

3. -3^8

3. _______________

base_______________

exponent___________

Objective 2   Use the product rule for exponents.

Review these examples for Objective 2:

3. Use the product rule for exponents to simplify each expression, if possible.

a. $8^4 \cdot 8^5$

$8^4 \cdot 8^5 = 8^{4+5}$

$= 8^9$

c. $m^7 \cdot m^8 \cdot m^9$

$m^7 \cdot m^8 \cdot m^9 = m^{7+8+9}$

$= m^{24}$

e. $5^2 + 5^3$

$5^2 + 5^3 = 25 + 125$

$= 150$

Now Try:

3. Use the product rule for exponents to simplify each expression, if possible.

a. $9^6 \cdot 9^7$

____

c. $m^{11} \cdot m^9 \cdot m^7$

____

e. $3^4 + 3^3$

____

f. $(-5x^4)(6x^9)$

$$(-5x^4)(6x^9) = (-5 \cdot 6) \cdot (x^4 \cdot x^9)$$
$$= -30x^{4+9}$$
$$= -30x^{13}$$

f. $(-6x^5)(3x^6)$

____

Objective 2 Practice Exercises

For extra help, see Example 3 on page 273 of your text.

Use the product rule to simplify each expression, if possible. Write each answer in exponential form.

4. $7^4 \cdot 7^3$

4. _____________

5. $(-2c^7)(-4c^8)$

5. _____________

6. $(3k^7)(-8k^2)(-2k^9)$

6. _____________

Objective 3   **Use the rule** $(a^m)^n = a^{mn}$.

Review these examples for Objective 3:
4. Use power rule (a) for exponents to simplify.

Now Try:
4. Use power rule (a) for exponents to simplify.

a. $(5^6)^3$

$$(5^6)^3 = 5^{6 \cdot 3}$$
$$= 5^{18}$$

a. $(7^2)^4$

____

c. $(x^3)^4$

$$(x^3)^4 = x^{3 \cdot 4}$$
$$= x^{12}$$

c. $(x^5)^6$

____

Objective 3 Practice Exercises

For extra help, see Example 4 on page 274 of your text.

Simplify each expression. Write all answers in exponential form.

7. $\left(7^3\right)^4$ 7. _______________

8. $-\left(v^4\right)^9$ 8. _______________

9. $\left[(-3)^3\right]^7$ 9. _______________

Objective 4   Use the rule $\left(ab\right)^m = a^m b^m$.

Review these examples for Objective 4: | **Now Try:**

5. Use power rule (b) for exponents to simplify.

5. Use power rule (b) for exponents to simplify.

a. $\left(5xy\right)^3$

a. $\left(4ab\right)^3$

$$\left(5xy\right)^3 = 5^3 x^3 y^3$$
$$= 125 x^3 y^3$$

____

d. $\left(-3^6\right)^3$

d. $\left(-7^2\right)^5$

$$\left(-3^6\right)^3 = (-1 \cdot 3^6)^3$$
$$= (-1)^3 \cdot (3^6)^3$$
$$= -1 \cdot 3^{18}$$
$$= -3^{18}$$

____

Objective 4 Practice Exercises

For extra help, see Example 5 on page 275 of your text.

Simplify each expression.

10. $\left(5r^3 t^2\right)^4$ 10. _______________

11. $\left(-0.2a^4 b\right)^3$ 11. _______________

12. $\left(-2w^3 z^7\right)^4$ 12. _______________

                Copyright © 2025 Pearson Education, Inc.

Objective 5   Use the rule $\left(\dfrac{a}{b}\right)^m = \dfrac{a^m}{b^m}$.

Review this example for Objective 5:
6. Use power rule (c) for exponents to simplify.

$$\left(\frac{1}{8}\right)^3$$

$$\left(\frac{1}{8}\right)^3 = \frac{1^3}{8^3} = \frac{1}{512}$$

Now Try:
6. Use power rule (c) for exponents to simplify.

$$\left(\frac{1}{4}\right)^5$$

____

Objective 5 Practice Exercises

For extra help, see Example 6 on page 276 of your text.

Simplify each expression.

13. $\left(-\dfrac{2x}{5}\right)^3$

13. _____________

14. $\left(\dfrac{xy}{z^2}\right)^4$

14. _____________

15. $\left(\dfrac{-2a}{b^2}\right)^7$

15. _____________

Objective 6   Use combinations of the rules for exponents.

Review these examples for Objective 6:
7. Simplify each expression.

a. $\left(\dfrac{3}{4}\right)^3 \cdot 3^2$

$$\left(\frac{3}{4}\right)^3 \cdot 3^2 = \frac{3^3}{4^3} \cdot \frac{3^2}{1}$$

$$= \frac{3^3 \cdot 3^2}{4^3 \cdot 1}$$

$$= \frac{3^{3+2}}{4^3}$$

$$= \frac{3^5}{4^3}, \quad \text{or} \quad \frac{243}{64}$$

Now Try:
7. Simplify each expression.

a. $\left(\dfrac{5}{2}\right)^3 \cdot 5^2$

____

d. $\left(-x^5y\right)^4\left(-x^6y^5\right)^3$

$\left(-x^5y\right)^4\left(-x^6y^5\right)^3$

$=\left(-1x^5y\right)^4\left(-1x^6y^5\right)^3$

$=(-1)^4\left(x^5\right)^4\left(y^4\right)\cdot(-1)^3\left(x^6\right)^3\left(y^5\right)^3$

$=(-1)^4\left(x^{20}\right)\left(y^4\right)\cdot(-1)^3\left(x^{18}\right)\left(y^{15}\right)$

$=(-1)^7x^{20+18}y^{4+15}$

$=-1x^{38}y^{19}$

$=-x^{38}y^{19}$

d. $\left(-x^5y\right)^3\left(-x^6y^5\right)^2$

____

Objective 6 Practice Exercises

For extra help, see Example 7 on pages 276–277 of your text.

Simplify. Write all answers in exponential form.

16. $\left(-x^3\right)^2\left(-x^5\right)^4$

16. __________________

17. $\left(2ab^2c\right)^5(ab)^4$

17. __________________

18. $\left(5x^2y^3\right)^7\left(5xy^4\right)^4$

18. __________________

Objective 7   Use the rules for exponents in a geometry application.

Review this example for Objective 7: | **Now Try:**

8.   Find the area of the figure.

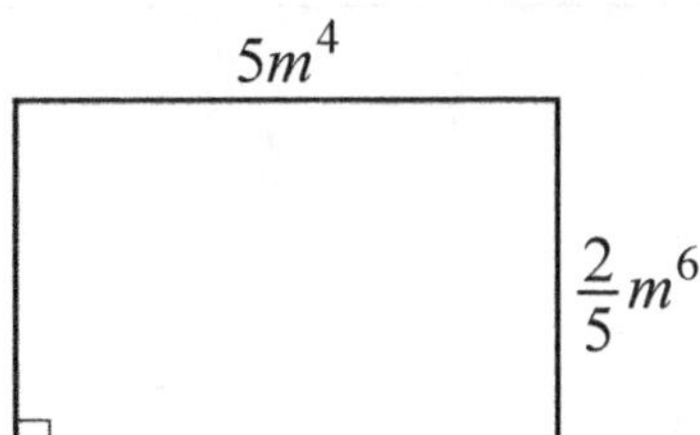

Use the formula for the area of a rectangle.
$$A = LW$$
$$A = \left(5m^4\right)\left(\frac{2}{5}m^6\right)$$
$$A = 5 \cdot \frac{2}{5} \cdot m^{4+6}$$
$$A = 2m^{10}$$

8.   Find the area of the figure.

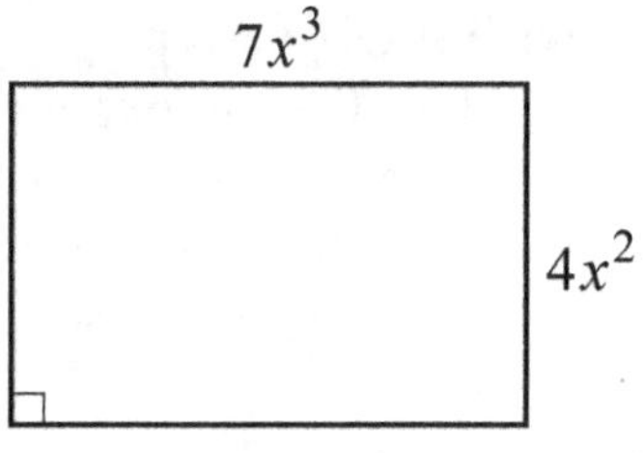

____

Objective 7 Practice Exercises

For extra help, see Example 8 on pages 277–278 of your text.

Find a polynomial that represents the area of each figure.

19.

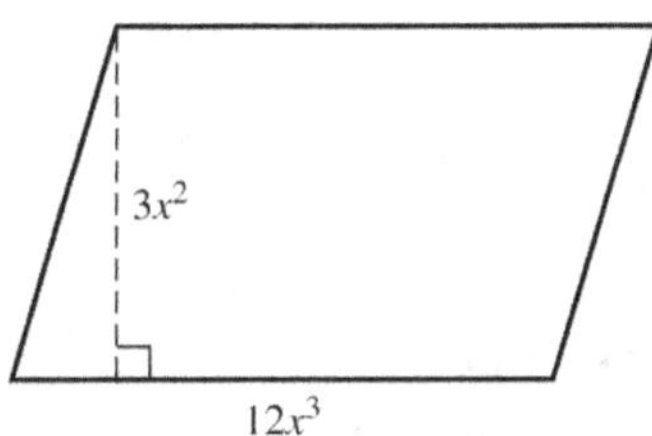

19. ___________________

20.

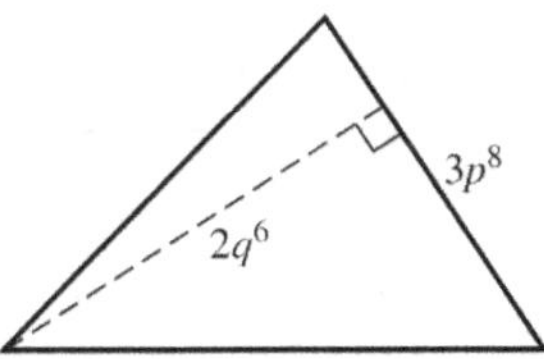

20. ___________________

21.

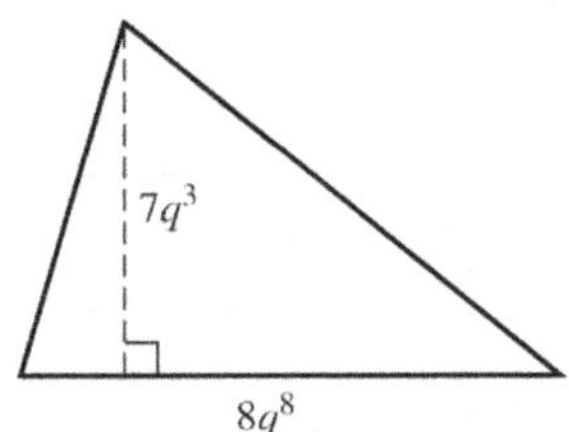

21. ___________________

Chapter 4 EXPONENTS AND POLYNOMIALS

4.2   Integer Exponents and the Quotient Rule

Learning Objectives
1   Use 0 as an exponent.
2   Use negative numbers as exponents.
3   Use the quotient rule for exponents.
4   Use combinations of the rules for exponents.

Key Terms

Use the vocabulary terms listed below to complete each statement in exercises 1−3.

exponent   **base**   **product rule for exponents**

power rule for exponents

1.   The statement "If m and n are any integers, then $\left(a^m\right)^n = a^{mn}$" is an example of the _______________________________.

2.   In the expression a^m, a is the _______________ and m is the _______________.

3.   The statement "If m and n are any integers, then $a^m \cdot a^n = a^{m+n}$" is an example of the _______________________________.

Objective 1   Use 0 as an exponent.

Review these examples for Objective 1:

1. Evaluate.

    a. $75^0 = 1$

    c. $-75^0 = -(1)$ or -1

    f. $(-9x)^0 = 1$ $(x \neq 0)$

    g. $3^0 + 12^0 = 1 + 1 = 2$

Now Try:

1. Evaluate.

    a. 88^0

    ____

    c. -88^0

    ____

    f. $(-88a)^0$ $(a \neq 0)$

    ____

    g. $5^0 - 16^0$

    ____

Name: Date:

Instructor: Section:

Objective 1 Practice Exercises

For extra help, see Example 1 on page 281 of your text.

Evaluate each expression.

1. -12^0

1. _______________

2. $-15^0 - (-15)^0$

2. _______________

3. $\dfrac{0^8}{8^0}$

3. _______________

Objective 2   Use negative numbers as exponents.

Review these examples for Objective 2:

2. Simplify by writing with positive exponents. Assume that all variables represent nonzero real numbers.

a. 4^{-3}

$$4^{-3} = \frac{1}{4^3}, \quad \text{or} \quad \frac{1}{64}$$

c. $\left(\dfrac{1}{3}\right)^{-3}$

$$\left(\frac{1}{3}\right)^{-3} = 3^3, \quad \text{or} \quad 27$$

d. $\left(\dfrac{2}{3}\right)^{-5}$

$$\left(\frac{2}{3}\right)^{-5} = \left(\frac{3}{2}\right)^5$$
$$= \frac{3^5}{2^5}$$
$$= \frac{243}{32}$$

Now Try:

2. Simplify by writing with positive exponents. Assume that all variables represent nonzero real numbers.

a. 3^{-3}

____

c. $\left(\dfrac{1}{5}\right)^{-2}$

____

d. $\left(\dfrac{3}{2}\right)^{-3}$

____

f. $5^{-1} - 3^{-1}$

$$5^{-1} - 3^{-1} = \frac{1}{5} - \frac{1}{3}$$

$$= \frac{3}{15} - \frac{5}{15}$$

$$= -\frac{2}{15}$$

i. $q^{-3} \quad (q \neq 0)$

$$q^{-3} = \frac{1}{q^3}$$

3. Simplify. Assume that all variables represent nonzero real numbers.

a. $\dfrac{3^{-4}}{7^{-2}} = \dfrac{7^2}{3^4}, \quad$ or $\quad \dfrac{49}{81}$

b. $a^{-6}b^4 = \dfrac{b^4}{a^6}$

c. $\dfrac{x^{-3}y}{4z^{-4}} = \dfrac{yz^4}{4x^3}$

f. $4^{-1} - 8^{-1}$

____

i. $p^{-5} \quad (p \neq 0)$

____

3. Simplify. Assume that all variables represent nonzero real numbers.

a. $\dfrac{6^{-2}}{5^{-3}}$

____

b. $x^{-7}y^2$

____

c. $\dfrac{p^{-3}q}{4r^{-5}}$

____

Objective 2 Practice Exercises

For extra help, see Examples 2–3 on pages 282–284 of your text.

Evaluate or simplify each expression, and write it using only positive exponents. Assume that all variables represent nonzero real numbers.

4. $-2k^{-4}$

4. ________________

5. $(m^2 n)^{-9}$

5. ________________

6. $\dfrac{2x^{-4}}{3y^{-7}}$

6. ________________

Objective 3   Use the quotient rule for exponents.

Review these examples for Objective 3:

4. Simplify. Assume that all variables represent nonzero real numbers.

a. $\dfrac{4^9}{4^6} = 4^{9-6} = 4^3 = 64$

d. $\dfrac{p^6}{p^{-4}} = p^{6-(-4)} = p^{10}$

f. $\dfrac{(x+7)^{-3}}{(x+7)^{-5}}$ $x \neq -7$

$\dfrac{(x+7)^{-3}}{(x+7)^{-5}} = (x+7)^{-3-(-5)}$

$= (x+7)^{-3+5}$

$= (x+7)^2$

h. $\dfrac{8x^{-4}y^3}{5^{-1}x^3y^{-4}}$

$\dfrac{8x^{-4}y^3}{5^{-1}x^3y^{-4}} = \dfrac{8 \cdot 5y^3y^4}{x^3x^4}$

$= \dfrac{40y^7}{x^7}$

Now Try:

4. Simplify. Assume that all variables represent nonzero real numbers.

a. $\dfrac{3^{18}}{3^{16}}$

____

d. $\dfrac{z^6}{z^{-4}}$

____

f. $\dfrac{(a-b)^{-8}}{(a-b)^{-10}}$ $a \neq b$

____

h. $\dfrac{9a^{-5}b^3}{4^{-1}a^3b^{-4}}$

____

Objective 3 Practice Exercises

For extra help, see Example 4 on page 285 of your text.

Use the quotient rule to simplify each expression, and write it using only positive exponents. Assume that all variables represent nonzero real numbers.

7. $\dfrac{4k^7m^{10}}{8k^3m^5}$

7. _______________

8. $\dfrac{a^4 b^3}{a^{-2} b^{-3}}$ 8. ______________

9. $\dfrac{3^{-1} m^{-4} p^6}{3^4 m^{-1} p^{-2}}$ 9. ______________

Objective 4   Use combinations of the rules for exponents.

Review these examples for Objective 4:	Now Try:

Review these examples for Objective 4:

5. Simplify each expression. Assume that all variables represent nonzero real numbers.

a. $\dfrac{\left(5^4\right)^2}{5^6}$

$$\frac{\left(5^4\right)^2}{5^6} = \frac{5^8}{5^6}$$

$$= 5^{8-6}$$

$$= 5^2$$

$$= 25$$

c. $(3a)^4 (3a)^2$

$$(3a)^4 (3a)^2 = (3a)^6$$

$$= 3^6 a^6$$

$$= 729 a^6$$

d. $\left(\dfrac{3x^4}{4}\right)^{-5}$

$$\left(\frac{3x^4}{4}\right)^{-5} = \left(\frac{4}{3x^4}\right)^{5}$$

$$= \frac{4^5}{3^5 x^{20}}$$

$$= \frac{1024}{243 x^{20}}$$

Now Try:

5. Simplify each expression. Assume that all variables represent nonzero real numbers.

a. $\dfrac{\left(6^3\right)^2}{6^5}$

____

c. $(5b)^3 (5b)^2$

____

d. $\left(\dfrac{2p^4}{3}\right)^{-5}$

____

e. $\dfrac{\left(k^2 m^{-3} n\right)^{-5}}{\left(3 k m^2 n^{-4}\right)^{-6}}$

$\dfrac{\left(k^2 m^{-3} n\right)^{-5}}{\left(3 k m^2 n^{-4}\right)^{-6}} = \dfrac{\left(k^2\right)^{-5}\left(m^{-3}\right)^{-5} n^{-5}}{3^{-6} k^{-6} \left(m^2\right)^{-6}\left(n^{-4}\right)^{-6}}$

$= \dfrac{k^{-10} m^{15} n^{-5}}{3^{-6} k^{-6} m^{-12} n^{24}}$

$= \dfrac{3^6 m^{15+12}}{k^{-6+10} n^{24+5}}$

$= \dfrac{729 m^{27}}{k^4 n^{29}}$

e. $\dfrac{\left(7 x y^{-2} z^3\right)^{-3}}{\left(x^{-4} y z^{-2}\right)^4}$

____

Objective 4 Practice Exercises

For extra help, see Example 5 on pages 286–287 of your text.

Simplify each expression, and write it using only positive exponents. Assume that all variables represent nonzero real numbers.

10. $\left(9xy\right)^7 \left(9xy\right)^{-8}$

10. _______________

11. $\dfrac{\left(a^{-1} b^{-2}\right)^{-4}\left(ab^2\right)^6}{\left(a^3 b\right)^{-2}}$

11. _______________

12. $\left(\dfrac{k^3 t^4}{k^2 t^{-1}}\right)^{-4}$

12. _______________

Chapter 4 EXPONENTS AND POLYNOMIALS

4.3   Scientific Notation

Learning Objectives
1   Express numbers in scientific notation.
2   Convert numbers in scientific notation to standard notation.
3   Use scientific notation in calculations.

Key Terms

Use the vocabulary terms listed below to complete each statement in exercises 1–3.

scientific notation   **quotient rule**   **power rule**

1.   A number written as $a \times 10^n$, where $1 \leq |a| < 10$ and n is an integer, is written in

_______________________________.

2.   The statement "If m and n are any integers and $b \neq 0$, then $\left(\dfrac{a}{b}\right)^m = \dfrac{a^m}{b^m}$" is an

example of the _______________________________.

3.   The statement "If m and n are any integers and $b \neq 0$, then $\dfrac{a^m}{a^n} = a^{m-n}$" is an

example of the _______________________________.

Objective 1   Express numbers in scientific notation.

Review these examples for Objective 1:

1.   Write each number in scientific notation.

    a. 84,300,000,000

Move the decimal point 10 places to the left.
$$84,300,000,000 = 8.43 \times 10^{10}$$

    b. 0.00573

The first nonzero digit is 5. Count the places.
Move the decimal point 3 places to the right.
$$0.00573 = 5.73 \times 10^{-3}$$

Now Try:

1.   Write each number in scientific notation.

    a. 47,710,000,000

____

    b. 0.0463

____

Objective 1 Practice Exercises

For extra help, see Example 1 on page 292 of your text.

Write each number in scientific notation.

1. 23,651

2. $-429,600,000,000$

3. -0.0002208

1. _________________

2. _________________

3. _________________

Objective 2   Convert numbers in scientific notation to standard notation.

Review these examples for Objective 2:

2. Write each number without exponents.

    a. 3.57×10^6

Move the decimal point 6 places to the right, and attach four zeros.

$$3.57 \times 10^6 = 3,570,000$$

    b. 8.98×10^{-3}

Move the decimal point 3 places to the left.

$$8.98 \times 10^{-3} = 0.00898$$

Now Try:

2. Write each number without exponents.

    a. 2.796×10^7

____

    b. 1.64×10^{-4}

____

Objective 2 Practice Exercises

For extra help, see Example 2 on page 293 of your text.

Write each number in standard notation.

4. -2.45×10^6

5. 6.4×10^{-3}

6. -4.02×10^4

4. _________________

5. _________________

6. _________________

Objective 3   Use scientific notation in calculations.

Review these examples for Objective 3:

3. Perform each calculation. Write answers in scientific notation and also without exponents.

   a. $\left(8\times10^4\right)\left(7\times10^3\right)$

$$\left(8\times10^4\right)\left(7\times10^3\right)=(8\times7)\left(10^4\times10^3\right)$$
$$=56\times10^7$$
$$=\left(5.6\times10^1\right)\times10^7$$
$$=5.6\times10^8$$
$$=560,000,000$$

   b. $\dfrac{6\times10^{-4}}{3\times10^2}$

$$\frac{6\times10^{-4}}{3\times10^2}=\frac{6}{3}\times\frac{10^{-4}}{10^2}$$
$$=2\times10^{-6}$$
$$=0.000002$$

4. The Sahara desert covers approximately 3.5×10^6 square miles. Its sand is, on average, 12 feet deep. Find the volume, in cubic feet, of sand in the Sahara. $\left(\text{Hint: }1\text{ mi}^2=5280^2\text{ ft}^2\right)$ Round your answer to two decimal places.

$$\left(3.5\times10^6\right)\left(5280^2\right)(12)$$
$$=97574400(12)\times10^6$$
$$=1170892800\times10^6$$
$$\approx1.17\times10^{15}\text{ ft}^3$$

The volume is 1.17×10^{15} cubic feet.

Now Try:

3. Perform each calculation. Write answers in scientific notation and also without exponents.

   a. $\left(9\times10^5\right)\left(3\times10^2\right)$

____

____

   b. $\dfrac{39\times10^{-3}}{13\times10^5}$

____

____

4. The Sahara desert covers approximately 3.5×10^6 square miles. Its sand is, on average, 12 feet deep. The volume of a single grain of sand is approximately 1.3×10^{-9} cubic feet. About how many grains of sand are in the Sahara?

____

Objective 3 Practice Exercises

For extra help, see Examples 3–5 on pages 293–294 of your text.

Perform the indicated operations, and write the answers in scientific notation.

7. $\left(2.3\times10^{4}\right)\times\left(1.1\times10^{-2}\right)$ 7. ________________

8. $\dfrac{9.39\times10^{1}}{3\times10^{3}}$ 8. ________________

Work the problem. Give answer in scientific notation.

9.   There are about 6×10^{23} atoms in a mole of atoms.   9. ________________
      How many atoms are there in 8.1×10^{-5} mole?

Chapter 4 EXPONENTS AND POLYNOMIALS

4.4   Adding, Subtracting, and Graphing Polynomials

Learning Objectives
1   Identify terms and coefficients.
2   Combine like terms.
3   Describe polynomials using appropriate vocabulary.
4   Evaluate polynomials.
5   Add and subtract polynomials.
6   Graph equations defined by polynomials of degree 2.

Key Terms

Use the vocabulary terms listed below to complete each statement in exercises 1−14.

term	**leading term**	**coefficient**	**like terms**
polynomial	**degree of a term**	**degree of a polynomial**	
monomial	**binomial**	**trinomial**	
descending powers		**parabola**	
vertex	**axis**	**line of symmetry**	

1. The _______________________ is the sum of the exponents on the variables in that term.

2. A polynomial in x is written in _______________________ if the exponents on x in its terms are decreasing order.

3. A _______________ is a number, a variable, or a product or quotient of a number and one or more variables raised to powers.

4. A polynomial with exactly three terms is called a _______________________.

5. A _______________________ is a term, or the sum of a finite number of terms with whole number exponents.

6. A polynomial with exactly one term is called a _______________________.

7. The _______________________ is the greatest degree of any term of the polynomial.

8. A _______________________ is a polynomial with exactly two terms.

9. Terms with exactly the same variables (including the same exponents) are called _______________________.

10. If a graph is folded on its_________________________________, the two sides coincide.

11. The ________________________________ of a parabola that opens upward or downward is the lowest or highest point on the graph.

12. The ________________________________ of a parabola that opens upward or downward is a vertical line through the vertex.

13. The graph of the quadratic equation $y = ax^2 + bx + c$ is called a

_____________________.

14. In the algebraic expression $5x^3 - 7x^2 - 8x + 9$, the number 5 is called the

_____________________ and $5x^3$ is called the _____________________.

Objective 1   Identify terms and coefficients.

Review this example for Objective 1:

1. For each expression, determine the number of terms and name the coefficients of the terms.

$$6 - 3x^4 - x^2$$

Rewrite the expression as $6x^0 - 3x^4 - 1x^2$.

There are three terms: $6, -3x^4$, and $-x^2$.

The coefficients are 6, –3, and –1.

Now Try:

1. For each expression, determine the number of terms and name the coefficients of the terms.

$$x^2 + 7 - 2x$$

____

Objective 1 Practice Exercises

For extra help, see Example 1 on page 299 of your text.

For each expression, determine the number of terms and name the coefficients of the terms.

1. $3x^2 - 2 + x$

1. _____________

2. $5 + 6z^3$

2. _____________

3. $8y - y^3 - 1$

3. _____________

Objective 2   Combine like terms.

Review this example for Objective 2:

2. Simplify the expression by combining like terms.

$$19m^3 + 6m + 5m^3$$

$$19m^3 + 6m + 5m^3 = (19 + 5)m^3 + 6m$$

$$= 24m^3 + 6m$$

Now Try:

2. Simplify the expression by combining like terms.

$$22m^2 + 15m^3 + 7m^2$$

____

Objective 2 Practice Exercises

For extra help, see Example 2 on page 300 of your text.

In each polynomial, combine like terms whenever possible. Write the result with descending powers.

4. $7z^3 - 4z^3 + 5z^3 - 11z^3$

4. ________________

5. $-1.3z^7 + 0.4z^7 + 2.6z^8$

5. ________________

6. $6c^3 - 9c^2 - 2c^2 + 14 + 3c^2 - 6c - 8 + 2c^3$

6. ________________

Objective 3   Describe polynomials using appropriate vocabulary.

Review these examples for Objective 3:

3. Simplify each polynomial, if possible, and write in descending powers of the variable. Then give the degree and tell whether the polynomial is a monomial, a binomial, a trinomial, or none of these.

a. $5x + 6x^3 - 7x - 2x^2 + 4x$

$$5x + 6x^3 - 7x - 2x^2 + 4x = 6x^3 - 2x^2 + 2x$$

The degree is 3. The simplified polynomial is a trinomial.

Now Try:

3. Simplify each polynomial, if possible, and write in descending powers of the variable. Then give the degree and tell whether the polynomial is a monomial, a binomial, a trinomial, or none of these.

a. $2x - 7x^2 - 6x + 3x^3 + 8x$

____

                          Copyright © 2025 Pearson Education, Inc.

b. $9y^3 - 7y^5 + 3y^3 + 2y^5$

$9y^3 - 7y^5 + 3y^3 + 2y^5 = -5y^5 + 12y^3$

The degree is 5. The simplified polynomial is a binomial.

b. $w^5 - 4w^2 + 3w^5 + w^2$

____

Objective 3 Practice Exercises

For extra help, see Example 3 on page 301 of your text.

For each polynomial, first simplify, if possible, and write the resulting polynomial in descending powers of the variable. Then give the degree of this polynomial, and tell whether it is a monomial, *a* binomial, *a* trinomial, *or* none of these.

7. $3n^8 - n^2 - 2n^8$

7. _______________

degree: _______________

type: _______________

8. $-d^2 + 3.2d^3 - 5.7d^8 - 1.1d^5$

8. _______________

degree: _______________

type: _______________

9. $-6c^4 - 6c^2 + 9c^4 - 4c^2 + 5c^5$

9. _______________

degree: _______________

type: _______________

Objective 4   Evaluate polynomials.

Review this example for Objective 4:

4.   Find the value of $4x^3 + 6x^2 - 5x - 5$ for

$x = 2$

$4x^3 + 6x^2 - 5x - 5$

$= 4(2)^3 + 6(2)^2 - 5(2) - 5$

$= 4(8) + 6(4) - 5(2) - 5$

$= 32 + 24 - 10 - 5$

$= 41$

Now Try:

4.   Find the value of
$5x^4 + 3x^2 - 9x - 7$ for
$x = 4$

____

Objective 4 Practice Exercises

For extra help, see Example 4 on page 302 of your text.

Find the value of each polynomial (a) *when x = –2 and* (b) *when x = 3.*

10. $3x^3 + 4x - 19$

                                                  10. a._______________

                                                          b._______________

11. $-4x^3 + 10x^2 - 1$

                                                  11. a._______________

                                                          b._______________

12. $x^4 - 3x^2 - 8x + 9$

                                                  12. a._______________

                                                          b._______________

Objective 5   Add and subtract polynomials.

Review these examples for Objective 5:   **Now Try:**

6.   Find each sum.   **6.**   Find each sum.

a. Add $5x^4 - 7x^3 + 9$ and $-3x^4 + 8x^3 - 7$ **a.** Add $15x^3 - 5x + 3$ and $-11x^3 + 6x + 9$

$$(5x^4 - 7x^3 + 9) + (-3x^4 + 8x^3 - 7)$$
$$= 5x^4 - 3x^4 - 7x^3 + 8x^3 + 9 - 7$$
$$= 2x^4 + x^3 + 2$$

____

b. $(5x^4 - 7x^2 + 6x) + (-3x^3 + 4x^2 - 7)$ **b.** $(8x^2 - 6x + 4) + (7x^3 - 8x - 5)$

$$(5x^4 - 7x^2 + 6x) + (-3x^3 + 4x^2 - 7)$$
$$= 5x^4 - 3x^3 - 7x^2 + 4x^2 + 6x - 7$$
$$= 5x^4 - 3x^3 - 3x^2 + 6x - 7$$

____

7. Perform the subtraction.

Subtract $8x^3 - 5x^2 + 8$ from $9x^3 + 6x^2 - 7$.

$(9x^3 + 6x^2 - 7) - (8x^3 - 5x^2 + 8)$

$\quad = (9x^3 + 6x^2 - 7) + (-8x^3 + 5x^2 - 8)$

$\quad = x^3 + 11x^2 - 15$

9. Perform the indicated operations to simplify the expression

$(5 - 2x + 9x^2) - (7 - 5x + 8x^2) + (6 + 3x - 5x^2)$

Rewrite, changing the subtraction to adding the opposite.

$(5 - 2x + 9x^2) - (7 - 5x + 8x^2) + (6 + 3x - 5x^2)$

$= (5 - 2x + 9x^2) + (-7 + 5x - 8x^2) + (6 + 3x - 5x^2)$

$= (-2 + 3x + x^2) + (6 + 3x - 5x^2)$

$= 4 + 6x - 4x^2$

10. Add or subtract as indicated.

$(3x^2y + 5xy + y^2) - (4x^2y + xy - 3y^2)$

Change the signs of the terms in the parentheses and add like terms vertically.

$$\begin{array}{r} 3x^2y + 5xy + y^2 \\ -4x^2y - xy + 3y^2 \\ \hline -x^2y + 4xy + 4y^2 \end{array}$$

7. Perform the subtraction.

Subtract $18x^3 + 4x - 6$ from $7x^3 - 3x - 5$.

____

9. Perform the indicated operations to simplify the expression

$(10 - 7x + 6x^2) - (5 - 11x + 3x^2)$

$+ (2 + 4x - 7x^2)$

____

10. Add or subtract as indicated.

$(7x^2y + 3xy + 4y^2)$

$- (6x^2y - xy + 4y^2)$

____

Objective 5 Practice Exercises

For extra help, see Examples 5–10 on pages 302–304 of your text.

Add or subtract as indicated.

13. $(3r^3 + 5r^2 - 6) + (2r^2 - 5r + 4)$

13. _______________

14. $\left(-8w^3 + 11w^2 - 12\right) - \left(-10w^2 + 3\right)$ **14.** _________________

15. $\left(2x^2 y + 2xy - 4xy^2\right) + \left(6xy + 9xy^2\right) - \left(9x^2 y + 5xy\right)$ **15.** _________________

Objective 6   Graph equations defined by polynomials of degree 2.

Review this example for Objective 6:

11. Graph the equation.

$$y = x^2 - 3$$

Find several ordered pairs. Let $x = 0$ to find the y-intercept.

$$y = x^2 - 3 = 0^2 - 3 = -3$$

This gives the ordered pair $(0, -3)$. Select several values for x and find the corresponding values for y. Plot the ordered pairs and join them with a smooth curve.

x	y
2	1
1	-2
0	-3
-1	-2
-2	1

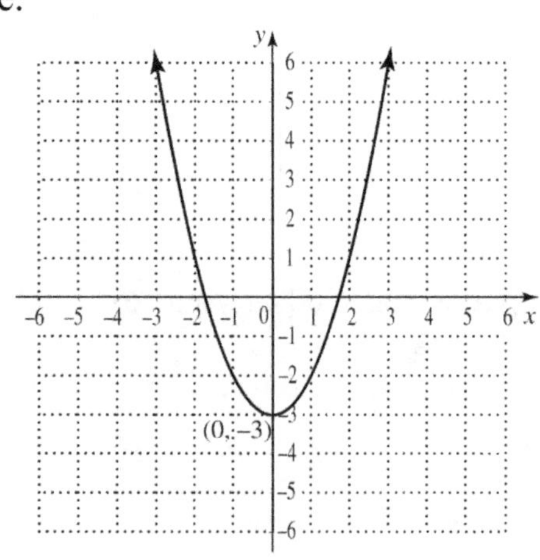

Now Try:

11. Graph the equation.

$$y = 9 - x^2$$

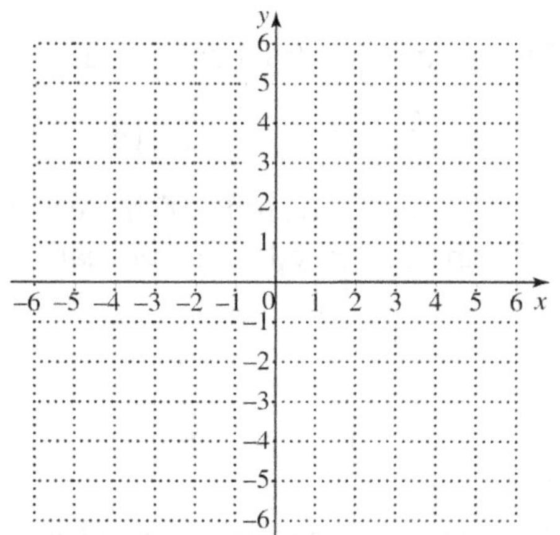

                Copyright © 2025 Pearson Education, Inc.

Name: Date:
Instructor: Section:

Objective 6 Practice Exercises

For extra help, see Example 11 on page 305 of your text.

Graph each equation.

16. $y = -x^2 - 1$

16.

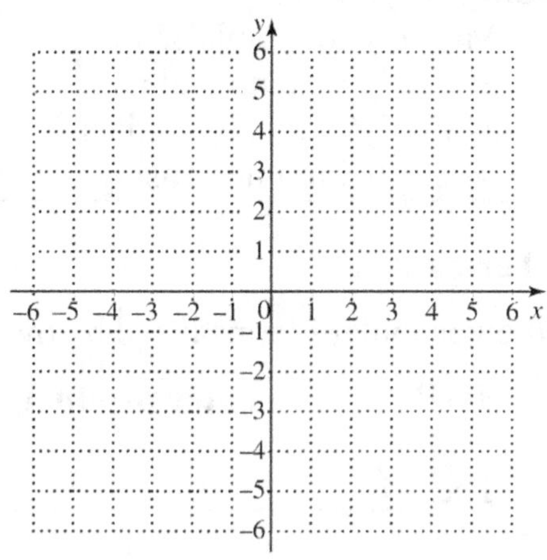

vertex: _______________________

17. $y = x^2 + 2$

17.

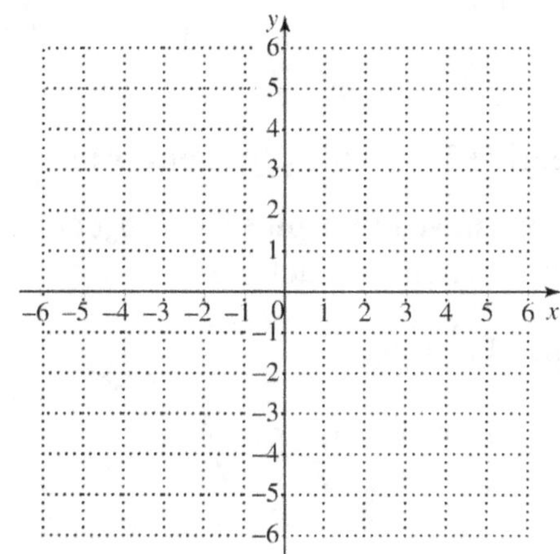

vertex: _______________________

Chapter 4 EXPONENTS AND POLYNOMIALS

4.5   Multiplying Polynomials

Learning Objectives
1   Multiply monomials.
2   Multiply a monomial and a polynomial.
3   Multiply two polynomials.
4   Multiply binomials using the FOIL method.

Key Terms

Use the vocabulary terms listed below to complete each statement in exercises 1−3.

FOIL   outer product   inner product

1.   The ______________________________ of $(2y-5)(y+8)$ is $-5y$.

2.   ____________________ is a shortcut method for finding the product of two binomials.

3.   The ______________________________ of $(2y-5)(y+8)$ is $16y$.

Objective 1   Multiply monomials.

Review these examples for Objective 1:

1.   Find each product.

   a. $4x^3(-5x)$

   $$4x^3(-5x) = 4(-5) \cdot x^3 x^1$$
   $$= -20x^{3+1}$$
   $$= -20x^4$$

   b. $2m^2 n^3 (7mn^2)$

   $$2m^2 n^3 (7mn^2) = 2(7) \cdot m^2 m^1 \cdot n^3 n^2$$
   $$= 14m^{2+1} n^{3+2}$$
   $$= 14m^3 y^5$$

Now Try:

1.   Find each product.

   a. $9a^4(-2a)$

   ____

   b. $3p^4 q^2 (5p^3 q^2)$

   ____

          Copyright © 2025 Pearson Education, Inc.

Objective 1 Practice Exercises

For extra help, see Example 1 on page 310 of your text.

Find each product.

1. $12p^3(3p^2)$

1. ______________________

2. $5a^4(2a^2)$

2. ______________________

3. $-6x^4(2x^3)$

3. ______________________

Objective 2   Multiply a monomial and a polynomial.

Review this example for Objective 2:

2. Find the product.

$$5x^2(7x+3)$$

Use the distributive property.
$$5x^2(7x+3)=5x^2(7x)+5x^2(3)$$
$$=35x^3+15x^2$$

Now Try:

2. Find the product.

$$8x^3(4x+8)$$

____

Objective 2 Practice Exercises

For extra help, see Example 2 on page 310 of your text.

Find each product.

4. $7z(5z^3+2)$

4. ______________________

5. $2m(3+7m^2+3m^3)$

5. ______________________

6. $-3y^2(2y^3+3y^2-4y+11)$

6. ______________________

Objective 3   Multiply two polynomials.

Review these examples for Objective 3:

3. Multiply $(x^2 + 6)(5x^3 - 4x^2 + 3x)$.

Multiply each term of the second polynomial by each term of the first.

$(x^2 + 6)(5x^3 - 4x^2 + 3x)$

$= x^2(5x^3) + x^2(-4x^2) + x^2(3x)$

$\quad + 6(5x^3) + 6(-4x^2) + 6(3x)$

$= 5x^5 - 4x^4 + 3x^3 + 30x^3 - 24x^2 + 18x$

$= 5x^5 - 4x^4 + 33x^3 - 24x^2 + 18x$

4. Multiply $(2x^3 + 7x^2 + 5x - 1)(4x + 6)$ vertically.

Write the polynomials vertically.

$$\begin{array}{r} 2x^3 + 7x^2 + 5x - 1 \\ 4x + 6 \\ \hline \end{array}$$

Begin by multiplying each term in the top row by 6.

$$\begin{array}{r} 2x^3 + 7x^2 + 5x - 1 \\ 4x + 6 \\ \hline 12x^3 + 42x^2 + 30x - 6 \end{array}$$

Now multiply each term in the top row by $4x$. Then add like terms.

$$\begin{array}{r} 2x^3 + 7x^2 + 5x - 1 \\ 4x + 6 \\ \hline 12x^3 + 42x^2 + 30x - 6 \\ 8x^4 + 28x^3 + 20x^2 - 4x \\ \hline 8x^4 + 40x^3 + 62x^2 + 26x - 6 \end{array}$$

The product is $8x^4 + 40x^3 + 62x^2 + 26x - 6$.

Now Try:

3. Multiply
$(x^3 + 9)(4x^4 - 2x^2 + x)$

____

4. Multiply
$(4x^3 - 3x^2 + 6x + 5)(7x - 3)$
vertically.

____

5. Find the product of $-16m^3 + 12m^2 + 4$ and $\frac{1}{4}m^2 + \frac{3}{4}$.

Multiply each term of the second polynomial by each term of the first.

$$\left(-16m^3 + 12m^2 + 4\right)\left(\frac{1}{4}m^2 + \frac{3}{4}\right)$$

$$= -16m^3\left(\frac{1}{4}m^2\right) - 16m^3\left(\frac{3}{4}\right) + 12m^2\left(\frac{1}{4}m^2\right)$$

$$\quad + 12m^2\left(\frac{3}{4}\right) + 4\left(\frac{1}{4}m^2\right) + 4\left(\frac{3}{4}\right)$$

$$= -12m^3 + 9m^2 + 3 - 5m^5 + 3m^4 + m^2$$

$$= -4m^5 + 3m^4 - 12m^3 + 10m^2 + 3$$

The product is $-4m^5 + 3m^4 - 12m^3 + 10m^2 + 3$.

5. Find the product of $12x^3 - 36x^2 + 6$ and $\frac{1}{6}x^2 + \frac{5}{6}$.

____

Objective 3 Practice Exercises

For extra help, see Examples 3–5 on page 311 of your text.

Find each product.

7. $(x+3)\left(x^2 - 3x + 9\right)$

7. __________________

8. $\left(2m^2 + 1\right)\left(3m^3 + 2m^2 - 4m\right)$

8. __________________

9. $\left(3x^2 + x\right)\left(2x^2 + 3x - 4\right)$

9. __________________

Objective 4   Multiply binomials using the FOIL method.

Review these examples for Objective 4:

6. Use the FOIL method to find the product $(x+7)(x-5)$.

Step 1 F Multiply the first terms: $x(x)=x^2$.

Step 2 O Find the outer product: $x(-5)=-5x$.

Step 3 I Find the inner product: $7(x)=7x$.
Add the outer and inner products mentally:
$$-5x+7x=2x$$
Step 4 L Multiply the last terms: $7(-5)=-35$.

The product $(x+7)(x-5)$ is $x^2+2x-35$.

7. Multiply $(7x-3)(4y+5)$.

First $7x(4y)=28xy$

Outer $7x(5)=35x$

Inner $-3(4y)=-12y$

Last $-3(5)=-15$

The product $(7x-3)(4y+5)$ is
$28xy+35x-12y-15$.

8. Find the product.

$(3k+7m)(2k+9m)$

$(3k+7m)(2k+9m)$
$=3k(2k)+3k(9m)+7m(2k)+7m(9m)$
$=6k^2+27km+14km+63m^2$
$=6k^2+41km+63m^2$

Now Try:

6. Use the FOIL method to find the product $(x+9)(x-6)$.

____

7. Multiply $(8y-7)(2x+9)$.

____

8. Find the product.

$(5k+8n)(3k+4n)$

____

Objective 4 Practice Exercises

For extra help, see Examples 6–8 on page 313 of your text.

Find each product.

10.　$(5a - b)(4a + 3b)$　　　　　　　　　　10. _______________

11.　$(3 + 4a)(1 + 2a)$　　　　　　　　　　11. _______________

12.　$(2m + 3n)(-3m + 4n)$　　　　　　　　12. _______________

Chapter 4 EXPONENTS AND POLYNOMIALS

4.6   Special Products

Learning Objectives
1   Square binomials.
2   Find the product of the sum and difference of two terms.
3   Find greater powers of binomials.

Key Terms

Use the vocabulary terms listed below to complete each statement in exercises 1–3.

conjugate   **binomial**   **difference of two squares**

1.   A polynomial with two terms is called a _______________________________.

2.   The _____________________ of $a + b$ is $a - b$.

3.   The product of $a + b$ and $a - b$ is a _____________________.

Objective 1   Square binomials.

Review these examples for Objective 1:
2.   Square each binomial.

    a. $(b-5)^2$

$$(b-5)^2 = b^2 - 2(b)(5) + 5^2$$
$$= b^2 - 10b + 25$$

    c. $(6x+3y)^2$

$$(6x+3y)^2 = (6x)^2 + 2(6x)(3y) + (3y)^2$$
$$= 36x^2 + 36xy + 9y^2$$

    e. $\left(6n+\dfrac{1}{4}\right)^2$

$$\left(6n+\frac{1}{4}\right)^2 = (6n)^2 + 2(6n)\left(\frac{1}{4}\right) + \left(\frac{1}{4}\right)^2$$
$$= 36n^2 + 3n + \frac{1}{16}$$

Now Try:
2.   Square each binomial.

    a. $(c-4)^2$

____

    c. $(2a+9k)^2$

____

    e. $\left(3p+\dfrac{1}{6}\right)^2$

____

                    Copyright © 2025 Pearson Education, Inc.

Objective 1 Practice Exercises

For extra help, see Examples 1–2 on pages 317–318 of your text.

Find each square by using the pattern for the square of a binomial.

1. $(7+x)^2$

1. _________________

2. $(2m-3p)^2$

2. _________________

3. $(4y-0.7)^2$

3. _________________

Objective 2   Find the product of the sum and difference of two terms.

Review these examples for Objective 2:

3. Find each product.

   a. $(x+5)(x-5)$

Use the rule for the product of the sum and difference of two terms.

$$(x+5)(x-5) = x^2 - 5^2$$
$$= x^2 - 25$$

b. $\left(\dfrac{3}{4}-y\right)\left(\dfrac{3}{4}+y\right)$

$$\left(\dfrac{3}{4}-y\right)\left(\dfrac{3}{4}+y\right) = \left(\dfrac{3}{4}\right)^2 - y^2$$
$$= \dfrac{9}{16} - y^2$$

4. Find each product.

   a. $(6x+w)(6x-w)$

$$(6x+w)(6x-w) = (6x)^2 - w^2$$
$$= 36x^2 - w^2$$

b. $3q\left(q^2+4\right)\left(q^2-4\right)$

First, multiply the conjugates.

$$3q\left(q^2+4\right)\left(q^2-4\right) = 3q\left(q^4-16\right)$$
$$= 3q^5 - 48q$$

Now Try:

3. Find each product.

   a. $(x+9)(x-9)$

____

b. $\left(\dfrac{5}{6}+a\right)\left(\dfrac{5}{6}-a\right)$

____

4. Find each product.

   a. $(11x-y)(11x+y)$

____

b. $4p\left(p^2+6\right)\left(p^2-6\right)$

____

Objective 2 Practice Exercises

For extra help, see Examples 3–4 on page 319 of your text.

Find each product by using the pattern for the sum and difference of two terms.

4. $(12+x)(12-x)$

4. _______________

5. $(8k+5p)(8k-5p)$

5. _______________

6. $\left(\frac{4}{7}t+2u\right)\left(\frac{4}{7}t-2u\right)$

6. _______________

Objective 3   Find greater powers of binomials.

Review these examples for Objective 3:

5. Find each product.

a. $(x+4)^3$

$(x+4)^3$

$= (x+4)^2(x+4)$

$= \left(x^2+8x+16\right)(x+4)$

$= x^3+8x^2+16x+4x^2+32x+64$

$= x^3+12x^2+48x+64$

b. $(5y-4)^4$

$(5y-4)^4$

$= (5y-4)^2(5y-4)^2$

$= \left(25y^2-40y+16\right)\left(25y^2-40y+16\right)$

$= 625y^4-1000y^3+400y^2$

$\quad -1000y^3+1600y^2-640y$

$\quad +400y^2-640y+256$

$= 625y^4-2000y^3+2400y^2-1280y+256$

Now Try:

5. Find each product.

a. $(x+6)^3$

____

b. $(3x-5)^4$

____

Name:

Instructor:

Date:

Section:

Objective 3 Practice Exercises

For extra help, see Example 5 on page 320 of your text.

Find each product.

7. $(a-3)^3$

7. ______________

8. $(j+3)^4$

8. ______________

9. $(4s+3t)^4$

9. ______________

Chapter 4 EXPONENTS AND POLYNOMIALS

4.7   Dividing Polynomials

| **Learning Objectives** |
| 1   Divide a polynomial by a monomial. |
| 2   Divide a polynomial by a polynomial. |
| 3   Apply polynomial division in a geometry problem. |

Key Terms

Use the vocabulary terms listed below to complete each statement in exercises 1−3.

      quotient   **dividend**   **divisor**

1. In the division $\dfrac{5x^5 - 10x^3}{5x^2} = x^3 - 2x$, the expression $5x^5 - 10x^3$ is the

_________________.

2. In the division $\dfrac{5x^5 - 10x^3}{5x^2} = x^3 - 2x$, the expression $x^3 - 2x$ is the __________.

3. In the division $\dfrac{5x^5 - 10x^3}{5x^2} = x^3 - 2x$, the expression $5x^2$ is the _____________.

Objective 1   Divide a polynomial by a monomial.

Review these examples for Objective 1:

1. Divide $6x^4 - 18x^3$ by $6x^2$.

$$\frac{6x^4 - 18x^3}{6x^2} = \frac{6x^4}{6x^2} - \frac{18x^3}{6x^2}$$

$$= x^2 - 3x$$

Check Multiply. $6x^2\left(x^2 - 3x\right) = 6x^4 - 18x^3$

2. Divide. $\dfrac{25a^6 - 15a^4 + 10a^2}{5a^3}$

Divide each term by $5a^3$.

$$\frac{25a^6 - 15a^4 + 10a^2}{5a^3} = \frac{25a^6}{5a^3} - \frac{15a^4}{5a^3} + \frac{10a^2}{5a^3}$$

$$= 5a^3 - 3a + \frac{2}{a}$$

Now Try:

1. Divide $20x^4 - 10x^2$ by $2x$.

____

2. Divide. $\dfrac{27n^5 - 36n^4 - 18n^2}{9n^3}$

____

3. Divide $-12x^4 + 15x^5 - 5x$ by $-5x$.

Write the polynomial in descending powers before dividing.

$$\frac{15x^5 - 12x^4 - 5x}{-5x} = \frac{15x^5}{-5x} - \frac{12x^4}{-5x} - \frac{5x}{-5x}$$

$$= -3x^4 + \frac{12}{5}x^3 + 1$$

Check $\;-5x\left(-3x^4 + \frac{12}{5}x^3 + 1\right)$

$$= -5x(-3x^4) - 5x\left(\frac{12}{5}x^3\right) - 5x(1)$$

$$= 15x^5 - 12x^4 - 5x$$

4. Divide
$-225x^5y^9 + 150x^3y^7 - 110x^2y^5 + 80xy^3 - 75y^2$
by $-25xy^2$.

$$\frac{-225x^5y^9 + 150x^3y^7 - 110x^2y^5 + 80xy^3 - 75y^2}{-25xy^2}$$

$$= \frac{-225x^5y^9}{-25xy^2} + \frac{150x^3y^7}{-25xy^2} - \frac{110x^2y^5}{-25xy^2} + \frac{80xy^3}{-25xy^2} - \frac{75y^2}{-25xy^2}$$

$$= 9x^4y^7 - 6x^2y^5 + \frac{22xy^3}{5} - \frac{16y}{5} + \frac{3}{x}$$

Check by multiplying the quotient by the divisor.

3. Divide $-8z^5 + 7z^6 - 10z - 6$ by $2z^2$.

____

4. Divide
$-80a^5b^3 - 160a^4b^2 + 120a^2b$
by $-40a^2b$.

____

Objective 1 Practice Exercises

For extra help, see Examples 1–4 on pages 323–324 of your text.

Perform each division.

1. $\dfrac{16a^5 - 24a^3}{8a^2}$

1. ________________

2. $\dfrac{12z^5 + 28z^4 - 8z^3 + 3z}{4z^3}$

2. ________________

3. $\dfrac{39m^4 - 12m^3 + 15}{-3m^2}$

3. ________________

                    183

Objective 2   Divide a polynomial by a polynomial.

Review these examples for Objective 2: | **Now Try:**

5. Divide $\dfrac{2x^2-11x+15}{x-3}$.

Step 1 $2x^2$ divided by x is $2x$

$2x(x-3)=2x^2-6x$

Step 2   Subtract. Bring down the next term.

Step 3 $-5x$ divided by x is -5.

$-5(x-3)=-5x+15$

Step 4   Subtract. The remainder is 0.

$$
\begin{array}{r}
2x\ -5 \\
x-3\overline{)2x^2-11x+15} \\
\underline{2x^2-\ 6x} \\
-5x+15 \\
\underline{-5x+15} \\
0
\end{array}
$$

$$\frac{2x^2-11x+15}{x-3}=2x-5$$

Check $(x-3)(2x-5)=2x^2-5x-6x+15$

$=2x^2-11x+15$

6. Divide $\dfrac{8x+9x^3-7-9x^2}{3x-1}$.

Write the dividend in descending powers as
$9x^3-9x^2+8x-7$.

Step 1 $9x^3$ divided by $3x$ is $3x^2$.

$3x^2(3x-1)=9x^3-3x^2$

Step 2   Subtract. Bring down the next term.

Step 3 $-6x^2$ divided by $3x$ is $-2x$.

$-2x(3x-1)=-6x^2+2x$

Step 4   Subtract. Bring down the next term.

Step 5 $6x$ divided by $3x$ is 2.

$2(3x-1)=6x-2$

Now Try:

5. Divide $\dfrac{4x^2-5x-6}{x-2}$.

____

6. Divide $\dfrac{-12x^2+10x^3-3-8x}{5x-1}$.

____

$$\begin{array}{r}
3x^2 - 2x + 2 \\
3x-1\overline{)9x^3 - 9x^2 + 8x - 7} \\
\underline{9x^3 - 3x^2} \\
-6x^2 + 8x \\
\underline{-6x^2 + 2x} \\
6x - 7 \\
\underline{6x - 2} \\
-5
\end{array}$$

$$\frac{9x^3 - 9x^2 + 8x - 7}{3x - 1} = 3x^2 - 2x + 2 + \frac{-5}{3x - 1}$$

Step 7 Multiply to check.

Check $\ \ (3x-1)\left(3x^2 - 2x + 2 + \dfrac{-5}{3x-1}\right)$

$$= (3x-1)(3x^2) + (3x-1)(-2x)$$
$$+ (3x-1)(2) + (3x-1)\left(\frac{-5}{3x-1}\right)$$
$$= 9x^3 - 3x^2 - 6x^2 + 2x + 6x - 2 - 5$$
$$= 9x^3 - 9x^2 + 8x - 7$$

7. Divide $x^3 - 64$ by $x - 4$.

Here the dividend is missing the x^2-term and the x-term. We use 0 as the coefficient for each missing term.

$$\begin{array}{r}
x^2 + 4x + 16 \\
x-4\overline{)x^3 + 0x^2 + 0x - 64} \\
\underline{x^3 - 4x^2} \\
4x^2 + 0x \\
\underline{4x^2 - 16x} \\
16x - 64 \\
\underline{16x - 64} \\
0
\end{array}$$

The remainder is 0. The quotient is $x^2 + 4x + 16$.

Check $\ \ (x-4)\left(x^2 + 4x + 16\right)$

$$= x^3 + 4x^2 + 16x - 4x^2 - 16x - 64$$
$$= x^3 - 64$$

7. Divide $x^3 - 1000$ by $x - 10$.

____

8. Divide $x^4 - 3x^3 + 7x^2 - 8x + 14$ by $x^2 + 2$.

Since $x^2 + 2$ is missing the x-term, we write it as $x^2 + 0x + 2$.

$$
\begin{array}{r}
x^2 - 3x + 5 \\
x^2 + 0x + 2 \overline{)\, x^4 - 3x^3 + 7x^2 - 8x + 14} \\
\underline{x^4 + 0x^3 + 2x^2} \\
-3x^3 + 5x^2 - 8x \\
\underline{-3x^3 + 0x^2 - 6x} \\
5x^2 - 2x + 14 \\
\underline{5x^2 + 0x + 10} \\
-2x + 4
\end{array}
$$

The quotient is $x^2 - 3x + 5 + \dfrac{-2x + 4}{x^2 + 2}$

The check shows that the quotient multiplied by the divisor gives the original dividend.

8. Divide $3x^4 + 5x^3 - 7x^2 - 12x + 9$ by $x^2 - 4$.

____

Objective 2 Practice Exercises

For extra help, see Examples 5–9 on pages 326–328 of your text.

Perform each division.

4. $\dfrac{-6x^2 + 23x - 20}{2x - 5}$

4. _______________

5. $\dfrac{6x^4 - 12x^3 + 13x^2 - 5x - 1}{2x^2 + 3}$

5. _______________

              Copyright © 2025 Pearson Education, Inc.

6. $\dfrac{2a^4 + 5a^2 + 3}{2a^2 + 3}$

6. _______________________

Objective 3   Apply polynomial division in a geometry problem.

Review this example for Objective 3:

10. The area of a rectangle is given by $12p^3 - 7p^2 + 5p - 1$ square units, and the width is $4p - 1$ units. What is the length of the rectangle?

For a rectangle, $A = LW$. Solving for L gives $L = \dfrac{A}{W}$. Divide the area, $12p^3 - 7p^2 + 5p - 1$ by the width $4p - 1$.

$$
\begin{array}{r}
3p^2 - p + 1 \\
4p-1\overline{)\,12p^3 - 7p^2 + 5p - 1} \\
\underline{12p^3 - 3p^2} \\
-4p^2 + 5p \\
\underline{-4p^2 + p} \\
4p - 1 \\
\underline{4p - 1} \\
0
\end{array}
$$

The length is $3p^2 - p + 1$ units.

Now Try:

10. The area of a rectangle is given by $6r^3 - 5r^2 + 16r - 5$ square units, and the width is $3r - 1$ units. What is the length of the rectangle?

____

Objective 3 Practice Exercises

For extra help, see Example 10 on page 329 of your text.

Work each problem.

7. The area of a parallelogram is given by
$4y^3 - 44y - 600$ square units, and the height is
$y - 6$ units. What is the base of the parallelogram?

7. _________________

8. The area of a parallelogram is given by
$3t^3 + 16t^2 - 32t - 64$ square units, and the base is
$t^2 + 4t - 16$ units. What is the height of the
parallelogram?

8. _________________

Chapter 5 FACTORING AND APPLICATIONS

5.1   Greatest Common Factors; Factoring by Grouping

<table>
<tr><td colspan="2">Learning Objectives</td></tr>
<tr><td>1</td><td>Find the greatest common factor of a list of numbers.</td></tr>
<tr><td>2</td><td>Find the greatest common factor of a list of variable terms.</td></tr>
<tr><td>3</td><td>Factor out the greatest common factor.</td></tr>
<tr><td>4</td><td>Factor by grouping.</td></tr>
</table>

Key Terms

Use the vocabulary terms listed below to complete each statement in exercises 1−5.

> **factor**   **factored form**   **greatest common factor (GCF)**
>
> **factoring**   **common factor**

1.   The process of writing a polynomial as a product is called ______________________.

2.   An expression is in ______________________________ when it is written as a product.

3.   The ______________________________ is the largest quantity that is a factor of each of a group of quantities.

4.   An integer that is a factor of two or more integers is a ______________________________ of those integers.

5.   An expression A is a ______________________ of an expression B if B can be divided by A with 0 remainder.

Objective 1   Find the greatest common factor of a list of numbers.

Review these examples for Objective 1:

1.   Find the greatest common factor for each list of numbers.

   a. 60, 45

   Write the prime factored form of each number.
   $$60 = 2 \cdot 2 \cdot 3 \cdot 5$$
   $$45 = 3 \cdot 3 \cdot 5$$
   Use each prime the least number of times it appears in all the factored forms.
   $$GCF = 3^1 \cdot 5^1 = 15$$

Now Try:

1.   Find the greatest common factor for each list of numbers.

   a. 18, 36, 42

   ____

b. 90, 36, 108

Write the prime factored form of each number.

$$90 = 2 \cdot 3 \cdot 3 \cdot 5$$

$$36 = 2 \cdot 2 \cdot 3 \cdot 3$$

$$108 = 2 \cdot 2 \cdot 3 \cdot 3 \cdot 3$$

There is one factor of 2 and two factors of 3.

$$\text{GCF} = 2^1 \cdot 3^2 = 18$$

c. 17, 18, 24

$$17 = 17$$

$$18 = 2 \cdot 3 \cdot 3$$

$$24 = 2 \cdot 2 \cdot 2 \cdot 3$$

There are no primes common to all three numbers, so the GCF is 1.

b. 32, 40, 72

____

c. 26, 27, 28

____

Objective 1 Practice Exercises

For extra help, see Example 1 on pages 344–345 of your text.

Find the greatest common factor for each list.

1.   84, 280, 112

1. _______________

2.   6, 8, 30

2. _______________

3.   9, 15, 52

3. _______________

Objective 2   Find the greatest common factor of a list of variable terms.

Review this example for Objective 2:

2.   Find the greatest common factor for the list of terms.

$$28m^4, \ 35m^6, \ 49m^9, \ 70m^5$$

$$28m^4 = 2 \cdot 2 \cdot 7 \cdot m^4$$

$$35m^6 = 5 \cdot 7 \cdot m^6$$

$$49m^9 = 7 \cdot 7 \cdot m^9$$

$$70m^5 = 2 \cdot 5 \cdot 7 \cdot m^5$$

Then, $\text{GCF} = 7m^4$.

Now Try:

2.   Find the greatest common factor for the list of terms.

$$54x^5, \ 48x^7, \ 42x^9, \ 30x^4$$

____

                Copyright © 2025 Pearson Education, Inc.

Objective 2 Practice Exercises

For extra help, see Example 2 on page 346 of your text.

Find the greatest common factor for each list.

4. $xy^2,\ x^2y,\ x^2y^3$

4. _______________

5. $6k^2m^4n^5,\ 8k^3m^7n^4,\ k^4m^8n^7$

5. _______________

6. $9xy^4,\ 72x^4y^7,\ 27xy^2,\ 108x^2y^5$

6. _______________

Objective 3   Factor out the greatest common factor.

Review these examples for Objective 3:

3.   Write in factored form by factoring out the greatest common factor.

$$12x^5 + 27x^4 - 15x^3$$

$$\text{GCF} = 3x^3$$
$$12x^5 + 27x^4 - 15x^3$$
$$= 3x^3(4x^2) + 3x^3(9x) + 3x^3(-5)$$
$$= 3x^3(4x^2 + 9x - 5)$$

Check Multiply the factored form.
$$3x^3(4x^2 + 9x - 5)$$
$$= 3x^3(4x^2) + 3x^3(9x) + 3x^3(-5)$$
$$= 12x^5 + 27x^4 - 15x^3$$

5.   Write in factored form by factoring out the greatest common factor.

a. $x(x+9) + 7(x+9)$

Factor out $x+9$.
$$x(x+9) + 7(x+9) = (x+9)(x+7)$$

Now Try:

3.   Write in factored form by factoring out the greatest common factor.
$$20y^4 - 12y^3 + 4y^2$$

____

5.   Write in factored form by factoring out the greatest common factor.

a. $y(y+8) + 4(y+8)$

____

b. $a^2(a+6)-7(a+6)$

Factor out $a+6$.

$a^2(a+6)-7(a+6)=(a+6)(a^2-7)$

b. $z^2(z+5)-11(z+5)$

____

Objective 3 Practice Exercises

For extra help, see Examples 3–5 on pages 346–348 of your text.

Factor out the greatest common factor or a negative common factor if the coefficient of the term of greatest degree is negative.

7. $20x^2+40x^2y-70xy^2$

7. __________________

8. $2a(x-2y)+9b(x-2y)$

8. __________________

9. $26x^8-13x^{12}+52x^{10}$

9. __________________

Objective 4   Factor by grouping.

Review these examples for Objective 4:

6. Factor by grouping.

  a. $7by+28y+b+4$

Group the terms, then factor each group.

$7by+28y+b+4$

$=(7by+28y)+(b+4)$

$=7y(b+4)+1(b+4)$

$=(b+4)(7y+1)$

Check   Use the FOIL method.

$(b+4)(7y+1)=7by+1b+4(7y)+4$

$=7by+28y+b+4$

Now Try:

6. Factor by grouping.

  a. $36x+4tx+9+t$

____

b. $3x^2 - 18x + 5xy - 30y$

$3x^2 - 18x + 5xy - 30y$
$$= \left(3x^2 - 18x\right) + \left(5xy - 30y\right)$$
$$= 3x(x-6) + 5y(x-6)$$
$$= (x-6)(3x+5y)$$
Check by multiplying using the FOIL method.

c. $r^3 + 3r^2 - 5r - 15$

$r^3 + 3r^2 - 5r - 15$
$$= \left(r^3 + 3r^2\right) + (-5r - 15)$$
$$= r^2(r+3) - 5(r+3)$$
$$= (r+3)\left(r^2 - 5\right)$$
Check by multiplying using the FOIL method.

7. Factor by grouping.

$18x^2 - 20y + 24x - 15xy$

Group the terms, then factor each group.
$18x^2 - 20y + 24x - 15xy$
$$= 2\left(9x^2 - 10y\right) + 3x(8 - 5y)$$
This does not lead to a common factor, so we try rearranging the terms.
$18x^2 - 20y + 24x - 15xy$
$$= 18x^2 - 15xy + 24x - 20y$$
$$= \left(18x^2 - 15xy\right) + (24x - 20y)$$
$$= 3x(6x - 5y) + 4(6x - 5y)$$
$$= (6x - 5y)(3x + 4)$$
Check Use the FOIL method.
$(6x - 5y)(3x + 4)$
$$= 18x^2 + 24x - 15xy - 20y$$
$$= 18x^2 - 20y + 24x - 15xy$$

b. $4x^2 - 28x + 5xy - 35y$

____

c. $x^3 + 7x^2 - 2x - 14$

____

7. Factor by grouping.

$56x^2 + 32x - 21xy - 12y$

____

Name: Date:
Instructor: Section:

Objective 4 Practice Exercises

For extra help, see Examples 6–7 on pages 348–350 of your text.

Factor each polynomial by grouping.

10. $15 - 5x - 3y + xy$ 10. _______________

11. $2x^2 - 14xy + xy - 7y^2$ 11. _______________

12. $3r^3 - 2r^2 s + 3s^2 r - 2s^3$ 12. _______________

Chapter 5 FACTORING AND APPLICATIONS

5.2   Factoring Trinomials

Learning Objectives
1   Factor trinomials with coefficient 1 for the second-degree term.
2   Factor out the greatest common factor first.

Key Terms

Use the vocabulary terms listed below to complete each statement in exercises 1−3.

       prime polynomial   **factoring**   **greatest common factor**

1.   _______________________ is the process of writing a polynomial as a product.

2.   The _______________________ of a polynomial is the greatest term that is a factor of all the terms in the polynomial.

3.   A _______________________ is a polynomial that cannot be factored using only integers.

Objective 1   Factor trinomials with coefficient 1 for the second-degree term.

Review these examples for Objective 1:

1.   Factor $m^2 + 8m + 15$.

Look for integers whose product is 15 and whose sum is 8. Only positive signs are needed.

Factors of 15	Sums of Factors
15, 1	$15 + 1 = 16$
5, 3	$5 + 3 = 8$

From the table, 5 and 3 are the required integers.

$m^2 + 8m + 15$ factors as $(m+5)(m+3)$

Check   Use the FOIL method.

$$(m+5)(m+3) = m^2 + 3m + 5m + 15$$
$$= m^2 + 8m + 15$$

Now Try:

1.   Factor $x^2 + 11x + 24$.

____

2. Factor $x^2 - 11x + 28$.

Look for integers whose product is 28 and whose sum is -11. Since the numbers have a positive product and a negative sum, we consider only pairs of negative integers.

Factors of 28	Sums of Factors
$-28, -1$	$-28 + (-1) = -29$
$-14, -2$	$-14 + (-2) = -16$
$-7, -4$	$-7 + (-4) = -11$

The required integers are -7 and -4.

$x^2 - 11x + 28$ factors as $(x - 7)(x - 4)$

Check   Use the FOIL method.

$$(x - 7)(x - 4) = x^2 - 4x - 7x + 28$$
$$= x^2 - 11x + 28$$

3. Factor $x^2 + 2x - 15$.

Look for integers whose product is -15 and whose sum is 2. To get a negative product, the pairs of integers must have different signs.

Factors of -15	Sums of Factors
$15, -1$	$15 + (-1) = 14$
$-15, 1$	$-15 + 1 = -14$
$5, -3$	$5 + (-3) = 2$

The required integers are 5 and -3.

$x^2 + 2x - 15$ factors as $(x + 5)(x - 3)$

Check   Use the FOIL method.

$$(x + 5)(x - 3) = x^2 - 3x + 5x - 15$$
$$= x^2 + 2x - 15$$

5. Factor the trinomial.

$$x^2 - 7x + 18$$

Look for integers whose product is 18 and whose sum is -7. Since the numbers have a positive product and a negative sum, we consider only pairs of negative integers.

2. Factor $y^2 - 12y + 35$.

____

3. Factor $p^2 + 6p - 27$.

____

5. Factor the trinomial.

$$m^2 - 7m + 5$$

____

Factors of 18	Sums of Factors
$-18, -1$	$-18 + (-1) = -19$
$-9, -2$	$-9 + (-2) = -11$
$-6, -3$	$-6 + (-3) = -9$

None of the pairs of integers has a sum of -7.

$x^2 - 7x + 18$ cannot be factored.

It is a prime polynomial.

6. Factor $x^2 - 6xy - 7y^2$.

Here, the coefficient of x in the middle term is $-6y$, so we need to find two expressions whose product is $-7y^2$ and whose sum is $-6y$.

Factors of $-7y^2$	Sums of Factors
$7y, -y$	$7y + (-y) = 6y$
$-7y, y$	$-7y + y = -6y$

$x^2 - 6xy - 7y^2$ factors as $(x - 7y)(x + y)$

Check   Use the FOIL method.

$$(x - 7y)(x + y) = x^2 + xy - 7xy - 7y^2$$
$$= x^2 - 6xy - 7y^2$$

6. Factor $p^2 - 5pq - 14q^2$.

____

Objective 1 Practice Exercises

For extra help, see Examples 1–6 on pages 354–357 of your text.

Factor completely. If a polynomial cannot be factored, write prime.

1. $r^2 + r + 3$

1. _______________

2. $x^2 - 11x + 28$

2. _______________

3. $x^2 - 8x - 33$

3. _______________

Name: Date:

Instructor: Section:

Objective 2   Factor out the greatest common factor first.

Review these examples for Objective 2:

7.   Factor each trinomial completely.

   a. $5x^5 - 45x^4 + 90x^3$

   There is no second-degree term. Look for a
   common factor.
   $$5x^5 - 45x^4 + 90x^3 = 5x^3(x^2 - 9x + 18)$$
   Now factor $x^2 - 9x + 18$. The integers –3
   and –6 have a product of 18 and a sum of –9.
   $5x^5 - 45x^4 + 90x^3$ factors as $5x^3(x - 6)(x - 3)$

   Check   Use the FOIL method.
   $$5x^3(x - 6)(x - 3) = 5x^3(x^2 - 3x - 6x + 18)$$
   $$= 5x^3(x^2 - 9x + 18)$$
   $$= 5x^5 - 45x^4 + 90x^3$$

   b. $-2y^7 - 8y^6 + 42y^5$

   The coefficient of the first term is negative, so
   we factor out $-2y^5$.
   $$-2y^7 - 8y^6 + 42y^5 = -2y^5(y^2 + 4y - 21)$$
   Now factor $y^2 + 4y - 21$. The integers 7
   and –3 have a product of –21 and a sum of 4.
   Thus, $-2y^7 - 8y^6 + 42y^5$ factors as
   $-2y^5(y + 7)(y - 3)$.

Now Try:

7.   Factor each trinomial
   completely.

   a. $7x^6 - 49x^5 + 70x^4$.

   ____

   b. $-3y^6 - 3y^5 + 60y^4$

   ____

Objective 2 Practice Exercises

For extra help, see Example 7 on page 357 of your text.

Factor completely. If a polynomial cannot be factored, write **prime**.

   4. $2n^4 - 16n^3 + 30n^2$ 4. ________________

   5. $2a^3b - 10a^2b^2 + 12ab^3$ 5. ________________

   6. $10k^6 + 70k^5 + 100k^4$ 6. ________________

                   Copyright © 2025 Pearson Education, Inc.

Chapter 5 FACTORING AND APPLICATIONS

5.3   More on Factoring Trinomials

> **Learning Objectives**
> 1   Factor trinomials by grouping when the coefficient of the second-degree term is not 1.
> 2   Factor trinomials using the FOIL method.

Key Terms

Use the vocabulary terms listed below to complete each statement in exercises 1–5.

> **coefficient**   **trinomial**   **FOIL**
>
> **outer product**   **inner product**

1. In the term $6x^2 y$, 6 is the ___________________________.

2. A polynomial with three terms is a ___________________________.

3. The ___________________________ of $(2y-5)(y+8)$ is $-5y$.

4. ___________________ is a shortcut method for finding the product of two binomials.

5. The ___________________________ of $(2y-5)(y+8)$ is $16y$.

Objective 1   Factor trinomials by grouping when the coefficient of the second-degree term is not 1.

Review these examples for Objective 1:

1. Factor $3x^2 + 11x + 10$.

   Look for two positive integers whose product is $3 \cdot 10 = 30$ and whose sum is 11.

   The integers are 5 and 6, since $5 \cdot 6 = 30$ and $5 + 6 = 11$.

   $$3x^2 + 11x + 10 = 3x^2 + 5x + 6x + 10$$
   $$= (3x^2 + 5x) + (6x + 10)$$
   $$= x(3x + 5) + 2(3x + 5)$$
   $$= (3x + 5)(x + 2)$$

   Check   Multiply $(3x+5)(x+2)$ to obtain $3x^2 + 11x + 10$.

Now Try:

1. Factor $5x^2 + 17x + 6$.

   ____

2. Factor each trinomial.

a. $8x^2 - 2x - 1$

We must find two integers with a product of $8(-1) = -8$ and a sum of –2. The integers are –4 and 2. We write the middle term as $-4x + 2x$.

$$8x^2 - 2x - 1 = 8x^2 - 4x + 2x - 1$$
$$= (8x^2 - 4x) + (2x - 1)$$
$$= 4x(2x - 1) + 1(2x - 1)$$
$$= (2x - 1)(4x + 1)$$

Check Multiply $(2x - 1)(4x + 1)$ to obtain $8x^2 - 2x - 1$.

b. $15z^2 + z - 2$

Look for two integers whose product is $15(-2) = -30$ and whose sum is 1.

The integers are 6 and –5.

$$15z^2 + z - 2 = 15z^2 - 5z + 6z - 2$$
$$= (15z^2 - 5z) + (6z - 2)$$
$$= 5z(3z - 1) + 2(3z - 1)$$
$$= (3z - 1)(5z + 2)$$

Check Multiply $(3z - 1)(5z + 2)$ to obtain $15z^2 + z - 2$.

c. $12r^2 + 5rs - 2s^2$

Two integers whose product is $12(-2) = -24$ and whose sum is 5 are 8 and –3. Rewrite the trinomial with four terms.

$$12r^2 + 5rs - 2s^2 = 12r^2 + 8rs - 3rs - 2s^2$$
$$= (12r^2 + 8rs) + (-3rs - 2s^2)$$
$$= 4r(3r + 2s) - s(3r + 2s)$$
$$= (3r + 2s)(4r - s)$$

Check Multiply $(3r + 2s)(4r - s)$ to obtain $12r^2 + 5rs - 2s^2$.

2. Factor each trinomial.

a. $14x^2 - 3x - 5$

____

b. $3m^2 - m - 14$

____

c. $10x^2 + xy - 3y^2$

____

3. Factor $100x^5 + 140x^4 - 15x^3$.

Factor out the greatest common factor, $5x^3$.
$$100x^5 + 140x^4 - 15x^3 = 5x^3\left(20x^2 + 28x - 3\right)$$

To factor $20x^2 + 28x - 3$, find two integers whose product is $20(-3) = -60$ and whose sum is 28. Factor 60 into prime factors.
$$60 = 2 \cdot 2 \cdot 3 \cdot 5$$
Combine the prime factors in pairs using one positive factor and one negative factor to get -60. The factors of 30 and -2 have the correct sum, 28.
$$100x^5 + 140x^4 - 15x^3$$
$$= 5x^3\left(20x^2 + 28x - 3\right)$$
$$= 5x^3\left(20x^2 + 30x - 2x - 3\right)$$
$$= 5x^3\left[\left(20x^2 + 30x\right) + \left(-2x - 3\right)\right]$$
$$= 5x^3\left[10x(2x + 3) - 1(2x + 3)\right]$$
$$= 5x^3(2x + 3)(10x - 1)$$

3. Factor $30x^5 + 87x^4 - 63x^3$.

____

Objective 1 Practice Exercises

For extra help, see Examples 1–3 on pages 361–362 of your text.

Factor each trinomial by grouping.

1. $8b^2 + 18b + 9$

1. _______________

2. $7a^2b + 18ab + 8b$

2. _______________

3. $10c^2 - 29ct + 21t^2$

3. _______________

Objective 2   Factor trinomials using the FOIL method.

Review these examples for Objective 2:	**Now Try:**

5.   Factor $6x^2 + 13x + 7$.

5.   Factor $15x^2 + 26x + 7$.

The number 6 has several possible pairs of factors, but 7 has only 1 and 7, or -1 and -7. We choose positive factors since all coefficients in the trinomial are positive.

$$(\underline{} + 7)(\underline{} + 1)$$

The possible pairs of $6x^2$ are $6x$ and x, or $3x$ and $2x$.

$$(3x + 7)(2x + 1)$$

gives middle term $3x + 14x = 17x$. Incorrect.

$$(2x + 7)(3x + 1)$$

gives middle term $2x + 21x = 23x$. Incorrect.

$$(6x + 7)(x + 1)$$

gives middle term $6x + 7x = 13x$. Correct.

$6x^2 + 13x + 7$ factors as $(6x + 7)(x + 1)$.

Check. Multiply $(6x + 7)(x + 1)$ to obtain $6x^2 + 13x + 7$.

6.   Factor $10x^2 - 19x + 7$.

6.   Factor $20x^2 - 13x + 2$.

Since 7 has only 1 and 7 or -1 and -7 as factors, it is better to begin by factoring 7. We need two negative factors because the product of two negative factors is positive and their sum is negative, as required.

We try -1 and -7.

$$(\underline{} - 1)(\underline{} - 7)$$

The factors of $10x^2$ are $10x$ and x, or $5x$ and $2x$.

$$(10x - 1)(x - 7)$$

has middle term $-70x - x = -71x$. Incorrect.

$$(5x - 1)(2x - 7)$$

has middle term $-35x - 2x = -37x$. Incorrect.

$$(2x - 1)(5x - 7)$$

has middle term $-14x - 5x = -19x$. Correct.

Thus $10x^2 - 19x + 7$ factors as $(2x - 1)(5x - 7)$.

7. Factor $6x^2 - x - 15$.

The integer 6 has several possible pairs of factors, as does -15. Since the constant term is negative, one positive factor and one negative factor of -15 are needed. Since the coefficient of the middle term is relatively small, it is wise to avoid large factors. We try $3x$ and $2x$ as factors of $6x^2$ and 5 and -3 as factors of -15.

$$(3x + 5)(2x - 3)$$
has middle term $-9x + 10x = x$. Incorrect.
$$(3x - 5)(2x + 3)$$
has middle term $9x - 10x = -x$. Correct.

$6x^2 - x - 15$ factors as $(3x - 5)(2x + 3)$.

8. Factor $18x^2 - 3xy - 28y^2$.

There are several factors of $18x^2$, including
$\quad$ $18x$ and x, $\;9x$ and $2x$, and $6x$ and $3x$.
There are many possible pairs of factors of $-28y^2$, including
$\quad$ $28y$ and $-y$, $\;-28y$ and y, $\;14y$ and $-2y$,
$\quad$ $-14y$ and $2y$, $\;7y$ and $-4y$, $\;-7y$ and $4y$.

Once again, since the coefficient of the middle term is relatively small, avoid the larger factors. Try the factors of $6x$ and $3x$, and $4y$ and $-7y$.
$$(6x + 4y)(3x - 7y) \quad \text{Incorrect}$$
The first binomial has a common factor of 2.
$$(6x - 7y)(3x + 4y)$$
has middle term $24xy - 21xy = 3xy$. Incorrect.
Interchange the signs of the last two terms.
$$(6x + 7y)(3x - 4y)$$
has middle term $-24xy + 21xy = -3xy$. Correct.

Thus, $18x^2 - 3xy - 28y^2$ factors as
$(6x + 7y)(3x - 4y)$

7. Factor $8x^2 + 2x - 21$.

____

8. Factor $24x^2 - 2xy - 15y^2$.

____

9. Factor the trinomial.

$$-105a^3 + 65a^2 - 10a$$

The common factor is $-5a$. Then use trial and error.

$$-105a^3 + 65a^2 - 10a = -5a\left(21a^2 - 13a + 2\right)$$
$$= -5a(3a-1)(7a-2)$$

Check.

$$-5a(3a-1)(7a-2) = -5a\left(21a^2 - 13a + 2\right)$$
$$= -105a^3 + 65a^2 - 10a$$

9. Factor the trinomial.

$$-18a^3 + 66a^2 - 60a$$

____

Objective 2 Practice Exercises

For extra help, see Examples 4–9 on pages 363–366 of your text.

Factor each trinomial completely.

4. $8q^2 + 10q + 3$

4. ______________

5. $3a^2 + 8ab + 4b^2$

5. ______________

6. $4c^2 + 14cd - 8d^2$

6. ______________

Chapter 5 FACTORING AND APPLICATIONS

5.4   Special Factoring Techniques

Learning Objectives
1   Factor a difference of squares.
2   Factor a perfect square trinomial.
3   Factor a difference of cubes.
4   Factor a sum of cubes.

Key Terms

Use the vocabulary terms listed below to complete each statement in exercises 1–4.

perfect square trinomial   difference

difference of cubes   sum of cubes

1.   A _________________________ is the result of a subtraction.

2.   The product $(x+y)(x^2-xy+y^2)$ results in a _____________________.

3.   The product $(x-y)(x^2+xy+y^2)$ results in a _____________________.

4.   A _____________________________ is a trinomial that can be factored as the square of a binomial.

Objective 1   Factor a difference of squares.

Review these examples for Objective 1:

1.   Factor the binomial, if possible.

$$x^2-64$$

$$x^2-64=x^2-8^2=(x+8)(x-8)$$

2.   Factor each difference of squares.

a. $36x^2-25$

$$36x^2-25=(6x)^2-5^2$$
$$=(6x+5)(6x-5)$$

b. $81y^2-49$

$$81y^2-49=(9y)^2-7^2$$
$$=(9y+7)(9y-7)$$

Now Try:

1.   Factor the binomial, if possible.

$$z^2-36$$

____

2.   Factor each difference of squares.

a. $4x^2-81$

____

b. $25t^2-49$

____

3. Factor completely.

  a. $28y^2 - 175$

$$28y^2 - 175 = 7\left(4y^2 - 25\right)$$
$$= 7\left[(2y)^2 - 5^2\right]$$
$$= 7(2y + 5)(2y - 5)$$

  b. $p^4 - 81$

$$p^4 - 81 = \left(p^2\right)^2 - 9^2$$
$$= \left(p^2 + 9\right)\left(p^2 - 9\right)$$
$$= \left(p^2 + 9\right)(p + 3)(p - 3)$$

3. Factor completely.

  a. $90x^2 - 490$

    ____

  b. $p^4 - 256$

    ____

Objective 1 Practice Exercises

For extra help, see Examples 1–3 on pages 370–371 of your text.

Factor each binomial completely. If a binomial prime, say so.

1. $x^2 - 49$

1. _______________

2. $81x^4 - 16$

2. _______________

3. $9x^2 + 16$

3. _______________

Objective 2   **Factor a perfect square trinomial.**

Review these examples for Objective 2:

4.   Factor $x^2 + 20x + 100$.

  The terms x^2 and 100 are perfect squares.
$$x^2 + 20x + 100 = (x + 10)^2$$
  Check the middle term. $2(x)(10) = 20x$
  The trinomial is a perfect square.

Now Try:

4.   Factor $p^2 + 16p + 64$.

    ____

5. Factor each trinomial.

   a. $36y^2 + 42y + 49$

The first and last terms are perfect squares.
$$36y^2 = (6y)^2 \quad \text{and} \quad 49 = 7^2$$
Twice the product of the first and last terms of the binomial is $2 \cdot (6y)(7) = 84y$, which is not the middle term of $36y^2 + 42y + 49$.
   It is a prime polynomial.

   b. $128z^3 + 192z^2 + 72z$

$128z^3 + 192z^2 + 72z$
$$= 8z(16z^2 + 24z + 9)$$
$$= 8z\left[(4z)^2 + 2(4z)(3) + 3^2\right]$$
$$= 8z(4z + 3)^2$$

   c. $49m^2 - 70m + 25$

$49m^2 - 70m + 25$
$$= (7m)^2 + 2(7m)(-5) + (-5)^2$$
$$= (7m - 5)^2$$

5. Factor each trinomial.

   a. $100y^2 - 70y + 49$

____

   b. $20x^3 + 100x^2 + 125x$

____

   c. $64m^2 + 48m + 9$

____

Objective 2 Practice Exercises

For extra help, see Examples 4–5 on pages 372–373 of your text.

Factor each trinomial completely. If the trinomial is prime, say so.

4. $z^2 - 26z + 169$

4. _______________

5. $9j^2 + 12j + 4$

5. _______________

6. $-12a^2 + 60ab - 75b^2$

6. _______________

Objective 3   Factor a difference of cubes.

Review these examples for Objective 3:

6.   Factor each binomial.

    a. $m^3 - 1000$

Use the pattern for a difference of cubes.
$$m^3 - 1000 = m^3 - 10^3$$
$$= (m-10)(m^2 + 10m + 100)$$

    b. $64p^3 - 125$

$$64p^3 - 125 = (4p^3) - 5^3$$
$$= (4p-5)\left[(4p)^2 + (4p)(5) + 5^2\right]$$
$$= (4p-5)(16p^2 + 20p + 25)$$

    c. $64y^3 + 1000x^6$

$$64y^3 + 1000x^6$$
$$= (4y)^3 + (10x^2)^3$$
$$= (4y + 10x^2)[(4y)^2 - (4y)(10x^2) + (10x^2)^2]$$
$$= (4y + 10x^2)(16y^2 - 40x^2y + 100x^4)$$

Now Try:

6.   Factor each binomial.

    a. $t^3 - 216$

    ____

    b. $27k^3 - y^3$

    ____

    c. $27x^3 + 343y^6$

    ____

Objective 3 Practice Exercises

For extra help, see Example 6 on page 375 of your text.

Factor each binomial completely.

7. $8a^3 - 125b^3$

7. _________________

8. $216x^3 - 8y^3$

8. _________________

9. $(m+n)^3 - (m-n)^3$

9. _________________

Objective 4   Factor a sum of cubes.

Review these examples for Objective 4:

7. Factor each binomial.

   a. $k^3 + 1000$

$$k^3 + 1000 = k^3 + 10^3$$
$$= (k+10)(k^2 - 10k + 100)$$

   b. $2m^3 + 250n^3$

$$2m^3 + 250n^3 = 2(m^3 + 125n^3)$$
$$= 2\left[m^3 + (5n)^3\right]$$
$$= 2(m+5n)\left[m^2 - m(5n) + (5n)^2\right]$$
$$= 2(m+5n)(m^2 - 5mn + 25n^2)$$

Now Try:

7. Factor each binomial.

   a. $216x^3 + 1$

   ____

   b. $6x^3 + 48y^3$

   ____

Objective 4 Practice Exercises

For extra help, see Example 7 on pages 375–376 of your text.

Factor each binomial completely.

10. $27r^3 + 8s^3$ 10. _____________

11. $8a^3 + 64b^3$ 11. _____________

12. $64x^3 + 343y^3$ 12. _____________

Chapter 5 FACTORING AND APPLICATIONS

5.5   Solving Quadratic Equations Using the Zero-Factor Property

---

Learning Objectives
1   Solve quadratic equations using the zero-factor property.
2   Solve other equations using the zero-factor property.

Key Terms

Use the vocabulary terms listed below to complete each statement in exercises 1–3.

> **quadratic equation**   **standard form**   **double solution**

1.   An equation written in the form $ax^2 + bx + c = 0$ is written in the
________________________ of a second-degree equation.

2.   Two factors are identical and both lead to the same solution, called a
________________________.

3.   An equation that can written in the form $ax^2 + bx + c = 0$, with $a \neq 0$, is a
________________________.

Objective 1   Solve quadratic equations using the zero-factor property.

Review these examples for Objective 1:

1.   Solve each equation.

   a. $(x+9)(5x-6)=0$

By the zero-factor property, either $x+9=0$ or $5x-6=0$.

$$x+9=0 \quad \text{or} \quad 5x-6=0$$
$$x=-9 \quad \text{or} \quad 5x=6$$
$$x=\frac{6}{5}$$

Check:
Let $x=-9$.
$$(x+9)(5x-6)=0$$
$$(-9+9)[5(-9)-6]\overset{?}{=}0$$
$$0(-51)\overset{?}{=}0$$
$$0=0 \ \text{True}$$

Now Try:

1.   Solve each equation.

   a. $(x+12)(4x-7)=0$

____

Let $x = \dfrac{6}{5}$.

$$(x+9)(5x-6) = 0$$

$$\left(\dfrac{6}{5}+9\right)\left[5\left(\dfrac{6}{5}\right)-6\right] \overset{?}{=} 0$$

$$\left(\dfrac{51}{5}\right)(6-6) \overset{?}{=} 0$$

$$0 = 0 \quad \text{True}$$

Both values check, so the solution set is

$$\left\{-9,\ \dfrac{6}{5}\right\}.$$

b. $x(8x-11) = 0$

Use the zero-factor property.

$$x = 0 \quad \text{or} \quad 8x - 11 = 0$$
$$8x = 11$$
$$x = \dfrac{11}{8}$$

Check these solutions by substituting each in the original equation. The solution set is $\left\{0,\ \dfrac{11}{8}\right\}$.

3. Solve $8p^2 + 30 = 46p$.

$$8p^2 + 30 = 46p$$
$$8p^2 - 46p + 30 = 0 \quad \text{Standard form}$$
$$2\left(4p^2 - 23p + 15\right) = 0 \quad \text{Factor out 2.}$$
$$4p^2 - 23p + 15 = 0 \quad \text{Divide each side by 2}$$
$$(4p - 3)(p - 5) = 0 \quad \text{Factor.}$$
$$4p - 3 = 0 \quad \text{or} \quad p - 5 = 0 \quad \text{Zero-factor}$$
$$p = \dfrac{3}{4} \quad \text{or} \quad p = 5 \quad \text{property}$$

The solution set is $\left\{\dfrac{3}{4},\ 5\right\}$.

b. $x(6x-11) = 0$

____

3. Solve $15p^2 + 36 = 57p$.

____

4. Solve the equation.

$$64m^2 - 49 = 0$$

$$64m^2 - 49 = 0$$
$$(8m + 7)(8m - 7) = 0$$
$$8m + 7 = 0 \quad \text{or} \quad 8m - 7 = 0$$
$$8m = -7 \quad \text{or} \quad 8m = 7$$
$$m = -\frac{7}{8} \quad \text{or} \quad m = \frac{7}{8}$$

The solution set is $\left\{-\dfrac{7}{8}, \dfrac{7}{8}\right\}$.

4. Solve the equation.

$$100x^2 - 9 = 0$$

____

Objective 1 Practice Exercises

For extra help, see Examples 1–5 on pages 383–386 of your text.

Solve each equation and check your solutions.

1. $2x^2 - 3x - 20 = 0$

1. _______________

2. $25x^2 = 20x$

2. _______________

3. $c(5c + 17) = 12$

3. _______________

Objective 2   Solve other equations using the zero-factor property.

Review these examples for Objective 2:

Now Try:

6. Solve $12z^3 - 3z = 0$.

$$12z^3 - 3z = 0$$

$$3z(4z^2 - 1) = 0$$

$$3z(2z + 1)(2z - 1) = 0$$

By an extension of the zero-factor property, we have

$$3z = 0 \quad \text{or} \quad 2z + 1 = 0 \quad \text{or} \quad 2z - 1 = 0$$

$$z = 0 \quad \text{or} \quad z = -\frac{1}{2} \quad \text{or} \quad z = \frac{1}{2}$$

Check by substituting each value in the original equation. The solution set is $\left\{-\frac{1}{2}, \ 0, \ \frac{1}{2}\right\}$.

6. Solve $3r^3 = 75r$.

____

7. Solve $(3x - 1)(x^2 - 13x + 36) = 0$.

$$(3x - 1)(x^2 - 13x + 36) = 0$$

$$(3x - 1)(x - 4)(x - 9) = 0$$

$$3x - 1 = 0 \quad \text{or} \quad x - 4 = 0 \quad \text{or} \quad x - 9 = 0$$

$$x = \frac{1}{3} \quad \text{or} \quad x = 4 \quad \text{or} \quad x = 9$$

Check by substituting each value in the original equation. The solution set is $\left\{\frac{1}{3}, \ 4, \ 9\right\}$.

7. Solve
$$(5x - 2)(x^2 - 11x + 18) = 0.$$

____

8. Solve $x(3x - 7) = (x - 1)^2 + 11$.

$$x(3x - 7) = (x - 1)^2 + 11$$

$$3x^2 - 7x = x^2 - 2x + 1 + 11$$

$$3x^2 - 7x = x^2 - 2x + 12$$

$$2x^2 - 5x - 12 = 0$$

$$(2x + 3)(x - 4) = 0$$

$$2x + 3 = 0 \quad \text{or} \quad x - 4 = 0$$

$$x = -\frac{3}{2} \quad \text{or} \quad x = 4$$

Check by substituting each value in the original equation. The solution set is $\left\{-\frac{3}{2}, \ 4\right\}$.

8. Solve $x(2x + 5) = (x + 2)^2 + 8$.

____

         213

Objective 2 Practice Exercises

For extra help, see Examples 6–8 on pages 387–388 of your text.

Solve each equation and check your solutions.

4. $x^3 + 2x^2 - 8x = 0$ 4. ______________________

5. $z^4 + 8z^3 - 9z^2 = 0$ 5. ______________________

6. $\left(y^2 - 5y + 6\right)\left(y^2 - 36\right) = 0$ 6. ______________________

Chapter 5 FACTORING AND APPLICATIONS

5.6   **Applications of Quadratic Equations**

Learning Objectives

1   Solve problems involving geometric figures.
2   Solve problems involving consecutive integers.
3   Solve problems by applying the Pythagorean theorem.
4   Solve problems using given quadratic models.

Key Terms

Use the vocabulary terms listed below to complete each statement in exercises 1−3.

     hypotenuse   **legs**   **consecutive**

1.   In a right triangle, the sides that form the right angle are the ________________.

2.   The longest side of a right triangle is the ________________________.

3.   ____________________ integers are integers that are next to each other on a number line.

Objective 1   Solve problems involving geometric figures.

Review this example for Objective 1:

1.   The length of a rectangle is three times its width. If the width was increased by 4 and the length remained the same, the resulting rectangle would have an area of 231 square inches. Find the dimensions of the original rectangle.

Step 1 Read the problem carefully. Find the dimensions of the original rectangle.

Step 2 Assign a variable.
     Let x = width.
       $3x$ = length
   $x + 4$ = new width
       $3x$ = length

Step 3 Write an equation. The area of the rectangle is given by $\text{Area} = \text{Length} \times \text{Width}$. Substitute 231 for area, $3x$ for length, and $x + 4$ for width.
$$231 = 3x(x + 4)$$

Now Try:

1.   Mr. Fixxall is building a box which will have a volume of 60 cubic meters. The height of the box will be 4 meters, and the length will be 2 meters more than the width. Find the width and length of the box.

                 ____

         215

Step 4 Solve.

$$231 = 3x(x+4)$$
$$231 = 3x^2 + 12x$$
$$3x^2 + 12x - 231 = 0$$
$$3(x^2 + 4x - 77) = 0$$
$$x^2 + 4x - 77 = 0$$
$$(x+11)(x-7) = 0$$
$$x + 11 = 0 \quad \text{or} \quad x - 7 = 0$$
$$x = -11 \quad \text{or} \quad x = 7$$

Step 5 State the answer. The solutions are –11 and 7. A rectangle cannot have a side of negative length, so we discard –11. The width is 7 inches. The length is $3(7) = 21$ inches.

Step 6 Check. The new width is 7 + 4 = 11. The new area of the rectangle is $11(21) = 231$ square inches.

Objective 1 Practice Exercises

For extra help, see Example 1 on page 391 of your text.

Solve each problem. Check your answers to be sure they are reasonable.

1. A book is three times as long as it is wide. Find the length and width of the book in inches if its area is numerically 128 more than its perimeter.

1. width________________

   length ____________

2. The area of a triangle is 42 square centimeters. The base is 2 centimeters less than twice the height. Find the base and height of the triangle.

2. base________________

   height ____________

3. The volume of a box is 192 cubic feet. If the length of the box is 8 feet and the width is 2 feet more than the height, find the height and width of the box.

3. height _____________

width_____________

Objective 2   Solve problems involving consecutive integers.

Review this example for Objective 2:

2. The product of the first and second of three consecutive integers is 2 more than 6 times the third integer. Find the integers.

Step 1 Read carefully. Note that the integers are consecutive integers.

Step 2 Assign a variable.
  Let x = the first integer.
  Then $x + 1$ = the second integer,
  and $x + 2$ = the third integer.

Step 3 Write an equation. The product of the first and second integer is 2 more than 6 times the third integer.
$$x(x+1) = 6(x+2)+2$$

Step 4 Solve.
$$x(x+1) = 6(x+2)+2$$
$$x^2 + x = 6x + 14$$
$$x^2 - 5x - 14 = 0$$
$$(x+2)(x-7) = 0$$
$$x+2 = 0 \quad \text{or} \quad x-7 = 0$$
$$x = -2 \quad \text{or} \quad x = 7$$

Step 5 State the answer. The value -2 and 7 each lead to a distinct answer.
If $x = -2$, then $x + 1 = -1$, and $x + 2 = 0$.
  The integers are $-2, -1, 0$.
If $x = 7$, then $x + 1 = 8$, and $x + 2 = 9$.
  The integers are $7, 8, 9$.

Now Try:

2. The product of the second and third of three consecutive integers is 2 more than 8 times the first integer. Find the integers.

____

Step 6 Check. The product of the first and second integers must equal 2 more than 6 times the third. Because

$$-2(-1) = 6(0) + 2 \text{ and } 7(8) = 6(9) + 2$$

are both true, both sets of consecutive integers satisfy the statement of the problem.

Objective 2 Practice Exercises

For extra help, see Example 2 on page 392 of your text.

Solve each problem.

4. Find all possible pairs of consecutive odd integers whose sum is equal to their product decreased by 47.

4. _______________

5. Find two consecutive positive even integers whose product is six more than three times its sum.

5. _______________

6. Find three consecutive positive odd integers such that four times the sum of all three equals 13 more than the product of the smaller two.

6. _______________

Name: Date:
Instructor: Section:

Objective 3   Solve problems by applying the Pythagorean theorem.

Review this example for Objective 3:

3.   Penny and Carla started biking from the same corner. Penny biked east and Carla biked south. When they were 26 miles apart, Carla had biked 14 miles further than Penny. Find the distance each biked.

Step 1 Read carefully. Find the two distances.

Step 2 Assign a variable.
   Let x = Penny's distance.
   Then $x + 14$ = Carla's distance.

Step 3 Write an equation. Substitute into the Pythagorean theorem.
$$a^2 + b^2 = c^2$$
$$x^2 + (x+14)^2 = 26^2$$

Step 4 Solve.
$$x^2 + x^2 + 28x + 196 = 676$$
$$2x^2 + 28x - 480 = 0$$
$$2(x^2 + 14x - 240) = 0$$
$$x^2 + 14x - 240 = 0$$
$$(x + 24)(x - 10) = 0$$
$$x + 24 = 0 \quad \text{or} \quad x - 10 = 0$$
$$x = -24 \quad \text{or} \quad x = 10$$

Step 5 State the answer. Since –24 cannot be a distance, 10 is the distance for Penny, and 10 + 14 = 24 is the distance for Carla.

Step 6 Check. Since $10^2 + 24^2 = 26^2$ is true, the answer is correct.

Now Try:

3.   A ladder is leaning against a building. The distance from the bottom of the ladder to the building is 8 feet less than the length of the ladder. How high up the side of the building is the top of the ladder if that distance is 4 feet less than the length of the ladder?

____

Objective 3 Practice Exercises

For extra help, see Example 3 on pages 393–394 of your text.

Solve each problem.

7.   A field is in the shape of a right triangle. The shorter leg measures 45 meters. The hypotenuse measures 45 meters less than twice the longer the leg. Find the dimensions of the lot.

7. _______________________

8.   A train and a car leave a station at the same time, the   **8.** car_______________
     train traveling due north and the car traveling west.
     When they are 100 miles apart, the train has traveled   train ______________
     20 miles farther than the car. Find the distance each
     has traveled.

9.   Two ships left a dock at the same time. When they   **9.** _______________
     were 25 miles apart, the ship that sailed due south
     had gone 10 miles less than twice the distance
     traveled by the ship that sailed due west. Find the
     distance traveled by the ship that sailed due south.

Objective 4   Solve problems using given quadratic models.

Review this example for Objective 4:

4. You throw a stone straight upward at 46 feet per
     second from a dock 6 feet above a lake. The
     height of the stone above the lake t seconds after
     it is thrown is given by $h = -16t^2 + 46t + 6$.
     How long will it take for the stone to reach a
     height of 39 feet?

Substitute 39 for h.

$$39 = -16t^2 + 46t + 6$$

Solve for t.

$$16t^2 - 46t + 33 = 0$$

$$(8t - 11)(2t - 3) = 0$$

$$8t - 11 = 0 \quad \text{or} \quad 2t - 3 = 0$$

$$t = \frac{11}{8} \quad \text{or} \quad t = \frac{3}{2}$$

Since we have found two acceptable answers,

Now Try:

4. A ball is dropped from the roof
     of a 19.6 meter high building. Its
     height h (in meters) t seconds
     later is given by the equation
     $h = -4.9t^2 + 19.6$. After how
     many second is the height 14.7
     meters?

     ____

the stone will be at height of 39 feet twice (once on its way up and once on its way down) —at $\frac{11}{8}$ sec or $\frac{3}{2}$ sec.

Objective 4 Practice Exercises

For extra help, see Examples 4–5 on pages 394–395 of your text.

Solve each problem.

10. If an object is propelled upward from a height of 16 feet with an initial velocity of 48 feet per second, its height h (in feet) t seconds later is given by the equation $h = -16t^2 + 48t + 16$.

   (a) After how many seconds is the height 52 feet?

   (b) After how many seconds is the height 48 feet?

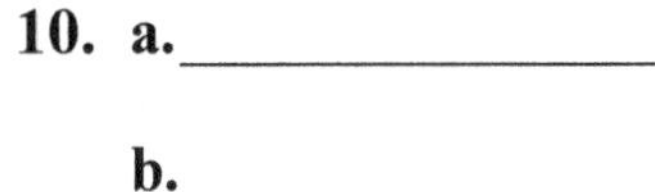

11. A company determines that its daily revenue R (in dollars) for selling x items is modeled by the equation $R = x(150 - x)$. How many items must be sold for its revenue to be \$4400?

11. ________________

12. If a ball is batted at an angle of 35°, the distance that the ball travels is given approximately by $D = 0.029v^2 + 0.021v - 1$, where v is the bat speed in miles per hour and D is the distance traveled in feet. Find the distance a batted ball will travel if the ball is batted with a velocity of 90 miles per hour. Round your answer to the nearest whole number.

12. ________________

Chapter 6 RATIONAL EXPRESSIONS AND APPLICATIONS

6.1　The Fundamental Property of Rational Expressions

Learning Objectives
1　Find the numerical value of a rational expression.
2　Find the values of the variable for which a rational expression is undefined.
3　Write rational expressions in lowest terms.
4　Recognize equivalent forms of rational expressions.

Key Terms

Use the vocabulary terms listed below to complete each statement in exercises 1–2.

    rational expression　　　　**lowest terms**

1.　The quotient of two polynomials with denominator not 0 is called a

    ________________________.

2.　A rational expression is written in ____________________ if the
greatest common factor of its numerator and denominator is 1.

Objective 1　Find the numerical value of a rational expression.

Review this example for Objective 1:

1.　Find the numerical value of $\dfrac{4x+8}{3x-6}$ for the value of x.

    $x = -2$

$$\frac{4x+8}{3x-6} = \frac{4(-2)+8}{3(-2)-6} = \frac{0}{-12} = 0$$

Now Try:

1.　Find the numerical value of $\dfrac{3x-7}{x+5}$ for the value of x.

    $x = -7$

    ____

Objective 1 Practice Exercises

For extra help, see Example 1 on page 412 of your text.

Find the numerical value of each expression when (a) $x = 4$ and (b) $x = -1$.

1.　$\dfrac{-3x+1}{2x+1}$

1. (a) ______________

    (b) ______________

2. $\dfrac{2x^2 - 4}{x^2 - 2}$

2. (a)______________

   (b)______________

3. $\dfrac{2x - 5}{2 - x - x^2}$

3. (a)______________

   (b)______________

Objective 2   Find the values of the variable for which a rational expression is undefined.

Review these examples for Objective 2:

2. Find any values of the variable for which each rational expression is undefined.

a. $\dfrac{3x + 8}{5x + 4}$

Step 1 Set the denominator equal to 0.
$$5x + 4 = 0$$

Step 2 Solve.
$$5x = -4$$
$$x = -\dfrac{4}{5}$$

Step 3 The given expression is undefined for $-\dfrac{4}{5}$, so $x \neq -\dfrac{4}{5}$.

b. $\dfrac{6r}{r^2 + 49}$

This denominator will not equal 0 for any value of r. There are no values for which this expression is undefined.

Now Try:

2. Find any values of the variable for which each rational expression is undefined.

a. $\dfrac{y + 6}{7y - 1}$

____

b. $\dfrac{12t^2}{t^2 + 100}$

____

c. $\dfrac{3m^2}{m^2 - 4m - 5}$

Set the denominator equal to 0.

$$m^2 - 4m - 5 = 0$$

$$(m + 1)(m - 5) = 0$$

$$m + 1 = 0 \quad \text{or} \quad m - 5 = 0$$

$$m = -1 \quad \text{or} \quad m = 5$$

The given expression is undefined for -1 and 5, so $m \neq -1$, $m \neq 5$.

c. $\dfrac{15m^2}{m^2 - m - 20}$

____

Objective 2 Practice Exercises

For extra help, see Example 2 on page 413 of your text.

Find any value(s) of the variable for which each rational expression is undefined. Write answers with $\neq$.

4. $\dfrac{4x^2}{x + 7}$

4. ______________

5. $\dfrac{2x^2}{x^2 + 4}$

5. ______________

6. $\dfrac{2y - 5}{2y^2 + 4y - 16}$

6. ______________

Objective 3   Write rational expressions in lowest terms.

Review these examples for Objective 3:

3. Write the expression in lowest terms.

$$\dfrac{15k^3}{3k^4}$$

Write k^3 as $k \cdot k \cdot k$ and k^4 as $k \cdot k \cdot k \cdot k$.

$$\dfrac{15k^3}{3k^4} = \dfrac{3 \cdot 5 \cdot k \cdot k \cdot k}{3 \cdot k \cdot k \cdot k \cdot k}$$

$$= \dfrac{5 \cdot (3 \cdot k \cdot k \cdot k)}{k \cdot (3 \cdot k \cdot k \cdot k)}$$

$$= \dfrac{5}{k}$$

Now Try:

3. Write the expression in lowest terms.

$$\dfrac{12k^5}{4k^8}$$

____

4.   Write each rational expression in lowest terms.

    a. $\dfrac{6x-18}{5x-15}$

$$\frac{6x-18}{5x-15}=\frac{6(x-3)}{5(x-3)}=\frac{6}{5}$$

    c. $\dfrac{m^2+5m-24}{3m^2-5m-12}$

$$\frac{m^2+5m-24}{3m^2-5m-12}=\frac{(m+8)(m-3)}{(3m+4)(m-3)}$$
$$=\frac{m+8}{3m+4}$$

5.   Write $\dfrac{3x-2y}{2y-3x}$ in lowest terms.

Factor -1 from the denominator.
$$\frac{3x-2y}{2y-3x}=\frac{3x-2y}{-1(-2y+3x)}$$
$$=\frac{3x-2y}{-1(3x-2y)}$$
$$=-1$$

4.   Write each rational expression in lowest terms.

    a. $\dfrac{7x-35}{9x-45}$

____

    c. $\dfrac{m^2-3m-54}{2m^2-15m-27}$

____

5.   Write $\dfrac{4y-5x}{5x-4y}$ in lowest terms.

____

Objective 3 Practice Exercises

For extra help, see Examples 3–6 on pages 414–417 of your text.

Write each rational expression in lowest terms. Assume that no values of any variable make any denominator zero.

7. $\dfrac{15ab^3c^9}{-24ab^2c^{10}}$

7. ________________

8. $\dfrac{16-x^2}{2x-8}$

8. ________________

                    Copyright © 2025 Pearson Education, Inc.

9. $\dfrac{9x^2 - 9x - 108}{2x - 8}$

9. _______________

Objective 4 Recognize equivalent forms of rational expressions.

Review this example for Objective 4:

7. Write four equivalent forms of the following rational expression.

$$-\dfrac{4x + 3}{x - 8}$$

If we apply the negative sign to the numerator, we obtain the first two equivalent forms.

$$\dfrac{-(4x + 3)}{x - 8} \quad \text{and} \quad \dfrac{-4x - 3}{x - 8}$$

If we apply the negative sign to the denominator, we obtain the last two equivalent forms.

$$\dfrac{4x + 3}{-(x - 8)} \quad \text{and} \quad \dfrac{4x + 3}{-x + 8}$$

Now Try:

7. Write four equivalent forms of the following rational expression.

$$-\dfrac{10x - 7}{4x - 3}$$

____

Objective 4 Practice Exercises

For extra help, see Example 7 on page 418 of your text.

Write four equivalent forms of the following rational expressions. Assume that no values of any variable make any denominator zero.

10. $-\dfrac{4x + 5}{3 - 6x}$

10. _______________

11. $\dfrac{2p - 1}{1 - 4p}$

11. _______________

12. $-\dfrac{2x - 3}{x + 2}$

12. _______________

Chapter 6 RATIONAL EXPRESSIONS AND APPLICATIONS

6.2   **Multiplying and Dividing Rational Expressions**

Learning Objectives
1   Multiply rational expressions.
2   Find reciprocals of rational expressions.
3   Divide rational expressions.

Key Terms

Use the vocabulary terms listed below to complete each statement in exercises 1−3.

        rational expression   **reciprocal**   **lowest terms**

1.   The _____________________ of the expression $\dfrac{4x-5}{x+2}$ is $\dfrac{x+2}{4x-5}$.

2.   A _______________________________ is the quotient of two polynomials with denominator not 0.

3.   A rational expression is written in _____________________________ when the numerator and denominator have no common terms.

Objective 1   Multiply rational expressions.

Review these examples for Objective 1: | **Now Try:**

1.   Multiply. Write each answer in lowest terms.

a. $\dfrac{7}{12}\cdot\dfrac{3}{14}$

$$\dfrac{7}{12}\cdot\dfrac{3}{14}=\dfrac{7\cdot3}{12\cdot14}$$

$$=\dfrac{7\cdot3}{2\cdot2\cdot3\cdot2\cdot7}$$

$$=\dfrac{1}{8}$$

b. $\dfrac{9}{x^2}\cdot\dfrac{x^3}{15}$

$$\dfrac{9}{x^2}\cdot\dfrac{x^3}{15}=\dfrac{9\cdot x^3}{x^2\cdot15}$$

$$=\dfrac{3\cdot3\cdot x\cdot x\cdot x}{3\cdot5\cdot x\cdot x}$$

$$=\dfrac{3x}{5}$$

Now Try:

1.   Multiply. Write each answer in lowest terms.

a. $\dfrac{5}{9}\cdot\dfrac{12}{25}$

____

b. $\dfrac{8}{x^3}\cdot\dfrac{x^2}{6}$

____

2. Multiply. Write the answer in lowest terms.

$$\frac{3x+2y}{5x}\cdot\frac{x^2}{(3x+2y)^2}$$

Multiply numerators, multiply denominators, factor, and identify the common factors.

$$\frac{3x+2y}{5x}\cdot\frac{x^2}{(3x+2y)^2}=\frac{(3x+2y)x^2}{5x(3x+2y)^2}$$

$$=\frac{(3x+2y)\cdot x\cdot x}{5x(3x+2y)(3x+2y)}$$

$$=\frac{x}{5(3x+2y)}$$

3. Multiply. Write the answer in lowest terms.

$$\frac{x^2+10x+9}{x^2-5x}\cdot\frac{x^2+3x-40}{x^2+9x+8}$$

$$\frac{x^2+10x+9}{x^2-5x}\cdot\frac{x^2+3x-40}{x^2+9x+8}$$

$$=\frac{\left(x^2+10x+9\right)\left(x^2+3x-40\right)}{\left(x^2-5x\right)\left(x^2+9x+8\right)}$$

$$=\frac{(x+9)(x+1)(x+8)(x-5)}{x(x-5)(x+1)(x+8)}$$

$$=\frac{x+9}{x}$$

The quotients $\frac{x+1}{x+1}$, $\frac{x+8}{x+8}$, and $\frac{x-5}{x-5}$ are all

equal to 1, justifying the final product $\frac{x+9}{x}$.

2. Multiply. Write the answer in lowest terms.

$$\frac{r-s}{6s}\cdot\frac{s^3}{(r-s)^2}$$

____

3. Multiply. Write the answer in lowest terms.

$$\frac{7x^2-7}{x^2+x-2}\cdot\frac{5x+10}{x^2+x}$$

____

Objective 1 Practice Exercises

For extra help, see Examples 1–3 on pages 423–424 of your text.

Multiply. Write each answer in lowest terms.

1. $\dfrac{8m^4n^3}{3}\cdot\dfrac{5}{4mn^2}$

1. ____________

2. $\dfrac{m^2-16}{m-3}\cdot\dfrac{9-m^2}{4-m}$

2. _______________

3. $\dfrac{3x+12}{6x-30}\cdot\dfrac{x^2-x-20}{x^2-16}$

3. _______________

Objective 2   Find reciprocals of rational expressions.

Review these examples for Objective 2:

4. Find the reciprocal of each rational expression.

a. $\dfrac{5a^2}{7b}$

$\dfrac{5a^2}{7b}$ has reciprocal $\dfrac{7b}{5a^2}$.

b. $\dfrac{p^2-16}{p^2-7p+12}$

$\dfrac{p^2-16}{p^2-7p+12}$ has reciprocal $\dfrac{p^2-7p+12}{p^2-16}$.

Now Try:

4. Find the reciprocal of each rational expression.

a. $\dfrac{2x^3}{7y}$

____

b. $\dfrac{q^2-25}{q^2-8q-15}$

____

Objective 2 Practice Exercises

For extra help, see Example 4 on page 424 of your text.

Find the reciprocal of each rational expression.

4. $\dfrac{8x^5}{9y^2}$

4. _______________

5. $\dfrac{13}{6z^2-2z+1}$

5. _______________

6. $\dfrac{r^2-3r+4}{r^2+2r}$

6. _______________

Objective 3   Divide rational expressions.

Review these examples for Objective 3:

5. Divide. Write each answer in lowest terms.

   a. $\dfrac{7}{9} \div \dfrac{4}{27}$

   Multiply the first expression by the reciprocal of the second.

   $$\dfrac{7}{9} \div \dfrac{4}{27} = \dfrac{7}{9} \cdot \dfrac{27}{4}$$
   $$= \dfrac{7 \cdot 27}{9 \cdot 4}$$
   $$= \dfrac{7 \cdot 3 \cdot 9}{9 \cdot 4}$$
   $$= \dfrac{21}{4}$$

   b. $\dfrac{y}{y-5} \div \dfrac{7y}{y+4}$

   $$\dfrac{y}{y-5} \div \dfrac{7y}{y+4} = \dfrac{y}{y-5} \cdot \dfrac{y+4}{7y}$$
   $$= \dfrac{y(y+4)}{(y-5)(7y)}$$
   $$= \dfrac{y+4}{7(y-5)}$$

7. Divide. Write the answer in lowest terms.

   $$\dfrac{m^2-16}{(m-5)(m-4)} \div \dfrac{(m+4)(m-5)}{-6m}$$

   Multiply by the reciprocal.

   $$\dfrac{m^2-16}{(m-5)(m-4)} \div \dfrac{(m+4)(m-5)}{-6m}$$
   $$= \dfrac{m^2-16}{(m-5)(m-4)} \cdot \dfrac{-6m}{(m+4)(m-5)}$$
   $$= \dfrac{-6m(m^2-16)}{(m-5)(m-4)(m+4)(m-5)}$$
   $$= \dfrac{-6m(m+4)(m-4)}{(m-5)(m-4)(m+4)(m-5)}$$
   $$= \dfrac{-6m}{(m-5)^2}, \text{ or } -\dfrac{6m}{(m-5)^2}$$

Now Try:

5. Divide. Write each answer in lowest terms.

   a. $\dfrac{9}{10} \div \dfrac{3}{25}$

   ____

   b. $\dfrac{y}{y+2} \div \dfrac{6y}{y-2}$

   ____

7. Divide. Write the answer in lowest terms.

   $$\dfrac{x^2-49}{(x-7)(x-3)} \div \dfrac{(x+7)(x-3)}{8x}$$

   ____

8. Divide. Write the answer in lowest terms.

$$\frac{x^2-25}{x^2-9} \div \frac{3x^2-15x}{3-x}$$

Multiply by the reciprocal.

$$\frac{x^2-25}{x^2-9} \div \frac{3x^2-15x}{3-x}$$

$$= \frac{x^2-25}{x^2-9} \cdot \frac{3-x}{3x^2-15x}$$

$$= \frac{\left(x^2-25\right)(3-x)}{\left(x^2-9\right)\left(3x^2-15x\right)}$$

$$= \frac{(x+5)(x-5)(3-x)}{(x+3)(x-3)(3x)(x-5)}$$

$$= \frac{-1(x+5)}{3x(x+3)}, \quad \text{or} \quad \frac{-x-5}{3x(x+3)}$$

8. Divide. Write the answer in lowest terms.

$$\frac{m^2-64}{m^2-81} \div \frac{5m^2+40m}{9-m}$$

____

Objective 3 Practice Exercises

For extra help, see Examples 5–8 on pages 425–426 of your text.

Divide. Write each answer in lowest terms.

7. $\dfrac{b-7}{16} \div \dfrac{7-b}{8}$

7. ________________

8. $\dfrac{m^2+2mn+n^2}{m^2+m} \div \dfrac{m^2-n^2}{m^2-1}$

8. ________________

9. $\dfrac{27-3k^2}{3k^2+8k-3} \div \dfrac{k^2-6k+9}{6k^2-19k+3}$

9. ________________

Chapter 6 RATIONAL EXPRESSIONS AND APPLICATIONS

6.3   Least Common Denominators

Learning Objectives
1   Find the least common denominator for a list of fractions.
2   Write equivalent rational expressions.

Key Terms

Use the vocabulary terms listed below to complete each statement in exercises 1−2.

least common denominator   **equivalent expressions**

1. $\dfrac{24x-8}{9x^2-1}$ and $\dfrac{8}{3x+1}$ are _________________________________.

2.   The simplest expression that is divisible by all denominators is called the

_________________________________.

Objective 1   Find the least common denominator for a list of fractions.

Review these examples for Objective 1:

1.   Find the LCD for the pair of fractions.

$$\frac{1}{35},\ \frac{11}{45}$$

Step 1 Factor each denominator into prime factors.

$$35 = 5 \cdot 7$$

$$45 = 3 \cdot 3 \cdot 5 = 3^2 \cdot 5$$

Step 2 List each denominator the greatest number of times it appears as a factor in any of the denominators.

The factor 3 appears two times, and the factors 5 and 7 each appear once.

Step 3 Multiply to get the LCD.

$$LCD = 3 \cdot 3 \cdot 5 \cdot 7$$

$$= 3^2 \cdot 5 \cdot 7$$

$$= 315$$

Now Try:

1.   Find the LCD for the pair of fractions.

$$\frac{5}{18},\ \frac{13}{24}$$

____

2. Find the LCD for $\dfrac{9}{28s^3}$ and $\dfrac{5}{42s^2}$.

Step 1

$$28s^3 = 2\cdot2\cdot7\cdot s^3$$

$$42s^2 = 2\cdot3\cdot7\cdot s^2$$

Step 2

Here s appears three times, 2 appears twice, and 3 and 7 each appear once.

Step 3

$$\text{LCD} = 2^2\cdot3\cdot7\cdot s^3 = 84s^3$$

3. Find the LCD for the fractions in each list.

a. $\dfrac{5}{7b}$, $\dfrac{8}{b^2-5b}$

$$7b = 7\cdot b$$

$$b^2 - 5b = b(b-5)$$

$$\text{LCD} = 7\cdot b(b-5) = 7b(b-5)$$

b. $\dfrac{6}{c^2-5c-6}$, $\dfrac{10}{c^2+3c-54}$, $\dfrac{4}{c^2-12c+36}$

$$c^2 - 5c - 6 = (c-6)(c+1)$$

$$c^2 + 3c - 54 = (c-6)(c+9)$$

$$c^2 - 12c + 36 = (c-6)^2$$

Use each factor the greatest number of times it appears as a factor.

$$\text{LCD} = (c+1)(c+9)(c-6)^2$$

c. $\dfrac{1}{a-8}$, $\dfrac{7}{8-a}$

$$-(a-8) = -a+8 = 8-a$$

Therefore either $8-a$ or $a-8$ can be used as the LCD.

2. Find the LCD for
$\dfrac{15}{40a^2}$ and $\dfrac{13}{24a^4}$.

____

3. Find the LCD for the fractions in each list.

a. $\dfrac{7}{9w}$, $\dfrac{13}{w^2-2w}$

____

b. $\dfrac{12}{b^2-16}$, $\dfrac{6}{b^2-3b-4}$, $\dfrac{9}{b^2-8b+16}$

____

c. $\dfrac{13}{p-14}$, $\dfrac{12}{14-p}$

____

Objective 1 Practice Exercises

For extra help, see Examples 1–3 on pages 430–431 of your text.

Find the least common denominator for each list of rational expressions.

1. $\dfrac{13}{36b^4}$, $\dfrac{17}{27b^2}$

1. ______________

                      Copyright © 2025 Pearson Education, Inc.

2. $\dfrac{-7}{a^2-2a}, \ \dfrac{3a}{2a^2+a-10}$ 2. _________________

3. $\dfrac{8}{w^3-9w}, \ \dfrac{4w}{w^2+w-6}$ 3. _________________

Objective 2   Write equivalent rational expressions.

Review these examples for Objective 2:

4. Write the rational expression as an equivalent expression with the indicated denominator.

$$\frac{7}{9}=\frac{?}{45}$$

Step 1 Factor both denominators.

$$\frac{7}{9}=\frac{?}{5\cdot 9}$$

Step 2 A factor of 5 is missing.

Step 3 Multiply $\dfrac{7}{9}$ by $\dfrac{5}{5}$.

$$\frac{7}{9}=\frac{7}{9}\cdot\frac{5}{5}=\frac{35}{45}$$

5. Write each rational expression as an equivalent expression with the indicated denominator.

a. $\dfrac{9}{5x+2}=\dfrac{?}{15x+6}$

Factor the denominator on the right.

$$\frac{9}{5x+2}=\frac{?}{3(5x+2)}$$

The missing factor is 3, so multiply by $\dfrac{3}{3}$.

$$\frac{9}{5x+2}\cdot\frac{3}{3}=\frac{27}{15x+6}$$

Now Try:

4. Write the rational expression as an equivalent expression with the indicated denominator.

$$\frac{13}{6}=\frac{?}{30}$$

____

5. Write each rational expression as an equivalent expression with the indicated denominator.

a. $\dfrac{19}{6c-5}=\dfrac{?}{24c-20}$

____

                    235

b. $\dfrac{7}{q^2 + 6q} = \dfrac{?}{q^3 + q^2 - 30q}$

b. $\dfrac{3}{z^2 - 7z} = \dfrac{?}{z^3 - 5z^2 - 14z}$

Factor the denominator in each rational expression.

$$\dfrac{7}{q(q+6)} = \dfrac{?}{q(q+6)(q-5)}$$

The missing factor is $(q-5)$, so multiply by $\dfrac{q-5}{q-5}$.

$$\dfrac{7}{q(q+6)} = \dfrac{7}{q(q+6)} \cdot \dfrac{q-5}{q-5}$$

$$= \dfrac{7(q-5)}{q^3 - q^2 - 30q}$$

$$= \dfrac{7q - 35}{q^3 - q^2 - 30q}$$

____

Objective 2 Practice Exercises

For extra help, see Examples 4–5 on pages 431–432 of your text.

Rewrite each rational expression with the indicated denominator. Give the numerator of the new fraction.

4. $\dfrac{5a}{8a - 3} = \dfrac{?}{6 - 16a}$

4. _______________

5. $\dfrac{3}{5r - 10} = \dfrac{?}{50r^2 - 100r}$

5. _______________

6. $\dfrac{3}{k^2 + 3k} = \dfrac{?}{k^3 + 10k^2 + 21k}$

6. _______________

Chapter 6 RATIONAL EXPRESSIONS AND APPLICATIONS

6.4 Adding and Subtracting Rational Expressions

Learning Objectives
1 Add rational expressions having the same denominator.
2 Add rational expressions having different denominators.
3 Subtract rational expressions.

Key Terms

Use the vocabulary terms listed below to complete each statement in exercises 1–2.

least common multiple **greatest common factor**

1. The _______________________________ of $2m^2 - 5m - 3$ and $2m - 6$ is $m - 3$.

2. The _______________________________ of $2m^2 - 5m - 3$ and $2m - 6$ is $2(m-3)(2m+1)$.

Objective 1   Add rational expressions having the same denominator.

Review these examples for Objective 1:

1. Add. Write each answer in lowest terms.

   a. $\dfrac{1}{8} + \dfrac{3}{8}$

   The denominators are the same, so add the numerators and keep the common denominator.

   $$\frac{1}{8} + \frac{3}{8} = \frac{1+3}{8}$$
   $$= \frac{4}{8}$$
   $$= \frac{4 \cdot 1}{4 \cdot 2}$$
   $$= \frac{1}{2}$$

   b. $\dfrac{4x}{x+2} + \dfrac{8}{x+2}$

   $$\frac{4x}{x+2} + \frac{8}{x+2} = \frac{4x+8}{x+2} = \frac{4(x+2)}{x+2} = 4$$

Now Try:

1. Add. Write each answer in lowest terms.

   a. $\dfrac{9}{20} + \dfrac{7}{20}$

   ____

   b. $\dfrac{2x^2}{x+4} + \dfrac{8x}{x+4}$

   ____

                    237

Objective 1 Practice Exercises

For extra help, see Example 1 on page 436 of your text.

Add. Write each answer in lowest terms.

1. $\dfrac{5}{3w^2}+\dfrac{7}{3w^2}$

1. _________________

2. $\dfrac{b}{b^2-4}+\dfrac{2}{b^2-4}$

2. _________________

3. $\dfrac{2x+3}{x^2+3x-10}+\dfrac{2-x}{x^2+3x-10}$

3. _________________

Objective 2   Add rational expressions having different denominators.

Review these examples for Objective 2:

2.   Add. Write each answer in lowest terms.

a. $\dfrac{5}{18}+\dfrac{7}{24}$

Step 1 Find the LCD.

$$18 = 2\cdot3\cdot3 = 2\cdot3^2$$

$$24 = 2\cdot2\cdot2\cdot3 = 2^3\cdot3$$

$$\text{LCD} = 2^3\cdot3^2 = 72$$

Step 2 Now write each expression as an equivalent expression with the LCD as the denominator.

$$\frac{5}{18}+\frac{7}{24} = \frac{5(4)}{18(4)}+\frac{7(3)}{24(3)}$$

$$= \frac{20}{72}+\frac{21}{72}$$

Step 3 Add the numerators.
Step 4 Write in lowest terms, if necessary.

$$= \frac{20+21}{72}$$

$$= \frac{41}{72}$$

Now Try:

2.   Add. Write each answer in lowest terms.

a. $\dfrac{9}{35}+\dfrac{8}{45}$

____

          Copyright © 2025 Pearson Education, Inc.

b. $\dfrac{5}{6z}+\dfrac{7}{9z}$

  Step 1 Find the LCD.
  $$6z = 2\cdot 3\cdot z$$
  $$9z = 3\cdot 3\cdot z = 3^2\cdot z$$
  $$\text{LCD} = 2\cdot 3^2\cdot z = 18z$$

  Step 2 Now write each expression as an equivalent expression with the LCD as the denominator.
  $$\frac{5}{6z}+\frac{7}{9z}=\frac{5(3)}{6z(3)}+\frac{7(2)}{9z(2)}$$
  $$=\frac{15}{18z}+\frac{14}{18z}$$

  Step 3 Add the numerators.
  Step 4 Write in lowest terms, if necessary.
  $$=\frac{15+14}{18z}$$
  $$=\frac{29}{18z}$$

b. $\dfrac{4}{9y}+\dfrac{2}{7y}$

  ____

4. Add. Write the answer in lowest terms.
  $$\frac{3x}{x^2-x-20}+\frac{5}{x^2-2x-15}$$

  The LCD is $(x+3)(x+4)(x-5)$.
  $$=\frac{3x}{(x+4)(x-5)}+\frac{5}{(x+3)(x-5)}$$
  $$=\frac{3x(x+3)}{(x+3)(x+4)(x-5)}+\frac{5(x+4)}{(x+3)(x+4)(x-5)}$$
  $$=\frac{3x(x+3)+5(x+4)}{(x+3)(x+4)(x-5)}$$
  $$=\frac{3x^2+9x+5x+20}{(x+3)(x+4)(x-5)}$$
  $$=\frac{3x^2+14x+20}{(x+3)(x+4)(x-5)}$$

4. Add. Write the answer in lowest terms.
  $$\frac{7x}{x^2-2x-8}+\frac{4}{x^2-3x-4}$$

  ____

Objective 2 Practice Exercises

For extra help, see Examples 2–5 on pages 437–438 of your text.

Add. Write each answer in lowest terms.

4. $\dfrac{7}{x-5}+\dfrac{4}{x+5}$

4. ______________

5. $\dfrac{3z}{z^2-4}+\dfrac{4z-3}{z^2-4z+4}$

5. _______________

6. $\dfrac{4z}{z^2+6z+8}+\dfrac{2z-1}{z^2+5z+6}$

6. _______________

Objective 3   Subtract rational expressions.

Review these examples for Objective 3:

8. Subtract. Write the answer in lowest terms.

$$\frac{5x}{x-7}-\frac{x-42}{7-x}$$

The denominators are opposites.

$$\frac{5x}{x-7}-\frac{x-42}{7-x}=\frac{5x}{x-7}-\frac{(x-42)(-1)}{(7-x)(-1)}$$

$$=\frac{5x}{x-7}-\frac{-x+42}{x-7}$$

$$=\frac{5x+x-42}{x-7}$$

$$=\frac{6x-42}{x-7}$$

$$=\frac{6(x-7)}{x-7}$$

$$=6$$

Now Try:

8. Subtract. Write the answer in lowest terms.

$$\frac{4x}{x-9}-\frac{3x-63}{9-x}$$

____

9. Subtract. Write the answer in lowest terms.

$$\frac{7x}{x^2-4x+4}-\frac{2}{x^2-4}$$

$$=\frac{7x}{(x-2)^2}-\frac{2}{(x+2)(x-2)}$$

The LCD is $(x+2)(x-2)^2$.

$$=\frac{7x}{(x-2)^2}-\frac{2}{(x+2)(x-2)}$$

$$=\frac{7x(x+2)}{(x+2)(x-2)^2}-\frac{2(x-2)}{(x+2)(x-2)^2}$$

$$=\frac{7x(x+2)-2(x-2)}{(x+2)(x-2)^2}$$

$$=\frac{7x^2+14x-2x+4}{(x+2)(x-2)^2}$$

$$=\frac{7x^2+12x+4}{(x+2)(x-2)^2}$$

9. Subtract. Write the answer in lowest terms.

$$\frac{8x}{x^2-10x+25}-\frac{3}{x^2-25}$$

____

Objective 3 Practice Exercises

For extra help, see Examples 6–10 on pages 439–441 of your text.

Subtract. Write each answer in lowest terms.

7. $\dfrac{z+2}{z-2}-\dfrac{z-2}{z+2}$

7. __________________

8. $\dfrac{-4}{x^2-4}-\dfrac{3}{4-2x}$

8. __________________

9. $\dfrac{m}{m^2-4}-\dfrac{1-m}{m^2+4m+4}$

9. __________________

Chapter 6 RATIONAL EXPRESSIONS AND APPLICATIONS

6.5   Complex Fractions

Learning Objectives
Learning Objectives
1   Define and recognize a complex fraction.
2   Simplify a complex fraction by writing it as a division problem (Method 1).
3   Simplify a complex fraction by multiplying numerator and denominator by the LCD (Method 2).
4   Simplify rational expressions with negative exponents.

Key Terms

Use the vocabulary terms listed below to complete each statement in exercises 1−2.

       complex fraction   **LCD**

1.   A __________________________ is a rational expression with one or more fractions in the numerator, denominator, or both.

2.   To simplify a complex fraction, multiply the numerator and denominator by the __________________________ of all the fractions within the complex fraction.

Objective 1   Define and recognize a complex fraction.

Objective 2   Simplify a complex fraction by writing it as a division problem (Method 1).

Review these examples for Objective 2:

1. Simplify the complex fraction.

$$\dfrac{8+\dfrac{4}{x}}{\dfrac{x}{6}+\dfrac{1}{12}}$$

Step 1 Write the numerator as a single fraction.

$$8+\frac{4}{x}=\frac{8}{1}+\frac{4}{x}=\frac{8x}{x}+\frac{4}{x}=\frac{8x+4}{x}$$

Do the same with each denominator.

$$\frac{x}{6}+\frac{1}{12}=\frac{x(2)}{6(2)}+\frac{1}{12}=\frac{2x}{12}+\frac{1}{12}=\frac{2x+1}{12}$$

Step 2 Write the equivalent complex fraction as a division problem.

$$\dfrac{\dfrac{8x+4}{x}}{\dfrac{2x+1}{12}}=\frac{8x+4}{x}\cdot\frac{2x+1}{12}$$

Now Try:

1. Simplify the complex fraction.

$$\dfrac{9+\dfrac{3}{x}}{\dfrac{x}{10}+\dfrac{1}{30}}$$

____

              Copyright © 2025 Pearson Education, Inc.

Step 3 Divide by multiplying by the reciprocal.

$$\frac{8x+4}{x} \div \frac{2x+1}{12} = \frac{8x+4}{x} \cdot \frac{12}{2x+1}$$

$$= \frac{4(2x+1)}{x} \cdot \frac{12}{2x+1}$$

$$= \frac{48}{x}$$

2. Simplify the complex fraction.

$$\frac{\dfrac{rs^2}{t^3}}{\dfrac{s^3}{r^3 t}}$$

Use the definition of division and then the fundamental property.

$$= \frac{rs^2}{t^3} \div \frac{s^3}{r^3 t}$$

$$= \frac{rs^2}{t^3} \cdot \frac{r^3 t}{s^3}$$

$$= \frac{r^4}{st^2}$$

3. Simplify the complex fraction.

$$\frac{\dfrac{30}{x+4} - 6}{\dfrac{4}{x+4} + 1} = \frac{\dfrac{30}{x+4} - \dfrac{6(x+4)}{x+4}}{\dfrac{4}{x+4} + \dfrac{1(x+4)}{x+4}}$$

$$= \frac{\dfrac{30 - 6(x+4)}{x+4}}{\dfrac{4 + 1(x+4)}{x+4}}$$

$$= \frac{\dfrac{30 - 6x - 24}{x+4}}{\dfrac{4 + x + 4}{x+4}}$$

$$= \frac{\dfrac{6 - 6x}{x+4}}{\dfrac{x+8}{x+4}}$$

$$= \frac{6 - 6x}{x+4} \cdot \frac{x+4}{x+8}$$

$$= \frac{6 - 6x}{x+8}$$

2. Simplify the complex fraction.

$$\frac{\dfrac{a^3 b^2}{c}}{\dfrac{a^5 b}{c^3}}$$

____

3. Simplify the complex fraction.

$$\frac{\dfrac{20}{x-5} - 9}{\dfrac{5}{x-5} + 2}$$

____

Name: Date:
Instructor: Section:

Objective 2 Practice Exercises

For extra help, see Examples 1–3 on pages 445–447 of your text.

Simplify each complex fraction by writing it as a division problem.

1. $\dfrac{\dfrac{49m^3}{18n^5}}{\dfrac{21m}{27n^2}}$

1. _______________________

2. $\dfrac{\dfrac{p}{2}-\dfrac{1}{3}}{\dfrac{p}{3}+\dfrac{1}{6}}$

2. _______________________

3. $\dfrac{3+\dfrac{4}{s}}{2s+\dfrac{2}{3}}$

3. _______________________

Objective 3   Simplify a complex fraction by multiplying numerator and denominator by the LCD (Method 2).

Review these examples for Objective 3:

4. Simplify the complex fraction.

$$\dfrac{12+\dfrac{4}{x}}{\dfrac{x}{5}+\dfrac{1}{15}}$$

Step 1 Find the LCD for all the denominators.
   The LCD for x, 5, and 15 is $15x$.
Step 2 Multiply the numerator and denominator
of the complex fraction by the LCD.

Now Try:

4. Simplify the complex fraction.

$$\dfrac{4+\dfrac{2}{x}}{\dfrac{x}{3}+\dfrac{1}{6}}$$

____

$$\frac{12+\dfrac{4}{x}}{\dfrac{x}{5}+\dfrac{1}{15}}=\frac{15x\left(12+\dfrac{4}{x}\right)}{15x\left(\dfrac{x}{5}+\dfrac{1}{15}\right)}$$

$$=\frac{180x+60}{3x^2+x}$$

$$=\frac{60(3x+1)}{x(3x+1)}$$

$$=\frac{60}{x}$$

5. Simplify the complex fraction.

$$\frac{\dfrac{9}{7n}-\dfrac{3}{n^2}}{\dfrac{8}{3n}+\dfrac{5}{6n^2}}$$

The LCD is $42n^2$.

$$=\frac{42n^2\left(\dfrac{9}{7n}-\dfrac{3}{n^2}\right)}{42n^2\left(\dfrac{8}{3n}+\dfrac{5}{6n^2}\right)}$$

$$=\frac{42n^2\left(\dfrac{9}{7n}\right)-42n^2\left(\dfrac{3}{n^2}\right)}{42n^2\left(\dfrac{8}{3n}\right)+42n^2\left(\dfrac{5}{6n^2}\right)}$$

$$=\frac{54n-126}{112n+35},\ \ \text{or}\ \ \frac{18(3n-7)}{7(16n+5)}$$

5. Simplify the complex fraction.

$$\frac{\dfrac{2}{9n}-\dfrac{2}{5n^2}}{\dfrac{4}{5n}+\dfrac{2}{3n^2}}$$

____

Objective 3 Practice Exercises

For extra help, see Examples 4–6 on pages 448–450 of your text.

Simplify each complex fraction by multiplying numerator and denominator by the least common denominator.

4. $\dfrac{\dfrac{9}{x^2}-1}{\dfrac{3}{x}-1}$

4. _______________

5. $\dfrac{\dfrac{x-2}{x+2}}{\dfrac{x}{x-2}}$

5. ___________

6. $\dfrac{\dfrac{6}{k+1}-\dfrac{5}{k-3}}{\dfrac{3}{k-3}+\dfrac{2}{k+2}}$

6. ___________

Objective 4   Simplify rational expressions with negative exponents.

Review this example for Objective 4:

7. Simplify the expression using only positive exponents in the answer.

$$\frac{3x^{-2}+5x^{-3}}{10x^{-1}+6}$$

Write with positive exponents. Use Method 2 to simplify.

$$\frac{3x^{-2}+5x^{-3}}{10x^{-1}+6}=\frac{\dfrac{3}{x^2}+\dfrac{5}{x^3}}{\dfrac{10}{x}+6}$$

$$=\frac{x^3\left(\dfrac{3}{x^2}+\dfrac{5}{x^3}\right)}{x^3\left(\dfrac{10}{x}+6\right)}$$

$$=\frac{x^3\cdot\dfrac{3}{x^2}+x^3\cdot\dfrac{5}{x^3}}{x^3\cdot\dfrac{10}{x}+x^3\cdot6}$$

$$=\frac{3x+5}{10x^2+6x^3}$$

Now Try:

7. Simplify the expression using only positive exponents in the answer.

$$\frac{1+x^{-1}}{(x+y)^{-1}}$$

____

Objective 4 Practice Exercises

For extra help, see Example 7 on pages 450–451 of your text.

Simplify each expression using only positive exponents in the answer.

7. $\dfrac{2x^{-1}+y^2}{z^{-3}}$ 7. _________________

8. $\left(a^{-2}-b^{-2}\right)^{-1}$ 8. _________________

9. $\dfrac{4x^{-2}}{2+6y^{-3}}$ 9. _________________

Chapter 6 RATIONAL EXPRESSIONS AND APPLICATIONS

6.6   Solving Equations with Rational Expressions

Learning Objectives	
1	Distinguish between operations with rational expressions and equations with terms that are rational expressions.
2	Solve equations with rational expressions.
3	Solve a formula for a specified variable.

Key Terms

Use the vocabulary terms listed below to complete each statement in exercises 1−2.

proposed solution   **extraneous solution**

1.   A solution that is not an actual solution of a given equation is called a(n)

_________________________.

2.   A value of the variable that appears to be a solution after both sides of an equation with rational expressions are multiplied by a variable expression is called a(n)

_________________________.

Objective 1   Distinguish between operations with rational expressions and equations with terms that are rational expressions.

Review these examples for Objective 1:

1.   Identify each of the following as an expression or an equation. Then simplify the expression or solve the equation.

a. $\dfrac{8}{9}x - \dfrac{5}{6}x$

This is a difference of two terms. It represents an expression since there is no equality symbol. The LCD is 18.

$$\frac{8}{9}x - \frac{5}{6}x = \frac{2 \cdot 8}{2 \cdot 9}x - \frac{3 \cdot 5}{3 \cdot 6}x$$

$$= \frac{16}{18}x - \frac{15}{18}x$$

$$= \frac{1}{18}x$$

Now Try:

1.   Identify each of the following as an expression or an equation. Then simplify the expression or solve the equation.

a. $\dfrac{4}{5}x - \dfrac{3}{10}x$

____

                 Copyright © 2025 Pearson Education, Inc.

b. $\dfrac{8}{9}x - \dfrac{5}{6}x = \dfrac{2}{3}$

Because there is an equality symbol, this is an equation to be solved. The LCD is 18.

$$18\left(\dfrac{8}{9}x - \dfrac{5}{6}x\right) = 18\left(\dfrac{2}{3}\right)$$

$$18\left(\dfrac{8}{9}x\right) - 18\left(\dfrac{5}{6}x\right) = 18\left(\dfrac{2}{3}\right)$$

$$16x - 15x = 12$$

$$x = 12$$

A check shows that the solution set is $\{12\}$.

b. $\dfrac{4}{5}x - \dfrac{3}{10}x = 7$

____

Objective 1 Practice Exercises

For extra help, see Example 1 on pages 454–455 of your text.

Identify each of the following as an expression or an equation. Then simplify the expression or solve the equation.

1. $\dfrac{3x}{5} - \dfrac{4x}{3} = \dfrac{22}{15}$

1. _______________

2. $\dfrac{4x}{5} - \dfrac{5x}{10}$

2. _______________

3. $\dfrac{2x}{5} + \dfrac{7x}{3}$

3. _______________

Objective 2   Solve equations with rational expressions.

Review this example for Objective 2:

4. Solve, and check the proposed solution.

$$\dfrac{x}{x-3} - \dfrac{3}{x-3} = 3$$

Note that x cannot equal 3, since 3 causes both denominators to equal 0.

Now Try:

4. Solve, and check the proposed solution.

$$\dfrac{8x}{x+1} + \dfrac{8}{x+1} = 4$$

____

Multiply by the LCD, $x-3$.

$$(x-3)\left(\frac{x}{x-3}-\frac{3}{x-3}\right)=(x-3)(3)$$

$$(x-3)\left(\frac{x}{x-3}\right)-(x-3)\left(\frac{3}{x-3}\right)=(x-3)(3)$$

$$x-3=3x-9$$

$$-2x-3=-9$$

$$-2x=-6$$

$$x=3$$

Check $\quad \dfrac{x}{x-3}-\dfrac{3}{x-3}=3$

$$\frac{3}{3-3}-\frac{3}{3-3}\overset{?}{=}3$$

$$\frac{3}{0}-\frac{3}{0}\overset{?}{=}3$$

Division by 0 is undefined.

Thus, the proposed solution 3 must be rejected, and the solution set is $\varnothing$.

7. Solve, and check the proposed solution(s).

$$\frac{6}{x^2-1}=1-\frac{3}{x+1}$$

$x\neq 1,-1$ or a denominator is 0.
Factor the denominator.

$$\frac{6}{(x+1)(x-1)}=1-\frac{3}{x+1}$$

The LCD is $(x+1)(x-1)$.

$$(x+1)(x-1)\frac{6}{(x+1)(x-1)}$$

$$=(x+1)(x-1)\left(1-\frac{3}{x+1}\right)$$

$$(x+1)(x-1)\frac{6}{(x+1)(x-1)}$$

$$=(x+1)(x-1)-(x+1)(x-1)\frac{3}{x+1}$$

$$6=(x+1)(x-1)-3(x-1)$$

$$6=x^2-1-3x+3$$

$$0=x^2-3x-4$$

$$0=(x+1)(x-4)$$

$$x+1=0 \quad \text{or} \quad x-4=0$$

$$x=-1 \quad \text{or} \quad x=4$$

7. Solve, and check the proposed solution(s).

$$\frac{x}{x+1}+\frac{4}{x}=\frac{4}{x^2+x}$$

____

Since -1 makes the original denominator equal 0, the proposed solution -1 is an extraneous value.

Check $\dfrac{6}{x^2-1}=1-\dfrac{3}{x+1}$

$$\dfrac{6}{4^2-1}\overset{?}{=}1-\dfrac{3}{4+1}$$

$$\dfrac{2}{5}=\dfrac{2}{5}\quad\text{True}$$

A check shows that $\{4\}$ is the solution set.

Objective 2 Practice Exercises

For extra help, see Examples 2–8 on pages 455–460 of your text.

Solve each equation and check your solutions.

4. $\dfrac{4}{n+2}-\dfrac{2}{n}=\dfrac{1}{6}$

4. _________________

5. $\dfrac{x}{3x+16}=\dfrac{4}{x}$

5. _________________

6. $\dfrac{-16}{n^2-8n+12}=\dfrac{3}{n-2}+\dfrac{n}{n-6}$

6. _________________

Objective 3   Solve a formula for a specified variable.

Review these examples for Objective 3: | **Now Try:**

9. Solve the following formula $r = \dfrac{s+w}{v}$ for w.

Isolate w. Multiply by v.

$$r = \frac{s+w}{v}$$

$$rv = s+w$$

$$rv - s = w$$

Check $r = \dfrac{s+w}{v}$

$$r \overset{?}{=} \frac{s+rv-s}{v}$$

$$r \overset{?}{=} \frac{rv}{v}$$

$$r = r \quad \text{True}$$

9. Solve the following formula
$a = \dfrac{b-c}{q}$ for b.

____

10. Solve the following formula $S = \dfrac{a_1}{1-r}$ for r.

Isolate r. Multiply by $1 - r$.

$$S = \frac{a_1}{1-r}$$

$$S(1-r) = a_1$$

$$S - rS = a_1$$

$$-rS = a_1 - S$$

$$r = \frac{a_1 - S}{-S}, \quad \text{or} \quad \frac{S - a_1}{S}$$

10. Solve the following formula
$w = \dfrac{x}{y+z}$ for y.

____

11. Solve the formula $\dfrac{1}{p} + \dfrac{1}{q} = \dfrac{1}{r}$ for p.

Isolate p, the specified variable. Multiply by the LCD, pqr.

$$pqr\left(\frac{1}{p} + \frac{1}{q}\right) = pqr\left(\frac{1}{r}\right)$$

$$pqr\left(\frac{1}{p}\right) + pqr\left(\frac{1}{q}\right) = pqr\left(\frac{1}{r}\right)$$

$$qr + pr = pq$$

$$qr = pq - pr$$

$$qr = p(q-r)$$

$$\frac{qr}{q-r} = p$$

11. Solve the formula $\dfrac{1}{x} = \dfrac{1}{y} + \dfrac{1}{z}$
for x.

____

                        Copyright © 2025 Pearson Education, Inc.

Objective 3 Practice Exercises

For extra help, see Examples 9–11 on pages 461–462 of your text.

Solve each formula for the specified variable.

7. $\dfrac{1}{f} = \dfrac{1}{d_0} + \dfrac{1}{d_1}$ for f 7. ________________

8. $m = \dfrac{y_2 - y_1}{x_2 - x_1}$ for y_1 8. ________________

9. $A = \dfrac{2pf}{b(q+1)}$ for q 9. ________________

Chapter 6 RATIONAL EXPRESSIONS AND APPLICATIONS

6.7   Applications of Rational Expressions

Learning Objectives
1   Solve problems about numbers.
2   Solve problems about distance, rate, and time.
3   Solve problems about work.

Key Terms

Use the vocabulary terms listed below to complete each statement in exercises 1–3.

      reciprocal   **numerator**   **denominator**

1.   In the fraction $\dfrac{x+5}{x-2}$, $x+5$ is the _______________________________.

2.   In the fraction $\dfrac{x+5}{x-2}$, $x-2$ is the _______________________________.

3.   The fraction $\dfrac{x+5}{x-2}$ is the _____________________ of the fraction $\dfrac{x-2}{x+5}$.

Objective 1   Solve problems about numbers.

Review this example for Objective 1:

1.   If a certain number is added to the numerator and twice that number is subtracted from the denominator of the fraction $\dfrac{3}{5}$, the result is equal to 5. Find the number.

Step 1 Read the problem carefully. We are trying to find a number.

Step 2 Assign a variable.
    Let $x =$ the number.

Step 3 Write an equation. The fraction $\dfrac{3+x}{5-2x}$ represents adding the number to the numerator and twice that number is subtracted from the denominator of the fraction $\dfrac{3}{5}$. The result is equal to 5.

$$\frac{3+x}{5-2x}=5$$

Now Try:

1.   If the same number is added to the numerator and denominator of the fraction $\dfrac{5}{9}$, the value of the resulting fraction is $\dfrac{2}{3}$. Find the number.

    ____

Step 4 Solve. Multiply by the LCD, $5-2x$.

$$(5-2x)\frac{3+x}{5-2x} = (5-2x)5$$

$$3+x = 25-10x$$

$$11x = 22$$

$$x = 2$$

Step 5 State the answer. The number is 2.

Step 6 Check the solution in the original problem. If 2 is added to the numerator, and twice 2 is subtracted from the denominator of $\frac{3}{5}$,

the result is $\frac{3+2}{5-2(2)} = \frac{5}{1}$, or 5, as required.

Objective 1 Practice Exercises

For extra help, see Example 1 on page 468 of your text.

Solve each problem. Check your answers to be sure they are reasonable.

1. If three times a number is subtracted from twice its
    reciprocal, the result is -1. Find the number.

    1. ______________

2. If two times a number is added to one-half of its
    reciprocal, the result is $\frac{13}{6}$. Find the number.

    2. ______________

3. The denominator of a fraction is 1 less than twice the
    numerator. If the numerator and the denominator are
    each increased by 3, the resulting fraction simplifies
    to $\frac{3}{4}$. Find the original fraction.

    3. ______________

Objective 2   Solve problems about distance, rate, and time.

Review this example for Objective 2:	Now Try:

Review this example for Objective 2:

2. A boat goes 6 miles per hour in still water. It takes as long to go 40 miles upstream as 80 miles downstream. Find the speed of the current.

Step 1 Read the problem carefully. Find the speed of the current.

Step 2 Assign a variable.
Let x = the speed of the current.
The rate of traveling upstream is $6 - x$.
The rate of traveling downstream is $6 + x$.

	d	r	t
Upstream	40	$6 - x$	$\dfrac{40}{6 - x}$
Downstream	80	$6 + x$	$\dfrac{80}{6 + x}$

The times are equal.

Step 3 Write an equation.
$$\frac{40}{6 - x} = \frac{80}{6 + x}$$

Step 4 Solve. The LCD is $(6 + x)(6 - x)$.
$$(6 + x)(6 - x)\frac{40}{6 - x} = (6 + x)(6 - x)\frac{80}{6 + x}$$
$$40(6 + x) = 80(6 - x)$$
$$240 + 40x = 480 - 80x$$
$$240 + 120x = 480$$
$$120x = 240$$
$$x = 2$$

Step 5 State the answer. The speed of the current is 2 miles per hour.

Step 6 Check.

Upstream: $\dfrac{40}{6 - 2} = 10$ hr

Downstream: $\dfrac{80}{6 + 2} = 10$ hr

The time upstream is the same as the time downstream, as required.

Now Try:

2. The Cuyahoga River has a current of 2 miles per hour. Ali can paddle 10 miles downstream in the time it takes her to paddle 2 miles upstream. How fast can Ali paddle?

____

Objective 2 Practice Exercises

For extra help, see Example 2 on pages 469–470 of your text.

Solve each problem.

4. A boat travels 15 miles per hour in still water. The boat travels 20 miles downstream in the same time it takes the boat to travel 10 miles upstream. How fast is the current?

    4. ________________

5. A ship goes 120 miles downriver in $2\frac{2}{3}$ hours less than it takes to go the same distance upriver. If the speed of the current is 6 miles per hour, find the speed of the ship.

    5. ________________

6. On Saturday, Pablo jogged 6 miles. On Monday, jogging at the same speed, it took him 30 minutes longer to cover 10 miles. How fast did Pablo jog?

    6. ________________

Objective 3   Solve problems about work.

Review this example for Objective 3:

3. Skip can paint a house in 8 hours and Phil can paint a house in 12 hours. How long will it take to paint a house if they work together?

Step 1 Read the problem carefully. Find the time working together.

Step 2 Assign a variable.
Let x = the number of hours working together.

The rate for Skip is $\frac{1}{8}$.

The rate for Phil is $\frac{1}{12}$.

Step 3 Write an equation. The sum of the fractional part for each multiplied by the time working together is the whole job.

$$\frac{1}{8}x + \frac{1}{12}x = 1$$

Step 4 Solve. The LCD is 24.

$$24\left(\frac{1}{8}x + \frac{1}{12}x\right) = 24(1)$$

$$24\left(\frac{1}{8}x\right) + 24\left(\frac{1}{12}x\right) = 24$$

$$3x + 2x = 24$$

$$5x = 24$$

$$x = \frac{24}{5} = 4\frac{4}{5}$$

Step 5 State the answer. Working together, it takes $4\frac{4}{5}$ hours to paint the house.

Step 6 Check. Substitute $\frac{24}{5}$ for x in the equation from Step 3.

$$\frac{1}{8}x + \frac{1}{12}x = 1$$

$$\frac{1}{8}\left(\frac{24}{5}\right) + \frac{1}{12}\left(\frac{24}{5}\right) = 1$$

$$\frac{3}{5} + \frac{2}{5} = 1 \quad \text{True}$$

Now Try:

3. Chuck can weed the garden in $\frac{1}{2}$ hour, but David takes 2 hours. How long does it take them to weed the garden if they work together?

____

Name: Date:
Instructor: Section:

Objective 3 Practice Exercises

For extra help, see Example 3 on pages 471–472 of your text.

Solve each problem.

7. Kelly can clean the house in 6 hours, but it takes
    Linda 4 hours. How long would it take them to clean
    the house if they worked together?

7. _____________________

8. Michael can type twice as fast as Sharon. Together
    they can type a certain job in 2 hours. How long
    would it take Michael to type the entire job by
    himself?

8. _____________________

Chapter 7 LINEAR EQUATIONS, GRAPHS, AND SYSTEMS

7.1 Review of Graphs and Slopes of Lines

Learning Objectives	
1	Plot ordered pairs.
2	Graph lines and find intercepts.
3	Graph equations of horizontal and vertical lines.
4	Find the midpoint of a line segment.
5	Find the slope of a line.
6	Graph a line given its slope and a point on the line.
7	Determine whether two lines are parallel, perpendicular, or neither using slope.
8	Solve problems involving average rate of change.

Key Terms

Use the vocabulary terms listed below to complete each statement in exercises 1−17.

ordered pair	origin	x-axis	y-axis
rectangular (Cartesian) coordinate system		plot	
components	coordinate	quadrant	graph of an equation
first-degree equation		linear equation in two variables	
y-intercept	x-intercept rise	run	slope

1. If a graph intersects the y-axis at k, then the ___________________ is $(0, k)$.

2. An equation that can be written in the form $Ax + By = C$, where A, B, and C are real numbers and A, $B \neq 0$, is called a ___________________________________.

3. Each number in an ordered pair represents a _______________________ of the corresponding point.

4. The axis lines in a coordinate system intersect at the ___________________.

5. If a graph intersects the x-axis at k, then the ___________________ is $(k, 0)$.

6. In a rectangular coordinate system, the horizontal number line is called the

    ___________________________.

7. A _______________________________ is one of the four regions in the plane determined by a rectangular coordinate system.

8. A pair of numbers written between parentheses in which order is important is called a(n) ___________________________.

9. In a rectangular coordinate system, the vertical number line is called the

    ___________________________.

10. The two numbers in an ordered pair are the _________________________ of the ordered pair.

11. To _________________ an ordered pair is to locate the corresponding point on a coordinate system.

12. Together, the x-axis and the y-axis form a _________________________________.

13. The _________________ of a line is the ratio of the change in y compared to the change in x when moving along the line from one point to another.

14. The _____________________________ is the set of points corresponding to all ordered pairs that satisfy the equation.

15. A _____________________________ has no term with a variable to a power greater than one.

16. The vertical change between two different points on a line is called the _____________________.

17. The horizontal change between two different points on a line is called the _____________________.

Objective 1   Plot ordered pairs.

For extra help, see page 490 of your text.

Objective 2   Graph lines and find intercepts.

Review these examples for Objective 2:

1. Find the x- and y-intercepts of $3x + y = 6$ and graph the equation.

To find the y-intercept, let $x = 0$.
To find the x-intercept, let $y = 0$.

$$3(0) + y = 6 \quad | \quad 3x + 0 = 6$$
$$0 + y = 6 \quad | \quad 3x = 6$$
$$y = 6 \quad | \quad x = 2$$

The intercepts are $(0, 6)$ and $(2, 0)$. To find a third point, as a check, we let $x = 1$.

$$3(1) + y = 6$$
$$3 + y = 6$$
$$y = 3$$

This gives the ordered pair $(1, 3)$.

Now Try:

1. Find the x- and y-intercepts of $5x - 2y = -10$ and graph the equation.

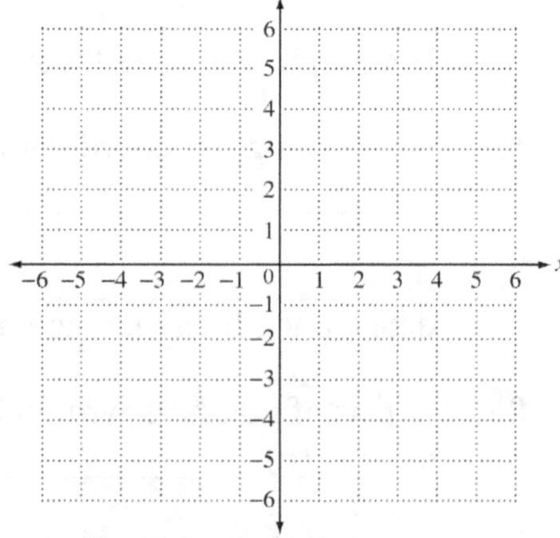

           Copyright © 2025 Pearson Education, Inc.

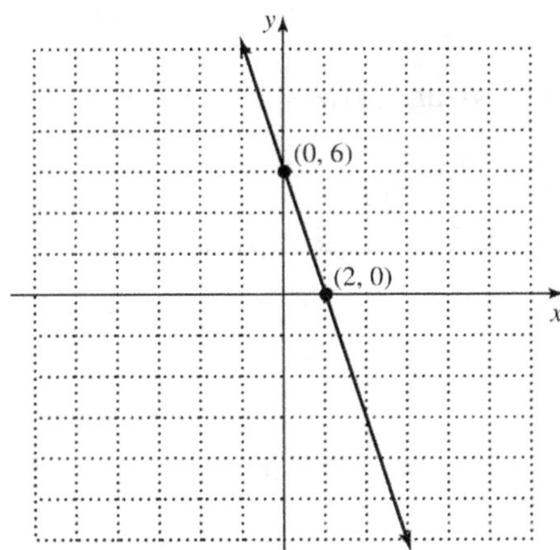

2. Graph $x + 5y = 0$.

To find the y-intercept, let $x = 0$.
To find the x-intercept, let $y = 0$.

$$0 + 5y = 0 \quad\mid\quad x + 5(0) = 0$$
$$5y = 0 \quad\mid\quad x + 0 = 0$$
$$y = 0 \quad\mid\quad x = 0$$

The x- and y-intercepts are the same point $(0, 0)$. We must select two other values for x or y to find two other points. We choose $y = 1$ and $y = -1$.

$$x + 5(1) = 0 \quad\mid\quad x + 5(-1) = 0$$
$$x + 5 = 0 \quad\mid\quad x - 5 = 0$$
$$x = -5 \quad\mid\quad x = 5$$

We use $(-5, 1)$, $(0, 0)$, and $(5, -1)$ to draw the graph.

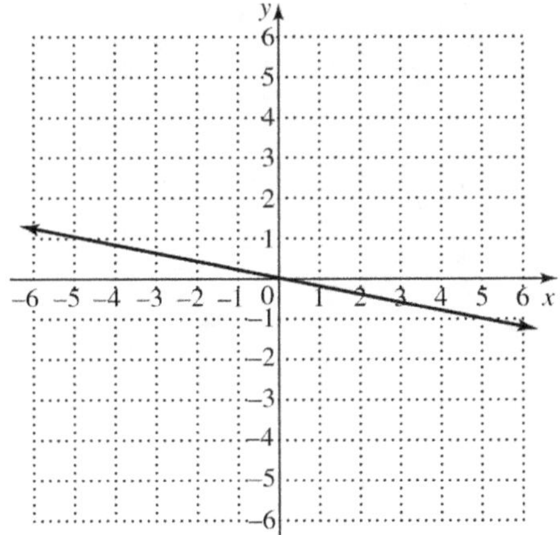

2. Graph $3x - y = 0$.

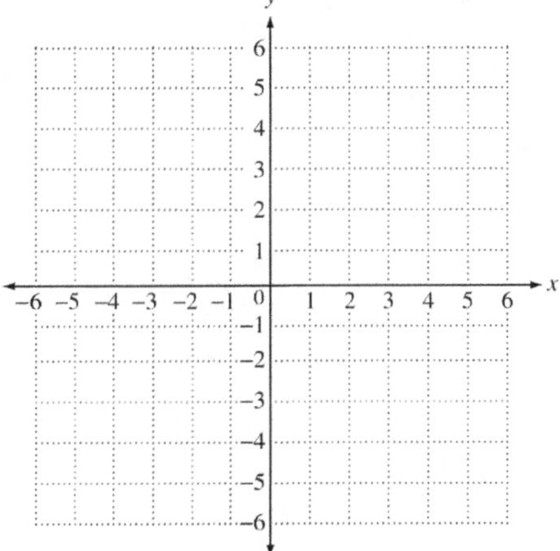

Objective 2 Practice Exercises

For extra help, see Examples 1–2 on pages 492–493 of your text.

Find the intercepts, then graph the equation.

1. $4x - y = 4$

1. _______________________

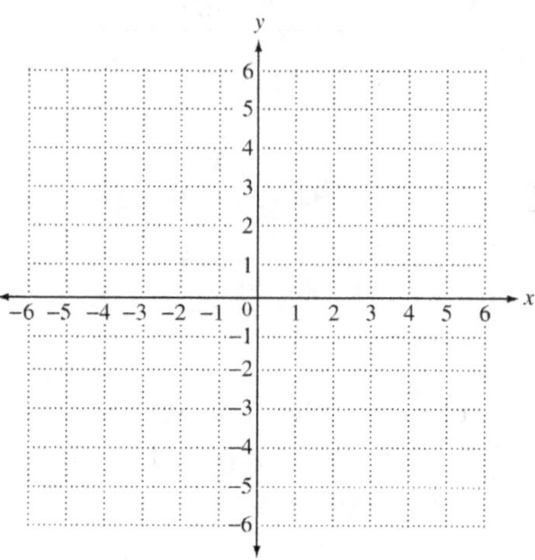

2. $2x - 3y = 6$

2. _______________________

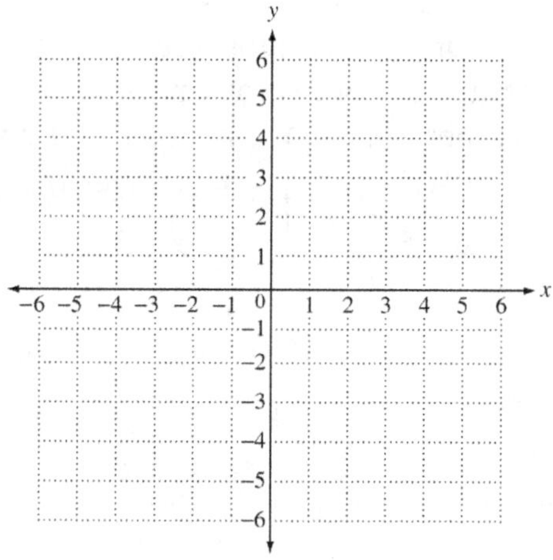

     Copyright © 2025 Pearson Education, Inc.

Objective 3   Graph equations of horizontal and vertical lines.

Review these examples for Objective 3:

3.   Graph each equation.

   a. $y = -2$

   For any value of x, y is always -2. Three ordered pairs that satisfy the equation are $(-4, -2)$, $(0, -2)$ and $(2, -2)$. Drawing a line through these points gives the horizontal line. The y-intercept is $(0, -2)$. There is no x-intercept.

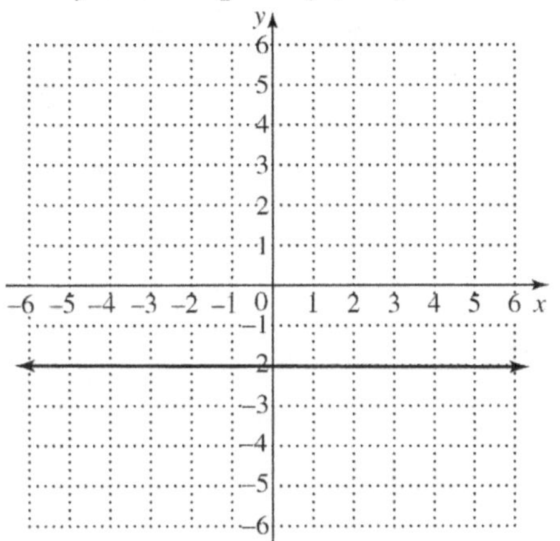

   b. $x + 4 = 0$

   First we subtract 4 from each side of the equation to get the equivalent equation $x = -4$. All ordered-pair solutions of this equation have x-coordinate -4.

   Three ordered pairs that satisfy the equation are $(-4, -1)$, $(-4, 0)$, and $(-4, 3)$. The graph is a vertical line. The x-intercept is $(-4, 0)$. There is no y-intercept.

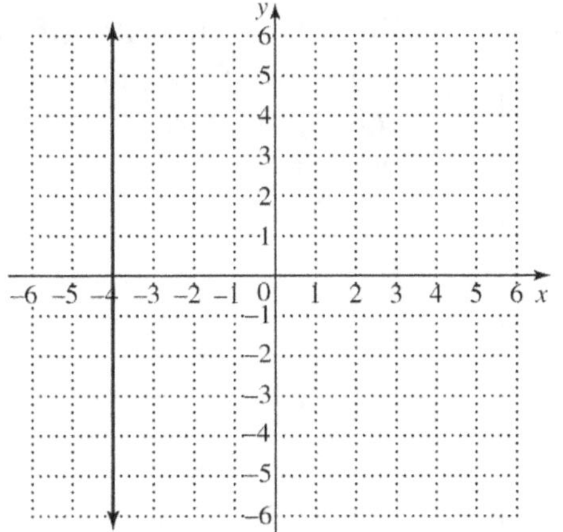

Now Try:

3.   Graph each equation.

   a. $y = 4$

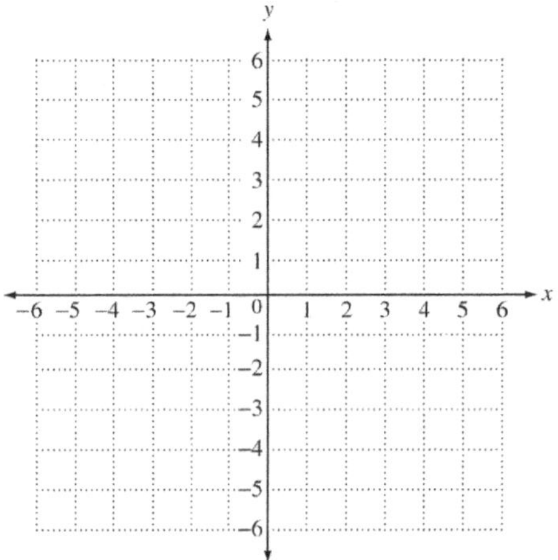

   b. $x = 0$

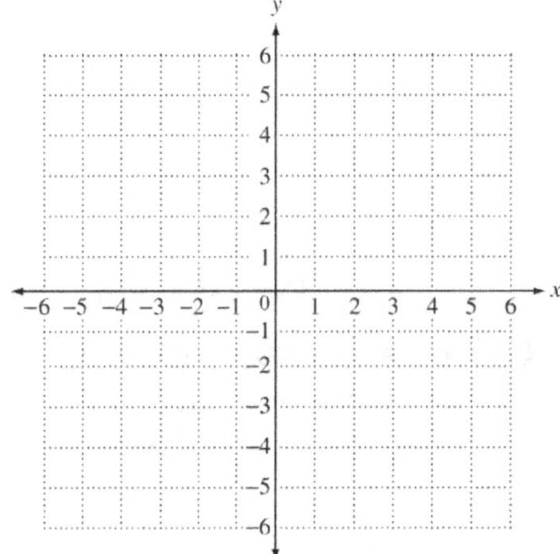

Name: Date:
Instructor: Section:

Objective 3 Practice Exercises

For extra help, see Example 3 on page 493 of your text.

Find the intercepts, and graph the line.

3. $x-1=0$ 3.

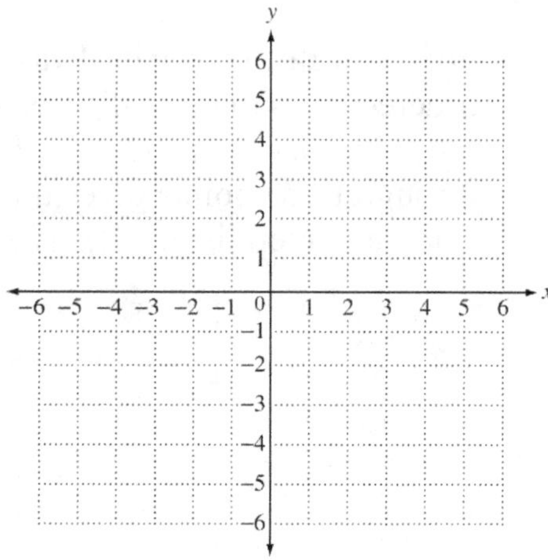

4. $y+3=0$ 4.

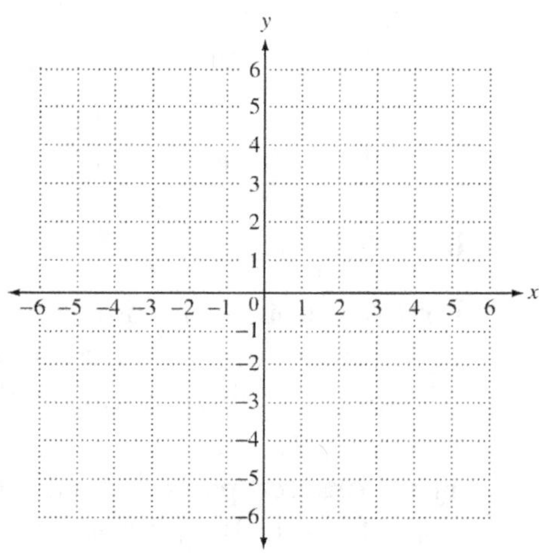

Objective 4   Find the midpoint of a line segment.

Review this example for Objective 4:

4. Find the coordinates of the midpoint of the line segment PQ with endpoints $P(8,-5)$ and $Q(4,-3)$.

$$P(8,-5)=(x_1,\,y_1) \text{ and } Q(4,-3)=(x_2,\,y_2)$$

$$\left(\frac{x_1+x_2}{2},\frac{y_1+y_2}{2}\right)=\left(\frac{8+4}{2},\frac{-5+(-3)}{2}\right)$$

$$=\left(\frac{12}{2},\frac{-8}{2}\right)$$

$$=(6,-4)$$

The midpoint of PQ is $(6,-4)$.

Now Try:

4. Find the coordinates of the midpoint of the line segment PQ with endpoints $P(7,-6)$ and $Q(3,2)$.

____

Objective 4 Practice Exercises

For extra help, see Example 4 on page 495 of your text.

Find the midpoint of each segment with the given endpoints.

5. $(-4, 8)$ and $(8, -4)$

5.   ______________

6. $(8, 5)$ and $(-3, -11)$

6.   ______________

7. $(-2.2, -9.3)$ and $(-8.4, 5.7)$

7.   ______________

Objective 5   Find the slope of a line.

Review this example for Objective 5:

5.   Find the slope of the line passing through $(-5, 4)$ and $(2, -6)$

Apply the slope formula.

$(x_1, y_1) = (-5,\ 4)$ and $(x_2, y_2) = (2, -6)$

$$\text{slope } m = \frac{y_2 - y_1}{x_2 - x_1} = \frac{-6 - 4}{2 - (-5)}$$

$$= \frac{-10}{7}, \text{or } -\frac{10}{7}$$

Now Try:

5.   Find the slope of the line passing through $(-6, 7)$ and $(3, -9)$

____

Objective 5 Practice Exercises

For extra help, see Examples 5–7 on pages 496–497 of your text.

Find the slope of the line through the given points.

8. $(4, 3)$ and $(3, 5)$

8.   ______________

9. $(5, -2)$ and $(2, 7)$

9.   ______________

Find the slope of the graph of the given line.

10. $x + 14y = 35$

10.   ______________

Objective 6   Graph a line given its slope and a point on the line.

Review this example for Objective 6:

8.   Graph the line passing through the point $(1,-3)$, with slope $-\dfrac{5}{2}$.

First, locate the point $(1, -3)$. Then write the slope $-\dfrac{5}{2}$ as

$$\text{slope } m = \frac{\text{change in } y \text{ (rise)}}{\text{change in } x \text{ (run)}} = \frac{5}{-2}.$$

Locate another point on the line by counting up 5 units from $(1,-3)$, and then to the left 2 units. Finally, draw the line through this new point, $(-1, 2)$.

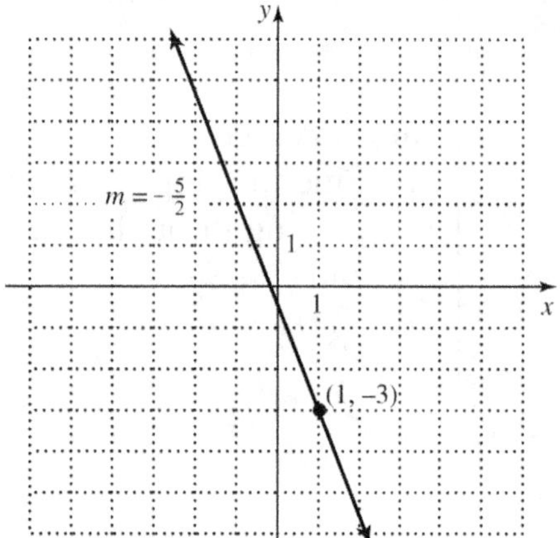

Now Try:

8.   Graph the line passing through the point $(2, 2)$, with slope $\dfrac{1}{3}$.

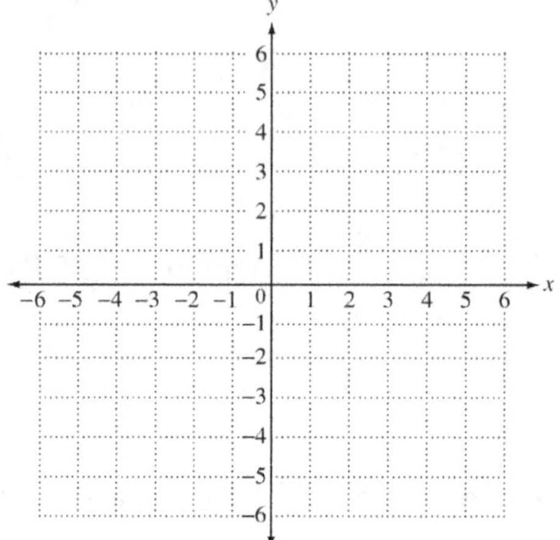

          Copyright © 2025 Pearson Education, Inc.

Objective 6 Practice Exercises

For extra help, see Example 8 on pages 498–499 of your text.

Graph the line passing through the given point and having the given slope.

10. $(4, -2)$; $m = -1$

10.

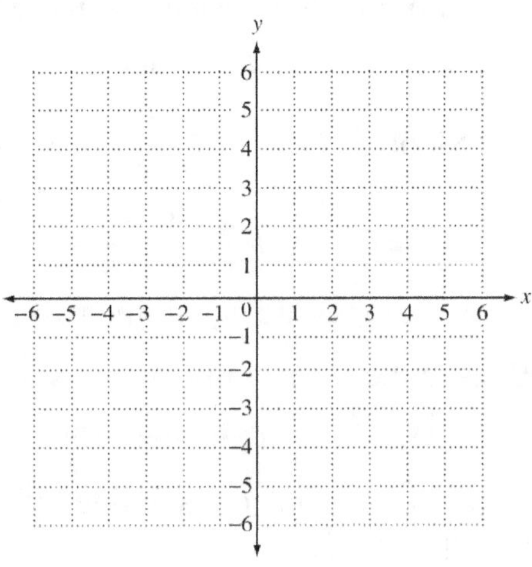

11. $(-3, -2)$; $m = \dfrac{2}{3}$

11.

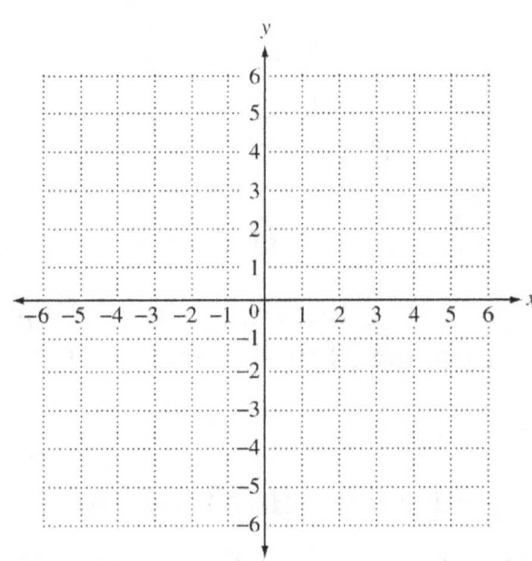

12. $(-3, -1)$; undefined slope

12.

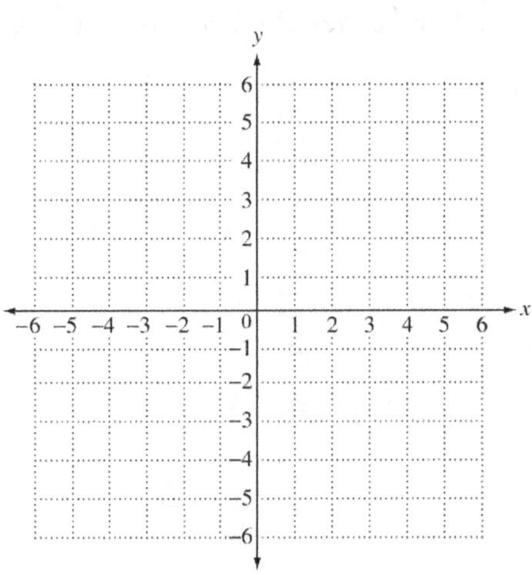

Objective 7   Determine whether two lines are parallel, perpendicular, or neither using slopes.

Review these examples for Objective 7:

9. Determine whether the lines L_1 passing through $(7, 6)$ and $(-4, 2)$ and L_2 passing through $(0,-4)$ and $(11, 0)$, are parallel.

Slope of L_1: $m_1 = \dfrac{2-6}{-4-7} = \dfrac{-4}{-11} = \dfrac{4}{11}$

Slope of L_2: $m_2 = \dfrac{0-(-4)}{11-0} = \dfrac{4}{11}$

The slopes are equal, the two lines are parallel.

10a. Are the lines with equations $x+3y=8$ and $-3x+y=5$ perpendicular?

Find the slope of each line by first solving each equation for y.

$3y=-x+8$ $\Big|$ $y=3x+5$

$y=-\dfrac{1}{3}x+\dfrac{8}{3}$

The slope is $-\dfrac{1}{3}$. $\Big|$ The slope is 3.

Check the product of the slopes: $-\dfrac{1}{3}(3)=-1$.

The two lines are perpendicular because the product of their slopes is -1.

Now Try:

9. Determine whether the lines L_1 passing through $(6, 3)$ and $(-4, 5)$ and L_2 passing through $(0, 3)$ and $(15, 0)$, are parallel.

____

10a. Are the lines with equations $9x-y=7$ and $x+9y=11$ perpendicular?

____

Objective 7 Practice Exercises

For extra help, see Examples 9–10 on page 500 of your text.

Decide whether the lines in each pair are **parallel,** **perpendicular,** *or* **neither.**

13. $y=-5x-2$

      $y=5x+11$

13. _______________

14. $-x+y=-7$

      $x-y=-3$

14. _______________

15. $2x + 2y = 7$

$2x - 2y = 5$

15. _________________

Objective 8 Solve problems involving average rate of change.

Review these examples for Objective 8:

11. A small company had the following sales during their first three years of operation.

Year	Sales
2021	$82,250
2022	$89,790
2023	$96,100

a. What was the rate of change in 2021–2022?
b. What was the rate of change in 2022–2023?
c. What was the rate of change in 2021–2023?

a. We use the ordered pairs (2021, 82,250) and (2022, 89,790).

$$\text{average rate of change} = \frac{89,790 - 82,250}{2022 - 2021}$$

$$= \frac{7540}{1} = 7540$$

This means sales increased by an average of $7540 from 2021 to 2022.

b. We use the ordered pairs (2022, 89,790) and (2023, 96,100).

$$\text{average rate of change} = \frac{96,100 - 89,790}{2023 - 2022}$$

$$= \frac{6310}{1} = 6310$$

This means sales increased by an average of $6310 from 2022 to 2023.

c. We use the ordered pairs (2021, 82,250) and (2023, 96,100).

$$\text{average rate of change} = \frac{96,100 - 82,250}{2023 - 2021}$$

$$= \frac{13,850}{2} = 6925$$

This means sales increased by an average of $6925 from 2021 to 2023.

Now Try:

11. A plane had an altitude of 8500 feet at 4:02 P.M. and 12,700 feet at 4:39 P.M. What was the average rate of change in the altitude in feet per minute?

____

12. Enrollment in a college was 11,500 two years ago, 10,975 last year, and 10,800 this year. What is the average rate of change in enrollment per year for this 3-year period?

We use the ordered pairs (1, 11,500) and (3, 10,800).

$$\text{average rate of change} = \frac{10,800 - 11,500}{3 - 1}$$

$$= \frac{-700}{2}$$

$$= -350$$

The enrollment decreases at a rate of 350 students per year.

12. A company had 44 employees during the first year of operation. During their eighth year, the company had 79 employees. What was the average rate of change in the number of employees per year?

____

Objective 8 Practice Exercises

For extra help, see Examples 11–12 on pages 501–502 of your text.

Solve each problem.

16. Suppose in 2018, the sales of a company were $1,625,000. In 2023, the company had sales of $2,250,000. Find the average rate of change in the sales per year.

16. ____________________

17. A state had a population of 755,000 in 2010 and a population of 809,000 in 2022. Find the average rate of change in population per year.

17. ____________________

18. Suppose a man's salary was $45,750 in 1995 and $60,000 in 2010. Find the average rate of change in the salary per year.

18. ____________________

                Copyright © 2025 Pearson Education, Inc.

Chapter 7 LINEAR EQUATIONS, GRAPHS, AND SYSTEMS

7.2   Review of Equations of Lines; Linear Models

Learning Objectives

1   Write an equation of a line given its slope and y-intercept.
2   Graph a line using its slope and y-intercept.
3   Write an equation of a line given its slope and a point on the line.
4   Write an equation of a line given two points on the line.
5   Write equations of horizontal and vertical lines.
6   Write an equation of a line parallel or perpendicular to a given line.
7   Write an equation of a line that models real data.

Key Terms

Use the vocabulary terms listed below to complete each statement in exercises 1–3.

slope-intercept form   point-slope form   standard form

1.   A linear equation in the form $y - y_1 = m(x - x_1)$ is written in

_______________________________.

2.   A linear equation in the form $Ax + By = C$ is written in

_______________________________.

3.   A linear equation in the form $y = mx + b$ is written in

_______________________________.

Objective 1   Write an equation of a line given its slope and y-intercept.

Review this example for Objective 1:

1.   Write an equation of the line with slope $\frac{5}{7}$ and y-intercept $(0, -6)$.

Here, $m = \frac{5}{7}$ and $b = -6$, so we can write the following equation.
$$y = mx + b$$
$$y = \frac{5}{7}x + (-6), \text{ or } y = \frac{5}{7}x - 6$$

Now Try:

1.   Write an equation of the line with slope $\frac{7}{9}$ and y-intercept $(0, 8)$.

____

Objective 1 Practice Exercises

For extra help, see Example 1 on page 508 of your text.

Write the slope-intercept form equation of the line with the given slope and y-intercept.

1. $m = \dfrac{3}{2};\ b = -\dfrac{2}{3}$

1. _____________________

2. $m = -7;\ b = -2$

2. _____________________

3. Slope: $-\dfrac{6}{5}$; y-intercept $\left(0,\ \dfrac{2}{5}\right)$

3. _____________________

Objective 2   Graph a line using its slope and y-intercept.

Review this example for Objective 2:

2. Graph the equation by using the slope and y-intercept.

$2x - 3y = 6$

Solve for y to write the equation in slope-intercept form.

$$2x - 3y = 6$$
$$-3y = -2x + 6$$
$$y = \frac{2}{3}x - 2$$

The y-intercept is (0,–2). Graph this point.

The slope is $\dfrac{2}{3}$. By definition,

$$\text{slope } m = \frac{\text{change in } y \text{ (rise)}}{\text{change in } x \text{ (run)}} = \frac{2}{3}$$

From the y-intercept, count up 2 units and to the right 3 units to obtain the point (3, 0).

Draw the line through the points (0,–2) and (3, 0) to obtain the graph.

Now Try:

2. Graph the equation by using the slope and y-intercept.

$2x - 3y = 0$

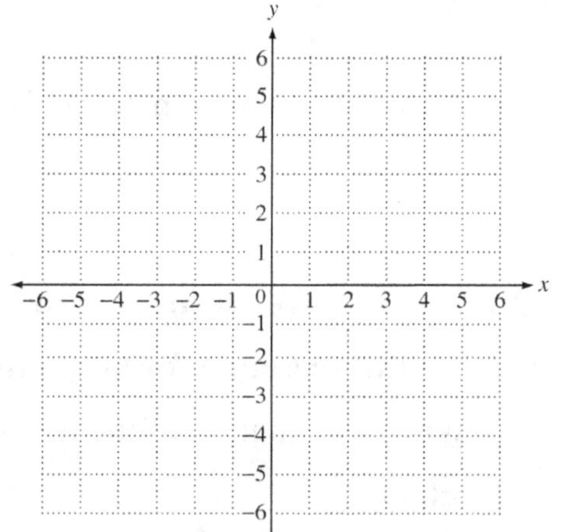

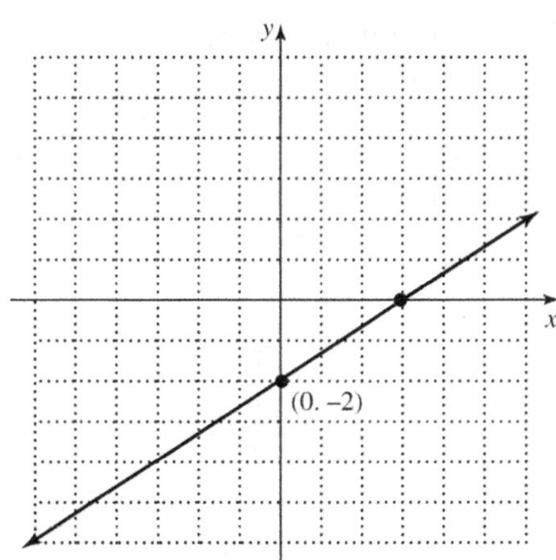

Objective 2 Practice Exercises

For extra help, see Example 2 on page 509 of your text.

Graph each equation by using the slope and y-intercept.

4. $4x - y = 4$

4.

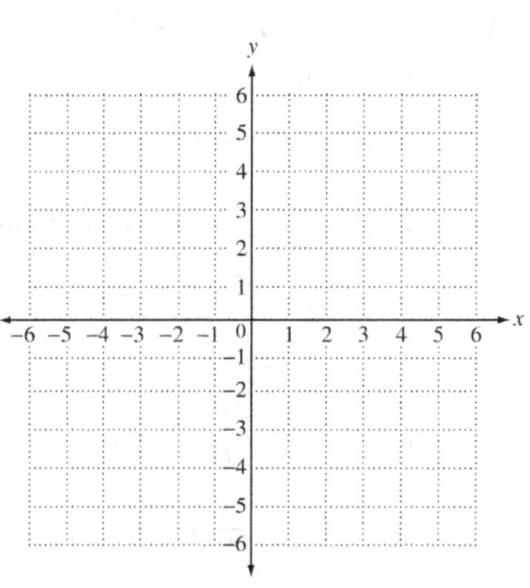

5. $y = -3x + 6$

5.

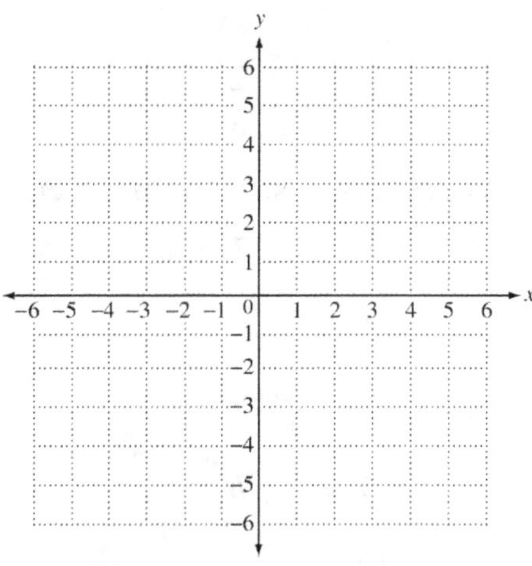

                              275

Graph the line passing through the given point and having the given slope.

6. $(-2,-2);\ m=0$

6.

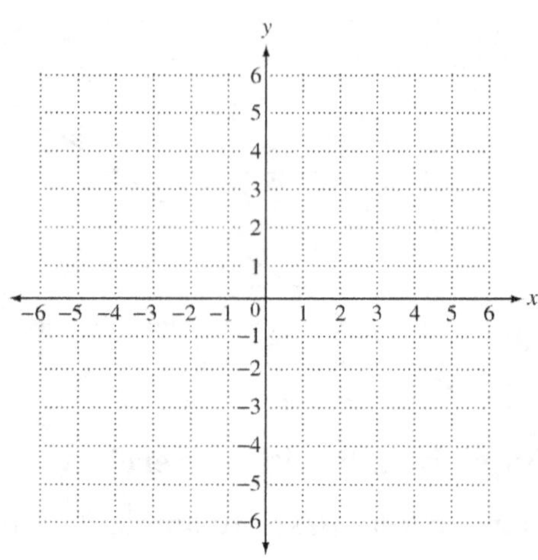

Objective 3   Write an equation of a line given its slope and a point on the line.

Review this example for Objective 3:

3. Write an equation of the line with slope $\frac{2}{3}$ passing through the point $(4,-7)$.

Method 1 Use point-slope form, with $(x_1,y_1)=(4,-7)$ and $m=\frac{2}{3}$.

$$y-y_1=m(x-x_1)$$

$$y-(-7)=\frac{2}{3}(x-4)$$

$$y+7=\frac{2}{3}(x-4)$$

$$3y+21=2x-8$$

$$3y=2x-29$$

$$y=\frac{2}{3}x-\frac{29}{3}$$

Method 2 Use slope-intercept form, with $(x_1,y_1)=(4,-7)$ and $m=\frac{2}{3}$.

$$y=mx+b$$

$$-7=\frac{2}{3}(4)+b$$

$$-7=\frac{8}{3}+b$$

$$-\frac{29}{3}=b,\ \text{or}\ b=-\frac{29}{3}$$

Now Try:

3. Write an equation of the line with slope $\frac{4}{5}$ passing through the point $(6,-4)$.

____

Knowing $m = \frac{2}{3}$ and $b = -\frac{29}{3}$ gives the

equation $y = \frac{2}{3}x - \frac{29}{3}$, same as Method 1.

Objective 3 Practice Exercises

For extra help, see Example 3 on page 510 of your text.

Write the equation in standard form of the line satisfying the given conditions.

7. $(-3,\ 4);\ m = -\frac{3}{5}$ 7. _______________

8. $(-4, -7);\ m = \frac{4}{3}$ 8. _______________

9. $(-1,\ 2);\ m = \frac{2}{3}$ 9. _______________

Objective 4   Write an equation of a line given two points on the line.

Review this example for Objective 4:

4.   Write the equation of the line passing through the point (6, 8) and (–3, 5). Give the final answer in slope-intercept form and then in standard form.

First, find the slope of the line.

$(x_1, y_1) = (6,\ 8)$ and $(x_2, y_2) = (-3,\ 5)$

$$\text{slope } m = \frac{y_2 - y_1}{x_2 - x_1} = \frac{5 - 8}{-3 - 6} = \frac{-3}{-9} = \frac{1}{3}$$

Now Try:

4.   Write the equation of the line passing through the point (7, 15) and (15, 9). Give the final answer in slope-intercept form and then in standard form.

____

Now use (x_1, y_1), here $(6, 8)$ and point-slope form.

$$y - y_1 = m(x - x_1)$$

$$y - 8 = \frac{1}{3}(x - 6)$$

$$y - 8 = \frac{1}{3}x - 2$$

$$y = \frac{1}{3}x + 6 \quad \text{Slope-intercept form}$$

$$3y = x + 18$$

$$-x + 3y = 18$$

$$x - 3y = -18 \qquad \text{Standard form}$$

Objective 4 Practice Exercises

For extra help, see Example 4 on page 511 of your text.

Write the equation in standard form of the line through the given points.

10. $(3, 7), (5, 4)$ **10.** ________________

11. $(2, -1), (5, -2)$ **11.** ________________

12. $(-1, -4), (-2, -3)$ **12.** ________________

Objective 5   Write equations of horizontal and vertical lines.

Review these examples for Objective 5:

5. Write an equation of the line passing through the point $(2, -2)$ that satisfies the given condition.

a. The line has slope 0.

Since the slope is 0, this is a horizontal line.
$$y = -2.$$

Now Try:

5. Write an equation of the line passing through the point $(-5, 5)$ that satisfies the given condition.

a. The line has slope 0.

____

b. The line has undefined slope.

This is a vertical line, since the slope is undefined.
$$x = 2$$

b. The line has undefined slope.

____

Objective 5 Practice Exercises

For extra help, see Example 5 on page 512 of your text.

Write the equation in standard form of the line through the given points.

13. $(-1, -7), (-1, 8)$

13. ________________

14. $(0, 2), (0, -6)$

14. ________________

15. $(4, -5), (8, -5)$

15. ________________

Objective 6   Write an equation of a line parallel or perpendicular to a given line.

Review these examples for Objective 6:

6.   Write an equation in slope-intercept form of the line passing through the point $(-4, 5)$ that satisfies the given condition.

a. The line is parallel to $5x + 2y = 10$.

First, find the slope of the given line.
$$5x + 2y = 10$$
$$2y = -5x + 10$$
$$y = -\frac{5}{2}x + 5$$

The slope is $-\frac{5}{2}$.

Use point-slope form with $\left(x_1, y_1\right) = (-4, 5)$ and $m = -\frac{5}{2}$.

Now Try:

6.   Write an equation in slope-intercept form of the line passing through the point $(-6, 8)$ that satisfies the given condition.

a. The line is parallel to $3x + 4y = 12$.

____

$$y - y_1 = m(x - x_1)$$
$$y - 5 = -\frac{5}{2}(x - (-4))$$
$$y - 5 = -\frac{5}{2}(x + 4)$$
$$y - 5 = -\frac{5}{2}x - 10$$
$$y = -\frac{5}{2}x - 5$$

b. The line is perpendicular to $5x + 2y = 10$.

From part (a), the line in slope-intercept form is
$y = -\frac{5}{2}x + 5$.

The line perpendicular to this line must have
slope $\frac{2}{5}$, the negative reciprocal of $-\frac{5}{2}$.

Use point-slope form with $(x_1, y_1) = (-4,\ 5)$ and
$m = \frac{2}{5}$.

$$y - y_1 = m(x - x_1)$$
$$y - 5 = \frac{2}{5}(x - (-4))$$
$$y - 5 = \frac{2}{5}(x + 4)$$
$$y - 5 = \frac{2}{5}x + \frac{8}{5}$$
$$y = \frac{2}{5}x + \frac{33}{5}$$

b. The line is perpendicular to
$3x + 4y = 12$.

____

Objective 6 Practice Exercises

For extra help, see Example 6 on pages 512–513 of your text.

Write the equation in standard form of the line satisfying the given conditions.

16.   parallel to $2x + 3y = -12$, through $(9, -3)$

16. ________________

17.   parallel to $4x - 3y = 8$, through $(-2, 3)$.

17. ________________

18.   perpendicular to $x - 3y = 0$, through $(-10, 2)$

18. ________________

Objective 7   Write an equation of a line that models real data.

Review these examples for Objective 7:

7.   The table and scatter graph shows the number of internet users in the world from 1998 to 2005, where year 0 represents 1998.

Year	Number of Internet Users (millions)
0	147
2	361
4	587
6	817
8	1093

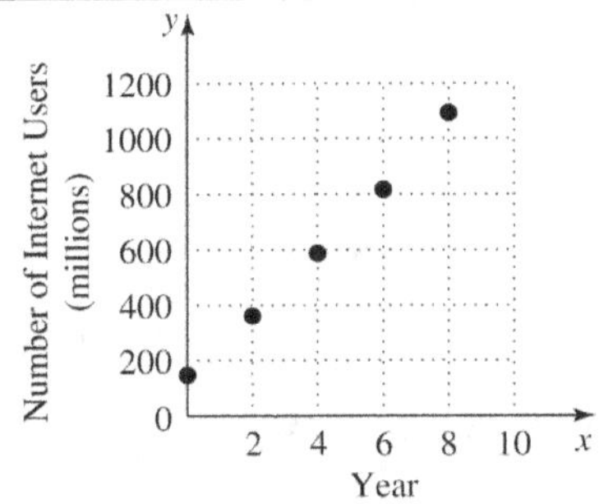

a. Find an equation that models the data.

The points appear to lie approximately in a straight line. y represents the number of internet users in year x. To find an equation of the line, we choose the ordered pairs (0, 147) and (8, 1093) from the table and find the slope of the line through these points.

$$(x_1, y_1) = (0,\ 147) \text{ and } (x_2, y_2) = (8,\ 1093)$$

$$\text{slope } m = \frac{y_2 - y_1}{x_2 - x_1} = \frac{1093 - 147}{8 - 0} = \frac{946}{8}$$

$$= 118.25$$

Now Try:

7.   The table and scatter graph shows the average annual telephone expenditures for residential and pay telephones from 2001 to 2006, where year 0 represents 2001.

Year	Annual Telephone Expenditures
0	$686
2	$620
3	$592
4	$570
5	$542

a. Find an equation that models the data.

____

Use the slope, 118.25, and the point (0, 147) in slope-intercept form.

$$y = mx + b$$

$$147 = 118.25(0) + b$$

$$147 = b$$

Thus, $m = 118.25$ and $b = 147$, so the equation of the line is $y = 118.25x + 147$.

b. Find and interpret the ordered pair associated with the equation for $x = 5$.

If $x = 5$, then

$$y = 118.25(5) + 147$$

$$= 591.25 + 147$$

$$= 738.25$$

In 2003, there were 738.25 million internet users.

b. Find the ordered pair associated with the equation for $x = 1$.

____

Objective 7 Practice Exercises

For extra help, see Examples 7–9 on pages 514–517 of your text.

Solve each problem.

19. To run a newspaper ad, there is a $25 set up fee plus a charge of $1.25 per line of type in the ad. Let x represent the number of lines in the ad so that y represents the total cost of the ad (in dollars).
   a. Write an equation in the form $y = mx + b$.
   b. Give three ordered pairs associated with the equation for x-values 0, 5, and 10.

19.

a. ___________________

b. ___________________

20. The table and scatter graph shows the U.S. municipal solid waste recycling percent since 1985, where year 0 represents 1985.

a. Find an equation that models the data.

b. Use the equation from part (a) to predict the percent of municipal solid waste recycling in the year 2015.

Year	Recycling Percent
0	10.1
5	16.2
10	26.0
15	29.1
20	32.5

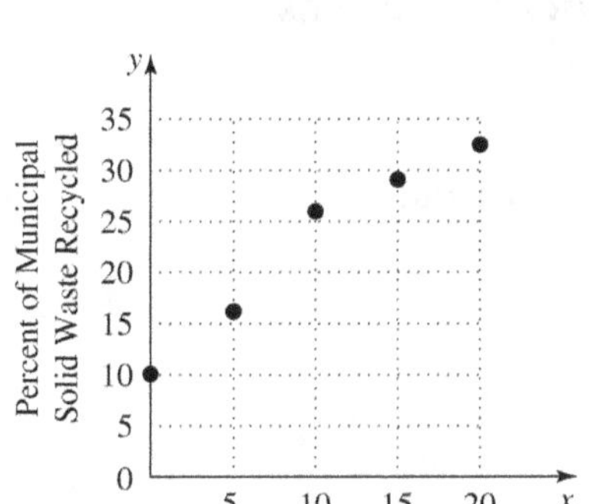

20.

a. _______________________

b. _______________________

Chapter 7 LINEAR EQUATIONS, GRAPHS, AND SYSTEMS

7.3   Solving Systems of Linear Equations by Graphing

Learning Objectives
1   Determine whether a given ordered pair is a solution of a system.
2   Solve linear systems by graphing.
3   Solve special systems by graphing.
4   Identify special systems without graphing.

Key Terms

Use the vocabulary terms listed below to complete each statement in exercises 1−7.

system of linear equations   **solution of the system**

solution set of the system   **consistent system**

inconsistent system   **independent equations**

dependent equations

1.   Equations of a system that have different graphs are called
     _______________________________.

2.   A system of equations with at least one solution is a
     _______________________________.

3.   The set of all ordered pairs that are solutions of a system is the
     ____.

4.   The ____ of linear equations is an ordered
     pair that makes all the equations of the system true at the same time.

5.   Equations of a system that have the same graph (because they are different forms
     of the same equation) are called _________________________________.

6.   A system with no solution is called a(n) _____________________________________.

7.   A(n) ____ consists of two or more linear
     equations with the same variables.

Name: Date:

Instructor: Section:

Objective 1   Determine whether a given ordered pair is a solution of a system.

Review this example for Objective 1:

1. Determine whether the ordered pair $(5, -2)$ is a solution of the system.

$$4x + 5y = 10$$

$$3x + 8y = 6$$

Again, substitute 5 for x and -2 for y in each equation.

$$
\begin{array}{c|c}
4x + 5y = 10 & 3x + 8y = 6 \\
4(5) + 5(-2) \stackrel{?}{=} 10 & 3(5) + 8(-2) \stackrel{?}{=} 6 \\
20 - 10 \stackrel{?}{=} 10 & 15 - 16 \stackrel{?}{=} 6 \\
10 = 10 \ \text{True} & \text{False} \ -1 = 6
\end{array}
$$

The ordered pair $(5, -2)$ is not a solution of this system because it does not satisfy the second equation.

Now Try:

1. Determine whether the ordered pair $(6, 5)$ is a solution of the system.

$$5x - 6y = 0$$

$$6x + 5y = 50$$

____

Objective 1 Practice Exercises

For extra help, see Example 1 on page 522 of your text.

Decide whether the given ordered pair is a solution of the given system.

1. $(2, -4)$

    $2x + 3y = 6$

    $3x - 2y = 14$

1. _______________

2. $(-3, -1)$

    $5x - 3y = -12$

    $2x + 3y = -9$

2. _______________

3. $(4, 0)$

    $4x + 3y = 16$

    $x - 4y = -4$

3. _______________

Objective 2   Solve linear systems by graphing.

Review this example for Objective 2:

2. Solve the system of equations by graphing both equations on the same axes.

$$6x - 5y = 4$$
$$2x - 5y = 8$$

Graph these equations by plotting several points for each line. To find the x-intercept, let $y = 0$. To find the y-intercept, let $x = 0$.

The tables show the intercepts and a check point for each graph.

$6x - 5y = 4$

x	y
0	$-\dfrac{4}{5}$
$\dfrac{2}{3}$	0
4	4

$2x - 5y = 8$

x	y
0	$\dfrac{8}{5}$
4	0
-6	-4

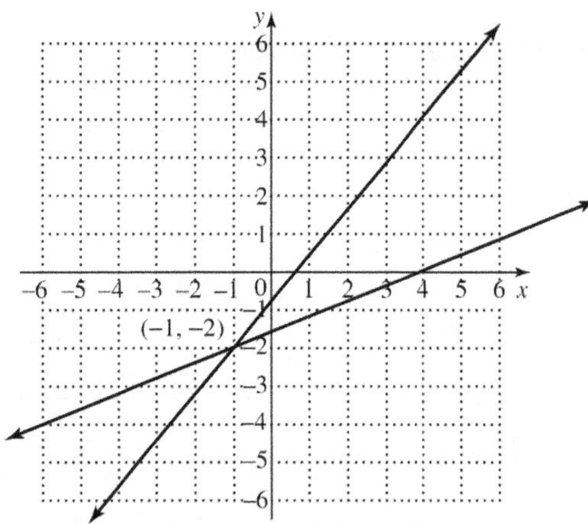

The lines suggest that the graphs intersect at the point $(-1, -2)$. We check by substituting -1 for x and -2 for y in both equations.

$$6x - 5y = 4 \qquad\qquad 2x - 5y = 8$$
$$6(-1) - 5(-2) \overset{?}{=} 4 \qquad 2(-1) - 5(-2) \overset{?}{=} 8$$
$$-6 + 10 \overset{?}{=} 4 \qquad\qquad -2 + 10 \overset{?}{=} 8$$
$$4 = 4 \ \text{True} \qquad\qquad \text{True} \ 8 = 8$$

Because $(-1, -2)$ satisfies both equations, the solution set of this system is $\{(-1, -2)\}$.

Now Try:

2. Solve the system of equations by graphing both equations on the same axes.

$$3x - y = -7$$
$$2x + y = -3$$

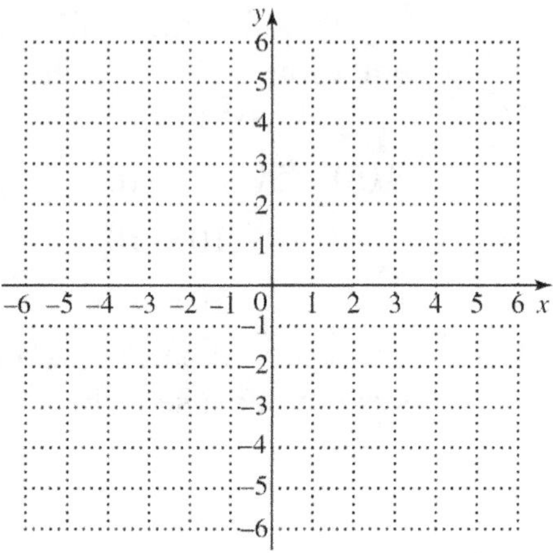

                Copyright © 2025 Pearson Education, Inc.

Objective 2 Practice Exercises

For extra help, see Example 2 on page 523 of your text.

Solve each system by graphing both equations on the same axes.

4. $x - 2y = 6$
        $2x + y = 2$

4. ______________________________

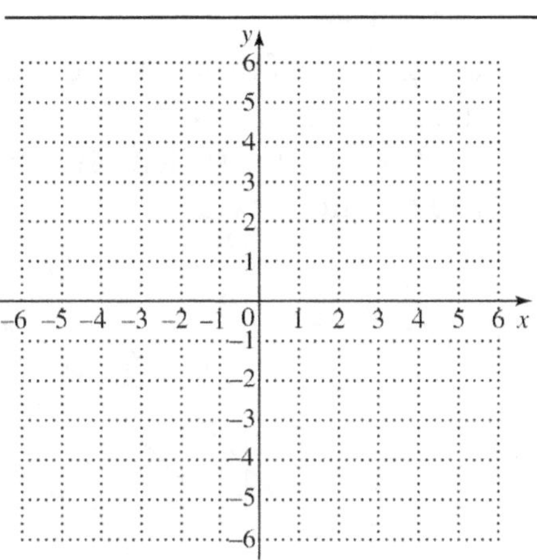

5. $2x = y$
        $5x + 3y = 0$

5. ______________________________

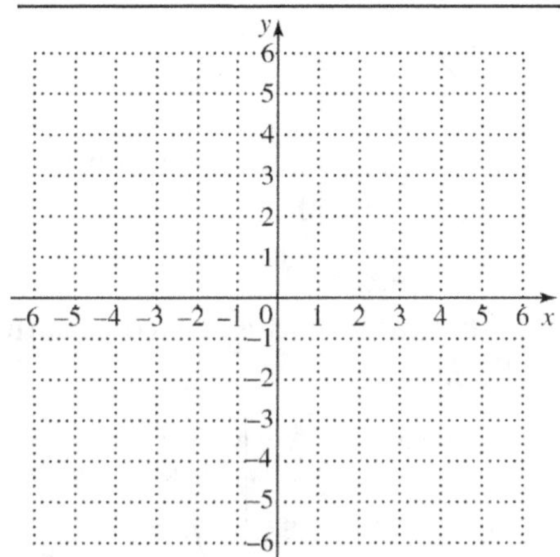

6. $3x + 2 = y$
        $2x - y = 0$

6. ______________________________

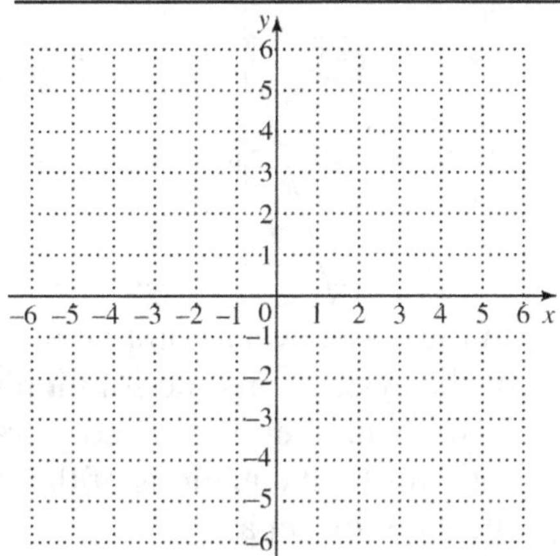

Objective 3   Solve special systems by graphing.

Review these examples for Objective 3:	**Now Try:**

3. Solve the system by graphing.
$$x - y = 1$$
$$x - y = -1$$

The graphs of these two equations are parallel and have no points in common. There is no solution for this system. It is inconsistent. The solution set is $\varnothing$.

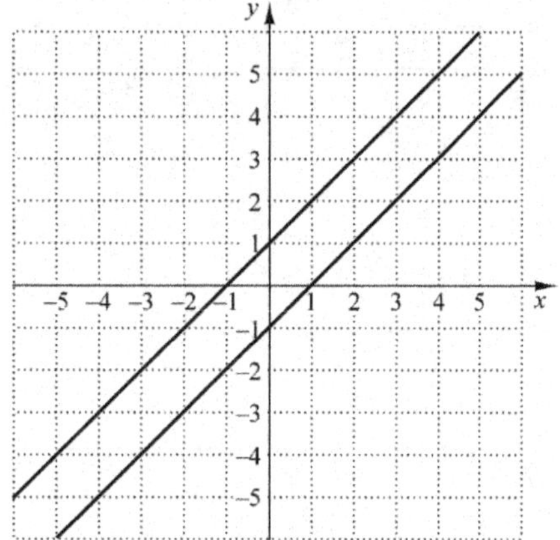

3. Solve the system by graphing.
$$x - 3y = 6$$
$$x - 3y = 4$$

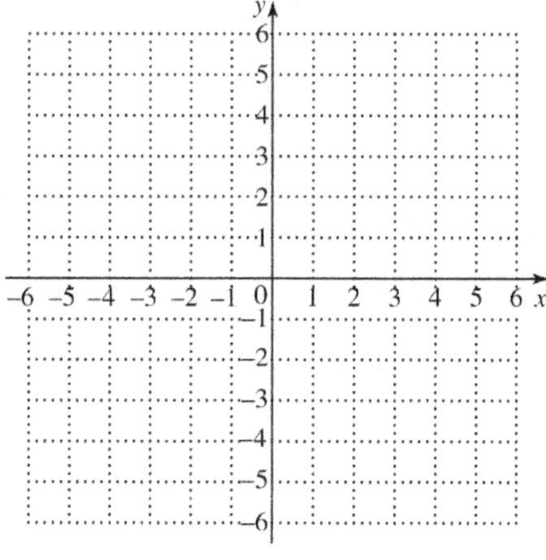

____

4. Solve the system by graphing.
$$3x - y = 0$$
$$2y = 6x$$

The graphs of these two equations are the same line.

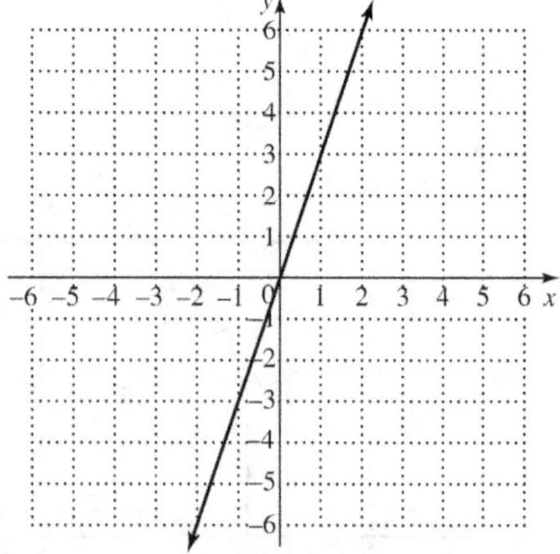

In this case, every point on the line is a solution of the system, and the solution set contains an infinite number of ordered pairs, each of which satisfies both equations of the system. We write the solution set as
$$\{(x, y)\mid 3x - y = 0\}.$$

4. Solve the system by graphing.
$$4x - 2y = 8$$
$$6x - 3y = 12$$

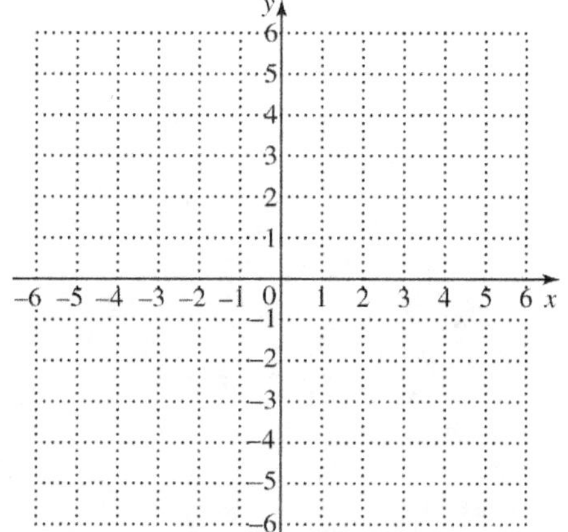

____

Name: ______________________   Date: ______________________

Instructor: ______________________   Section: ______________________

Objective 3 Practice Exercises

For extra help, see Examples 3–4 on pages 524–525 of your text.

*Solve each system of equations by graphing both equations on the same axes. If the two equations produce parallel lines, write **no solution**. If the two equations produce the same line, write **infinite number of solutions**.*

7. $8x + 4y = -1$

   $4x + 2y = 3$

7. _______________________________

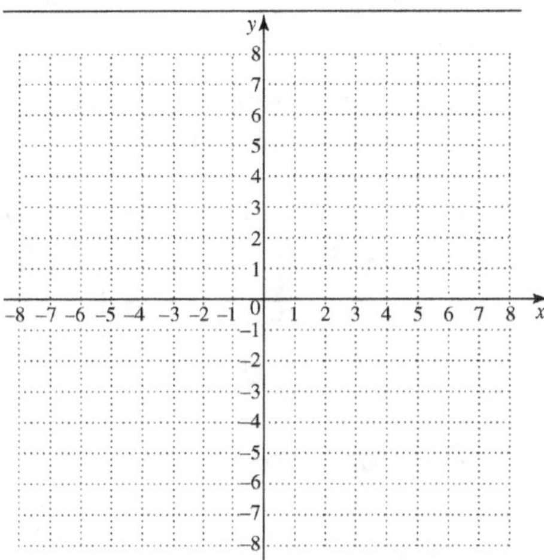

8. $-3x + 2y = 6$

   $-6x + 4y = 12$

8. _______________________________

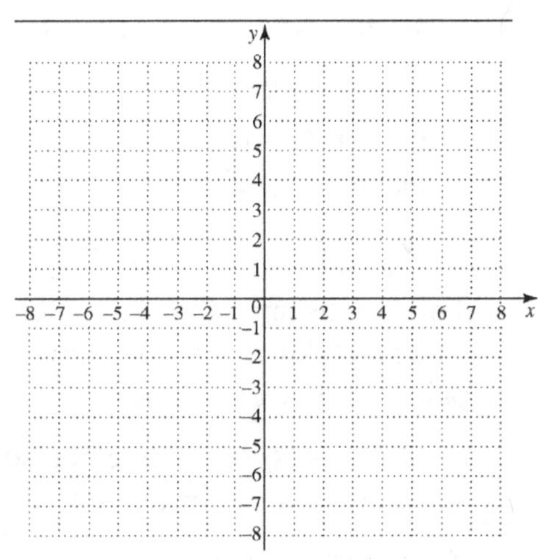

Objective 4   Identify special systems without graphing.

Review these examples for Objective 4:	**Now Try:**

Review these examples for Objective 4:

5. Given the linear system,

$$3x - 4y = 12$$

$$2x - 3y = 6$$

answer the following questions without graphing.

a. Is the system inconsistent, are the equations dependent, or neither?

Write each equation in slope-intercept form.

$$
\begin{array}{l|l}
3x - 4y = 12 & 2x - 3y = 6 \\
\quad -4y = -3x + 12 & \quad -3y = -2x + 6 \\
\quad y = \dfrac{3}{4}x - 3 & \quad y = \dfrac{2}{3}x - 2
\end{array}
$$

The slopes are different. This system is not inconsistent nor is it dependent. The system is neither.

b. Is the graph a pair of intersecting lines, a pair of parallel lines, or one line?

From part (a), we have written the system in slope-intercept form.

$$y = \frac{3}{4}x - 3 \qquad y = \frac{2}{3}x - 2$$

The graphs are neither parallel nor the same line, since the slopes are different. The system is a pair of intersecting lines.

c. Does the system have one solution, no solution, or an infinite number of solutions?

From part (a), we have written the system in slope-intercept form.

$$y = \frac{3}{4}x - 3 \qquad y = \frac{2}{3}x - 2$$

Since the slopes are different, the graph is neither parallel nor the same line, and therefore cannot have no solution nor an infinite number of solutions. This system has exactly one solution.

Now Try:

5. Given the linear system,

$$x - 6y = 4$$

$$6x - y = 9$$

answer the following questions without graphing.

a. Is the system inconsistent, are the equations dependent, or neither?

____

b. Is the graph a pair of intersecting lines, a pair of parallel lines, or one line?

____

c. Does the system have one solution, no solution, or an infinite number of solutions?

____

Name: Date:

Instructor: Section:

Objective 4 Practice Exercises

For extra help, see Example 5 on pages 526–527 of your text.

Without graphing, answer the following equations for each linear system.
(a) Is the system inconsistent, are the equations dependent, or neither?
(b) Is the graph a pair of intersecting lines, a pair of parallel lines, or one line?
(c) Does the system have one solution, no solution, or an infinite number of solutions?

9. $y = 2x + 1$

     $3x - y = 7$

9. (a)___________________

     (b)_______________

     (c) ______________

10. $-2x + y = 4$

     $-4x + 2y = -2$

10. (a)_______________

     (b)_______________

     (c) ______________

11. $4x + 3y = 12$

     $-12x = -36 + 9y$

11. (a)_______________

     (b)_______________

     (c) ______________

Chapter 7 LINEAR EQUATIONS, GRAPHS, AND SYSTEMS

7.4   Solving Systems of Linear Equations by Substitution

Learning Objectives
1   Solve linear systems by substitution.
2   Solve special systems by substitution.
3   Solve linear systems with fractions and decimals.

Key Terms

Use the vocabulary terms listed below to complete each statement in exercises 1−4.

substitution   **ordered pair**   **inconsistent system**

dependent system

1.   The solution of a linear system of equations is written as a(n) _________________.

2.   When one expression is replaced by another, _________________ is being used.

3.   A system of equations in which all solutions of the first equation are also solutions of the second equation is a(n) _________________________.

4.   A system of equations that has no common solution is called a(n)

                 _________________________.

Objective 1   Solve linear systems by substitution.

Review these examples for Objective 1:

1. Solve the system by the substitution method.
$$2x + 5y = 22 \quad (1)$$
$$y = 4x \quad (2)$$

Equation (2) is already solved for y. We substitute $4x$ for y in equation (1).
$$2x + 5y = 22$$
$$2x + 5(4x) = 22$$
$$2x + 20x = 22$$
$$22x = 22$$
$$x = 1$$

Find the value of y by substituting 1 for x in either equation. We use equation (2).
$$y = 4x$$
$$y = 4(1) = 4$$

We check the solution $(1, 4)$ by substituting 1 for

Now Try:

1. Solve the system by the substitution method.
$$x + y = 7$$
$$y = 6x$$

                        ____

x and 4 for y in both equations.

$$2x+5y=22 \qquad\qquad y=4x$$
$$2(1)+5(4)\overset{?}{=}22 \qquad\qquad 4\overset{?}{=}4(1)$$
$$2+20\overset{?}{=}22 \qquad\qquad \text{True } 4=4$$
$$22=22 \ \text{True}$$

Since $(1, 4)$ satisfies both equations, the solution set of the system is $\{(1, 4)\}$.

2. Solve the system by the substitution method.

$$4x+5y=13 \qquad (1)$$
$$x=-y+2 \quad (2)$$

Equation (2) gives x in terms of y. We substitute $-y+2$ for x in equation (1).

$$4x+5y=13$$
$$4(-y+2)+5y=13$$
$$-4y+8+5y=13$$
$$y+8=13$$
$$y=5$$

Find the value of x by substituting 5 for y in either equation. We use equation (2).

$$x=-y+2$$
$$x=-5+2=-3$$

We check the solution $(-3, 5)$ by substituting -3 for x and 5 for y in both equations.

$$4x+5y=13 \qquad\qquad x=-y+2$$
$$4(-3)+5(5)\overset{?}{=}13 \qquad\qquad -3\overset{?}{=}-5+2$$
$$-12+25\overset{?}{=}13 \qquad\qquad \text{True } -3=-3$$
$$13=13 \ \text{True}$$

Both results are true, so the solution set of the system is $\{(-3, 5)\}$.

3. Solve the system by the substitution method.

$$4x=5-y \quad (1)$$
$$7x+3y=15 \qquad (2)$$

Step 1 Solve one of the equations for x or y. Solve equation (1) for y to avoid fractions.

$$4x=5-y$$
$$y+4x=5$$
$$y=-4x+5$$

2. Solve the system by the substitution method.

$$2x+3y=6$$
$$x=5-y$$

____

3. Solve the system by the substitution method.

$$2x+7y=2$$
$$3y=2-x$$

____

Step 2 Now substitute $-4x+5$ for y in equation (2).
$$7x+3y=15$$
$$7x+3(-4x+5)=15$$

Step 3 Solve the equation from Step 2.
$$7x-12x+15=15$$
$$-5x+15=15$$
$$-5x=0$$
$$x=0$$

Step 4 Equation (1) solved for y is $y=-4x+5$. Substitute 0 for x.
$$y=-4(0)+5=5$$

Step 5 Check that $(0, 5)$ is the solution.

$$
\begin{array}{c|c}
4x=5-y & 7x+3y=15 \\
4(0)\overset{?}{=}5-5 & 7(0)+3(5)\overset{?}{=}15 \\
0=0 \ \text{True} & \text{True} \ 15=15
\end{array}
$$

Since both results are true, the solution set of the system is $\{(0, 5)\}$.

Objective 1 Practice Exercises

For extra help, see Examples 1–3 on pages 531–534 of your text.

Solve each system by the substitution method. Check each solution.

1. $\quad 3x+2y=14$

$\qquad y=x+2$

1. ______________________

2. $\quad x+\ y=9$

$\qquad 5x-2y=-4$

2. ______________________

3. $\quad 3x-21=y$

$\qquad y+2x=-1$

3. ______________________

Objective 2   Solve special systems by substitution.

Review these examples for Objective 2:	**Now Try:**

Review these examples for Objective 2:

4. Use substitution to solve the system.
$$x = 7 - 3y \quad (1)$$
$$5x + 15y = 1 \qquad (2)$$

Because equation (1) is already solved for x, we substitute $7 - 3y$ for x in equation (2).
$$5x + 15y = 1$$
$$5(7 - 3y) + 15y = 1$$
$$35 - 15y + 15y = 1$$
$$35 = 1 \quad \text{False}$$

A false result, here $35 = 1$, means that the equations in the system have graphs that are parallel lines. The system is inconsistent and has no solution, so the solution set is $\varnothing$.

5. Use substitution to solve the system.
$$14x - 7y = 21 \quad (1)$$
$$-2x + y = -3 \quad (2)$$

Begin by solving equation (2) for y to get $y = 2x - 3$. Substitute $2x - 3$ for y in equation (1).
$$14x - 7y = 21$$
$$14x - 7(2x - 3) = 21$$
$$14x - 14x + 21 = 21$$
$$0 = 0$$

This true result means that every solution of one equation is also a solution of the other, so the system has an infinite number of solutions. The solution set is $\{(x, y) \mid 14x - 7y = 21\}$.

Now Try:

4. Use substitution to solve the system.
$$5x - 10y = 8$$
$$x = 2y + 5$$

____

5. Use substitution to solve the system.
$$5x + 4y = 20$$
$$-10x + 40 = 8y$$

____

Objective 2 Practice Exercises

For extra help, see Examples 4–5 on pages 534–535 of your text.

Solve each system by the substitution method. Use set-builder notation for dependent equations.

4.
$$y = -\frac{1}{3}x + 5$$
$$3y + x = -9$$

4. __________

                    295

5. $\frac{1}{2}x + 3 = y$

   $6 = -x + 2y$

5. _______________

6. $4x + 3y = 2$

   $8x + 6y = 6$

6. _______________

Objective 3   Solve linear systems with fractions and decimals.

Review these examples for Objective 3:

6. Solve the system by the substitution method.

$$\frac{1}{2}x - y = 3 \quad (1)$$

$$\frac{1}{5}x + \frac{1}{2}y = \frac{3}{10} \quad (2)$$

Clear equation (1) of fractions by multiplying each side by 2.

$$2\left(\frac{1}{2}x - y\right) = 2(3)$$

$$2\left(\frac{1}{2}x\right) - 2y = 2(3)$$

$$x - 2y = 6$$

Clear equation (2) of fractions by multiplying each side by 10.

$$10\left(\frac{1}{5}x + \frac{1}{2}y\right) = 10\left(\frac{3}{10}\right)$$

$$10\left(\frac{1}{5}x\right) + 10\left(\frac{1}{2}y\right) = 10\left(\frac{3}{10}\right)$$

$$2x + 5y = 3$$

The given system of equations has been simplified to an equivalent system.

$$x - 2y = 6 \quad (3)$$

$$2x + 5y = 3 \quad (4)$$

To solve the system by substitution, solve equation (3) for x.

$$x - 2y = 6$$

$$x = 2y + 6$$

Now substitute the result for x in equation (4).

Now Try:

6. Solve the system by the substitution method.

$$x + \frac{1}{2}y = \frac{1}{2}$$

$$\frac{1}{2}x + \frac{1}{5}y = 0$$

____

$$2x + 5y = 3$$
$$2(2y + 6) + 5y = 3$$
$$4y + 12 + 5y = 3$$
$$9y + 12 = 3$$
$$9y = -9$$
$$y = -1$$

Substitute -1 for y in $x = 2y + 6$ (equation (3) solved for x).
$$x = 2(-1) + 6 = 4$$
Check $(4, -1)$ in both of the original equations. The solution set is $\{(4, -1)\}$.

7. Solve the system by the substitution method.
$$0.4x + 2.5y = 8 \qquad (1)$$
$$-0.1x + 3.5y = 14.5 \quad (2)$$

Clear each equation of decimals, by multiplying by 10.
$$0.4x + 2.5y = 8$$
$$10(0.4x + 2.5y) = 10(8)$$
$$10(0.4x) + 10(2.5y) = 10(8)$$
$$4x + 25y = 80$$

$$-0.1x + 3.5y = 14.5$$
$$10(-0.1x + 3.5y) = 10(14.5)$$
$$10(-0.1x) + 10(3.5y) = 10(14.5)$$
$$-x + 35y = 145$$

Now solve the equivalent system of equations by substitution.
$$4x + 25y = 80 \qquad (3)$$
$$-x + 35y = 145 \quad (4)$$

Equation (4) can be solved for x.
$$x = 35y - 145$$

Substitute this result for x in equation (3).
$$4x + 25y = 80$$
$$4(35y - 145) + 25y = 80$$
$$140y - 580 + 25y = 80$$
$$165y - 580 = 80$$
$$165y = 660$$
$$y = 4$$

7. Solve the system by the substitution method.
$$0.6x + 0.8y = -2.2$$
$$0.7x - 0.1y = 2.6$$

———————

Since equation (4) solved for x is $x = 35y - 145$, substitute 4 for y.
$$x = 35(4) - 145 = -5$$
Check $(-5, 4)$ in both of the original equations.
The solution set is $\{(-5, 4)\}$.

Objective 3 Practice Exercises

For extra help, see Examples 6–7 on pages 535–537 of your text.

Solve each system by the substitution method. Check each solution.

7. $\quad \dfrac{5}{4}x - y = -\dfrac{1}{4}$

$\qquad -\dfrac{7}{8}x + \dfrac{5}{8}y = 1$

7. _______________

8. $\quad \dfrac{1}{4}x + \dfrac{3}{8}y = -3$

$\qquad \dfrac{5}{6}x - \dfrac{3}{7}y = -10$

8. _______________

9. $\quad 0.6x + 0.8y = 1$

$\qquad 0.4y = 0.5 - 0.3x$

9. _______________

            Copyright © 2025 Pearson Education, Inc.

Chapter 7 LINEAR EQUATIONS, GRAPHS, AND SYSTEMS

7.5   Solving Systems of Linear Equations by Elimination

Learning Objectives
1 Solve linear systems by elimination.
2 Multiply when using the elimination method.
3 Use an alternative method to find the second value in a solution.
4 Solve special systems by elimination.

Key Terms

Use the vocabulary terms listed below to complete each statement in exercises 1–3.

addition property of equality   elimination method   substitution

1. Using the addition property to solve a system of equations is called the

    ______________________________.

2. The ______________________________ states that the same added quantity to
    each side of an equation results in equal sums.

3. ______________________________ is being used when one expression is replaced by
    another.

Objective 1   Solve linear systems by elimination.

Review this example for Objective 1:

1. Solve the system by the elimination method.
$$x+y=6 \quad (1)$$
$$-x+y=4 \quad (2)$$

Add the equations vertically.
$$x+y=6 \quad (1)$$
$$-x+y=4 \quad (2)$$
$$\overline{2y=10}$$
$$y=5$$

To find the x-value, substitute 5 for y in either of
the two equations of the system. We choose
equation (1).
$$x+y=6$$
$$x+5=6$$
$$x=1$$

Now Try:

1. Solve the system by the
    elimination method.
$$x+y=11$$
$$x-y=5$$

    ____

Check the solution $(1, 5)$, by substituting 1 for x and 5 for y in both equations of the given system.

$$x+y=6 \qquad\qquad -x+y=4$$
$$1+5 \overset{?}{=} 6 \qquad\qquad -1+5 \overset{?}{=} 4$$
$$6=6 \ \text{True} \qquad \text{True} \ 4=4$$

Since both results are true, the solution set of the system is $\{(1, 5)\}$.

Objective 1 Practice Exercises

For extra help, see Examples 1–2 on pages 539–541 of your text.

Solve each system by the elimination method. Check your answers.

1. $x-4y=-4$

    $-x+y=-5$

1. ____________________

2. $2x-y=10$

    $3x+y=10$

2. ____________________

3. $x-3y=5$

    $-x+4y=-5$

3. ____________________

Name: Date:

Instructor: Section:

Objective 2   Multiply when using the elimination method.

Review this example for Objective 2: | **Now Try:**

4. Solve the system by the elimination method.

$$3x + 8y = -2 \quad (1)$$

$$2x + 7y = 2 \quad (2)$$

To eliminate x, multiply equation (1) by 2 and multiply equation (2) by −3. Then add.

$$6x + 16y = -4$$
$$\underline{-6x - 21y = -6}$$
$$-5y = -10$$
$$y = 2$$

Find the value of x by substituting 2 for y in either equation (1) or (2).

$$2x + 7y = 2$$

$$2x + 7(2) = 2$$

$$2x + 14 = 2$$

$$2x = -12$$

$$x = -6$$

Check that the solution set of the system is $\{(-6, 2)\}$.

Now Try:

4. Solve the system by the elimination method.

$$3x + 4y = 24$$

$$4x + 3y = 11$$

____

Objective 2 Practice Exercises

For extra help, see Examples 3–4 on pages 541–542 of your text.

Solve each system by the elimination method. Check your answers.

4. $6x + 7y = 10$

    $2x - 3y = 14$

    4. __________________

5. $8x + 6y = 10$

    $4x - \ y = 1$

    5. __________________

6. $6x + y = 1$

    $3x - 4y = 23$

    6. __________________

Objective 3   Use an alternative method to find the second value in a solution.

Review this example for Objective 3:

5. Solve the system by the elimination method.

$$6x = 7 - 3y \quad (1)$$

$$8x - 5y = 5 \quad (2)$$

Write equation (1) in standard form.

$$6x + 3y = 7 \quad (3)$$

$$8x - 5y = 5 \quad (4)$$

One way to proceed is to eliminate y by multiplying each side of equation (3) by 5 and each side of equation (4) by 3, and then adding.

$$30x + 15y = 35$$
$$\underline{24x - 15y = 15}$$
$$54x \qquad = 50$$

$$x = \frac{50}{54} \text{ or } \frac{25}{27}$$

Substituting $\frac{25}{27}$ for x in one of the given

equations would give y, but the arithmetic would be complicated. Instead, solve for y by starting again with the original equations in standard form, and eliminating x.

Multiply equation (3) by 4 and equation (4) by -3.

$$24x + 12y = 28$$
$$\underline{-24x + 15y = -15}$$
$$27y = 13$$

$$y = \frac{13}{27}$$

The solution set is $\left\{ \left(\frac{25}{27},\ \frac{13}{27} \right) \right\}$.

Objective 3 Practice Exercises

For extra help, see Example 5 on page 543 of your text.

Solve each system by the elimination method. Check your answers.

7. $4x - 3y - 20 = 0$

       $6x + 5y + 8 = 0$

Now Try:

5. Solve the system by the elimination method.

$$8x = 5y + 1$$

$$6x - 8y = -2$$

            ____

7. _________________

8. $6x = 16 - 7y$

      $4x = 3y + 26$

8. ________________

9. $2x = 14 + 4y$

      $6y = -5x + 3$

9. ________________

Objective 4   Solve special systems by elimination.

Review these examples for Objective 4:

6. Solve each system by the elimination method.

a. $7x + y = 9$ (1)

    $-14x - 2y = -18$ (2)

Multiply each side of equation (1) by 2.

$$14x + 2y = 18$$
$$\underline{-14x - 2y = -18}$$
$$0 = 0 \quad \text{True}$$

A true statement occurs when the equations are equivalent. This indicates that every solution of one equation is also a solution of the other. The solution set is $\{(x, y) \mid 7x + y = 9\}$.

b. $5x + 10y = 9$ (1)

    $3x + 6y = 8$ (2)

Multiply each side of equation (1) by 3 and each side of equation (2) by –5.

$$15x + 30y = 27$$
$$\underline{-15x - 30y = -40}$$
$$0 = -13 \quad \text{False}$$

The false statement $0 = -13$ indicates that the system has solution set $\varnothing$.

Now Try:

6. Solve each system by the elimination method.

a. $9x - 7y = 5$

    $18x = 14y + 10$

____

b. $2x + 6y = 5$

    $5x + 15y = 8$

____

Objective 4 Practice Exercises

For extra help, see Example 6 on pages 543–544 of your text.

Solve each system by the elimination method. Use set-builder notation for dependent equations. Check your answers.

10. $12x - 8y = 3$

         $6x - 4y = 6$

10. _________________

11. $2x + 4y = -6$

         $-x - 2y = 3$

11. _________________

12. $15x + 6y = 9$

         $10x + 4y = 18$

12. _________________

Chapter 7 LINEAR EQUATIONS, GRAPHS, AND SYSTEMS

7.6   Systems of Linear Equations in Three Variables

Learning Objectives
1   Understand the geometry of systems of three equations in three variables.
2   Solve linear systems (with three equations and three variables) by elimination.
3   Solve linear systems (with three equations and three variables) in which some of the equations have missing terms.
4   Solve special systems.

Key Terms

Use the vocabulary terms listed below to complete each statement in exercises 1–5.

ordered triple   **focus variable**   **working equation**

inconsistent system   **dependent system**

1.   The solution of a linear system of equations in three variables is written as a(n)

_________________________.

2.   A system of equations in which all solutions of the first equation are also
       solutions of the second equation is a(n) _______________________________.

3.   A system of equations that has no common solution is called a(n)

_________________________.

4.   The first variable to be eliminated is called a(n) _________________________.

5.   The focus variable will always be present in the _________________________ .

**Objective 1   Understand the geometry of systems of three equations in three
                 variables.**

Objective 1 Practice Exercises

For extra help, see pages 549–549 of your text.

Answer each question.

1.   If a system of linear equations in three variables has   1.   ________________
      a single solution, how do the planes that are the
      graphs of the equations intersect?

2.   If a system of linear equations in three variables has   2.   ________________
      no solution, how do the planes that are the graphs of
      the equations intersect?

Objective 2   Solve linear systems (with three equations and three variables) by elimination.

Review this example for Objective 2:

1. Solve the system.

$$x - 2y + 5z = -7 \quad (1)$$
$$2x + 3y - 4z = 14 \quad (2)$$
$$3x - 5y + z = 7 \quad (3)$$

Step 1: Since x in equation (1) has coefficient 1, choose x as the focus variable and (1) as the working equation.

Step 2: Multiply working equation (1) by -2 and add the result to equation (2) to eliminate focus variable x.

$$-2x + 4y - 10z = 14 \quad \text{Multiply (1) by } -2$$
$$\underline{2x + 3y - 4z = 14 \quad (2)}$$
$$7y - 14z = 28 \quad (4)$$

Step 3: Multiply working equation (1) by -3 and add the result to equation (3) to eliminate focus variable x.

$$-3x + 6y - 15z = 21 \quad \text{Multiply (1) by } -3$$
$$\underline{3x - 5y + z = 7 \quad (3)}$$
$$y - 14z = 28 \quad (5)$$

Step 4: Write equations (4) and (5) as a system, then solve the system.

$$7y - 14z = 28 \quad (4)$$
$$y - 14z = 28 \quad (5)$$

We will eliminate z.

$$-7y + 14z = -28 \quad \text{Multiply (4) by } -1$$
$$\underline{y - 14z = 28 \quad (5)}$$
$$-6y = 0 \quad \text{Add.}$$
$$y = 0 \quad \text{Divide by } -6.$$

Substitute 0 for y in either equation to find z.

$$y - 14z = 28 \quad (5)$$
$$0 - 14z = 28 \quad \text{Let } y = 0.$$
$$-14z = 28$$
$$z = -2$$

Now Try:

1. Solve the system.

$$2x + y + 2z = -1$$
$$3x - y + 2z = -6$$
$$3x + y - z = -10$$

____

         Copyright © 2025 Pearson Education, Inc.

Step 5: Now substitute $y = 0$ and $z = -2$ in working equation (1) to find the value of the remaining variable, focus variable x.

$$x - 2y + 5z = -7 \quad (1)$$
$$x - 2(0) + 5(-2) = -7 \quad \text{Let } y = 0 \text{ and } z = -2.$$
$$x - 10 = -7$$
$$x = 3$$

Step 6: It appears that the ordered triple $(3, 0, -2)$ is the solution of the system. We must check that the solution satisfies all three original equations of the system.

$$x - 2y + 5z = -7 \quad (1)$$
$$3 - 2(0) + 5(-2) \overset{?}{=} -7$$
$$3 - 0 - 10 \overset{?}{=} -7$$
$$-7 = -7$$

Because $(3, 0, -2)$ also satisfies equations (2) and (3), the solution set is $\{(3, 0, -2)\}$.

Objective 2 Practice Exercises

For extra help, see Example 1 on pages 550–551 of your text.

Solve each system of equations.

3. $\quad x + y + z = 2$
    $\quad x - y + z = -2$
    $\quad x - y - z = -4$

3. _______________

4. $\quad 2x + y - z = 9$
    $\quad x + 2y + z = 3$
    $\quad 3x + 3y - z = 14$

4. _______________

5. $2x - 5y + 2z = 30$

$\quad x + 4y + 5z = -7$

$\quad \dfrac{1}{2}x - \dfrac{1}{4}y + z = 4$

5. ________________

Objective 3   Solve linear systems (with three equations and three variables) in which some of the equations have missing terms.

Review this example for Objective 3:

2. Solve the system.

$$3x \qquad - 4z = -23 \quad (1)$$
$$\quad y + 5z = 24 \quad (2)$$
$$x - 3y \qquad = 2 \quad (3)$$

Since equation (1) is missing the variable y, one way to begin is to eliminate y again, using equations (2) and (3).

$$3y + 15z = 72 \quad \text{Multiply (2) by 3}$$
$$\underline{x - 3y \qquad = 2 \quad (3)}$$
$$x \qquad + 15z = 74 \quad (4)$$

Now solve the system composed of equations (1) and (4).

$$3x - 4z = -23 \quad (1)$$
$$\underline{-3x - 45z = -222 \quad \text{Multiply (4) by } -3}$$
$$-49z = -245$$
$$z = 5$$

Substitute 5 for z in (1) and solve for x.

$$3x - 4z = -23 \quad (1)$$
$$3x - 4(5) = -23 \quad \text{Let } z = 5.$$
$$3x - 20 = -23$$
$$3x = -3$$
$$x = -1$$

Now Try:

2. Solve the system.

$$x + 5y \qquad = -23$$
$$\quad 4y - 3z = -29$$
$$2x \qquad + 5z = 19$$

____

Substitute 5 for z in (2) and solve for y.

$$y + 5z = 24 \quad (2)$$
$$y + 5(5) = 24 \quad \text{Let } z = 5.$$
$$y + 25 = 24$$
$$y = -1$$

A check verifies that the solution set is $\{(-1, -1, 5)\}$.

Objective 3 Practice Exercises

For extra help, see Example 2 on pages 551–552 of your text.

Solve each system of equations.

6. $\quad \begin{aligned} 7x \qquad\; + z &= -1 \\ 3y - 2z &= 8 \\ 5x + y \qquad\; &= 2 \end{aligned}$

6. _________________

7. $\quad \begin{aligned} 2x + 5y \qquad &= 18 \\ 3y + 2z &= 4 \\ \tfrac{1}{4}x - y \qquad &= -1 \end{aligned}$

7. _________________

8. $\quad \begin{aligned} 5x \qquad\;\; - 2z &= 8 \\ 4y + 3z &= -9 \\ \tfrac{1}{2}x + \tfrac{2}{3}y \qquad &= -1 \end{aligned}$

8. _________________

Objective 4   Solve special systems.

Review these examples for Objective 4:

3. Solve the system.
$$x - y + z = 7 \quad (1)$$
$$2x + 5y - 4z = 2 \quad (2)$$
$$-x + y - z = 4 \quad (3)$$

Since x in equation (1) has coefficient 1, choose x as the focus variable and (1) as the working equation. Using equations (1) and (3), we have
$$\begin{array}{l} x - y + z = 7 \quad (1) \\ \underline{-x + y - z = 4 \quad (3)} \\ \qquad\quad 0 = 11 \quad \text{False} \end{array}$$

The resulting false statement indicates that equations (1) and (3) have no common solution. Thus, the system is inconsistent and the solution set is $\varnothing$. The graph of this system would show that three planes are parallel to each other as shown below.

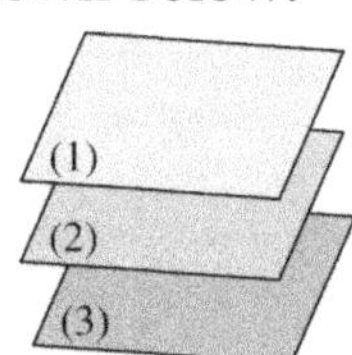

4. Solve the system.
$$3x - 2y + 5z = 4 \quad (1)$$
$$-6x + 4y - 10z = -8 \quad (2)$$
$$\frac{3}{2}x - y + \frac{5}{2}z = 2 \quad (3)$$

Multiplying each side of equation (1) by -2 gives equation (2). Multiplying each side of equation (1) by $\frac{1}{2}$ gives equation (3). Thus, the equations are dependent, and all three equations have the same graph as shown below. The solution set is written
$$\{(x, y, z) \mid 3x - 2y + 5z = 4\}.$$

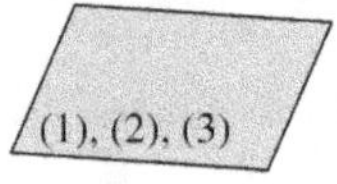

Now Try:

3. Solve the system.
$$-4x - 2y + z = -19$$
$$-6x + 2y - 6z = -8$$
$$-4x + 2y - 5z = -6$$

\underline{\hspace{3cm}}

4. Solve the system.
$$x - 5y + 2z = 0$$
$$-x + 5y - 2z = 0$$
$$\frac{1}{2}x - \frac{5}{2}y + z = 0$$

\underline{\hspace{3cm}}

5. Solve the system.
$$3x + 2y + z = 3 \quad (1)$$
$$-6x - 4y - 2z = -6 \quad (2)$$
$$x + \frac{2}{3}y + \frac{1}{3}z = 4 \quad (3)$$

Multiplying each side of equation (1) by -2 gives equation (2), so these two equations are dependent. Multiplying each side of equation (1) by $\frac{1}{3}$ does not give equation (3). Instead, we

obtain two equations with the same coefficients, but with different constant terms. The graphs of

5. Solve the system.
$$3x + 2y + z = 3$$
$$-6x - 4y - 2z = -6$$
$$9x + 6y + 3z = 1$$

____

Objective 4 Practice Exercises

For extra help, see Examples 3–5 on pages 552–553 of your text.

Solve each system of equations.

9.
$$8x - 7y + 2z = 1$$
$$3x + 4y - z = 6$$
$$-8x + 7y - 2z = 5$$

9. ________________

10.
$$3x - 2y + 4z = 5$$
$$-3x + 2y - 4z = -5$$
$$\frac{3}{2}x - y + 2z = \frac{5}{2}$$

10. ________________

11.
$$-x + 5y - 2z = 3$$
$$2x - 10y + 4z = -6$$
$$-3x + 15y - 6z = 9$$

11. ________________

Chapter 7 LINEAR EQUATIONS, GRAPHS, AND SYSTEMS

7.7   Applications of Systems of Linear Equations

Learning Objectives
1	Solve geometry problems using two variables.
2	Solve money problems using two variables.
3	Solve mixture problems using two variables
4	Solve distance-rate-time problems using two variables.
5	Solve problems with three variables using a system of three equations.

Key Terms

Use the vocabulary terms listed below to complete each statement in exercises 1−2.

elimination method   **substitution**

1. Using the addition property to solve a system of equations is called the

   _______________________________.

2. _______________________________ is being used when one expression is replaced by another.

Objective 1   Solve geometry problems using two variables.

Review this example for Objective 1:

1. The length of a rectangular field is 5 m more than the width. Find the length and width if the perimeter is 70 m.

   Step 1 Read the problem. We must find the dimensions of the field.

   Step 2 Assign variables. Let L = length and W = width.

   Step 3 Write a system of equations. We use the equation for perimeter.
   $$2L + 2W = 70$$
   A second equation uses the information given.
   $$L = W + 5$$
   The system of equations is
   $$2L + 2W = 70 \quad (1)$$
   $$L = W + 5 \quad\quad (2)$$

   Step 4 Solve the system. Since equation (2) is solved for L, we use substitution.

Now Try:

1. The length of a rectangle is 7 ft more than the width. The perimeter is 54 ft. Find the dimensions of the rectangle.

   ____

$$2L + 2W = 70$$
$$2(W + 5) + 2W = 70$$
$$2W + 10 + 2W = 70$$
$$4W + 10 = 70$$
$$4W = 60$$
$$W = 15$$

Let $W = 15$ in equation (2) to find L.
$$L = 15 + 5 = 20$$

Step 5 State the answer. The length is 20 m and the width is 15 m.

Step 6 Check.
$$2(20) + 2(15) = 70$$
$$20 = 15 + 5$$

The answer is correct.

Objective 1 Practice Exercises

For extra help, see Example 1 on pages 557–558 of your text.

Solve each problem.

1. The side of a square is 5 centimeters shorter than the side of an equilateral triangle. The perimeter of the square is 7 centimeters less than the perimeter of the triangle. Find the lengths of a side of the square and of a side of the triangle.

    1. square: __________

    triangle: __________

2. The perimeter of a rectangle is 96 inches. If the width were tripled, the width would be 36 inches more than the length. Find the length and width of the rectangle.

    2. length: __________

    width: __________

3. The perimeter of a triangle is 70 centimeters. Two sides of the triangle have the same length. The third side is 7 centimeters longer than either of the equal sides. Find the length of the equal sides of the triangle.

3. ______________

Objective 2   Solve money problems using two variables.

Review this example for Objective 2:

2. The total receipts for a basketball game were $4690.50. There were 723 tickets sold, some for children and some for adults. If the adult tickets cost $9.50 and the children's tickets cost $4, how many of each type were there?

Step 1 Read the problem. There are two unknowns.

Step 2 Assign variables.
   Let a = the number of adult tickets sold.
   Let c = the number of child tickets sold.

Step 3 Write a system of equations. We write one equation using the total number of tickets.
   $a + c = 723$
We write another equation using the cost.
   $9.50a + 4c = 4690.50$
The system of equations is
   $a + c = 723$ (1)
   $9.50a + 4c = 4690.50$ (2)

Step 4 Solve the system. To eliminate c, multiply equation (1) by –4, and add.
   $-4a - 4c = -2892$ Multiply (1) by -4.

   $9.5a + 4c = 4690.5$ (2)
   ____
   $5.5a \quad\quad = 1798.5$ Add.

   $a = 327$ Divide by 5.5.

To find the value of c, let a=327 in equation (1).
   $327 + c = 723$

   $c = 396$

Now Try:

2. The Garden Center ordered 6 ounces of marigold seed and 8 ounces of carnation seed, paying $214.54. They later ordered another 12 ounces of marigold seed and 18 ounces of carnation seed, paying $464.28. Find the price per ounce for each type of seed.

marigold   ______________

carnation   ______________

Step 5 State the answer. The number of adult tickets sold is 327 and the number of child tickets sold is 396.

Step 6 Check.

$$327 + 396 = 723$$

$$9.50(327) + 4(396) = 4690.50$$

The answer is correct.

Objective 2 Practice Exercises

For extra help, see Example 2 on page 559 of your text.

Solve each problem.

4. Pablo has some $10-bills and some $20-bills. The total value of the money is $650, with a total of 40 bills. How many of each are there?

4. $10-bills _________

   $20-bills _________

5. Big Giant Super Market will sell 5 large jars and 2 small jars of their peanut butter for $36. They will also sell 2 large jars and 5 small jars for $27. What is the price of each jar?

5. small _________

   large _________

6. A taxi charges a flat rate plus a certain charge per mile. A trip of 7 miles costs $5.30, while a trip of 3 miles costs $3.70. Find the flat rate and the charge per mile.

6. flat rate _________

   per mile _________

Objective 3   Solve mixture problems using two variables.

Review this example for Objective 3:

3. A 75% solution will be mixed with a 55% solution to get 70 liters of 63% solution? How many liters of the 55% and 75% solutions should be used?

Step 1 Read the problem. There are two solution strengths. We are looking for an "in between" strength.

Step 2 Assign variables.
    Let x = the number of liters of 75% solution.
    Let y = the number of liters of 55% solution.

Step 3 Write a system of equations.
Write one equation using the total amount.
    $x + y = 70$
Write each percent as a decimal and multiply each solution by its concentration.
    $0.75x + 0.55y = 0.63(70)$
The system of equations is
    $x + y = 70$ (1)

    $0.75x + 0.55y = 44.1$ (2)

Step 4 Solve the system. Multiply equation (2) by 100. Multiply equation (1) by −55 to eliminate y.

$$-55x - 55y = -3850 \quad \text{Multiply (1) by } -55.$$
$$\underline{75x + 55y = 4410} \quad \text{Multiply (2) by 100.}$$
$$20x = 560 \quad\quad \text{Add.}$$
$$x = 28 \quad\quad\quad \text{Divide by 20.}$$

Substitute the 28 for x in equation (1) to find the value of y.
    $28 + y = 70$

          $y = 42$

Step 5 State the answer. The desired mixture will contain 28 liters of 75% solution and 42 liters of 55% solution.

Step 6 Check.
Total amount: $28 + 42 = 70$
Total concentration: $0.75(28) + 0.55(42) = 44.1$
The answer is correct.

Now Try:

3. How many liters of water should be added to 25% antifreeze solution to get 30 liters of a 20% solution? How many liters of 25% solution are needed?

water   ______________

25% solution ______________

Objective 3 Practice Exercises

For extra help, see Example 3 on pages 560–561 of your text.

Solve each problem.

7. Jorge wishes to make 150 pounds of coffee blend that can be sold for $8 per pound. The blend will be a mixture of coffee worth $6 per pound and coffee worth $12 per pound. How many pounds of each kind of coffee should be used in the mixture?

7.
$6 coffee______________

$12 coffee_____________

8. Bags of coffee worth $90 a bag must be mixed with coffee worth $75 a bag to get 50 bags worth $87 a bag. How many bags of each are needed?

8.
$90 coffee_____________

$75 coffee_____________

9. A pharmacist wants to add water to a solution that contains 80% medicine. She wants to obtain 12 oz. of a solution that is 20% medicine. How much water and how much of the 80% solution should she use?

9.
water_________________

80% solution __________

Objective 4   Solve distance-rate-time problems using two variables.

Review this example for Objective 4:

4. A train travels 600 kilometers in the same time that a truck travels 520 kilometers. Find the speed of the train and the truck if the train's average speed is 8 kilometers per hour faster than the truck's.

Step 1 Read the problem. We need to find the rate of each vehicle.

Step 2 Assign variables.
Let x = the rate of the train.
Let y = the rate of the truck.

	d	r	t
Train	600	x	$\dfrac{600}{x}$
Truck	520	y	$\dfrac{520}{y}$

Step 3 Write a system of equations. From comparing the two speeds we have an equation.
$$x = y + 8$$
Since both vehicles travel for the same time, we have a second equation.
$$\frac{600}{x} = \frac{520}{y}$$
Multiplying both sides by xy, we have
$$600y = 520x.$$
The system of equations is
$$x = y + 8 \qquad (1)$$
$$600y = 520x \qquad (2)$$

Step 4 Solve the system. We solve the system by substitution. Replace x with $y + 8$ in equation (2).
$$600y = 520(y + 8)$$
$$600y = 520y + 4160$$
$$80y = 4160$$
$$y = 52$$
Because $x = y + 8$,
$$x = 52 + 8 = 60.$$

Step 5 State the answer. The train's rate is 60 km per hr. The truck's rate is 52 km per hr.

Now Try:

4. Ashley walks 10 miles in the same time that Taylor walks 6 miles. If Ashley walks 1 mile per hour less than twice Taylor's rate, what is the rate at which each walks?

Ashley _______________

Taylor _______________

Step 6 Check.

Train: $\dfrac{600}{60} = 10$ hr   Truck: $\dfrac{520}{52} = 10$ hr

The rate of the train is 8 km more than the rate of the truck.

Objective 4 Practice Exercises

For extra help, see Examples 4–5 on pages 561–563 of your text.

Solve each problem.

10. Two cars start together and travel in the same direction, one going twice as fast as the other. At the end of 3 hours, they are 96 miles apart. How fast is each traveling?

10.
slower________________

faster________________

11. Travis and his sister Kate jog to school daily. Travis jogs at 9 miles per hour, and Kate jogs at 5 miles per hour. When Travis reaches school, Kate is $\frac{1}{2}$ mile from the school. How far do Travis and Kate live from their school? How long does it take Travis to jog to school?

11. distance___________

time____________

Objective 5   Solve problems with three variables using a system of three equations.

Review this example for Objective 5:

7. Lee has some \$5, \$10, and \$20-bills. He has a total of 51 bills, worth \$795. The number of \$5-bills is 25 less than the number of \$20-bills. Find the number of each type of bill he has.

Step 1: Read the problem again. There are three unknowns.

Step 2: Assign variables.
Let x = the number of \$5 bills,
let y = the number of \$10 bills,
let z = the number of \$20 bills.

Step 3: Write a system of three equations.
There are a total of 51 bills, so
$$x + y + z = 51 \quad (1)$$
The bills amounted to \$795, so
$$5x + 10y + 20z = 795 \quad (2)$$
The number of \$5-bills is 25 less than the number of \$20-bills, so $x = z - 25$ or
$$x - z = -25 \quad (3)$$
The system is
$$x \; + y + \; z = \; 51 \quad (1)$$
$$5x + 10y + 20z = 795 \quad (2)$$
$$x \qquad - \; z = -25 \quad (3)$$

Step 4: Solve the system.
Eliminate y.

$$-10x - 10y - 10z = -510 \quad \text{Multiply (1) by } -10$$
$$\underline{5x + 10y + 20z = \; 795 \quad (2)}$$
$$-5x \qquad + 10z = \; 285 \quad (4)$$

Solve the system consisting of equations (3) and (4).

$$5x \; - \; 5z = -125 \quad \text{Multiply (3) by } 5$$
$$\underline{-5x + 10z = \; 285 \quad (4)}$$
$$5z = \; 160$$
$$z = 32$$

Substitute 32 for z in equation (3) and solve for x.

$$x - z = -25 \quad (3)$$
$$x - 32 = -25 \quad \text{Let } z = 32.$$
$$x = 7$$

Now Try:

7. The manager of the Sweet Candy Shop wishes to mix candy worth \$4 per pound, \$6 per pound, and \$10 per pound to get 100 pounds of a mixture worth \$7.60 per pound. The amount of \$10 candy must equal the total amounts of the \$4 and the \$6 candy. How many pounds of each must be used?

\$4 candy _____________

\$6 candy _____________

\$10 candy _____________

                     Copyright © 2025 Pearson Education, Inc.

Substitute 7 for x and 32 for z in equation (1) and solve for y.

$$x + y + z = 51 \quad (1)$$
$$7 + y + 32 = 51 \quad \text{Let } x = 7,\ z = 32.$$
$$y + 39 = 51$$
$$y = 12$$

Step 5: State the answer. Lee has 7 \$5-bills, 12 \$10-bills, and 32 \$20-bills.

Step 6: Check that the total value of the bills is \$795 and that the number of \$5-bills is 25 less than the number of \$20-bills.

Objective 5 Practice Exercises

For extra help, see Examples 6–7 on pages 564–566 of your text.

Solve each problem involving three unknowns.

12. Julie has \$80,000 to invest. She invests part at 5%, one fourth this amount at 6%, and the balance 7%. Her total annual income from interest is \$4700. Find the amount invested at each rate.

12. 5% _______________

6% _______________

7% _______________

13. A merchant wishes to mix gourmet coffee selling for \$8 per pound, \$10 per pound, and \$15 per pound to get 50 pounds of a mixture that can be sold for \$11.70 per pound. The amount of the \$8 coffee must be 3 pounds more than the amount of the \$10 coffee. Find the number of pounds of each that must be used.

13. \$8/lb _______________

\$10/lb _______________

\$15/lb _______________

14. A boy scout troop is selling popcorn. There are three different kinds of popcorn in three different arrangements. Arrangement I contains 1 bag of cheddar cheese popcorn, 2 bags of caramel popcorn, and 3 bags of microwave popcorn. Arrangement II contains 3 bags of cheddar cheese popcorn, 1 bag of caramel popcorn, and 2 bags of microwave popcorn. Arrangement III contains 2 bags of cheddar cheese popcorn, 3 bags of caramel popcorn, and 1 bag of microwave popcorn. Jim needs 28 bags of cheddar cheese popcorn, 22 bags of caramel popcorn, and 22 bags of microwave popcorn to give as stocking stuffers for Christmas. How many of each arrangement should he buy?

14. I ____________

II ____________

III ____________

Chapter 8   INEQUALITIES AND ABSOLUTE VALUE

8.1   Review of Linear Inequalities in One Variable

Learning Objectives
1   Review inequalities and interval notation.
2   Solve linear inequalities using the addition property.
3   Solve linear inequalities using the multiplication property.
4   Solve linear inequalities with three parts.

Key Terms

Use the vocabulary terms listed below to complete each statement in exercises 1−4.

interval   **interval notation**   **inequality**

linear inequality in one variable

1.   The ___________________________ for $a \leq x < b$ is $[a, b)$.

2.   A(n) ______________________ can be written in the form $Ax + B > C$, $Ax + B \geq C$, $Ax + B < C$, or $Ax + B \leq C$, where A, B, and C are real numbers with $A \neq 0$.

3.   An algebraic expression related by $>$, $\geq$, $<$, or $\leq$ is called a(n) _______________.

4.   An ___________________________ is a portion of a number line.

Objective 1   Review inequalities and interval notation.

For extra help, see pages 590–591 of your text.

Objective 2   Solve linear inequalities using the addition property.

Review this example for Objective 2:	Now Try:
1. Solve $x - 6 < -11$ and graph the solution set.	1. Solve $x + 3 < -1$ and graph the solution set.

Review this example for Objective 2:

1. Solve $x - 6 < -11$ and graph the solution set.

$$x - 6 < -11$$
$$x - 6 + 6 < -11 + 6$$
$$x < -5$$

A check confirms that $(-\infty, -5)$, graphed below, is the solution set.

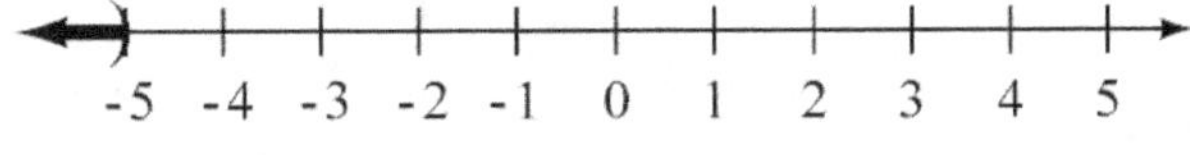

Now Try:

1. Solve $x + 3 < -1$ and graph the solution set.

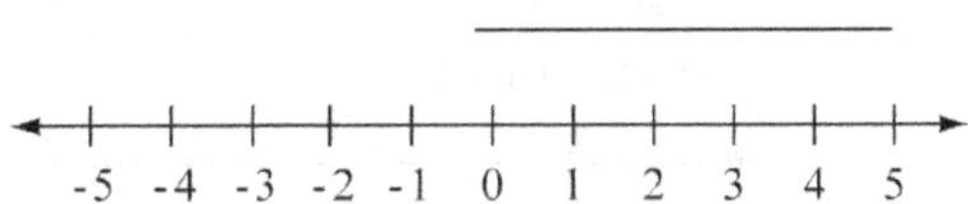

Objective 2 Practice Exercises

For extra help, see Example 1 on page 591 of your text.

Solve each inequality, giving its solution set in both interval and graph forms. Check your answers.

1. $5a + 3 \leq 6$

1.   __________________________

2. $6 + 3x < 4x + 4$

2.   __________________________

3. $3 + 5p \leq 4p + 3$

3.   __________________________

Objective 3   Solve linear inequalities using the multiplication property.

Review these examples for Objective 3:

2.   Solve the inequality, and graph the solution set.

$$6x < -24$$

We divide each side by 6.

$$6x < -24$$

$$\frac{6x}{6} < \frac{-24}{6}$$

$$x < -4$$

The graph of the solution set $(-\infty, -4)$, is shown below.

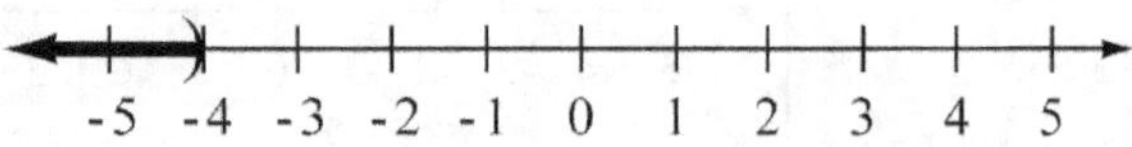

Now Try:

2.   Solve the inequality, and graph the solution set.

$$8x \leq -40$$

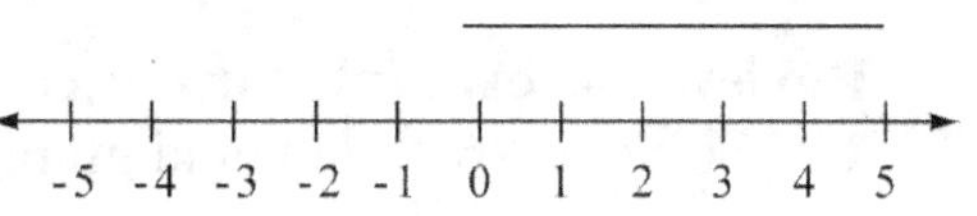

            Copyright © 2025 Pearson Education, Inc.

3. Solve $-5(x+3)+3 \geq 8-3x$, and graph the solution set.

Step 1 $\quad -5(x+3)+3 \geq 8-3x$

$\qquad\qquad -5x-15+3 \geq 8-3x$

$\qquad\qquad\quad -5x-12 \geq 8-3x$

Step 2 $\quad -5x-12+3x \geq 8-3x+3x$

$\qquad\qquad\quad\quad -2x-12 \geq 8$

$\qquad\qquad -2x-12+12 \geq 8+12$

$\qquad\qquad\qquad\quad -2x \geq 20$

Step 3 $\qquad\qquad \dfrac{-2x}{-2} \leq \dfrac{20}{-2}$

$\qquad\qquad\qquad\quad x \leq -10$

The solution set is $(-\infty, -10]$. The graph is shown below.

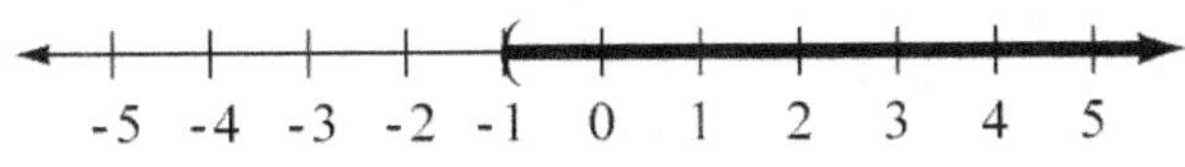

4. Solve $-\dfrac{2}{5}(r-4)-\dfrac{1}{3} < \dfrac{1}{3}(4-r)$, and graph the solution set.

Step 1 $\quad 15\left[-\dfrac{2}{5}(r-4)-\dfrac{1}{3}\right] < 15\left[\dfrac{1}{3}(4-r)\right]$

$\qquad\quad 15\left[-\dfrac{2}{5}(r-4)\right]-15\left(\dfrac{1}{3}\right) < 15\left[\dfrac{1}{3}(4-r)\right]$

$\qquad\qquad\qquad -6(r-4)-5 < 5(4-r)$

$\qquad\qquad\qquad -6r+24-5 < 20-5r$

$\qquad\qquad\qquad\quad -6r+19 < 20-5r$

Step 2 $\quad -6r+19+5r < 20-5r+5r$

$\qquad\qquad\qquad -r+19 < 20$

$\qquad\qquad -r+19-19 < 20-19$

$\qquad\qquad\qquad\qquad -r < 1$

Step 3 $\qquad\qquad -1(-r) > -1(1)$

$\qquad\qquad\qquad\qquad r > -1$

The solution set is $(-1, \infty)$. The graph is shown below.

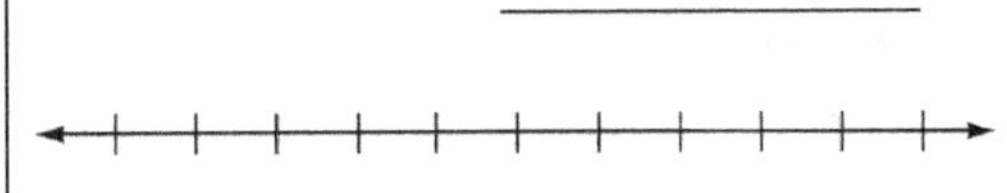

---

3. Solve $-6(x+4)+1 \geq 9-2x$, and graph the solution set.

____

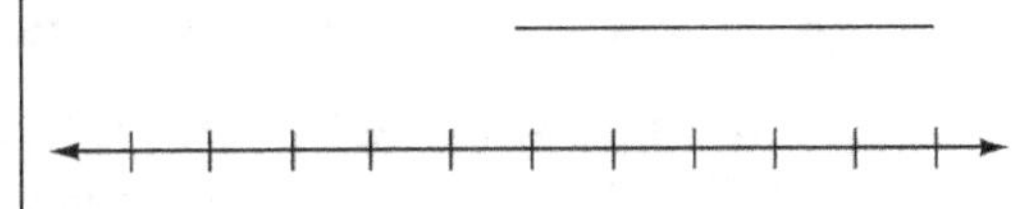

4. Solve $-\dfrac{1}{2}(r-5)+\dfrac{1}{3} < \dfrac{1}{3}(5-r)$, and graph the solution set.

____

Name: ________________________   Date: ________________

Instructor: ___________________   Section: _____________

Objective 3 Practice Exercises

For extra help, see Examples 2–4 on pages 592–594 of your text.

Solve each inequality, giving its solution set in both interval and graph forms. Check your answers.

4. $\quad -2s < 4$

4. ______________________

5. $\quad 4k \geq -16$

5. ______________________

6. $\quad -9m \geq -36$

6. ______________________

Objective 4   Solve linear inequalities with three parts.

Review this example for Objective 4:

5. Solve $3 \leq 4x - 5 < 7$ and graph the solution set.

$$3 \leq -4x - 5 < 7$$
$$3 + 5 \leq -4x < 7 + 5$$
$$8 \leq -4x < 12$$
$$-\frac{8}{4} \geq -\frac{4x}{4} > -\frac{12}{4}$$
$$-2 \geq x > -3$$
$$-3 < x \leq -2$$

The solution set is $(-3, -2]$. The graph is shown below.

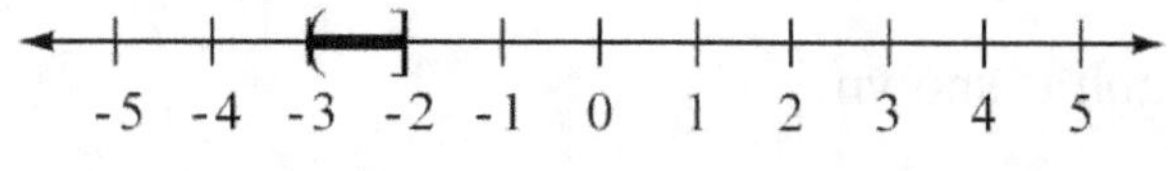

Now Try:

5. Solve $8 \leq -6x - 4 < 20$ and graph the solution set.

____

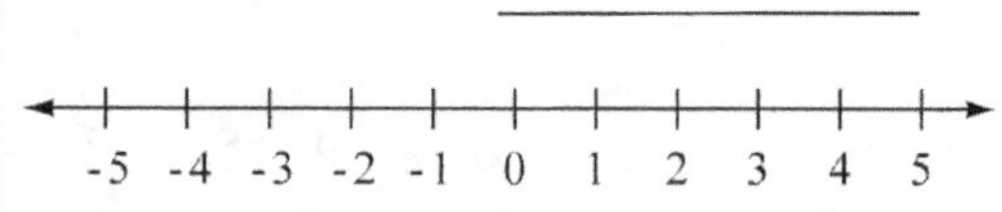

Objective 4 Practice Exercises

For extra help, see Examples 5–6 on page 595 of your text.

Solve each inequality, giving its solution set in both interval and graph forms. Check your answers.

7. $7 < 2x + 3 \leq 13$ **7.** _______________________

8. $-17 \leq 3x - 2 < -11$ **8.** _______________________

9. $1 < 3z + 4 < 19$ **9.** _______________________

Chapter 8 INEQUALITIES AND ABSOLUTE VALUE

8.2   Set Operations and Compound Inequalities

Learning Objectives
1   Recognize set intersection and union.
2   Find the intersection of two sets.
3   Solve compound inequalities with the word *and*.
4   Find the union of two sets.
5   Solve compound inequalities with the word *or*.

Key Terms

Use the vocabulary terms listed below to complete each statement in exercises 1−3.

  intersection   **compound inequality**   **union**

1.   The _________________________ of two sets, A and B, is the set of elements that belong to either A or B or both.

2.   A ____ is formed by joining two inequalities with a connective word such as *and* or *or*.

3.   The _________________________ of two sets, A and B, is the set of elements that belong to both A and B.

Objective 2   Find the intersection of two sets.

Review this example for Objective 2:

1.   Let $A = \{6, 7, 8, 9\}$ and $B = \{6, 8, 10\}$. Find $A \cap B$.

The set $A \cap B$ contains those elements that belong to both A and B.

$$A \cap B = \{6, 7, 8, 9\} \cap \{6, 8, 10\}$$
$$= \{6, 8\}$$

Now Try:

1.   Let $A = \{20, 30, 40, 50\}$ and $B = \{30, 50, 70\}$. Find $A \cap B$.

____

Objective 2 Practice Exercises

For extra help, see Example 1 on page 598 of your text.

Let $A = \{0, 1, 2, 3, 4, 5\}$, $B = \{2, 4, 6, 8, 10\}$, $C = \{1, 3, 5, 7, 9\}$, and $D = \{0, 2, 4\}$.
Specify each set.

1. $A \cap D$

1.   _______________

2. $B \cap C$ 2. _______________

3. $A \cap C$ 3. _______________

Objective 3   Solve compound inequalities with the word *and*.

Review these examples for Objective 3:

2. Solve $x+6 \le 15$ and $x+3 \ge 7$, and graph the solution set.

Step 1 Solve each inequality individually.

$$x+6 \le 15 \qquad \text{and} \qquad x+3 \ge 7$$
$$x+6-6 \le 15-6 \quad \text{and} \quad x+3-3 \ge 7-3$$
$$x \le 9 \qquad \text{and} \qquad x \ge 4$$

Step 2 The solution set of the compound inequality includes all numbers that satisfy both inequalities from Step 1.

The graphs of $x \le 9$ and $x \ge 4$ are shown below.

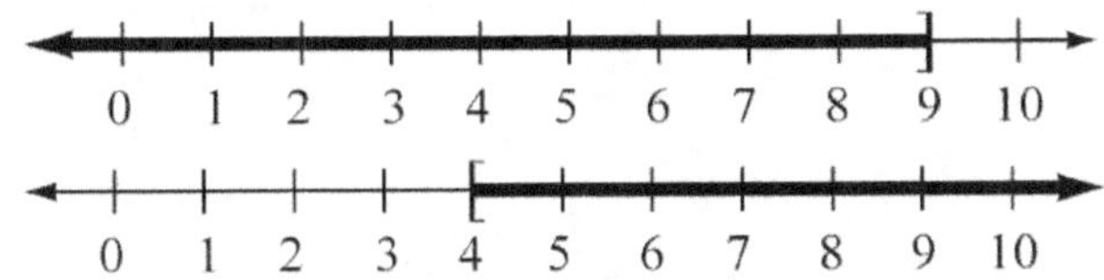

The intersection of these two graphs is the solution set [4, 9].

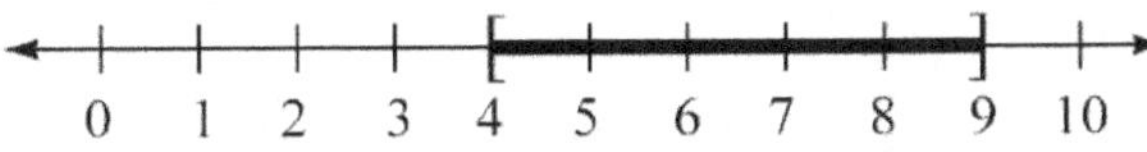

3. Solve $-5x-6 > 8$ and $6x-3 \le -21$, and graph the solution set.

Step 1 Solve each inequality individually.

$$-5x-6 > 8 \qquad \text{and} \quad 6x-3 \le -21$$
$$-5x > 14 \qquad \text{and} \qquad 6x \le -18$$
$$x < -\frac{14}{5} \quad \text{and} \qquad x \le -3$$

The graphs of $x < -\dfrac{14}{5}$ and $x \le -3$ are shown

Now Try:

2. Solve $x+4 \le 20$ and $x-3 \ge 9$ and graph the solution set.

____

3. Solve $-8x+6 > 30$ and $4x-7 \le 11$, and graph the solution set.

____

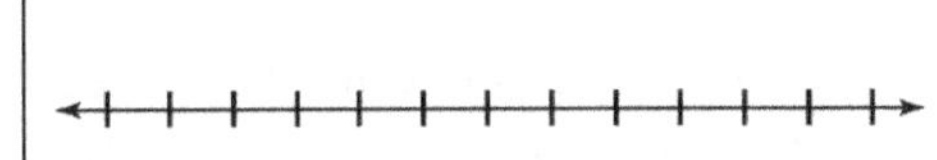

                    329

below.   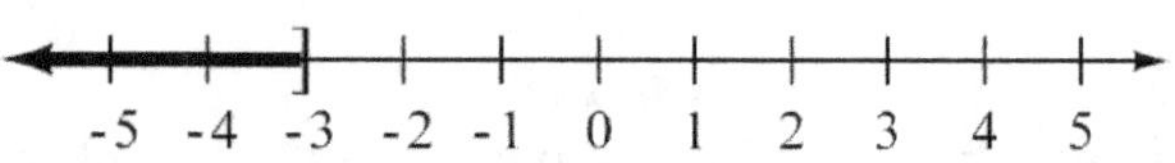

$Step\ 2$ Now find all the values of x that are less than $-\dfrac{14}{5}$ and also less than or equal to -3. The solution set is $(-\infty, -3]$.

Objective 3 Practice Exercises

For extra help, see Examples 2–4 on pages 599–600 of your text.

For each compound inequality, give the solution set in both interval and graph forms.

4. $1 - 2s \le 7$ and $2s + 7 \ge 11$

4.   ______________________

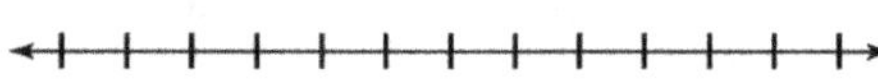

5. $3x + 2 < 11$ and $2 - 3x \le 14$

5.   ______________________

6. $5t > 0$ and $5t + 4 \le 9$

6.   ______________________

Objective 4   Find the union of two sets.

Review this example for Objective 4:

5.   Let $A = \{6,\ 7,\ 8,\ 9\}$ and $B = \{6,\ 8,\ 10\}$. Find $A \cup B$.

The set $A \cup B$ contains those elements that belong to either A or B (or both).

$$A \cup B = \{6,\ 7,\ 8,\ 9\} \cup \{6,\ 8,\ 10\}$$
$$= \{6,\ 7,\ 8,\ 9,\ 10\}$$

Now Try:

5.   Let $A = \{20,\ 30,\ 40,\ 50\}$ and $B = \{30,\ 50,\ 70\}$. Find $A \cup B$.

____

          Copyright © 2025 Pearson Education, Inc.

Objective 4 Practice Exercises

For extra help, see Example 5 on page 600 of your text.

Let $A = \{0, 1, 2, 3, 4, 5\}$, $B = \{2, 4, 6, 8, 10\}$, $C = \{1, 3, 5, 7, 9\}$, $D = \{0, 2, 4\}$, and $E = \{0\}$. Specify each set.

7. $A \cup D$ 7. ______________

8. $B \cup C$ 8. ______________

9. $A \cup E$ 9. ______________

Objective 5   Solve compound inequalities with the word *or*.

Review these examples for Objective 5:

6. Solve the compound inequality, and graph the solution set.

$$7x - 6 < 4x \quad \text{or} \quad -4x \le -16$$

Step 1 Solve each inequality individually.

$$7x - 6 < 4x \quad \text{or} \quad -4x \le -16$$
$$3x < 6$$
$$x < 2 \quad \text{or} \quad x \ge 4$$

The graphs of $x < 2$ and $x \ge 4$ are shown below.

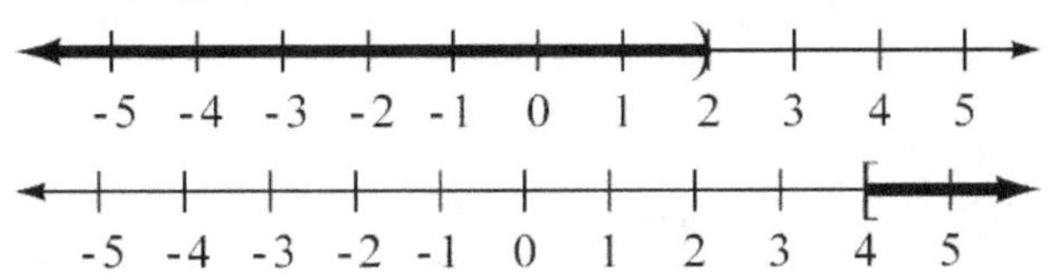

Step 2 Since the inequalities are joined with *or*, find the union of the two solution sets. The union is shown below. $(-\infty, 2) \cup [4, \infty)$

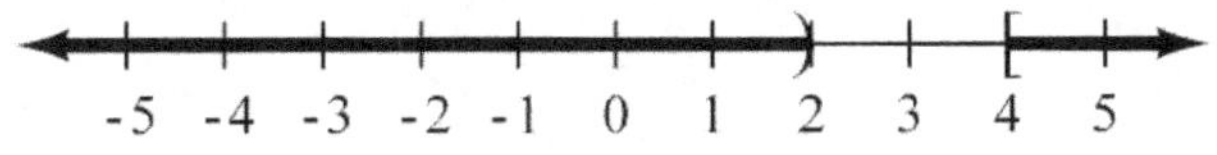

Now Try:

6. Solve the compound inequality, and graph the solution set.

$$5x - 6 < 2x \quad \text{or} \quad -7x \le -35$$

____

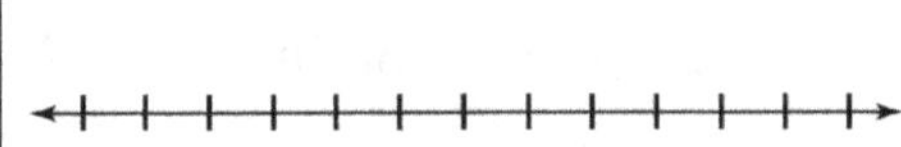

8. Solve the compound inequality, and graph the solution set.
$$-5x+4 \geq -16 \text{ or } 6x-11 \geq -29$$

Solve each inequality.
$$-5x+4 \geq -16 \quad \text{or} \quad 6x-11 \geq -29$$
$$-5x \geq -20 \quad \text{or} \quad 6x \geq -18$$
$$x \leq 4 \quad \text{or} \quad x \geq -3$$

Graph each inequality.

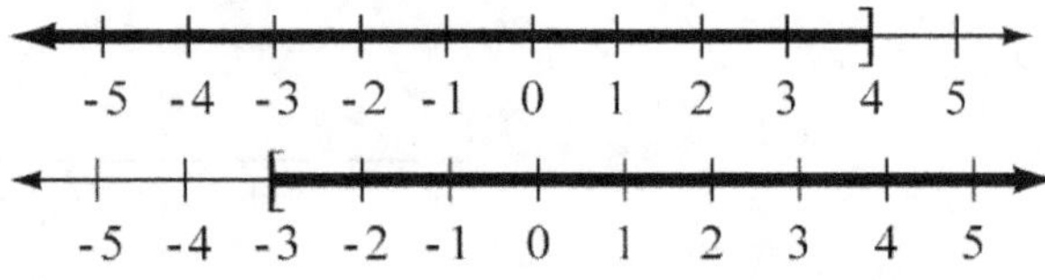

The solution is the union of the two sets,
$(-\infty, \infty)$.

8. Solve the compound inequality, and graph the solution set.
$$-9x+3 \geq -15 \text{ or } 8x-7 \geq -15$$

____

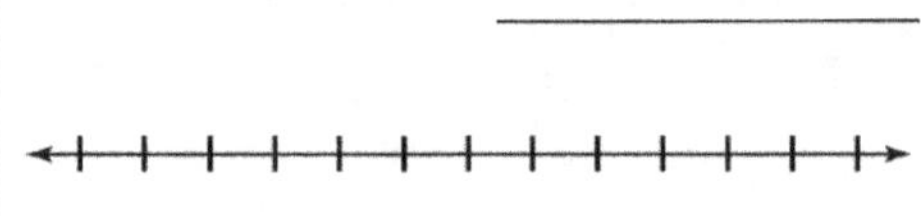

Objective 5 Practice Exercises

For extra help, see Examples 6–9 on pages 601–603 of your text.

For each compound inequality, give the solution set in both interval and graph forms.

10. $q + 3 > 7 \text{ or } q + 1 \leq -3$

10. ____________________

11. $3 > 4m + 2 \text{ or } 4m - 3 \geq -2$

11. ____________________

12. $2r + 4 \geq 8 \text{ or } 4r - 3 < 1$

12. ____________________

Chapter 8   INEQUALITIES AND ABSOLUTE VALUE

8.3   Absolute Value Equations and Inequalities

Learning Objectives

1.   Use the distance definition of absolute value.
2.   Solve equations of the form $|ax + b| = k$, for $k > 0$.
3.   Solve inequalities of the form $|ax + b| < k$ and of the form $|ax + b| > k$, for $k > 0$.
4.   Solve absolute value equations that involve rewriting.
5.   Solve equations of the form $|ax + b| = |cx + d|$.
6.   Solve special cases of absolute value equations and inequalities.
7.   Solve an application involving relative error.

Key Terms

Use the vocabulary terms listed below to complete each statement in exercises 1−2.

absolute value equation   **absolute value inequality**

1.   An _______________________________________ is an equation that
     involves the absolute value of a variable expression.

2.   An _______________________________________ is an inequality that
     involves the absolute value of a variable expression.

Objective 2   Solve equations of the form $|ax + b| = k$, for $k > 0$.

Review this example for Objective 2:

1.   Solve $|3x + 2| = 14$. Graph the solution set.

This is Case 1.

$$3x + 2 = 14 \quad \text{or} \quad 3x + 2 = 14$$
$$3x = -16 \quad \text{or} \quad 3x = 12$$
$$x = -\frac{16}{3} \quad \text{or} \quad x = 4$$

Check

$$\text{Let } x = -\frac{16}{3} \qquad \text{Let } x = 4$$

$$3\left(-\frac{16}{3}\right) + 2 \overset{?}{=} -14 \ \bigg| \ 3(4) + 2 \overset{?}{=} 14$$

$$-16 + 2 \overset{?}{=} -14 \ \bigg| \ 12 + 2 \overset{?}{=} 14$$

$$-14 = -14 \ \bigg| \ 14 = 14$$

The check confirms that the solution set is
$\left\{-\dfrac{16}{3},\ 4\right\}$.

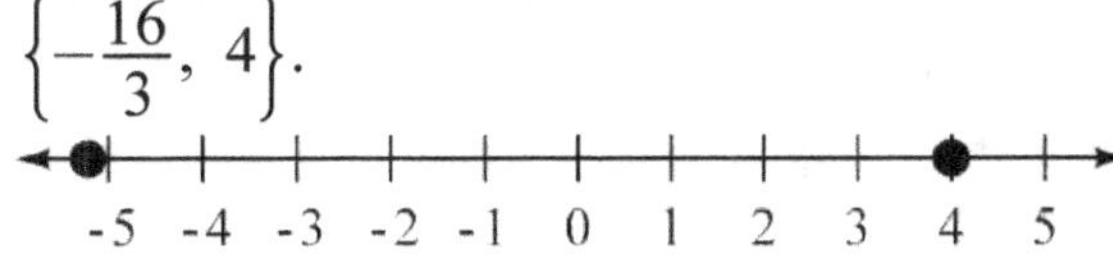

Now Try:

1.   Solve $|5x + 4| = 11$. Graph the solution set.

____

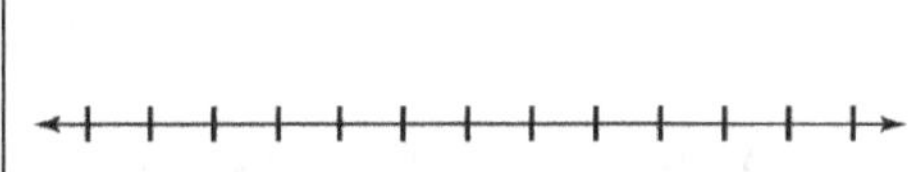

Objective 2 Practice Exercises

For extra help, see Example 1 on pages 608–609 of your text.

Solve each equation.

1. $|2x+3|=10$

1. _______________

2. $|5r-15|=0$

2. _______________

3. $\left|\dfrac{1}{2}x-3\right|=4$

3. _______________

Objective 3 **Solve inequalities of the form $|ax + b| < k$ and of the form $|ax + b| > k$, for $k > 0$.**

Review these examples for Objective 3:

2. Solve $|4x+2|>10$. Graph the solution set.

This is Case 2.
$$4x+2<-10 \quad \text{or} \quad 4x+2>10$$
$$4x<-12 \quad \text{or} \quad 4x>8$$
$$x<-3 \quad \text{or} \quad x>2$$
A check confirms that the solution set is $(-\infty,-3)\cup(2, \infty)$.

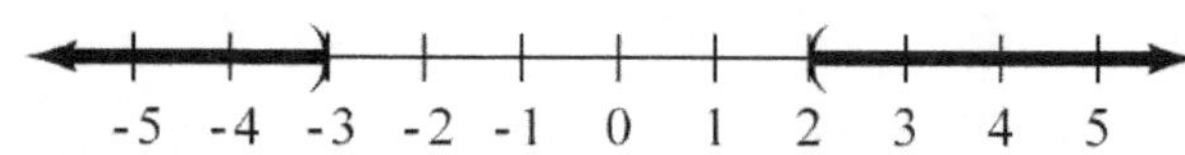

3. Solve $|10x+5|<25$. Graph the solution set.

This is Case 3.
$$-25<10x+5<25$$
$$-30< 10x <20$$
$$-3< x <2$$
A check confirms that the solution set is $(-3, 2)$.

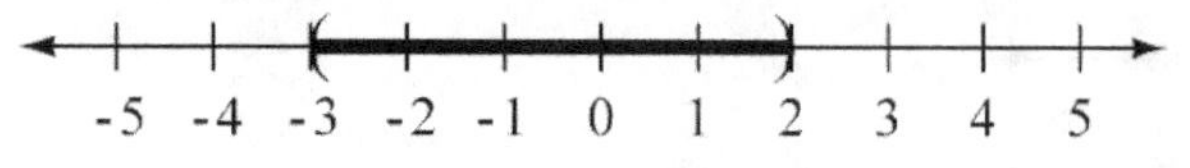

Now Try:

2. Solve $|8x+4|>12$. Graph the solution set.

____

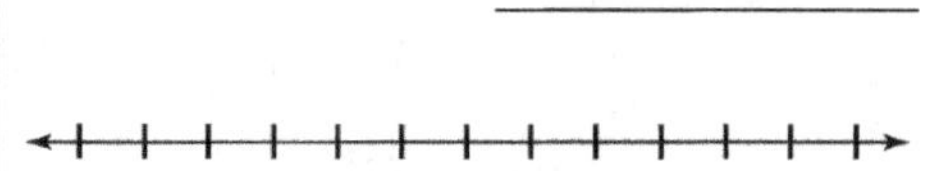

3. Solve $|6x+3|<15$. Graph the solution set.

____

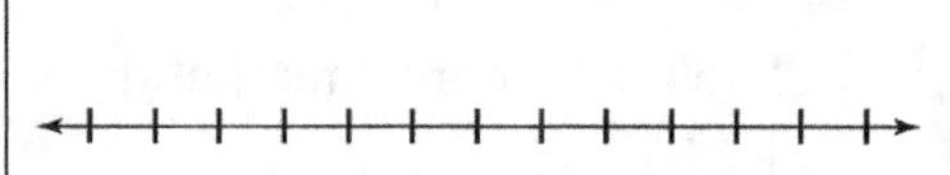

              Copyright © 2025 Pearson Education, Inc.

Name: _______________________________   Date: _______________

Instructor: ___________________________   Section: ____________

Objective 3 Practice Exercises

For extra help, see Examples 2–4 on pages 609–610 of your text.

Solve each inequality and graph the solution set.

4. $|x-2|>8$

4. _________________________

5. $|2r-9|\geq 23$

5. _________________________

6. $|5r+2|<18$

6. _________________________

Objective 4   Solve absolute value equations that involve rewriting.

Review this example for Objective 4:

5. Solve $|x-7|+3=15$.

Isolate the absolute value.
$$|x-7|+3=15$$
$$|x-7|+3-3=15-3$$
$$|x-7|=12$$
Now solve, using Case 1.
$$x-7=-12 \quad \text{or} \quad x-7=12$$
$$x=-5 \quad \text{or} \quad x=19$$
Check

Let $x=-5$ Let $x=19$

$$|-5-7|+3\overset{?}{=}15 \quad\Big|\quad |19-7|+3\overset{?}{=}15$$
$$|-12|+3\overset{?}{=}15 \quad\Big|\quad |12|+3\overset{?}{=}15$$
$$15=15 \quad\Big|\quad 15=15$$
The check confirms that the solution set is $\{-5, 19\}$.

Now Try:

5. Solve $|x-3|+4=11$.

____

6. Solve each inequality.

  a. $|x-7|+2 \geq 15$

$$|x-7|+2 \geq 15$$
$$|x-7| \geq 13$$
$$x-7 \leq -13 \quad \text{or} \quad x-7 \geq 13$$
$$x \leq -6 \quad \text{or} \quad x \geq 20$$

The solution set is $(-\infty, -6] \cup [20, \infty)$.

  b. $|x-7|+2 \leq 15$

$$|x-7|+2 \leq 15$$
$$|x-7| \leq 13$$
$$-13 \leq x-7 \leq 13$$
$$-6 \leq x \leq 20$$

The solution set is [–6, 20].

6. Solve each inequality.

  a. $|x+9|-2 \geq 11$

      ____

  b. $|x+9|-2 \leq 11$

      ____

Objective 4 Practice Exercises

For extra help, see Examples 5–6 on page 611 of your text.

Solve each equation.

7. $|2w-1|+7=12$

    7. ___________________

8. $\left|2-\dfrac{1}{2}x\right|-5=18$

    8. ___________________

9. $|4t+3|+8=10$

    9. ___________________

　　　　Copyright © 2025 Pearson Education, Inc.

Objective 5   Solve equations of the form $|ax + b| = |cx + d|$.

Review this example for Objective 5: | **Now Try:**

7.   Solve $|z+5| = |3z-9|$.

7.   Solve $|z+7| = |2z+8|$.

$$z+5 = -(3z-9) \quad \text{or} \quad z+5 = 3z-9$$
$$z+5 = -3z+9 \quad \text{or} \quad z+14 = 3z$$
$$4z = 4 \quad\quad \text{or} \quad\quad 14 = 2z$$
$$z = 1 \quad\quad\quad \text{or} \quad\quad\quad z = 7$$

Check

Let $z = 1$ Let $z = 7$

$$|1+5| \overset{?}{=} |3(1)-9| \quad\quad |7+5| \overset{?}{=} |3(7)-9|$$

$$|6| \overset{?}{=} |-6| \quad\quad\quad |12| \overset{?}{=} |21-9|$$

$$6 = 6 \quad\quad\quad\quad 12 = 12$$

The check confirms that the solution set is $\{1, 7\}$.

Objective 5 Practice Exercises

For extra help, see Example 7 on page 612 of your text.

Solve each problem.

10. $|y+5| = |3y+1|$

10.   ______________

11. $|2p-4| = |7-p|$

11.   ______________

12. $|3x-2| = |5x+8|$

12.   ______________

Objective 6   Solve special cases of absolute value equations and inequalities.

Review these examples for Objective 6:	**Now Try:**

8.   Solve each equation.

   a. $|9x-1|=-13$

The absolute value of an expression can never be negative, so there are no solutions for this equation. The solution set is $\varnothing$.

   b. $|8x-12|=0$

$$8x-12=0$$
$$8x=12$$
$$x=\frac{3}{2}$$

The solution set is $\left\{\frac{3}{2}\right\}$.

9.   Solve each inequality.

   a. $|x|\geq-12$

The absolute value of a number is always greater than or equal to 0. Thus, $|x|\geq-12$ is always true, and the solution set is $(-\infty,\ \infty)$.

   b. $|x-11|+7<2$

$$|x-11|+7<2$$
$$|x-11|<-5$$

There is no number whose absolute value is less than -5, so this inequality has no solution. The solution set is $\varnothing$.

   c. $|2x-8|-5\leq-5$

$$|2x-8|-5\leq-5$$
$$|2x-8|\leq0$$

The value of $|2x-8|$ will never be less than zero. However, $|2x-8|$ will equal 0.

$$2x-8=0$$
$$2x=8$$
$$x=4$$

The solution set is $\{4\}$.

Now Try:

8.   Solve each equation.

   a. $|6x+13|=-5$

____

   b. $|7x+35|=0$

____

9.   Solve each inequality.

   a. $|x|\geq-2$

____

   b. $|x+12|+6<3$

____

   c. $|5x-20|+7\leq7$

____

Objective 6 Practice Exercises

For extra help, see Examples 8–9 on pages 612–613 of your text.

Solve each problem.

13. $\left|7+\dfrac{1}{2}x\right|=0$

13. _________________

14. $|m-2|\geq -1$

14. _________________

15. $|k+5|\leq -2$

15. _________________

Objective 7   Solve an application involving relative error.

Review this example for Objective 7:

10. Suppose a machine filling 16.9 oz water bottles is set for a relative error that is no greater than 0.025 oz. How many ounces may a filled water bottle contain?

$$\left|\frac{16.9-x}{16.9}\right|\leq 0.025$$

$$-0.025\leq \frac{16.9-x}{16.9}\leq 0.025$$

$$-0.4225\leq 16.9-x\leq 0.4225$$

$$-17.3225\leq\ -x\ \leq -16.4775$$

$$17.3225\geq\ \ x\ \ \geq 16.4775$$

$$16.4775\leq\ \ x\ \ \leq 17.3225$$

The bottle may contain between 16.4775 and 17.3225 oz, inclusive.

Now Try:

10. Suppose a machine filling 16.9 oz water bottles is set for a relative error that is no greater than 0.05 oz. How many ounces may a filled water bottle contain?

____

Objective 7 Practice Exercises

For extra help, see Example 10 on page 613 of your text.

Determine the number of ounces a filled 16.9 oz water bottle may contain for the given relative error.

16. no greater than 0.04 oz

16. ______________________

17. no greater than 0.015 oz

17. ______________________

18. no greater than 0.03 oz

18. ______________________

Chapter 8   INEQUALITIES AND ABSOLUTE VALUE

8.4   Linear Inequalities and Systems in Two Variables

Learning Objectives
1   Graph linear inequalities in two variables.
2   Solve systems of linear inequalities by graphing.

Key Terms

Use the vocabulary terms listed below to complete each statement in exercises 1–4.

> **linear inequality in two variables**   **system of linear inequalities**
>
> **boundary line**   **solution set of a system of linear inequalities**

1.   In the graph of a linear inequality, the _________________ separates the region that satisfies the inequality from the region that does not satisfy the inequality.

2.   All ordered pairs that make all inequalities of the system true at the same time is called the _________________________________.

3.   A _________________________________ contains two or more linear inequalities (and no other kinds of inequalities).

4.   An inequality that can be written in the form $Ax + B < C$, $Ax + B > C$, $Ax + B \leq C$, or $Ax + B \geq C$ is called a _________________________.

Objective 1   Graph linear inequalities in two variables.

Review these examples for Objective 1:

1.   Graph $3x - 2y \leq 6$.

Step 1 Solve this inequality for y.
$$3x - 2y \leq 6$$
$$-2y \leq -3x + 6$$
$$y \geq \frac{3}{2}x - 3$$

Step 2 Graph the related equation $y = \frac{3}{2}x - 3$,

which has slope $\frac{3}{2}$ and y-intercept $(0,-3)$. Use the geometric interpretation of slope to find the second point, $(2, 0)$. Draw the line through these two points as a *solid* boundary line because of the $\geq$ symbol.

Now Try:

1.   Graph $2x + 5y \leq -8$.

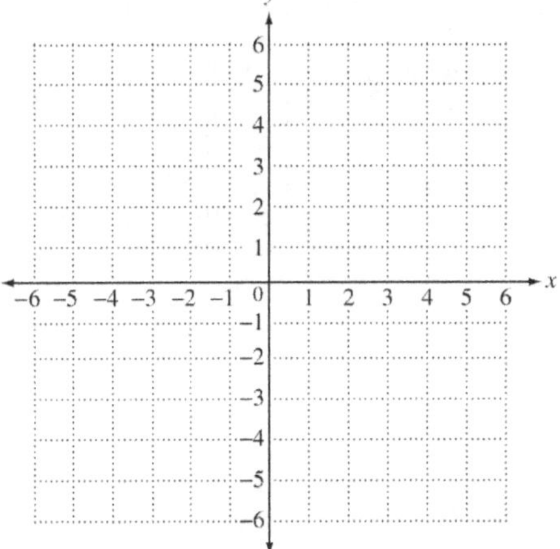

Step 3 Shade the appropriate region. Because the inequality symbol is $\geq$, we shade the region *above* the line.

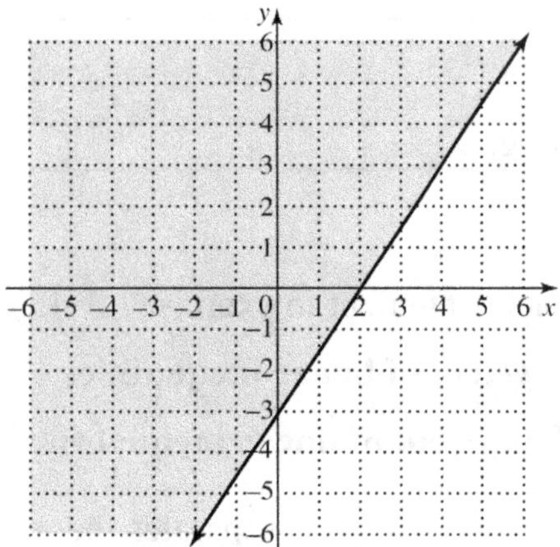

2.   Graph $y \geq 3x$.

Step 1 The inequality is already solved for y.

Step 2 Graph the equation. This line passes through the origin, $(0, 0)$, and its slope is 3. Draw the boundary line as *solid* because of the $\geq$ symbol.

Step 3 Shade the region *above* the boundary line because of the $\geq$ symbol.

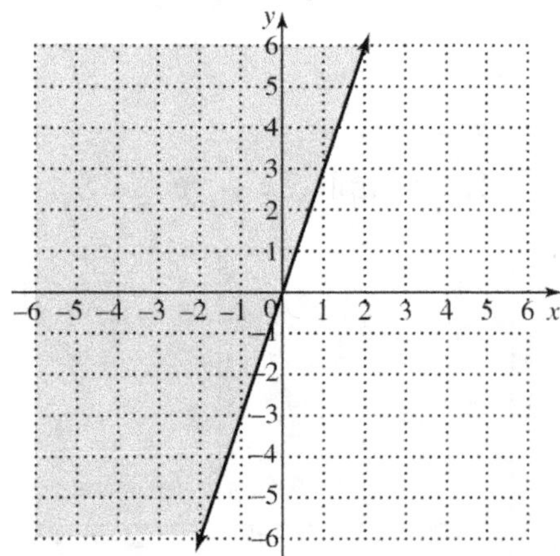

2.   Graph $y \geq x$.

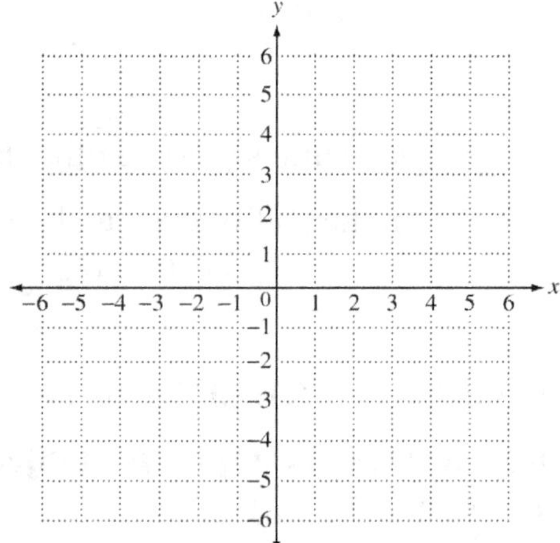

          Copyright © 2025 Pearson Education, Inc.

3. Graph $x - 4 \leq -1$.

First, solve the inequality for x.

$x \leq 3$

Now graph the line $x = 3$, a vertical line through the point $(3, 0)$. Use a solid line, and choose $(0, 0)$ as a test point.

$0 \leq 3$ True

Because $0 \leq 3$ is true, we shade the region containing $(0, 0)$.

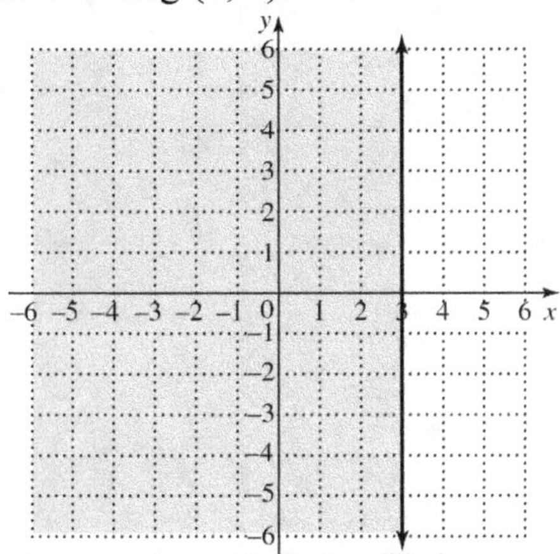

3. Graph $y \geq -1$.

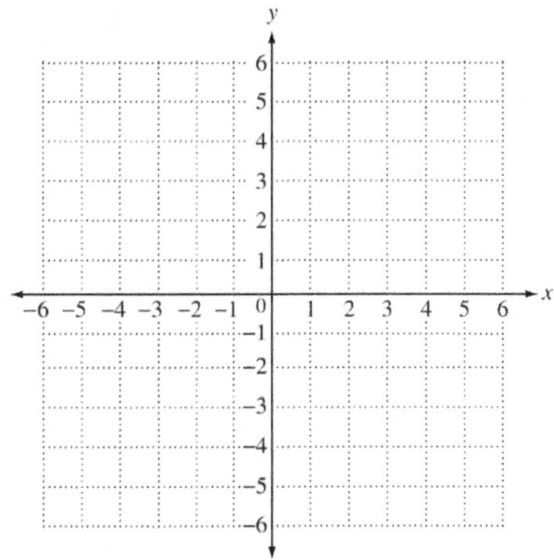

Objective 1 Practice Exercises

For extra help, see Examples 1–3 on pages 620–621 of your text.

Graph each linear inequality.

1. $x - y < 5$

1.

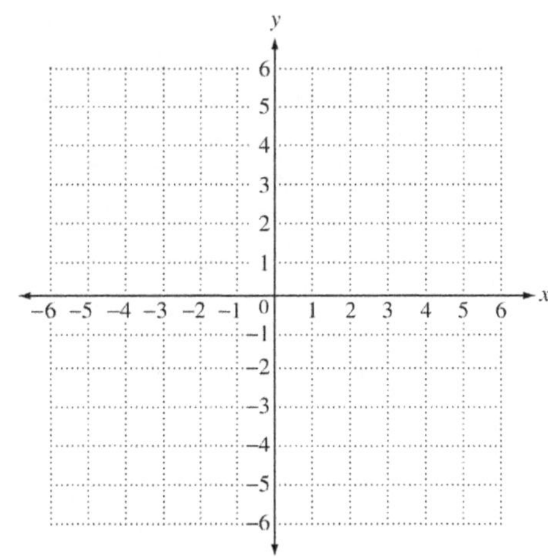

2. $2x + 3y \geq 6$ **2.**

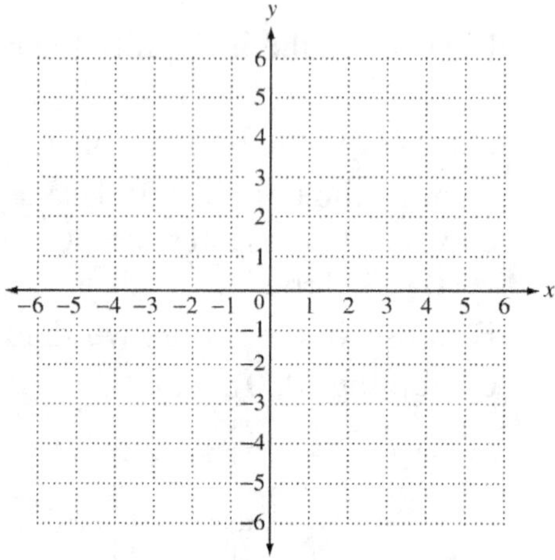

3. $x \leq 4y$ **3.**

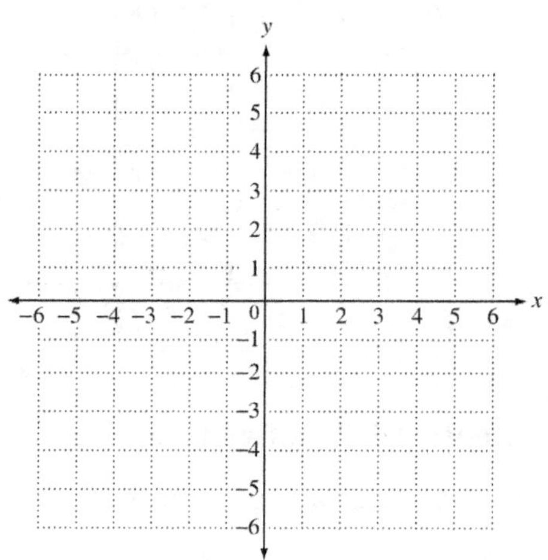

          Copyright © 2025 Pearson Education, Inc.

Objective 2 Solve systems of linear inequalities by graphing.

Review this example for Objective 2:

4. Graph the solution set of the system.

$$x + y \leq 3$$
$$5x - y \geq 5$$

Step 1 To graph each linear inequality, first solve for y.

$x + y \leq 3$	$5x - y \geq 5$
$y \leq -x + 3$	$-y \geq -5x + 5$
	$y \leq 5x - 5$

To graph $y \leq -x + 3$, graph the solid boundary line $y = -x + 3$ using the intercepts (0, 3) and (3, 0). Shade the region *below* the line.

To graph $y \leq 5x - 5$, graph the solid boundary line $y = 5x - 5$ using the intercepts (0, –5) and (1, 0). Shade the region *below* the line.

Step 2 The solution set of this system includes all points in the intersection (overlap) of the graph of the two inequalities. This intersection is the gray shaded region and portions of the two boundary lines that surround it.

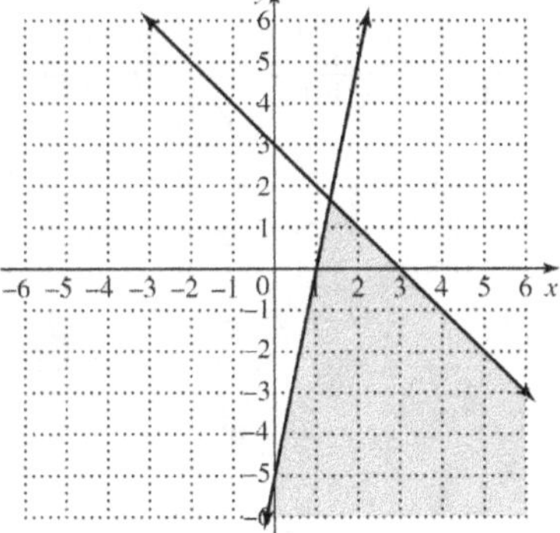

Now Try:

4. Graph the solution set of the system.

$$3x - y \leq 3$$
$$x + y \leq 0$$

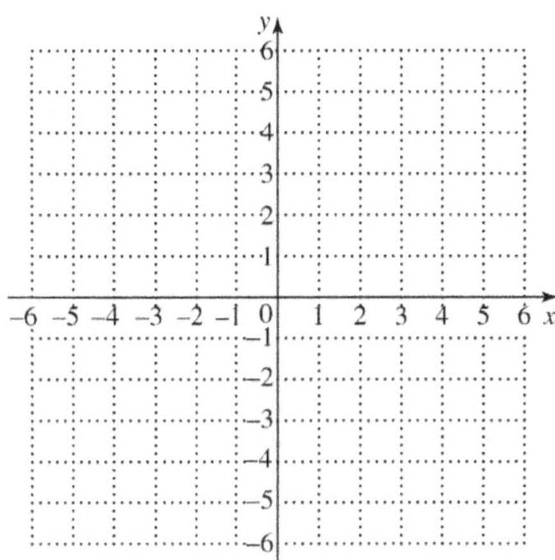

Objective 2 Practice Exercises

For extra help, see Examples 4–5 on pages 622–623 of your text.

Graph the solution of each system of linear inequalities.

4. $4x + 5y \leq 20$

       $y \leq x + 3$

4.

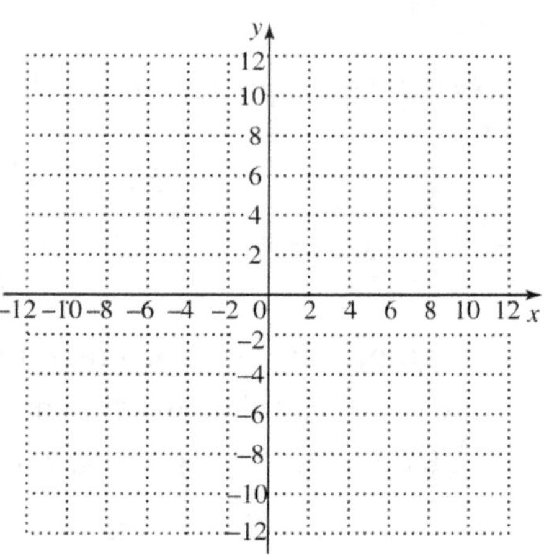

5. $x < 2y + 3$

       $0 < x + y$

5.

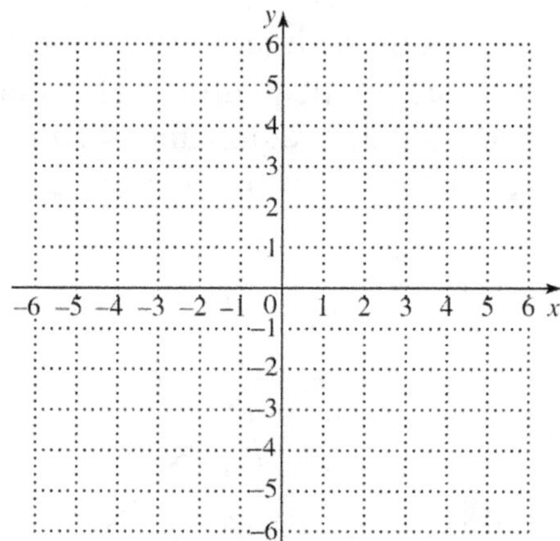

6. $y < 4$

       $x \geq -3$

6.

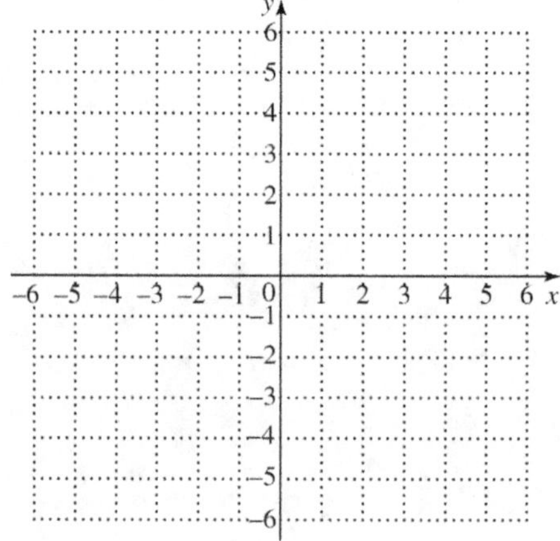

Chapter 9   RELATIONS AND FUNCTIONS

9.1   Introduction to Relations and Functions

Learning Objectives
1   Define and identify relations and functions.
2   Find the domain and range.
3   Identify functions defined by graphs and equations.

Key Terms

Use the vocabulary terms listed below to complete each statement in exercises 1−6.

dependent variable	independent variable	relation
function	domain	range

1.   The ___________________ of a relation is the set of second components (y-values) of the ordered pairs of the relation.

2.   A ___________________ is a set of ordered pairs of real numbers.

3.   If the quantity y depends on x, then y is called the ___________________ in a relation between x and y.

4.   The ___________________ of a relation is the set of first components (x-values) of the ordered pairs of the relation.

5.   A ___________________ is a set of ordered pairs in which each value of the first component, x, corresponds to exactly one value of the second component, y.

6.   If the quantity y depends on x, then x is called the ___________________ in a relation between x and y.

Objective 1   Define and identify relations and functions.

Review these examples for Objective 1:

1.   Write the relation as a set of ordered pairs.

Number of Hours Worked	Paycheck Amount (in dollars)
8	96
16	192
24	288
32	384

{(8, 96), (16, 192), (24, 288), (32, 384)}

Now Try:

1.   Write the relation as a set of ordered pairs.

Number of Hours Fishing	Number of Fish Caught
1	3
2	4
3	6
5	7

____

2. Determine whether each relation defines a function.

   a. $G = \{(-5, -2), (-2, 6), (6, 8), (8, 11), (11, 11)\}$

Relation G is a function. Although the last two ordered pairs have the same y-value, this does not violate the definition of a function.

   b. $H = \{(-9, 2), (-6, 2), (-6, 9)\}$

In relation H, the last two ordered pairs have the same x-value pair with different y-values. H is a relation, but not a function.

2. Determine whether each relation defines a function.

   a. $\{(10, 1), (100, 2), (70, 2)\}$

   ____

   b. $\{(0, 2), (2, 6), (6, 3), (0, 7)\}$

   ____

Objective 1 Practice Exercises

For extra help, see Examples 1–2 on pages 638–639 of your text.

Write the relation as a set of ordered pairs.

1.

x	y
1	3
1	4
2	−1
3	7

1. _______________

Decide whether each relation is a function.

2. $\{(2, -2,), (3, -3), (4, -4)\}$

2. _______________

3. $\{(3, 4), (5, 2), (4, 3), (5, 3), (-2, 2)\}$

3. _______________

Objective 2   Find the domain and range.

Review these examples for Objective 2:

3. Give the domain and range of each relation. Tell whether the relation defines a function.

$\{(15, 2), (20, 3), (6, 10), (-1, 2)\}$

The domain is the set of x-values $\{-1, 6, 15, 20\}$. The range is the set of y-values $\{2, 3, 10\}$. The relation is a function, because each x-value corresponds to exactly one y-value.

Now Try:

3. Give the domain and range of each relation. Tell whether the relation defines a function.
$\{(13, -1), (13, -2), (13, 4)\}$

   domain: _______________

   range: _______________

   ____

4. Give the domain and range of each relation.

a.

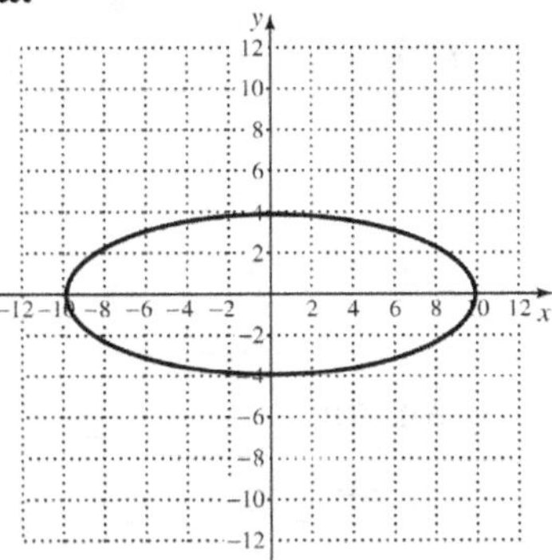

The *x*-values of the points on the graph include all numbers between −10 and 10, inclusive. The *y*-values include all numbers between −4 and 4, inclusive.
The domain is [−10, 10]. The range is [−4, 4].

b.

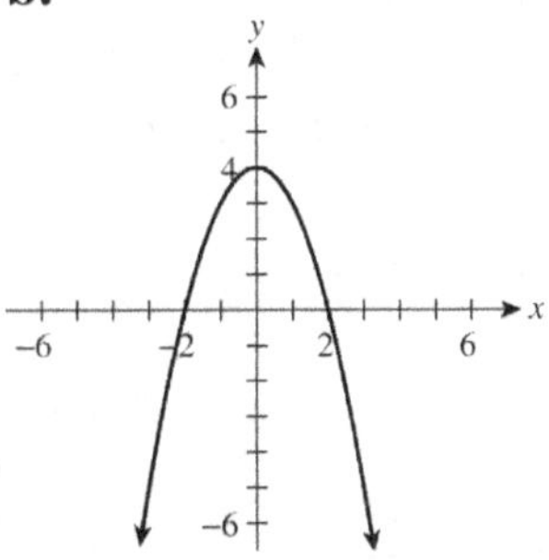

The graph extends indefinitely left and right, as well as downward. The domain is $(-\infty, \infty)$.
Because there is a greatest *y*-value, 4, the range includes all numbers less than or equal to 4, written $(-\infty, 4]$.

Objective 2 Practice Exercises

For extra help, see Examples 3–4 on pages 641–642 of your text.

Decide whether the relation is a function, and give the domain and range of the relation.

4. $\{(5,\ 2),\ (3,-1),\ (1,-3),\ (-1,-5)\}$

---

4. Give the domain and range of each relation.

a.

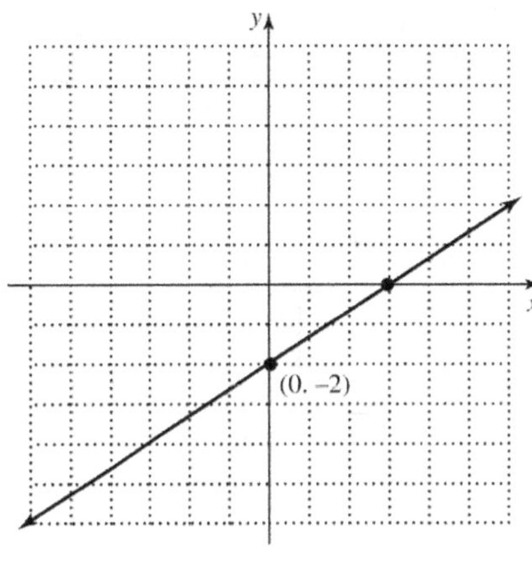

____

____

b.

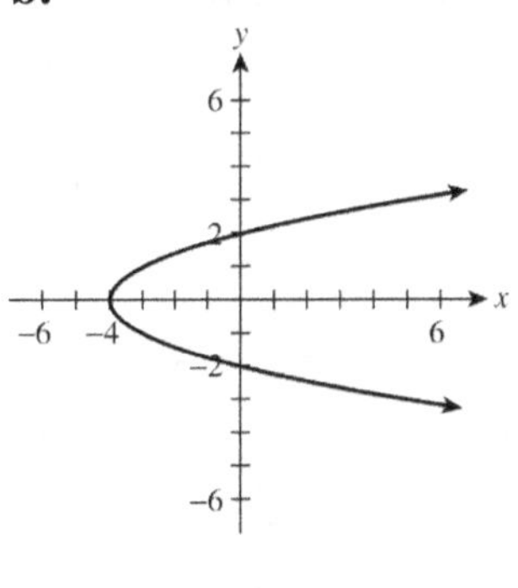

____

____

4. _____________________

domain:________________

range: _________________

5.

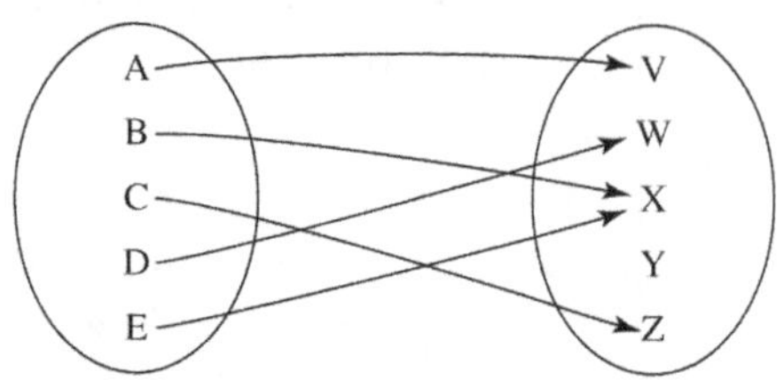

5. _______________

domain: _______________

range: _______________

6.

x	y
1	3
2	-1
-1	4
1	4

6. _______________

domain: _______________

range: _______________

Objective 3   Identify functions defined by graphs and equations.

Review these examples for Objective 3:

5. Use the vertical line test to determine whether the relation graphed is a function.

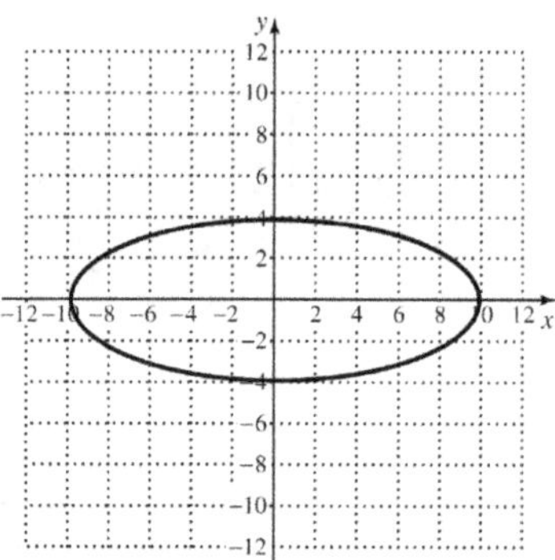

The graph is not a function.

6. Decide whether the relation defines y as a function of x. Give the domain.

$$y = 2x - 4$$

Each x value corresponds to just one y-value and the relation defines a function. Since x can be any real number, the domain is $(-\infty, \infty)$.

Now Try:

5. Use the vertical line test to determine whether the relation graphed is a function.

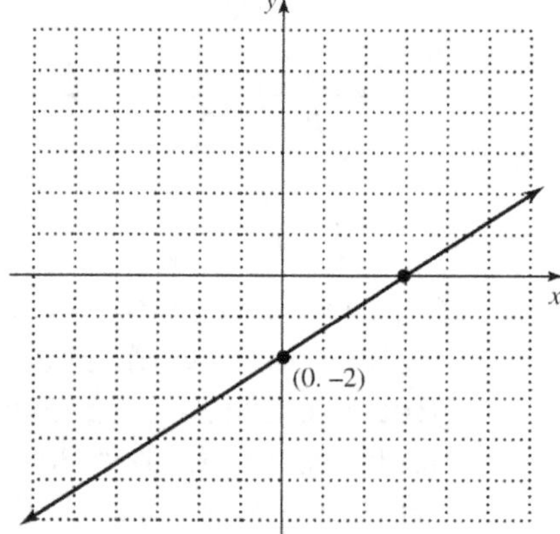

____

6. Decide whether the relation defines y as a function of x. Give the domain.

$$y = 3x - 1$$

____

____

Name: Date:
Instructor: Section:

Objective 3 Practice Exercises

For extra help, see Examples 5–6 on pages 643–645 of your text.

Use the vertical line test to determine whether the relation graphed is a function.

7.

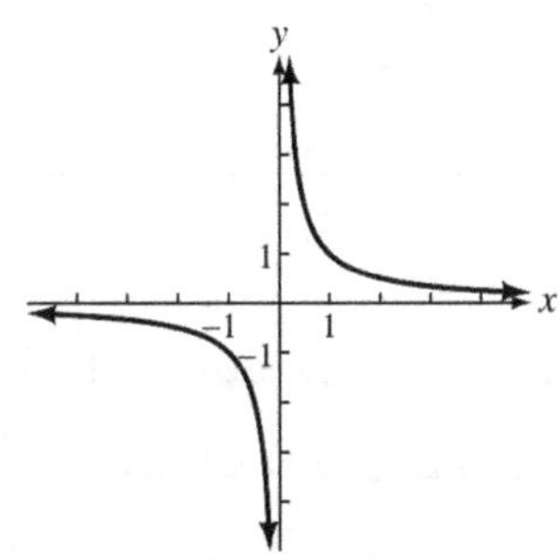

7. _______________________

Decide whether each equation defines y as a function of x. Give the domain.

8. $y^2 = x + 1$

8. _______________________

9. $y = \dfrac{3}{x + 6}$

9. _______________________

Chapter 9 RELATIONS AND FUNCTIONS

9.2   Function Notation and Linear Functions

Learning Objectives
1   Use function notation.
2   Graph linear and constant functions.

Key Terms

Use the vocabulary terms listed below to complete each statement in exercises 1–3.

     function notation   **linear function**   **constant function**

1.   A function defined by an equation of the form $f(x) = ax + b$, for real numbers a and b, is a ________________________________.

2.   ________________________________ $f(x)$ represents the value of the function at x, that is, the y-value that corresponds to x.

3.   A ________________________________ is a linear function of the form $f(x) = b$, for a real number b.

Objective 1   Use function notation.

Review these examples for Objective 1:	**Now Try:**
1. Let $f(x) = 7x - 3$. Evaluate the function f for the following.	1. Let $f(x) = 8x - 7$. Evaluate the function f for the following.
$\quad x = 4$	$\quad x = 3$
Start with the given function. Replace x with 4.	
$\quad f(x) = 7x - 3$	
$\quad f(4) = 7(4) - 3$	______________
$\quad f(4) = 28 - 3$	
$\quad f(4) = 25$	
Thus, $f(4) = 25$.	
2. Let $f(x) = -x^2 + 6x - 8$. Find the following.	2. Let $f(x) = -x^2 - 7x + 10$. Find the following.
a. $f(5)$	**a.** $f(-6)$
Replace x with 5.	______________

$$f(x) = -x^2 + 6x - 8$$
$$f(5) = -5^2 + 6 \cdot 5 - 8$$
$$f(5) = -25 + 30 - 8$$
$$f(5) = -3$$

b. $f(p)$

Replace x with p.
$$f(x) = -x^2 + 6x - 8$$
$$f(p) = -p^2 + 6p - 8$$

b. $f(c)$

____

3. Let $g(x) = 5x + 6$. Find and simplify $g(n+8)$.

Replace x with $n + 8$.
$$g(x) = 5x + 6$$
$$g(n+8) = 5(n+8) + 6$$
$$g(n+8) = 5n + 40 + 6$$
$$g(n+8) = 5n + 46$$

3. Let $g(x) = 4x - 7$. Find and simplify $g(a-1)$.

____

4. For the function, find $f(5)$.

$$f = \{(7, -27),\ (5, -25),\ (3, -23),\ (1, -21)\}$$

From the ordered pair $(5, -25)$, we have $f(5) = -25$.

4. For the function, find $f(-6)$.

$$f = \{(-2,\ 11),\ (-4,\ 17),$$
$$(-6,\ 21),\ (-8,\ 24)\}$$

____

5. Refer to the function graphed below.

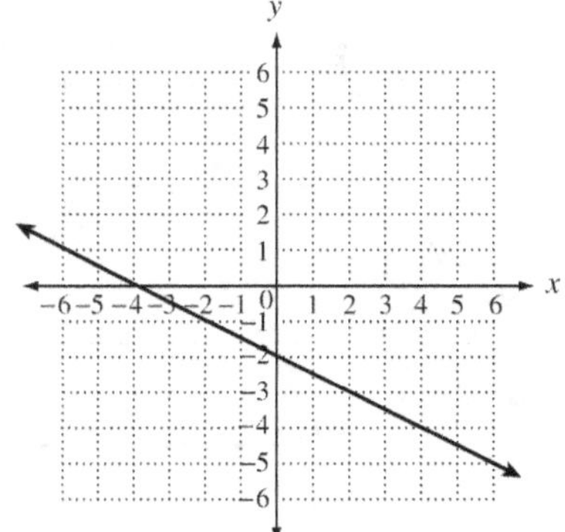

5. Refer to the function graphed below.

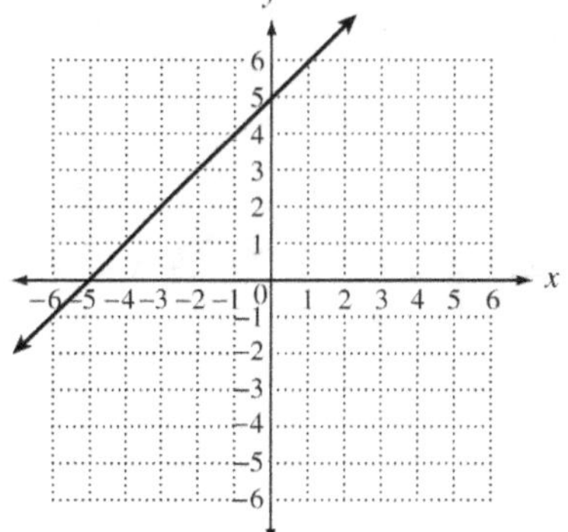

a. Find $f(-2)$.

Locate -2 on the x-axis. Moving down to the graph of f and over to the y-axis gives -1 for the corresponding y-value. Thus, $f(-2) = -1$.

b. For what value of x is $f(x) = -5$?

a. Find $f(-2)$.

____

b. For what value of x is

Locate -5 on the y-axis. Moving across the graph of f and up to the x-axis gives $x = 6$.
Thus, $f(6) = -5$.

$f(x) = 2?$

____

6. Write the equation using function notation $f(x)$. Then find $f(-5)$.

$x - 5y = 8$

Step 1 $x - 5y = 8$

$$-5y = -x + 8$$

$$y = \frac{1}{5}x - \frac{8}{5}$$

Step 2 $f(x) = \frac{1}{5}x - \frac{8}{5}$

$$f(-5) = \frac{1}{5}(-5) - \frac{8}{5}$$

$$f(-5) = -\frac{13}{5}$$

6. Write the equation using function notation $f(x)$. Then find $f(-3)$.

$2x + 3y = 7$

____

Objective 1 Practice Exercises

For extra help, see Examples 1–6 on pages 649–652 of your text.

For each function f, find (a) $f(-2)$*, (b)* $f(0)$*, and (c)* $f(-x)$*.*

1. $f(x) = 3x - 7$

1. a._______________

b._______________

c._______________

2. $f(x) = 2x^2 + x - 5$

2. a._______________

b._______________

c._______________

3. $f(x) = 9$

3. a._______________

b._______________

c._______________

Objective 2   Graph linear and constant functions.

Review this example for Objective 2:

7. Graph the function $f(x) = -2x - 3$. Give the domain and range.

The graph of the function has slope -2 and y-intercept -3. To graph this function, plot the y-intercept $(0, -3)$ and use the definition of slope as $\dfrac{\text{rise}}{\text{run}}$ to find a second point on the line. Since the slope is -2, move down two units and right one unit to the point $(1, -5)$. Draw the straight line through the points to obtain the graph. The domain and range are both $(-\infty, \infty)$.

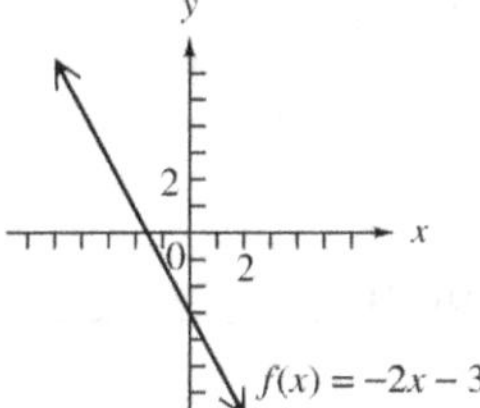

Now Try:

7. Graph the function $f(x) = \dfrac{1}{2}x + \dfrac{1}{2}$. Give the domain and range.

____

____

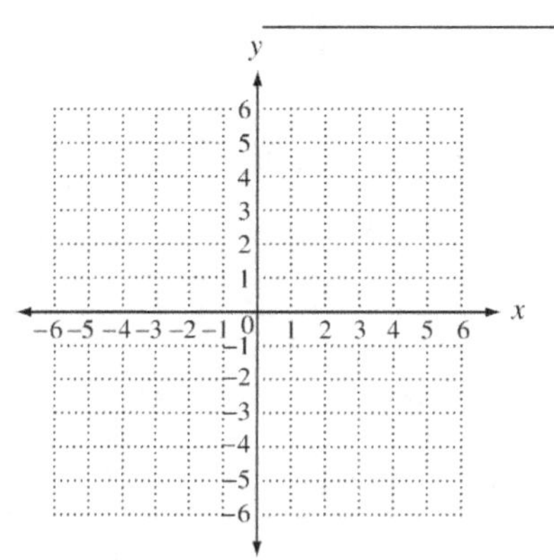

Objective 2 Practice Exercises

For extra help, see Example 7 on page 653 of your text.

Graph each function. Give the domain and range.

4. $2x - y = -2$

4. domain __________

range ____________

5. $y + \dfrac{1}{2}x = -2$

5. domain _________

range __________

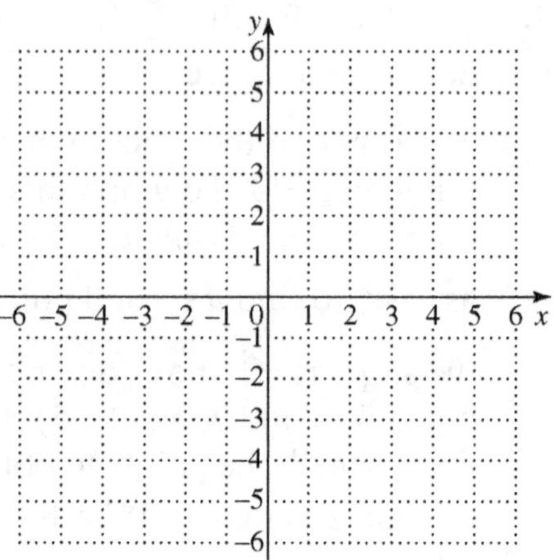

6. $y = 2$

6. domain _________

range __________

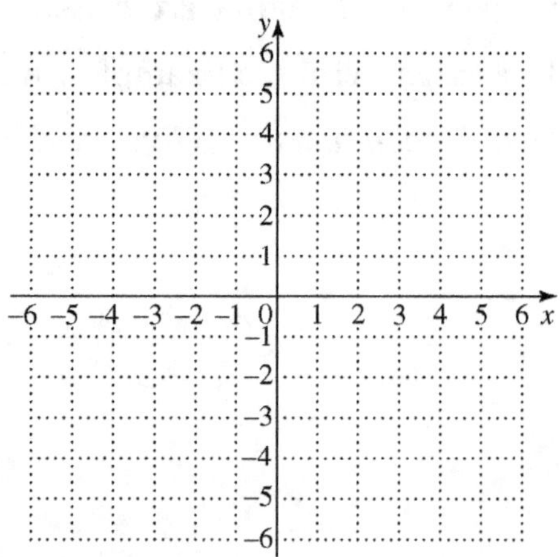

Chapter 9 RELATIONS AND FUNCTIONS

9.3   Polynomial Functions, Graphs, Operations, and Composition

Learning Objectives	
1	Recognize and evaluate polynomial functions.
2	Graph basic polynomial functions.
3	Perform operations on polynomial functions.
4	Find the composition of functions.

Key Terms

Use the vocabulary terms listed below to complete each statement in exercises 1–6.

    polynomial function of degree n   **identity function**

    squaring function   **cubing function**

    composite function   **composition**

1. The polynomial function defined by $f(x) = x^3$ is called the

   __________________________.

2. The function $g(f(x))$ is a ____________________________.

3. A function defined by $f(x) = a_n x^n + a_{n-1} x^{n-1} + \cdots + a_1 x + a_0$, where $a_n \neq 0$ and n is a whole number is a _____________________________________.

4. The polynomial function defined by $f(x) = x^2$ is called the

   __________________________.

5. The simplest polynomial function is the _____________________________ defined by $f(x) = x$.

6. If f and g are functions, then the ____________________________ of g and f is defined by $(g \circ f)(x) = g(f(x))$ for all x in the domain of f such that $f(x)$ is in the domain of g.

       357

Objective 1   Recognize and evaluate polynomial functions.

Review this example for Objective 1:	**Now Try:**

Review this example for Objective 1:

1. Let $f(x) = 6x^3 - 6x + 1$. Find $f(-3)$.

Substitute -3 for x.

$$f(-3) = 6(-3)^3 - 6(-3) + 1$$
$$= 6(-27) - 6(-3) + 1$$
$$= -162 + 18 + 1$$
$$= -143$$

Now Try:

1. Let $p(x) = -x^4 + 3x^2 - x + 7$. Find $p(2)$.

____

Objective 1 Practice Exercises

For extra help, see Example 1 on page 658 of your text.

For each polynomial function, find (a) $f(-2)$ and (b) $f(3)$.

1. $f(x) = -x^2 - x - 5$

1. (a) _______________

(b) _______________

2. $f(x) = 2x^2 + 3x - 5$

2. (a) _______________

(b) _______________

3. $f(x) = 3x^4 - 5x^2$

3. (a) _______________

(b) _______________

Objective 2   Graph basic polynomial functions.

Review these examples for Objective 2:

2.   Graph each function. Give the domain and range.

a. $f(x) = -3x + 2$

Plot the points and join them with a straight line.
The domain and range are both $(-\infty,\ \infty)$.

x	$f(x) = -3x + 2$
-1	5
0	2
1	-1
2	-4
3	-7

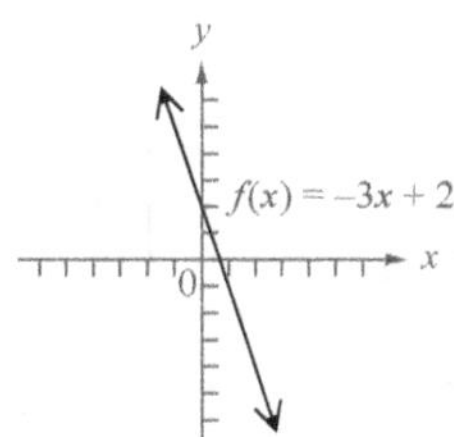

b. $g(x) = -2x^2$

The graph of $g(x)$ has the same shape as that of
$f(x) = x^2$ but is narrower and opens downward.
The domain is $(-\infty,\ \infty)$. The range is
$(-\infty,\ 0]$.

x	$g(x) = -2x^2$
-2	-8
-1	-2
0	0
1	-2
2	-8

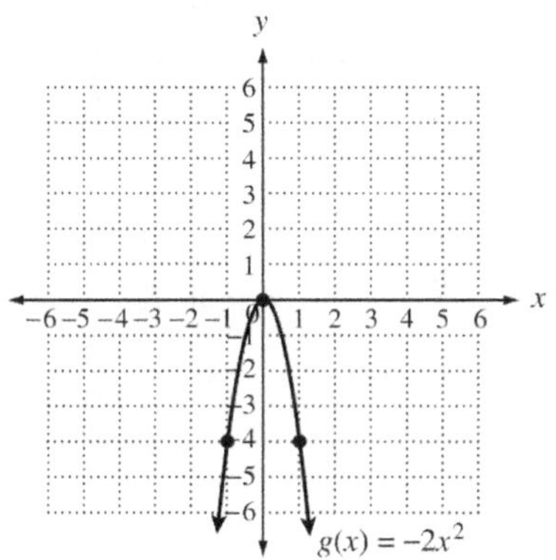

Now Try:

2.   Graph each function. Give the domain and range.

a. $f(x) = -2x - 3$

____

____

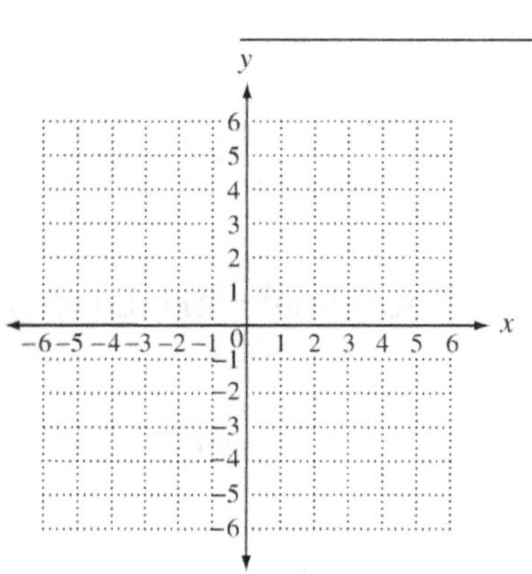

b. $f(x) = x^2 - 1$

____

____

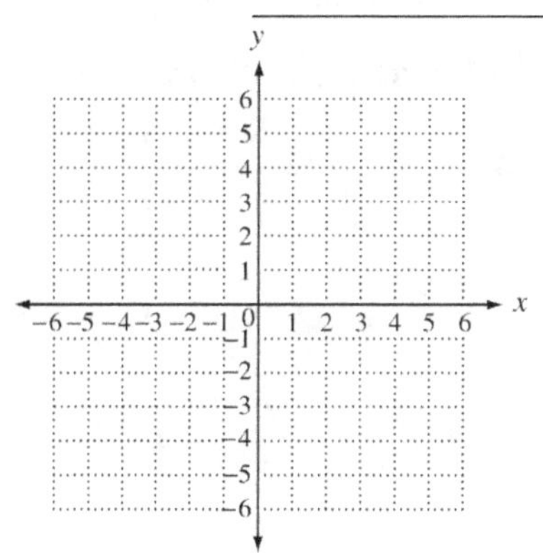

c. $f(x) = x^3 + 3.$

For this function, cube the input and add 3 to the result. The graph is the cubing function shifted 3 units up.

x	$f(x) = x^3 + 3$
-2	-5
-1	2
0	3
1	4
2	11

The domain and range is $(-\infty,\ \infty)$.

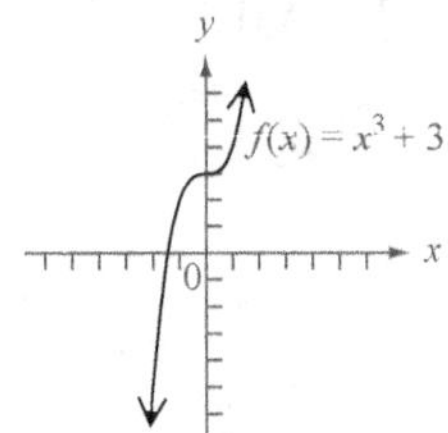

c. $f(x) = -x^3 + 1.$

____

____

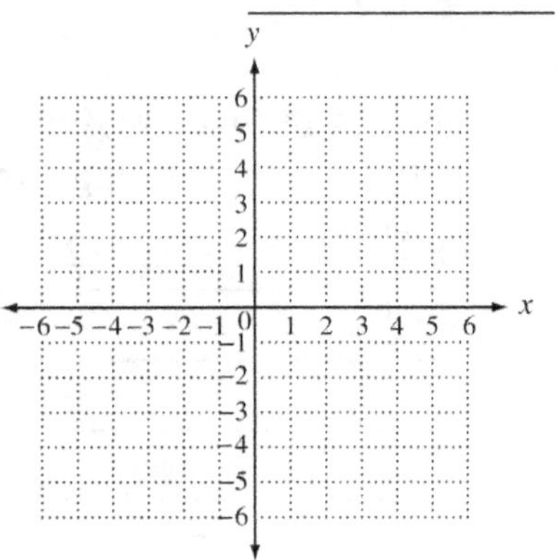

Objective 2 Practice Exercises

For extra help, see Example 2 on page 660 of your text.

Graph each function. Give the domain and range.

4. $f(x) = \frac{1}{2}x + \frac{1}{2}$

4. domain __________

     range __________

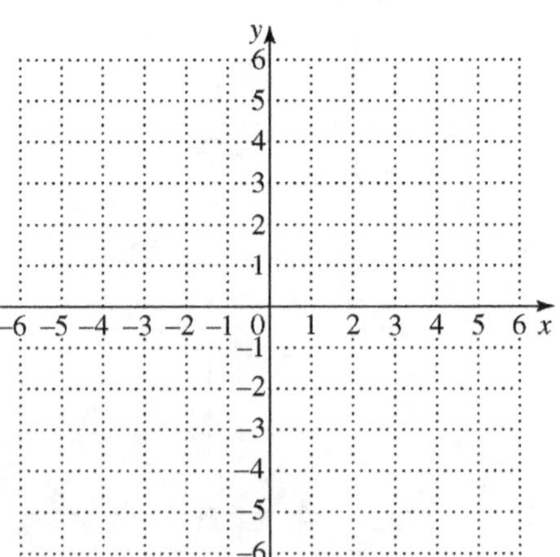

5. $f(x) = -2x^2$

5. domain _________

   range __________

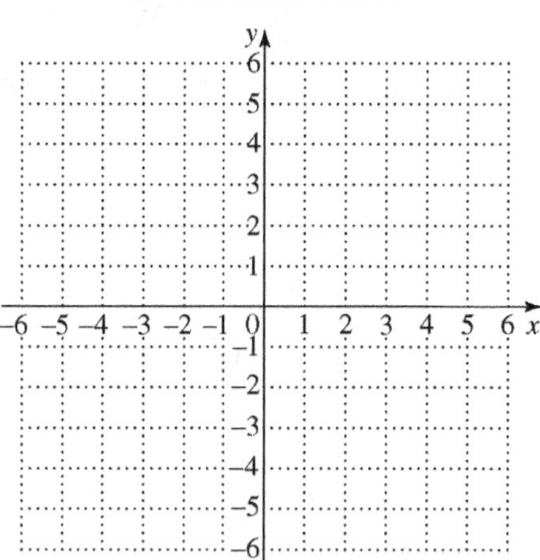

6. $f(x) = x^3 + 2$

6. domain _________

   range __________

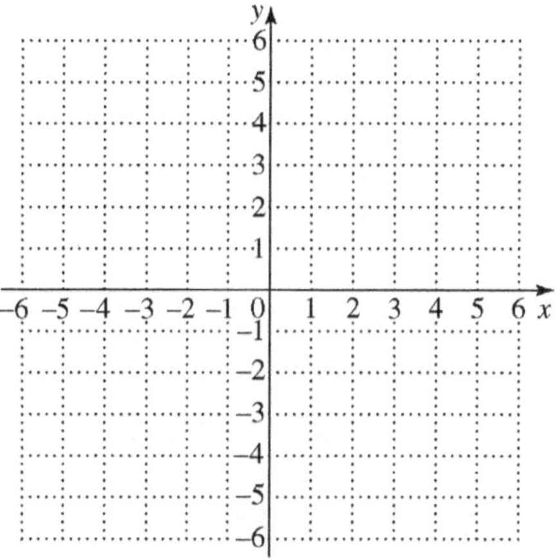

Objective 3   Perform operations on polynomial functions.

Review these examples for Objective 3:

3. For $f(x) = 2x^2 + 4x - 5$ and
$g(x) = -x^2 + 3x - 8$, find each of the following.

a. $(f + g)(x)$

$$(f + g)(x) = f(x) + g(x)$$
$$= (2x^2 + 4x - 5) + (-x^2 + 3x - 8)$$
$$= x^2 + 7x - 13$$

b. $(f - g)(x)$

$$(f - g)(x) = f(x) - g(x)$$
$$= (2x^2 + 4x - 5) - (-x^2 + 3x - 8)$$
$$= (2x^2 + 4x - 5) + (x^2 - 3x + 8)$$
$$= 3x^2 + x + 3$$

Now Try:

3. For $f(x) = 6x^2 - 7x + 12$ and
$g(x) = -3x^2 + x + 9$, find each
of the following.

a. $(f + g)(x)$

____

b. $(f - g)(x)$

____

5. For $f(x) = 5x + 2$ and $g(x) = 3x^2 + 4x$, find
$(fg)(x)$ and $(fg)(-2)$.

$$(fg)(x) = f(x) \cdot g(x)$$
$$= (5x + 2)(3x^2 + 4x)$$
$$= 15x^3 + 20x^2 + 6x^2 + 8x$$
$$= 15x^3 + 26x^2 + 8x$$
$$(fg)(-2) = 15(-2)^3 + 26(-2)^2 + 8(-2)$$
$$= -120 + 104 - 16$$
$$= -32$$

5. For $f(x) = 6x + 5$ and
$g(x) = 7x^2 + 2x$, find
$(fg)(x)$ and $(fg)(-3)$.

____

6. For $f(x) = 4x^2 - 17x - 15$ and $g(x) = x - 5$,
find $\left(\dfrac{f}{g}\right)(x)$ and $\left(\dfrac{f}{g}\right)(-2)$.

6. For $f(x) = 2x^2 - 3x - 20$ and
$g(x) = x - 4$, find $\left(\dfrac{f}{g}\right)(x)$ and

$$\left(\frac{f}{g}\right)(x) = \frac{f(x)}{g(x)} = \frac{4x^2 - 17x - 15}{x - 5}$$

$$\left(\frac{f}{g}\right)(-2).$$

$$\begin{array}{r} 4x + 3 \\ x - 5 \overline{) \, 4x^2 - 17x - 15} \\ \underline{4x^2 - 20x} \\ 3x - 15 \\ \underline{3x - 15} \\ 0 \end{array}$$

____

$$\left(\frac{f}{g}\right)(x) = 4x + 3, \quad x \neq 5$$

$$\left(\frac{f}{g}\right)(-2) = 4(-2) + 3 = -5$$

Objective 3 Practice Exercises

For extra help, see Examples 3–6 on pages 662–663 of your text.

For the pair of functions, find (a) $(f + g)(x)$ and (b) $(f - g)(x)$.

7. $f(x) = 2x^2 + 4x - 5, \ g(x) = -x^2 + 3x - 8$ 7. a._________________

                                                              b._________________

For the pair of functions, find (a) $(fg)(x)$ and (b) $(fg)(-1)$.

8. $f(x) = 3x^2 + 2, \ g(x) = -5x$ 8. a._________________

                                                              b._________________

For the pair of functions, find the quotient $\left(\frac{f}{g}\right)(x)$ and give any x-values that are not in the domain of the quotient function.

9. $f(x) = 4x^2 - 11x - 45, \ g(x) = x - 5$ 9. _________________

Objective 4   Find the composition of functions.

Review these examples for Objective 4:

7. Let $f(x)=3x-1$ and $g(x)=x^2+2$. Find $(f\circ g)(5)$.

$$(f\circ g)(5)=f\big(g(5)\big)$$
$$=f(5^2+2)$$
$$=f(27)$$
$$=3(27)-1$$
$$=80$$

9. Let $f(x)=3x-1$ and $g(x)=x^2+2$. Find the following.

$$(g\circ f)(2)$$

$$(g\circ f)(2)=g\big(f(2)\big)$$
$$=\big(f(2)\big)^2+2$$
$$=(3(2)-1)^2+2$$
$$=5^2+2$$
$$=27$$

10. Every item at the last day of an estate sale is 40% off its original price. After 3 PM, there is an additional discount of 75% off the sale price.

a. Write a function $g(x)$ that computes 40% off the original price.
$$g(x)=x-0.40x$$
$$g(x)=0.60x$$

b. Write a function $f(x)$ that computes 75% off the original price.
$$f(x)=x-0.75x$$
$$f(x)=0.25x$$

c. Find the composite function $(f\circ g)(x)$.
$$(f\circ g)(x)=f\big(g(x)\big)=f(0.60x)$$
$$=0.25(0.60x)$$
$$=0.15x$$

Now Try:

7. Let $f(x)=-3x-3$ and $g(x)=x^2-5$. Find $(f\circ g)(-3)$.

____

9. Let $f(x)=-3x-3$ and $g(x)=x^2-5$. Find the following.

$$(g\circ f)(2)$$

____

10. Every item at an estate sale is 60% off its original price. After noon, there is an extra discount of 70% off the sale price.

a. Write a function $g(x)$ that computes 55% off the original price.

____

b. Write a function $f(x)$ that computes 70% off the original price.

____

c. Find the composite function $(f\circ g)(x)$.

____

d. What is the final price of a $200 recliner purchased after 3 PM?

$$(f \circ g)(x) = 0.15x$$

$$(f \circ g)(200) = 0.15(200) = \$30$$

The sale price is $54.

d. What is the final price of a $400 daybed purchased after noon?

____

Objective 4 Practice Exercises

For extra help, see Examples 7–10 on pages 664–666 of your text.

Find the following.

10. Let $f(x) = 4x - 3$ and $g(x) = 2x^2 - 1$. Find the following.

    a. $(f \circ g)(2)$

    b. $(g \circ f)(-1)$

    c. $(f \circ g)(x)$

10. **a.** _______________

    b. _______________

    c. _______________

11. Let $f(x) = \dfrac{1}{x}$ and $g(x) = 3x^2 - 4x + 1$. Find the following.

    a. $(f \circ g)(2)$

    b. $(g \circ f)\left(\dfrac{1}{3}\right)$

    c. $(g \circ f)(x)$

11. **a.** _______________

    b. _______________

    c. _______________

Chapter 9   RELATIONS AND FUNCTIONS

9.4   Variation

Learning Objectives
1   Write an equation expressing direct variation.
2   Find the constant of variation, and solve direct variation problems.
3   Solve inverse variation problems.
4   Solve joint variation problems.
5   Solve combined variation problems.

Key Terms

Use the vocabulary terms listed below to complete each statement in exercises 1–4.

varies directly   **varies inversely**

constant of variation   **varies jointly**

1.   In the equations for direct and inverse variation, k is the __________________.

2.   If there exists a real number k such that $y = \dfrac{k}{x}$, then y __________________ as x.

3.   If there exists a real number k such that $y = kx$, then y __________________ as x.

4.   If there exists a real number k such that $y = kxz$,
       then y __________________ as x and z.

Objective 1   Write an equation expressing direct variation.

For extra help, see page 672 of your text.

Objective 2   Find the constant of variation, and solve direct variation problems.

Review these examples for Objective 2:

1.   If 12 gallons of gasoline cost $34.68, how much does 1 gallon of gasoline cost? Write the variation equation.

Let g represent the number of gallons of gasoline and let C represent the total cost of the gasoline. Then the variation equation is $C = kg$.

$$C = kg$$

$$34.68 = 12k$$

$$2.89 = k$$

The cost per gallon is $2.89.
The variation equation is $C = 2.89g$.

Now Try:

1.   One week a manufacturer sold 1200 items for a total profit of $30,000. What was the profit for one item. Write the variation equation.

       ____

       ____

2. A person's weight on the moon varies directly with the person's weight on Earth. A 120-pound person would weigh about 20 pounds on the moon. How much would a 150-pound person weigh on the moon?

If m represents the person's weight on the moon and w represents the person's weight on Earth.

$$m = kw$$

$$20 = k \cdot 120 \quad \text{Let } m = 20 \text{ and } w = 120.$$

$$\frac{20}{120} = \frac{1}{6} = k \quad \text{Solve for } k; \text{ lowest terms}$$

Now, substitute $\frac{1}{6}$ for k and 150 for w in the variation equation.

$$m = kw$$

$$m = \frac{1}{6} \cdot 150 = 25$$

A 150-pound person will weigh 25 pounds on the moon.

3. The surface area of a sphere varies directly as the square of its radius. If the surface area of a sphere with a radius of 12 inches is 576π square inches, find the surface area of a sphere with a radius of 3 inches.

Step 1 A represents the surface area and r represents the radius.

$$A = kr^2$$

Step 2 Find the value of k when A is 576π and r is 12.

$$A = kr^2$$

$$576\pi = k \cdot 12^2$$

$$576\pi = 144k$$

$$\frac{576\pi}{144} = 4\pi = k$$

Step 3 Rewrite the variation equation.

$$A = 4\pi r^2$$

Step 4 Let $r = 3$ to find the surface area.

$$A = 4\pi \cdot 3^2 = 4\pi \cdot 9 = 36\pi$$

The surface area of a sphere with a radius of 3 inches is 36π square inches.

2. The pressure exerted by a certain liquid at a given point varies directly as the depth of the point beneath the surface of the liquid. The pressure at 10 feet is 50 pounds per square inch (psi). What is the pressure at 25 feet?

____

3. The area of a circle varies directly as the square of the radius. A circle with a radius of 5 centimeters has an area of 78.5 square centimeters. Find the area if the radius changes to 7 centimeters.

____

Objective 2 Practice Exercises

For extra help, see Examples 1–3 on pages 673–674 of your text.

Find the constant of variation, and write a direct variation equation.

1. $y = 13.75$ when $x = 55$

1. ___________________

Solve each problem.

2.   The circumference of a circle varies directly as the radius. A circle with a radius of 7 centimeters has a circumference of 43.96 centimeters. Find the circumference of the circle if the radius changes to 11 centimeters.

2. ___________________

3.   The force required to compress a spring varies directly as the change in length of the spring. If a force of 20 newtons is required to compress a spring 2 centimeters in length, how much force is required to compress a spring of length 10 centimeters?

3. ___________________

Objective 3   Solve inverse variation problems.

Review these examples for Objective 3:

4.   For a specified distance, time varies inversely with speed. If Ramona walks a certain distance on a treadmill in 40 minutes at 4.2 miles per hour, how long will it take her to walk the same distance at 3.5 miles per hour?

Let t = time and s = speed.
Since t varies inversely as s, there is a constant k such that $t = \dfrac{k}{s}$. Recall that $40 \text{ min} = \dfrac{40}{60} \text{ hr}$.

$$t = \frac{k}{s}$$

$$\frac{40}{60} = \frac{k}{4.2}$$

$$2.8 = k$$

Now Try:

4.   The length of a violin string varies inversely with the frequency of its vibrations. A 10-inch violin string vibrates at a frequency of 512 cycles per second. Find the frequency of an 8-inch string.

____

Now use $t = \dfrac{k}{s}$ to find the value of t

when $s = 3.5$.

$$t = \frac{2.8}{3.5} = \frac{4}{5}$$

It takes $\dfrac{4}{5}$ hr, or 48 min to walk the same

distance.

5. With constant power, the resistance used in a simple electric circuit varies inversely as the square of the current. If the resistance is 120 ohms when the current is 12 amps, find the resistance if the current is reduced to 9 amps.

Let R represent resistance (in ohms) and

$I =$ current (in amps). Then $R = \dfrac{k}{I^2}$.

First, we solve for the constant of variation by substituting 120 for R and 12 for I.

$$120 = \frac{k}{12^2}$$

$$k = 120 \cdot 12^2$$

Now use the value for k and 9 for I to find R.

$$R = \frac{120 \cdot 12^2}{9^2} \approx 213.3$$

The resistance is about 213.3 ohms when the current is 9 amps.

5. If y varies inversely as x^3, and $y = 9$ when $x = 2$, find y when $x = 4$.

____

Objective 3 Practice Exercises

For extra help, see Examples 4–5 on pages 675–676 of your text.

Solve each problem.

4. The illumination produced by a light source varies inversely as the square of the distance from the source. If the illumination produced 4 feet from a light source is 75 footcandles, find the illumination produced 9 feet from the same source.

4. _______________

5. The weight of an object varies inversely as the square of its distance from the center of Earth. If an object 8000 miles from the center of Earth weighs 90 pounds, find its weight when it is 12,000 miles from the center of Earth.

5. _______________

6. The speed of a pulley varies inversely as its diameter. One kind of pulley, with a diameter of 3 inches, turns at 150 revolutions per minute. Find the speed of a similar pulley with diameter of 5 inches.

6. _______________

Objective 4   Solve joint variation problems.

Review this example for Objective 4:

6. For a fixed interest rate, interest varies jointly as the principal and the time in years. If $5000 invested for 4 years earns $900, how much interest will $6000 invested for 3 years earn at the same interest rate?

Let I = the interest, p = the principal, and t = the time in years. Then, $I = kpt$.

$$I = kpt$$

$$900 = k \cdot 5000 \cdot 4 \quad \text{Substitute given values.}$$

$$\frac{900}{20,000} = \frac{9}{200} = k$$

Now use $k = \dfrac{9}{200}$.

$$I = \frac{9}{200} \cdot 6000 \cdot 3 = 810$$

$6000 invested for three years will earn $810 in interest.

Now Try:

6. The strength of a rectangular beam varies jointly as its width and the square of its depth. If the strength of a beam 2 inches wide by 10 inches deep is 1000 pounds per square inch, what is the strength of a beam 4 inches wide and 8 inches deep?

____

Objective 4 Practice Exercises

For extra help, see Example 6 on page 677 of your text.

Solve each problem.

7. Suppose d varies jointly as f^2 and g^2, and $d = 384$ when $f = 3$ and $g = 8$. Find d when $f = 6$ and $g = 2$.

7. _______________________

8. The work w (in joules) done when lifting an object is jointly proportional to the product of the mass m (in kg) of the object and the height h (in meter) the object is lifted. If the work done when a 120 kg object is lifted 1.8 meters above the ground is 2116.8 joules, how much work is done when lifting a 100kg object 1.5 meters above the ground?

8. _______________________

9. The absolute temperature of an ideal gas varies jointly as its pressure and its volume. If the absolute temperature is 250° when the pressure is 25 pounds per square centimeter and the volume is 50 cubic centimeters, find the absolute temperature when the pressure is 50 pounds per square centimeter and the volume is 75 cubic centimeters.

9. _______________________

Objective 5   Solve combined variation problems.

Review this example for Objective 5:	**Now Try:**

Review this example for Objective 5:

7. The number of hours h that it takes w workers to assemble x machines varies directly as the number of machines and inversely as the number of workers. If four workers can assemble 12 machines in four hours, how many workers are needed to assemble 36 machines in eight hours?

The variation equation is $h = \dfrac{kx}{w}$.

To find k, let $h = 4$, $x = 12$, and $w = 4$.

$$4 = \frac{k \cdot 12}{4}$$

$$k = \frac{4 \cdot 4}{12}$$

$$k = \frac{4}{3}$$

Now find w when $x = 36$ and $h = 8$.

$$8 = \frac{\frac{4}{3} \cdot 36}{w}$$

$$8w = 48$$

$$w = 6$$

Eight workers are needed to assemble 36 machines in eight hours.

Now Try:

7. The volume of a gas varies directly as its temperature and inversely as its pressure. The volume of a gas at 85° C at a pressure of 12 kg/cm^2 is 300 cm^3. What is the volume when the pressure is 20 kg/cm^2 and the temperature is 30° C?

____

Objective 5 Practice Exercises

For extra help, see Example 7 on page 677 of your text.

Solve each problem.

10. The volume of a gas varies inversely as the pressure and directly as the temperature. If a certain gas occupies a volume of 1.3 liters at 300 K and a pressure of 18 kilograms per square centimeter, find the volume at 340 K and a pressure of 24 kilograms per square centimeter.

10. _______________

11. The time required to lay a sidewalk varies directly as its length and inversely as the number of people who are working on the job. If three people can lay a sidewalk 100 feet long in 15 hours, how long would it take two people to lay a sidewalk 40 feet long?

11. ______________________

12. When an object is moving in a circular path, the centripetal force varies directly as the square of the velocity and inversely as the radius of the circle. A stone that is whirled at the end of a string 50 centimeters long at 900 centimeters per second has a centripetal force of 3,240,000 dynes. Find the centripetal force if the stone is whirled at the end of a string 75 centimeters long at 1500 centimeters per second.

12. ______________________

Chapter 10 ROOTS, RADICALS, AND ROOT FUNCTIONS

10.1   Radical Expressions and Graphs

Learning Objectives
1   Find square roots.
2   Decide whether a given root is rational, irrational, or not a real number.
3   Find cube, fourth, and other roots.
4   Graph functions defined by radical expressions.
5   Find nth roots of nth powers.
5   Use a calculator to find roots.

Key Terms

Use the vocabulary terms listed below to complete each statement in exercises 1−12.

square root	**principal square root**	**negative square root**
radicand	**radical**	**perfect square**
cube root	**fourth root**	**index (order)**
radical expression	**square root function**	**cube root function**

1.   5 is the ___________________ of 625.

2.   A ___________________ is a radical sign and the number or expression that appears under it.

3.   The ___________________ of a positive number with even index n is the positive nth root of the number.

4.   A ___________________ is the number or expression that appears inside a radical sign.

5. $f(x) = \sqrt[3]{x}$ is called the ___________________.

6.   A number with a rational square root is called a ___________________.

7.   The symbol $-\sqrt{\ \ }$ is used for the ___________________ of a number.

8.   In a radical of the form $\sqrt[n]{a}$, the number n is the ___________________.

9.   The number b is a ___________________ of a if $b^3 = a$.

10.   The domain and range of $f(x) = \sqrt{x}$, ___________________, are both $[0, \infty)$.

11.   The ___________________ of a number is a number that, when multiplied by itself, gives the original number.

12.   A ___________________ is an algebraic expression containing a radical.

Objective 1   Find square roots.

Review these examples for Objective 1:

2. Find each square root.

a. $\sqrt{121}$

$11^2 = 121$, so $\sqrt{121} = 11$.

b. $-\sqrt{576}$

This is the negative square root of 576.
Since $\sqrt{576} = 24$, then $-\sqrt{576} = -24$.

c. $\sqrt{\dfrac{16}{25}}$

$\sqrt{\dfrac{16}{25}} = \dfrac{4}{5}$

3. Find the square of each radical expression.

a. $\sqrt{17}$

The square of $\sqrt{17}$ is $\left(\sqrt{17}\right)^2 = 17$.

b. $-\sqrt{31}$

$\left(-\sqrt{31}\right)^2 = 31$

c. $\sqrt{w^2 + 3}$

$\left(\sqrt{w^2 + 3}\right)^2 = w^2 + 3$

Now Try:

2. Find each square root.

a. $\sqrt{169}$

____

b. $\sqrt{1681}$

____

c. $\sqrt{\dfrac{9}{49}}$

____

3. Find the square of each radical expression.

a. $\sqrt{19}$

____

b. $-\sqrt{37}$

____

c. $\sqrt{n^2 + 5}$

____

Objective 1 Practice Exercises

For extra help, see Examples 1–3 on pages 692–693 of your text.

Find all square roots of each number.

1. 625

1. _______________

2. $\dfrac{121}{196}$

2. _______________

Find the square root.

3. $\sqrt{\dfrac{900}{49}}$

3. _______________

Objective 2   Decide whether a given root is rational, irrational, or not a real number.

Review these examples for Objective 2:

4. Tell whether each square root is rational, irrational, or not a real number.

   a. $\sqrt{5}$

   Because 5 is not a perfect square, $\sqrt{5}$ is irrational.

   b. $\sqrt{81}$

   81 is a perfect square, 9^2, so $\sqrt{81} = 9$ is a rational number.

   c. $\sqrt{-16}$

   There is no real number whose square is -16. Therefore, $\sqrt{-16}$ is not a real number.

Now Try:

4. Tell whether each square root is rational, irrational, or not a real number.

   a. $\sqrt{11}$

   ____

   b. $\sqrt{100}$

   ____

   c. $\sqrt{-9}$

   ____

Objective 2 Practice Exercises

For extra help, see Example 4 on page 694 of your text.

Tell whether each square root is rational, irrational, *or* not a real number.

   4. $\sqrt{72}$ **4.** ___________

   5. $\sqrt{-36}$ **5.** ___________

   6. $\sqrt{6400}$ **6.** ___________

Objective 3   Find cube, fourth, and other roots.

Review these examples for Objective 3:

5. Find each cube root.

   a. $\sqrt[3]{64}$

   Because $4^3 = 64$, $\sqrt[3]{64} = 4$.

   b. $\sqrt[3]{-64}$

   $\sqrt[3]{-64} = -4$, because $(-4)^3 = -64$.

Now Try:

5. Find each cube root.

   a. $\sqrt[3]{125}$

   ____

   b. $\sqrt[3]{-125}$

   ____

c. $\sqrt[3]{729}$

$\sqrt[3]{729} = 9,$ because $9^3 = 729.$

6. Find each root.

   a. $\sqrt[4]{81}$

$\sqrt[4]{81} = 3,$ because 3 is positive and $3^4 = 81.$

   b. $-\sqrt[4]{81}$

From part (a), $\sqrt[4]{81} = 3,$ so the negative root is $-\sqrt[4]{81} = -3.$

   c. $\sqrt[4]{-81}$

For a real number fourth root, the radicand must be nonnegative. There is no real number that equals $\sqrt[4]{-81}.$

   d. $-\sqrt[5]{1024}$

$-\sqrt[5]{1024} = -4$

   e. $\sqrt[5]{-1024}$

$\sqrt[5]{-1024} = -4,$ because $(-4)^5 = -1024.$

c. $\sqrt[3]{343}$

____

6. Find each root.

   a. $\sqrt[4]{625}$

____

   b. $-\sqrt[4]{625}$

____

   c. $\sqrt[4]{-625}$

____

   d. $-\sqrt[5]{3125}$

____

   e. $\sqrt[5]{-3125}$

____

Objective 3 Practice Exercises

For extra help, see Examples 5–6 on pages 695–696 of your text.

Find each root.

7. $\sqrt[3]{-64}$ **7.** _______________

8. $\sqrt[4]{256}$ **8.** _______________

9. $\sqrt[7]{-1}$ **9.** _______________

Objective 4   Graph functions defined by radical expressions.

Review these examples for Objective 3:

Now Try:

7. Graph each function by creating a table of values. Give the domain and the range.

a. $f(x) = \sqrt{x-1}$

Create a table of values.

x	$f(x) = \sqrt{x-1}$
1	$\sqrt{1-1} = 0$
5	$\sqrt{5-1} = 2$
10	$\sqrt{10-1} = 3$

For the radicand to be nonnegative, we must have $x-1 \geq 0$ or $x \geq 1$. Therefore, the domain is $[1, \infty)$. Function values are nonnegative, so the range is $[0, \infty)$.

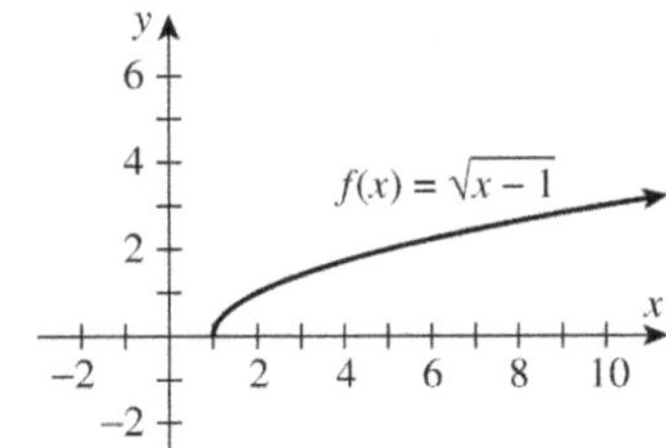

b. $f(x) = \sqrt[3]{x} + 1$

Create a table of values.

x	$f(x) = \sqrt[3]{x} + 1$
-8	$\sqrt[3]{-8} + 1 = -1$
-1	$\sqrt[3]{-1} + 1 = 0$
0	$\sqrt[3]{0} + 1 = 1$
1	$\sqrt[3]{1} + 1 = 2$
8	$\sqrt[3]{8} + 1 = 3$

Both the domain and range are $(-\infty, \infty)$.

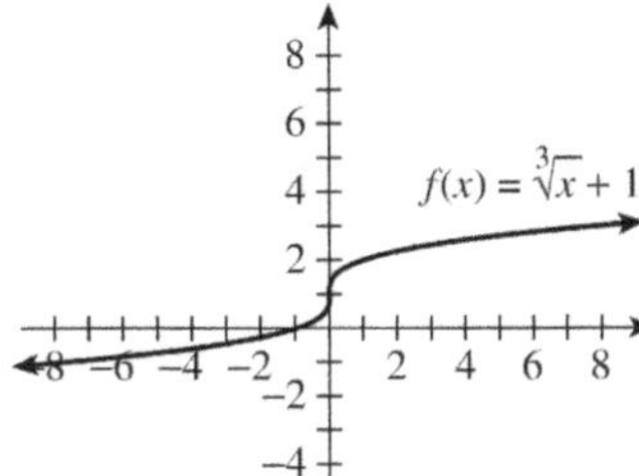

7. Graph each function by creating a table of values. Give the domain and the range.

a. $f(x) = \sqrt{x} - 1$

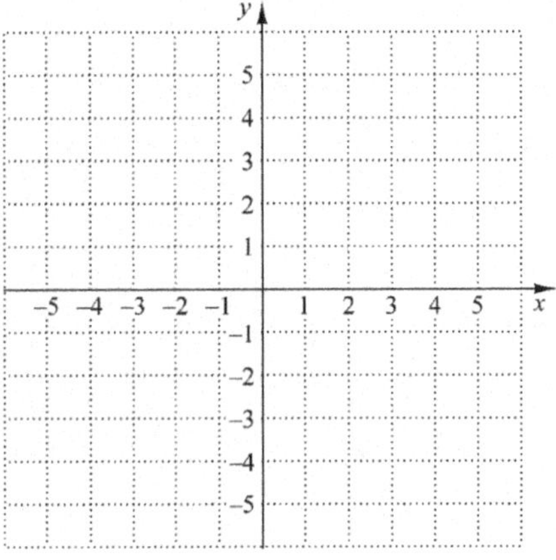

domain: _______________

range: _______________

b. $f(x) = \sqrt[3]{x} + 1$

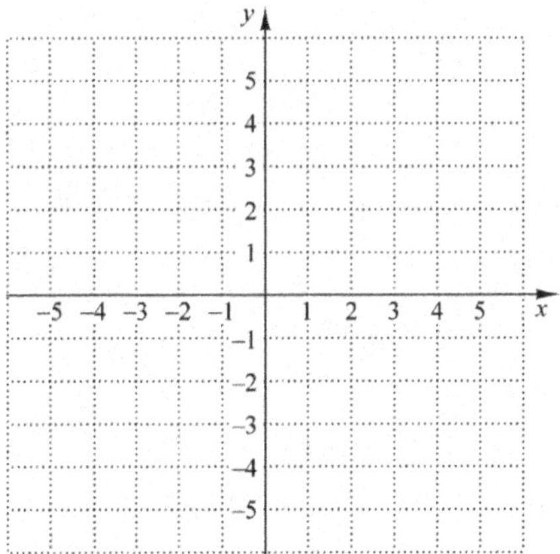

domain: _______________

range: _______________

Objective 4 Practice Exercises

For extra help, see Example 7 on page 697 of your text.

Graph each function and give its domain and its range.

10. $f(x) = \sqrt{x} + 2$

10.

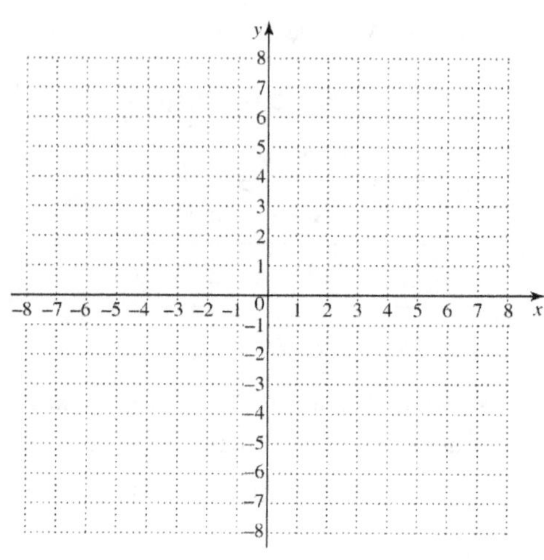

domain: _______________________

range: _______________________

11. $f(x) = \sqrt[3]{x} - 2$

11.

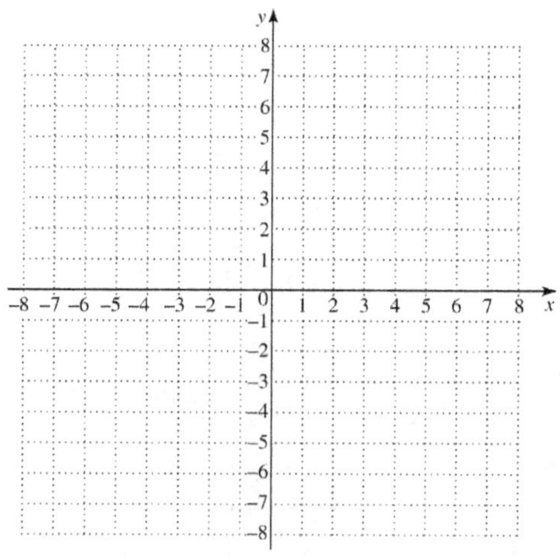

domain: _______________________

range: _______________________

Copyright © 2025 Pearson Education, Inc.

12. $f(x) = \sqrt[3]{x} + 2$ **12.**

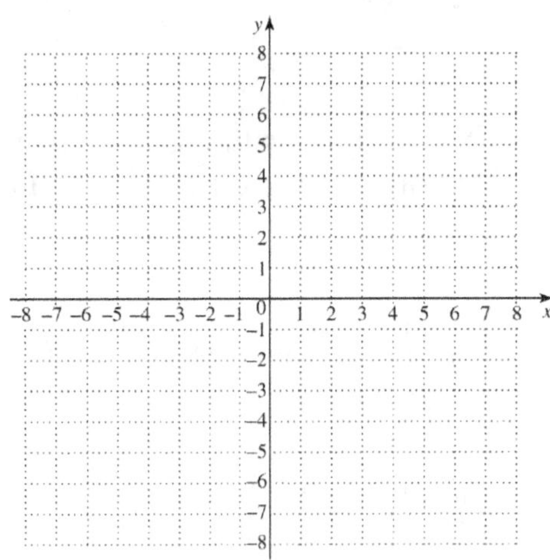

domain: ___________________________

range: ___________________________

Objective 5   Find *n*th roots of *n*th powers.

Review these examples for Objective 4:

8.   Find each square root. In part (d), m is a real number.

   a. $\sqrt{33^2}$

    $\sqrt{33^2} = |33| = 33$

   d. $\sqrt{(-m)^2}$

    $\sqrt{(-m)^2} = |-m| = |m|$

9.   Simplify each root.

   a. $\sqrt[4]{(-5)^4}$

   n is even. Use absolute value.

   $\sqrt[4]{(-5)^4} = |-5| = 5$

   b. $\sqrt[5]{(-3)^5}$

   n is odd.

   $\sqrt[5]{(-3)^5} = -3$

   c. $-\sqrt[6]{(-8)^6}$

   n is even. Use absolute value.

   $-\sqrt[6]{(-8)^6} = -|-8| = -8$

Now Try:

8.   Find each square root. In part (d), n is a real number.

   a. $\sqrt{73^2}$

   ____

   d. $\sqrt{(-n)^2}$

   ____

9.   Simplify each root.

   a. $\sqrt[8]{(-4)^8}$

   ____

   b. $\sqrt[7]{(-5)^7}$

   ____

   c. $-\sqrt[4]{(-2)^4}$

   ____

                    381

d. $-\sqrt{r^{12}}$

$$-\sqrt{r^{12}} = -\left|r^6\right| = -r^6$$

No absolute value bars are needed here since r^6 is nonnegative for any real number value of r.

e. $\sqrt[5]{s^{20}}$

$$\sqrt[5]{s^{20}} = s^4, \text{ because } s^{20} = (s^4)^5.$$

f. $\sqrt[4]{x^{20}}$

$$\sqrt[4]{x^{20}} = \left|x^5\right|$$

d. $-\sqrt{x^8}$

____

e. $\sqrt[3]{w^{30}}$

____

f. $\sqrt[6]{x^{30}}$

____

Objective 5 Practice Exercises

For extra help, see Examples 8–9 on page 698 of your text.

Simplify each root.

13. $\sqrt{(-9)^2}$

13. _______________

14. $-\sqrt[5]{x^5}$

14. _______________

15. $-\sqrt[4]{x^{16}}$

15. _______________

Objective 6   Use a calculator to find roots.

Review these examples for Objective 5:

10. Use a calculator to approximate each radical to three decimal places.

a. $\sqrt{12}$

Use the square root key on a calculator.
$$\sqrt{12} \approx 3.464$$

b. $-\sqrt{596}$

$$-\sqrt{596} \approx -24.413$$

c. $\sqrt[3]{61}$

$$\sqrt[3]{61} \approx 3.936$$

d. $\sqrt[4]{6902}$

$$\sqrt[4]{6902} \approx 9.115$$

Now Try:

10. Use a calculator to approximate each radical to three decimal places.

a. $\sqrt{14}$

____

b. $-\sqrt{678}$

____

c. $\sqrt[3]{431}$

____

d. $\sqrt[4]{8142}$

____

11. The minimum speed at which a driver is traveling when a car skids to a stop can be estimated by the formula $s = \sqrt{30\,fd}$, where s is the speed of the car in miles per hour when the brakes are applied, f is a constant drag factor, and d is the length of the skid marks in feet. Suppose that a car skids to a stop leaving a skid of 90 feet. Assume that the drag factor is 0.7. How fast was the driver going? Give your answer to the nearest tenth.

Start with the given formula. Substitute for f and d, and use a calculator.

$$s = \sqrt{30\,fd}$$

$$s = \sqrt{30(0.7)(90)}$$

$$s \approx 43.5$$

The driver was traveling at about 43.5 miles per hour at the start of the skid.

11. Use the formula at the left to find out how fast a driver was traveling if the skid mark is 20 feet and the drag factor is 0.9. Give your answer to the nearest tenth.

11. _______________

Objective 6 Practice Exercises

For extra help, see Examples 10–11 on pages 699–700 of your text.

Use a calculator to find a decimal approximation for each radical. Give the answer to the nearest thousandth.

16. $\sqrt[3]{701}$

16. _______________

17. $-\sqrt{990}$

17. _______________

18. The time t in seconds for one complete swing of a simple pendulum, where L is the length of the pendulum in feet is $t = 2\pi\sqrt{\dfrac{L}{32}}$. Find the time of a complete swing of a 4-ft pendulum to the nearest tenth of a second.

18. _______________

Chapter 10 ROOTS, RADICALS, AND ROOT FUNCTIONS

10.2   Rational Exponents

Learning Objectives
1   Use exponential notation for nth roots.
2   Define and use expressions of the form $a^{m/n}$.
3   Convert between radicals and rational exponents.
4   Use the rules for exponents with rational exponents.

Key Terms

Use the vocabulary terms listed below to complete each statement in exercises 1–3.

product rule for exponents   **quotient rule for exponents**

power rule for exponents

1. $\left(x^2 y^3\right)^4 = x^8 y^{12}$ is an example of the _________________________.

2. $w^5 w^3 = w^8$ is an example of the _________________________.

3. $\dfrac{z^6}{z^4} = z^2$ is an example of the _________________________.

Objective 1   Use exponential notation for nth roots.

Review these examples for Objective 1:

1.   Evaluate each exponential.

   a. $8^{1/3}$

$$8^{1/3} = \sqrt[3]{8} = 2$$

   b. $25^{1/2}$

$$25^{1/2} = \sqrt{25} = 5$$

   c. $-81^{1/4}$

$$-81^{1/4} = -\sqrt[4]{81} = -3$$

   d. $(-81)^{1/4}$

$(-81)^{1/4} = \sqrt[4]{-81}$ is not a real number.

   e. $(-8)^{1/3}$

$$(-8)^{1/3} = \sqrt[3]{-8} = -2$$

Now Try:

1.   Evaluate each exponential.

   a. $216^{1/3}$

   ____

   b. $121^{1/2}$

   ____

   c. $-1024^{1/10}$

   ____

   d. $(-1024)^{1/10}$

   ____

   e. $(-243)^{1/5}$

   ____

f. $\left(\dfrac{1}{125}\right)^{1/3}$ **f.** $\left(\dfrac{1}{16}\right)^{1/4}$

$$\left(\frac{1}{125}\right)^{1/3} = \sqrt[3]{\frac{1}{125}} = \frac{1}{5}$$

____

Objective 1 Practice Exercises

For extra help, see Example 1 on page 705 of your text.

Evaluate each exponential.

1. $-256^{1/4}$ **1.** _____________

2. $16^{1/2}$ **2.** _____________

3. $(-3375)^{1/3}$ **3.** _____________

Objective 2 Define and use expressions of the form $a^{m/n}$.

Review these examples for Objective 2:

2. Evaluate each polynomial.

 a. $100^{3/2}$

$$100^{3/2} = (100^{1/2})^3 = 10^3 = 1000$$

 b. $64^{2/3}$

$$64^{2/3} = (64^{1/3})^2 = 4^2 = 16$$

 c. $-729^{5/6}$

$$-729^{5/6} = -(729^{1/6})^5 = -(3)^5 = -243$$

 d. $(-125)^{2/3}$

$$(-125)^{2/3} = [(-125)^{1/3}]^2 = (-5)^2 = 25$$

 e. $(-729)^{5/6}$

$(-729)^{5/6} = [(-729)^{1/6}]^5$ is not a real number

since $(-729)^{1/6}$ is not a real number.

Now Try:

2. Evaluate each polynomial.

 a. $27^{2/3}$

____

 b. $16^{3/4}$

____

 c. $-36^{3/2}$

____

 d. $(-27)^{2/3}$

____

 e. $(-36)^{3/2}$

____

         385

3. Evaluate each exponential.

    a. $625^{-3/4}$

$$625^{-3/4} = \frac{1}{625^{3/4}} = \frac{1}{(625^{1/4})^3} = \frac{1}{\left(\sqrt[4]{625}\right)^3}$$

$$= \frac{1}{5^3} = \frac{1}{125}$$

    b. $36^{-3/2}$

$$36^{-3/2} = \frac{1}{36^{3/2}} = \frac{1}{\left(\sqrt{36}\right)^3} = \frac{1}{6^3} = \frac{1}{216}$$

    c. $\left(\dfrac{81}{16}\right)^{-3/4}$

$$\left(\frac{81}{16}\right)^{-3/4} = \left(\frac{16}{81}\right)^{3/4} = \left(\sqrt[4]{\frac{16}{81}}\right)^3 = \left(\frac{2}{3}\right)^3 = \frac{8}{27}$$

3. Evaluate each exponential.

    a. $32^{-2/5}$

    ____

    b. $125^{-4/3}$

    ____

    c. $\left(\dfrac{27}{64}\right)^{-2/3}$

    ____

Objective 2 Practice Exercises

For extra help, see Examples 2–3 on pages 705–706 of your text.

Evaluate each exponential.

4. $-81^{5/4}$

5. $36^{5/2}$

6. $\left(\dfrac{125}{27}\right)^{-2/3}$

4. ________________

5. ________________

6. ________________

Objective 3   Convert between radicals and rational exponents.

Review these examples for Objective 3:

4. Write each radical as an exponential. Assume that all variables represent positive real numbers. Use the definition that takes the root first.

    a. $17^{1/2}$

$$17^{1/2} = \sqrt{17}$$

    b. $10^{5/6}$

$$10^{5/6} = \left(\sqrt[6]{10}\right)^5$$

Now Try:

4. Write each radical as an exponential. Assume that all variables represent positive real numbers. Use the definition that takes the root first.

    a. $23^{1/3}$

    ____

    b. $21^{3/4}$

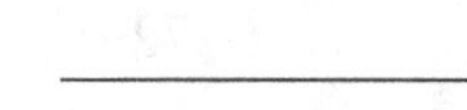

c. $8x^{3/4}$

$$8x^{3/4} = 8\left(\sqrt[4]{x}\right)^3$$

d. $2x^{2/5} - (4x)^{5/6}$

$$2x^{2/5} - (4x)^{5/6} = 2\left(\sqrt[5]{x}\right)^2 - \left(\sqrt[6]{4x}\right)^5$$

e. $x^{-4/5}$

$$x^{-4/5} = \frac{1}{x^{4/5}} = \frac{1}{\left(\sqrt[4]{x}\right)^5}$$

f. $\left(x^3 + y^2\right)^{1/5}$

$$\left(x^3 + y^2\right)^{1/5} = \sqrt[5]{x^3 + y^2}$$

c. $5x^{5/4}$

____

d. $(2x)^{4/3} - 3x^{2/5}$

____

e. $x^{-3/2}$

____

f. $\left(x^2 - y^2\right)^{1/4}$

____

5. Write each radical as an exponential and simplify. Assume that all variables represent positive real numbers.

a. $\sqrt{14}$

$$\sqrt{14} = 14^{1/2}$$

b. $\sqrt{7^4}$

$$\sqrt{7^4} = 7^{4/2} = 7^2 = 49$$

c. $\sqrt[3]{3^6}$

$$\sqrt[3]{3^6} = 3^{6/3} = 3^2 = 9$$

d. $\sqrt[7]{w^7}$

$$\sqrt[7]{w^7} = w^{7/7} = w^1 = w$$

5. Write each radical as an exponential and simplify. Assume that all variables represent positive real numbers.

a. $\sqrt{22}$

____

b. $\sqrt{10^4}$

____

c. $\sqrt[5]{3^{10}}$

____

d. $\sqrt[8]{m^8}$

____

Objective 3 Practice Exercises

For extra help, see Examples 4–5 on pages 707–708 of your text.

Write with radicals. Assume that all variables represent positive real numbers.

7. $4y^{2/5} + (5x)^{1/5}$

7. _______________

8. $\left(2x^4 - 3y^2\right)^{-4/3}$ 8. ________________

Simplify the radical by rewriting it with a rational exponent. Write answer in radical form if necessary. Assume that variables represent positive real numbers.

9. $\sqrt[8]{a^2}$ 9. ________________

Objective 4   Use the rules for exponents with rational exponents.

Review these examples for Objective 4:

6. Write with only positive exponents. Assume that all variables represent positive real numbers.

a. $13^{4/5} \cdot 13^{1/2}$

$$13^{4/5} \cdot 13^{1/2} = 13^{4/5 + 1/2} = 13^{13/10}$$

b. $\dfrac{8^{3/4}}{8^{1/4}}$

$$\frac{8^{3/4}}{8^{1/4}} = 8^{3/4 - 1/4} = 8^{1/2}$$

c. $\dfrac{\left(n^{7/4} w^{1/2}\right)^2}{w^{3/4}}$

$$\frac{\left(n^{7/4} w^{1/2}\right)^2}{w^{3/4}} = \frac{\left(n^{7/4}\right)^2 \left(w^{1/2}\right)^2}{w^{3/4}}$$

$$= \frac{n^{7/2} w^1}{w^{3/4}}$$

$$= n^{7/2} w^{1 - 3/4}$$

$$= n^{7/2} w^{1/4}$$

Now Try:

5. Write with only positive exponents. Assume that all variables represent positive real numbers.

a. $5^{3/4} \cdot 5^{7/4}$

____

b. $\dfrac{a^{4/5}}{a^{2/3}}$

____

c. $\dfrac{\left(x^{1/3} y^{2/3}\right)^6}{y^{1/2}}$

____

d. $\dfrac{a^{-1/3}b^{2/3}}{(a^3b)^{1/3}}$

$$\dfrac{a^{-1/3}b^{2/3}}{(a^3b)^{1/3}} = \dfrac{a^{-1/3}b^{2/3}}{a^{3/3}b^{1/3}}$$

$$= a^{-1/3-1}b^{2/3-1/3}$$

$$= a^{-4/3}b^{1/3}$$

$$= \dfrac{b^{1/3}}{a^{4/3}}$$

e. $\left(\dfrac{c^6x^3}{c^{-2}x^{1/2}}\right)^{-3/4}$

$$\left(\dfrac{c^6x^{1/2}}{c^{-2}x^3}\right)^{-3/4} = \left(c^{6-(-2)}x^{1/2-3}\right)^{-3/4}$$

$$= \left(c^8x^{-5/2}\right)^{-3/4}$$

$$= \left(c^8\right)^{-3/4}\left(x^{-5/2}\right)^{-3/4}$$

$$= c^{-6}x^{15/8}$$

$$= \dfrac{x^{15/8}}{c^6}$$

f. $a^{3/4}(a^{2/3}-a^{1/2})$

$$a^{3/4}(a^{2/3}-a^{1/2}) = a^{3/4}a^{2/3} - a^{3/4}a^{1/2}$$

$$= a^{3/4+2/3} - a^{3/4+1/2}$$

$$= a^{17/12} - a^{5/4}$$

7. Write all radicals as exponentials, and then apply the rules for rational exponents. Leave answers in exponential form. Assume that all variables represent positive real numbers.

a. $\sqrt[4]{x^3}\cdot\sqrt[5]{x}$

$$\sqrt[4]{x^3}\cdot\sqrt[5]{x} = x^{3/4}\cdot x^{1/5}$$

$$= x^{3/4+1/5}$$

$$= x^{15/20+4/20}$$

$$= x^{19/20}$$

d. $\dfrac{c^{2/3}d^{-1/3}}{(c^3d)^{1/3}}$

____

e. $\left(\dfrac{x^{-1}y^{2/3}}{x^{1/3}y^{1/2}}\right)^{-3/2}$

____

f. $r^{1/2}(r^{2/3}-r^{8/3})$

____

7. Write all radicals as exponentials, and then apply the rules for rational exponents. Leave answers in exponential form. Assume that all variables represent positive real numbers.

a. $\sqrt[6]{x^3}\cdot\sqrt[3]{x^2}$

____

b. $\dfrac{\sqrt[3]{y^5}}{\sqrt{y^3}}$

$$\frac{\sqrt[3]{y^5}}{\sqrt{y^3}} = \frac{y^{5/3}}{y^{3/2}} = y^{5/3-3/2} = y^{1/6}$$

c. $\sqrt[6]{ab} \cdot \sqrt{b}$

$$\sqrt[6]{ab} \cdot \sqrt{b} = (ab)^{1/6} \cdot b^{1/2}$$
$$= a^{1/6} b^{1/6} b^{1/2}$$
$$= a^{1/6} b^{1/6+1/2}$$
$$= a^{1/6} b^{4/6}$$
$$= a^{1/6} b^{2/3}$$

d. $\sqrt{\sqrt[4]{y^3}}$

$$\sqrt{\sqrt[4]{y^3}} = \sqrt{y^{3/4}} = \left(y^{3/4}\right)^{1/2} = y^{3/8}$$

b. $\dfrac{\sqrt[4]{y^5}}{\sqrt[3]{y^2}}$

____

c. $\sqrt[4]{c} \cdot \sqrt{cd}$

____

d. $\sqrt[3]{\sqrt[4]{x^3}}$

____

Objective 4 Practice Exercises

For extra help, see Examples 6–7 on pages 708–710 of your text.

Use the rules of exponents to simplify each expression. Write all answers with positive exponents. Assume that variables represent positive real numbers.

10. $y^{7/3} \cdot y^{-4/3}$

10. __________

11. $\dfrac{a^{2/3} \cdot a^{-1/3}}{\left(a^{-1/6}\right)^3}$

11. __________

12. $\dfrac{\left(x^{-3} y^2\right)^{2/3}}{\left(x^2 y^{-5}\right)^{2/5}}$

12. __________

Chapter 10 ROOTS, RADICALS, AND ROOT FUNCTIONS

10.3   Simplifying Radicals, the Distance Formula, and Circles

Learning Objectives
1   Use the product rule for radicals.
2   Use the quotient rule for radicals.
3   Simplify radicals.
4   Simplify products and quotients of radicals with different indexes.
5   Use the Pythagorean theorem.
6   Use the distance formula.
7   Find an equation of a circle given its center and radius.

Key Terms

Use the vocabulary terms listed below to complete each statement in exercises 1−6.

     index   **radicand**   **hypotenuse**   **legs**

     radius   **circle**   **center**

1.   In a right triangle, the side opposite the right angle is called the

     _______________________________.

2.   In the expression $\sqrt[4]{x^2}$, the "4" is the _______________________ and x^2 is

     the _______________________________.

3.   In a right triangle, the sides that form the right angle are called the

     _______________________.

4.   A(n) _______________________ is the set of all points in a plane that lie a

     fixed distance from a fixed point.

5.   A fixed point such that every point on a circle is a fixed distance from it is the

     _______________________.

6.   The distance from the center of a circle to a point on the circle is called the

     _______________________.

Objective 1   Use the product rule for radicals.

Review these examples for Objective 1:

1. Multiply. Assume that all variables represent positive real numbers.

   a. $\sqrt{13} \cdot \sqrt{5}$

   $\sqrt{13} \cdot \sqrt{5} = \sqrt{13 \cdot 5} = \sqrt{65}$

Now Try:

1. Multiply. Assume that all variables represent positive real numbers.

   a. $\sqrt{2} \cdot \sqrt{7}$

   ____

b. $\sqrt{5}\cdot\sqrt{2r}$

$\sqrt{5}\cdot\sqrt{2r}=\sqrt{5\cdot 2r}=\sqrt{10r}$

c. $\sqrt{5x}\cdot\sqrt{2yz}$

$\sqrt{5x}\cdot\sqrt{2yz}=\sqrt{10xyz}$

2. Multiply. Assume that all variables represent positive real numbers.

   a. $\sqrt[4]{2}\cdot\sqrt[4]{2x}$

$\sqrt[4]{2}\cdot\sqrt[4]{2x}=\sqrt[4]{2\cdot 2x}=\sqrt[4]{4x}$

   b. $\sqrt[3]{8x}\cdot\sqrt[3]{2y^2}$

$\sqrt[3]{8x}\cdot\sqrt[3]{2y^2}=\sqrt[3]{8x\cdot 2y^2}=\sqrt[3]{16xy^2}$

   c. $\sqrt[5]{6r^2}\cdot\sqrt[5]{4r^2}$

$\sqrt[5]{6r^2}\cdot\sqrt[5]{4r^2}=\sqrt[5]{6r^2\cdot 4r^2}=\sqrt[5]{24r^4}$

   d. $\sqrt[5]{2}\cdot\sqrt[4]{6}$

$\sqrt[5]{2}\cdot\sqrt[4]{6}$ cannot be simplified using the product rule for radicals, because the indexes (5 and 4) are different.

b. $\sqrt{5x}\cdot\sqrt{7}$

____

c. $\sqrt{3}\cdot\sqrt{11mn}$

____

2. Multiply. Assume that all variables represent positive real numbers.

   a. $\sqrt[3]{3}\cdot\sqrt[3]{7}$

____

   b. $\sqrt[3]{7x}\cdot\sqrt[3]{5y}$

____

   c. $\sqrt[5]{4w}\cdot\sqrt[5]{2w^3}$

____

   d. $\sqrt{3}\cdot\sqrt[3]{64}$

____

Objective 1 Practice Exercises

For extra help, see Examples 1–2 on page 713 of your text.

Multiply. Assume that variables represent positive real numbers.

1. $\sqrt{7x}\cdot\sqrt{6t}$

   1. _______________

2. $\sqrt[5]{6r^2t^3}\cdot\sqrt[5]{4r^2t}$

   2. _______________

3. $\sqrt{3}\cdot\sqrt[3]{7}$

   3. _______________

Objective 2   Use the quotient rule for radicals.

Review these examples for Objective 2:

3. Simplify. Assume that all variables represent positive real numbers.

a. $\sqrt{\dfrac{64}{9}}$

$$\sqrt{\dfrac{64}{9}} = \dfrac{\sqrt{64}}{\sqrt{9}} = \dfrac{8}{3}$$

b. $\sqrt{\dfrac{5}{16}}$

$$\sqrt{\dfrac{5}{16}} = \dfrac{\sqrt{5}}{\sqrt{16}} = \dfrac{\sqrt{5}}{4}$$

c. $\sqrt[3]{-\dfrac{27}{8}}$

$$\sqrt[3]{-\dfrac{27}{8}} = \dfrac{\sqrt[3]{-27}}{\sqrt[3]{8}} = \dfrac{-3}{2} = -\dfrac{3}{2}$$

d. $\sqrt[5]{-\dfrac{a^3}{243}}$

$$\sqrt[5]{-\dfrac{a^3}{243}} = \dfrac{\sqrt[5]{a^3}}{\sqrt[5]{-243}} = \dfrac{\sqrt[5]{a^3}}{-3} = -\dfrac{\sqrt[5]{a^3}}{3}$$

e. $\sqrt{\dfrac{z^4}{36}}$

$$\sqrt{\dfrac{z^4}{36}} = \dfrac{\sqrt{z^4}}{\sqrt{36}} = \dfrac{z^2}{6}$$

Now Try:

3. Simplify. Assume that all variables represent positive real numbers.

a. $\sqrt{\dfrac{36}{49}}$

____

b. $\sqrt{\dfrac{13}{81}}$

____

c. $\sqrt[3]{-\dfrac{343}{125}}$

____

d. $\sqrt[3]{-\dfrac{a^6}{125}}$

____

e. $\sqrt[4]{\dfrac{m}{81}}$

____

Objective 2 Practice Exercises

For extra help, see Example 3 on page 714 of your text.

Simplify each radical. Assume that variables represent positive real numbers.

4. $\sqrt[3]{\dfrac{27}{8}}$

4. _________________

5. $\sqrt[5]{\dfrac{7x}{32}}$

5. _________________

6. $\sqrt[3]{-\dfrac{x^9}{216}}$ **6.** _______________

Objective 3   Simplify radicals.

Review these examples for Objective 3:

4. Simplify.

  a. $\sqrt{90}$

$$\sqrt{90} = \sqrt{9 \cdot 10}$$
$$= \sqrt{9} \cdot \sqrt{10}$$
$$= 3\sqrt{10}$$

  b. $\sqrt{288}$

$$\sqrt{288} = \sqrt{144 \cdot 2}$$
$$= \sqrt{144} \cdot \sqrt{2}$$
$$= 12\sqrt{2}$$

  c. $\sqrt{35}$

No perfect square (other than 1) divides into 35, so $\sqrt{35}$ cannot be simplified further.

  d. $\sqrt[3]{81}$

$$\sqrt[3]{81} = \sqrt[3]{27 \cdot 3} = \sqrt[3]{27} \cdot \sqrt[3]{3} = 3\sqrt[3]{3}$$

  e. $-\sqrt[4]{3125}$

$$-\sqrt[4]{3125} = -\sqrt[4]{5^5} = \sqrt[4]{5^4 \cdot 5}$$
$$= -\sqrt[4]{5^4} \cdot \sqrt[4]{5}$$
$$= -5\sqrt[4]{5}$$

5. Simplify. Assume that all variables represent positive real numbers.

  a. $\sqrt{81x^3}$

$$\sqrt{81x^3} = \sqrt{9^2 \cdot x^2 \cdot x} = 9x\sqrt{x}$$

Now Try:

4. Simplify.

  a. $\sqrt{84}$

  ____

  b. $\sqrt{162}$

  ____

  c. $\sqrt{95}$

  ____

  d. $\sqrt[3]{256}$

  ____

  e. $-\sqrt[5]{512}$

  ____

5. Simplify. Assume that all variables represent positive real numbers.

  a. $\sqrt{100y^3}$

  ____

b. $\sqrt{56x^7y^6}$

$$\sqrt{56x^7y^6} = \sqrt{4\cdot 14\cdot \left(x^3\right)^2 \cdot x\cdot \left(y^3\right)^2}$$
$$= 2x^3y^3\sqrt{14x}$$

c. $\sqrt[3]{-270b^4c^8}$

$$\sqrt[3]{-270b^4c^8} = \sqrt[3]{\left(-27b^3c^6\right)\left(10bc^2\right)}$$
$$= \sqrt[3]{-27b^3c^6}\cdot\sqrt[3]{10bc^2}$$
$$= -3bc^2\sqrt[3]{10bc^2}$$

d. $-\sqrt[6]{448a^7b^7}$

$$-\sqrt[6]{448a^7b^7} = -\sqrt[6]{\left(64a^6b^6\right)\left(7ab\right)}$$
$$= -\sqrt[6]{64a^6b^6}\cdot\sqrt[6]{7ab}$$
$$= -2ab\sqrt[6]{7ab}$$

6. Simplify. Assume that all variables represent positive real numbers.

   a. $\sqrt[24]{5^4}$

$$\sqrt[24]{5^4} = \left(5^4\right)^{1/24} = 5^{4/24} = 5^{1/6} = \sqrt[6]{5}$$

   b. $\sqrt[12]{x^8}$

$$\sqrt[12]{x^8} = \left(x^8\right)^{1/12} = x^{8/12} = x^{2/3} = \sqrt[3]{x^2}$$

   c. $\sqrt[6]{x^{15}}$

$$\sqrt[6]{x^{15}} = \left(x^{15}\right)^{1/6} = x^{5/2} = \sqrt{x^5} = \sqrt{x^4}\cdot\sqrt{x}$$
$$= x^2\sqrt{x}$$

b. $\sqrt{48m^5r^9}$

____

c. $\sqrt[3]{-32n^7t^5}$

____

d. $-\sqrt[4]{405x^3y^9}$

____

6. Simplify. Assume that all variables represent positive real numbers.

   a. $\sqrt[12]{11^9}$

____

   b. $\sqrt[30]{z^{24}}$

____

   c. $\sqrt[6]{w^{21}}$

____

Objective 3 Practice Exercises

For extra help, see Examples 4–6 on pages 715–716 of your text.

Simplify each radical. Assume that variables represent positive real numbers.

7. $\sqrt[42]{x^{28}}$

7. _______________

8. $\sqrt{8x^3y^6z^{11}}$ 8. _______________

9. $\sqrt[3]{1250a^5b^7}$ 9. _______________

Objective 4   Simplify products and quotients of radicals with different indexes.

Review this example for Objective 4:

7. Simplify $\sqrt{3}\cdot\sqrt[5]{6}$.

Because the different indexes, 2 and 5, have least common multiple index of 10, we use rational exponents to write each radical as a tenth root.

$$\sqrt{3} = 3^{1/2} = 3^{5/10} = \sqrt[10]{3^5} = \sqrt[10]{243}$$

$$\sqrt[5]{6} = 6^{1/5} = 6^{2/10} = \sqrt[10]{6^2} = \sqrt[10]{36}$$

$$\sqrt{3}\cdot\sqrt[5]{6} = \sqrt[10]{243}\cdot\sqrt[10]{36}$$

$$= \sqrt[10]{243\cdot 36}$$

$$= \sqrt[10]{8748}$$

Now Try:

7. Simplify $\sqrt[3]{3}\cdot\sqrt[6]{7}$.

____

Objective 4 Practice Exercises

For extra help, see Example 7 on page 717 of your text.

Simplify each radical. Assume that variables represent positive real numbers.

10. $\sqrt{r}\cdot\sqrt[3]{r}$ 10. _______________

11. $\sqrt[4]{2}\cdot\sqrt[8]{7}$ 11. _______________

12. $\sqrt{3}\cdot\sqrt[5]{64}$ 12. _______________

                        Copyright © 2025 Pearson Education, Inc.

Objective 5   Use the Pythagorean theorem.

Review this example for Objective 5:

8. Use the Pythagorean theorem to find the length of the unknown side of the triangle.

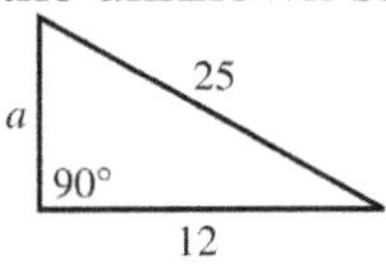

$$a^2 + b^2 = c^2$$

$$12^2 + b^2 = 25^2$$

$$144 + b^2 = 625$$

$$b^2 = 481$$

$$b = \sqrt{481}$$

The length of the side is $\sqrt{481}$.

Now Try:

8. Use the Pythagorean theorem to find the length of the unknown side of the triangle.

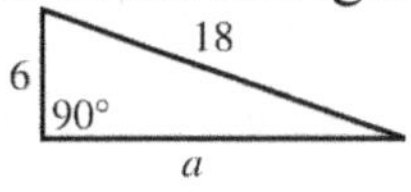

____

Objective 5 Practice Exercises

For extra help, see Example 8 on page 718 of your text.

Find the unknown length in each right triangle. Simplify the answer if necessary.

13.

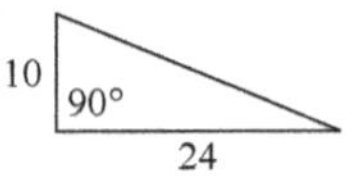

13. ___________________

14.

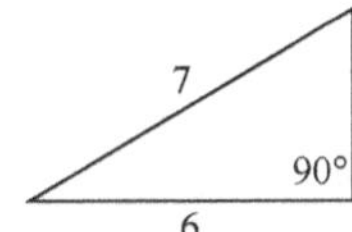

14. ___________________

15.

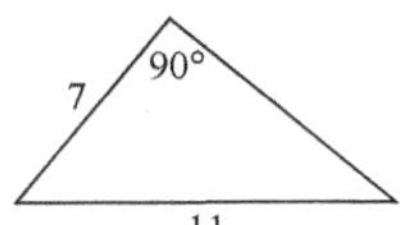

15. ___________________

Objective 6   Use the distance formula.

Review this example for Objective 6:

9. Find the distance between the points $(2,-2)$ and $(-6, 1)$.

   Use the distance formula. Let $(x_1, y_1) = (2, -2)$ and $(x_2, y_2) = (-6, 1)$.

   $$d = \sqrt{(x_2 - x_1)^2 + (y_2 - y_1)^2}$$
   $$= \sqrt{(-6-2)^2 + [1-(-2)]^2}$$
   $$= \sqrt{(-8)^2 + 3^2}$$
   $$= \sqrt{64+9}$$
   $$= \sqrt{73}$$

Now Try:

9. Find the distance between the points $(-1,-2)$ and $(-4, 3)$.

   ____

Objective 6 Practice Exercises

For extra help, see Example 9 on page 719 of your text.

Find the distance between each pair of points.

16. $(3, 4)$ and $(-1,-2)$

16. _______________

17. $(-2,-3)$ and $(-5, 1)$

17. _______________

18. $(4, 2)$ and $(3,-1)$

18. _______________

Objective 7 Find an equation of a circle given its center and radius.

Review these examples for Objective 7:

10. Find an equation of the circle with center $(0, 0)$ and radius 2, and graph it.

If the point (x, y) is on the circle, then the distance from (x, y) to the center $(0, 0)$ is 2.

$$\sqrt{(x_2 - x_1)^2 + (y_2 - y_1)^2} = d$$

$$\sqrt{(x - 0)^2 + (y - 0)^2} = 2$$

$$\left(\sqrt{x^2 + y^2}\right)^2 = 2^2$$

$$x^2 + y^2 = 4$$

The equation of this circle is $x^2 + y^2 = 4$.

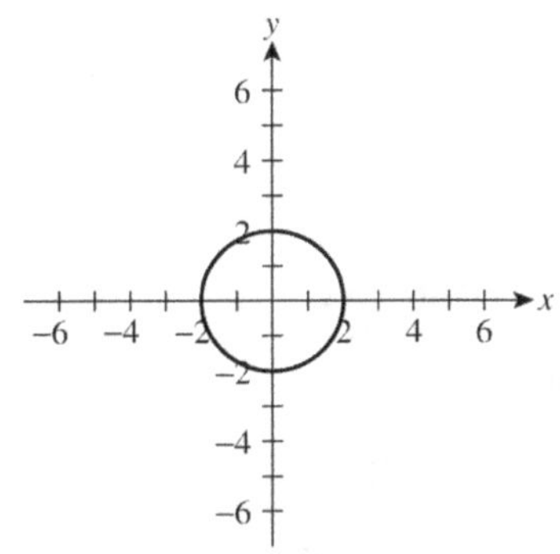

11. Find an equation of the circle with center $(-3, 2)$ and radius 3, and graph it.

$$\sqrt{(x_2 - x_1)^2 + (y_2 - y_1)^2} = d$$

$$\sqrt{(x - (-3))^2 + (y - 2)^2} = 3$$

$$\left(\sqrt{(x + 3)^2 + (y - 2)^2}\right)^2 = 3^2$$

$$(x + 3)^2 + (y - 2)^2 = 9$$

To graph the circle, plot the center $(-3, 2)$, then move three units right, left, up, and down from the center, plotting the points $(0, 2)$, $(-3, 5)$, $(-6, 2)$, and $(-3, -1)$. Draw a smooth curve through the points.

Now Try:

10. Find an equation of the circle with and center $(0, 0)$ and radius 5, and graph it.

____

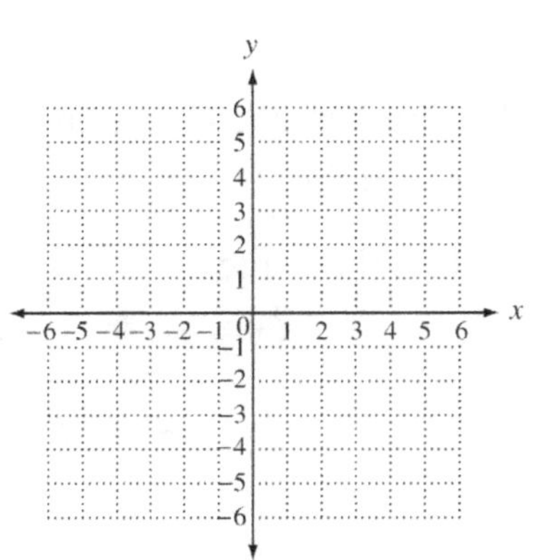

11. Find an equation of the circle with center $(-5, 4)$ and radius 4, and graph it.

____

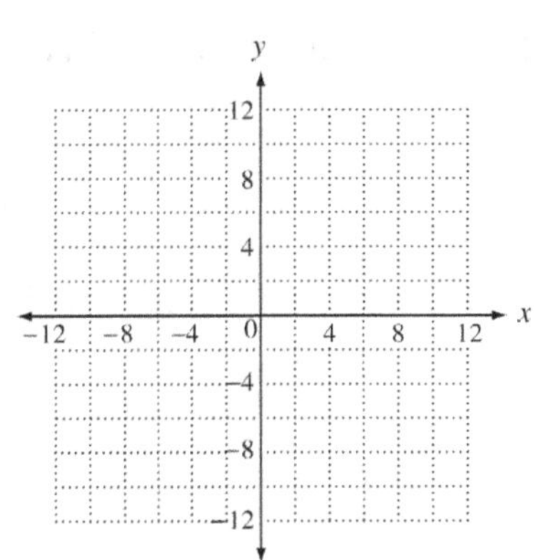

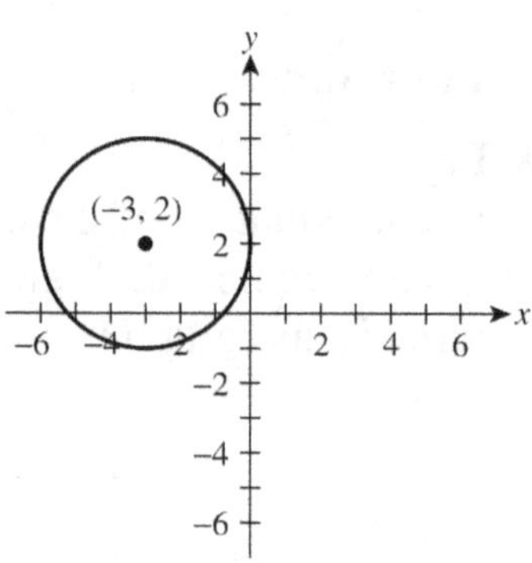

12. Write an equation of a circle with center $(-2,-4)$ and radius $2\sqrt{5}$.

Use the center-radius form.

$$(x-h)^2 + (y-k)^2 = r^2$$

$$[x-(-2)]^2 + [y-(-4)]^2 = \left(2\sqrt{5}\right)^2$$

$$(x+2)^2 + (y+4)^2 = 20$$

12. Write an equation of a circle with center $(3,-1)$ and radius $\sqrt{6}$.

____

Objective 7 Practice Exercises

For extra help, see Examples 10–12 on pages 720–721 of your text.

Find the equation of a circle satisfying the given conditions.

19.   center: $(3,-4)$; radius: 5

19. ______________

20.   center: $(-2,-2)$; radius: 3

20. ______________

21.   center: $(0, 3)$; radius: $\sqrt{2}$

21. ______________

Chapter 10 ROOTS, RADICALS, AND ROOT FUNCTIONS

10.4   Adding and Subtracting Radical Expressions

Learning Objectives
1   Simplify radical expressions involving addition and subtraction.

Key Terms

Use the vocabulary terms listed below to complete each statement in exercises 1–2.

like radicals   unlike radicals

1.   The expressions $2\sqrt{2}$ and $6\sqrt[3]{2}$ are _________________________________ .

2.   The expressions $2\sqrt{2}$ and $7\sqrt{2}$ are _________________________________ .

Objective 1   Simplify radical expressions involving addition and subtraction.

Review these examples for Objective 1:

1.   Add or subtract to simplify each radical expression.

a. $5\sqrt{3}+7\sqrt{3}$

$$5\sqrt{3}+7\sqrt{3}=(5+7)\sqrt{3}$$
$$=12\sqrt{3}$$

b. $3\sqrt{2}-6\sqrt{2}$

$$3\sqrt{2}-6\sqrt{2}=(3-6)\sqrt{2}$$
$$=-3\sqrt{2}$$

c. $3\sqrt{13}+5\sqrt{52}$

$$3\sqrt{13}+5\sqrt{52}=3\sqrt{13}+5\sqrt{4}\sqrt{13}$$
$$=3\sqrt{13}+5\cdot2\sqrt{13}$$
$$=3\sqrt{13}+10\sqrt{13}$$
$$=(3+10)\sqrt{13}$$
$$=13\sqrt{13}$$

Now Try:

1.   Add or subtract to simplify each radical expression.

a. $6\sqrt{5}+2\sqrt{5}$

____

b. $4\sqrt{3}-8\sqrt{3}$

____

c. $3\sqrt{54}-5\sqrt{24}$

____

d. $\sqrt{48x} - \sqrt{12x}, \; x \geq 0$

$$\sqrt{48x} - \sqrt{12x} = \sqrt{16} \cdot \sqrt{3x} - \sqrt{4} \cdot \sqrt{3x}$$
$$= 4\sqrt{3x} - 2\sqrt{3x}$$
$$= (4 - 2)\sqrt{3x}$$
$$= 2\sqrt{3x}$$

e. $7\sqrt{3} - 6\sqrt{21}$

The radicands differ and are already simplified, so this expression cannot be simplified further.

2. Simplify. Assume that all variables represent positive real numbers.

a. $7\sqrt[4]{32} - 9\sqrt[4]{2}$

$$7\sqrt[4]{32} - 9\sqrt[4]{2} = 7\sqrt[4]{16} \cdot \sqrt[4]{2} - 9\sqrt[4]{2}$$
$$= 7 \cdot 2 \cdot \sqrt[4]{2} - 9\sqrt[4]{2}$$
$$= 14\sqrt[4]{2} - 9\sqrt[4]{2}$$
$$= (14 - 9)\sqrt[4]{2}$$
$$= 5\sqrt[4]{2}$$

b. $6\sqrt[3]{27x^5 r} + 2x\sqrt[3]{x^2 r}$

$$6\sqrt[3]{27x^5 r} + 2x\sqrt[3]{x^2 r}$$
$$= 6 \cdot \sqrt[3]{27x^3} \cdot \sqrt[3]{x^2 r} + 2x\sqrt[3]{x^2 r}$$
$$= 18x\sqrt[3]{x^2 r} + 2x\sqrt[3]{x^2 r}$$
$$= (18x + 2x)\sqrt[3]{x^2 r}$$
$$= 20x\sqrt[3]{x^2 r}$$

d. $3\sqrt{18z} + 2\sqrt{8z}, \; z \geq 0$

____

e. $3\sqrt{7} + 2\sqrt{6}$

____

2. Simplify. Assume that all variables represent positive real numbers.

a. $7\sqrt[3]{54} - 6\sqrt[3]{128}$

____

b. $\sqrt[4]{32y^2 z^5} + 3z\sqrt[4]{2y^2 z}$

____

c. $3\sqrt{40x^5} + 5\sqrt[3]{48x^5}$

$3\sqrt{40x^5} + 5\sqrt[3]{48x^5}$

$= 3\cdot\sqrt{4x^4 \cdot 10x} + 5\sqrt[3]{8x^3 \cdot 6x^2}$

$= 3\cdot\sqrt{4x^4}\cdot\sqrt{10x} + 5\sqrt[3]{8x^3}\cdot\sqrt[3]{6x^2}$

$= 3\cdot 2x^2 \cdot\sqrt{10x} + 5\cdot 2x\cdot\sqrt[3]{6x^2}$

$= 6x^2\sqrt{10x} + 10x\sqrt[3]{6x^2}$

3. Simplify. Assume that all variables represent positive real numbers.

a. $\dfrac{\sqrt{32}}{3} + \dfrac{\sqrt{8}}{\sqrt{18}}$

$\dfrac{\sqrt{32}}{3} + \dfrac{\sqrt{8}}{\sqrt{18}} = \dfrac{\sqrt{16\cdot 2}}{3} + \dfrac{\sqrt{4\cdot 2}}{\sqrt{9\cdot 2}}$

$= \dfrac{4\sqrt{2}}{3} + \dfrac{2\sqrt{2}}{3\sqrt{2}}$

$= \dfrac{4\sqrt{2}}{3} + \dfrac{2}{3}$

$= \dfrac{4\sqrt{2}+2}{3}$

b. $\sqrt[3]{\dfrac{81}{y^6}} + 5\sqrt[3]{\dfrac{27}{y^3}}$

$\sqrt[3]{\dfrac{81}{y^6}} + 5\sqrt[3]{\dfrac{27}{y^3}} = \dfrac{\sqrt[3]{81}}{\sqrt[3]{y^6}} + 5\dfrac{\sqrt[3]{27}}{\sqrt[3]{y^3}}$

$= \dfrac{3\sqrt[3]{3}}{y^2} + 5\left(\dfrac{3}{y}\right)$

$= \dfrac{3\sqrt[3]{3}}{y^2} + \dfrac{15}{y}$

$= \dfrac{3\sqrt[3]{3}+15y}{y^2}$

c. $2\sqrt[3]{54x^7} + 2\sqrt{27x^7}$

____

3. Simplify. Assume that all variables represent positive real numbers.

a. $\sqrt{\dfrac{10}{18}} + \dfrac{\sqrt{15}}{\sqrt{27}}$

____

b. $\sqrt[3]{\dfrac{216}{w^6}} + \sqrt{\dfrac{121}{w^4}}$

____

Objective 1 Practice Exercises

For extra help, see Examples 1–3 on pages 727–729 of your text.

Add or subtract. Assume that all variables represent positive real numbers.

1. $\sqrt{100x} - \sqrt{9x} + \sqrt{25x}$

1. ______________________

2. $2\sqrt[3]{16r} + \sqrt[3]{54r} - \sqrt[3]{16r}$

2. ______________________

3. $\sqrt[3]{\dfrac{y^7}{125}} + y^2 \sqrt[3]{\dfrac{y}{27}}$

3. ______________________

Chapter 10 ROOTS, RADICALS, AND ROOT FUNCTIONS

10.5   Multiplying and Dividing Radical Expressions

Learning Objectives
1   Multiply radical expressions.
2   Rationalize denominators with one radical term.
3   Rationalize denominators with binomials involving radicals.
4   Write radical quotients in lowest terms.

Key Terms

Use the vocabulary terms listed below to complete each statement in exercises 1–2.

rationalizing the denominator   **conjugate**

1.   The ________________________ of $a + b$ is $a - b$.

2.   The process of removing radicals from the denominator so that the denominator contains only rational quantities is called ________________________________.

Objective 1   Multiply radical expressions.

Review these examples for Objective 1:

2.   Multiply, using the FOIL method.

a. $\left(\sqrt{6} + \sqrt{5}\right)\left(\sqrt{6} - \sqrt{5}\right)$

This is the difference of squares.

$$\left(\sqrt{6} + \sqrt{5}\right)\left(\sqrt{6} - \sqrt{5}\right) = \left(\sqrt{6}\right)^2 - \left(\sqrt{5}\right)^2$$
$$= 6 - 5$$
$$= 1$$

b. $\left(5 - \sqrt{3}\right)\left(\sqrt{2} + \sqrt{5}\right)$

$$\left(5 - \sqrt{3}\right)\left(\sqrt{2} + \sqrt{5}\right)$$
$$= 5 \cdot \sqrt{2} + 5 \cdot \sqrt{5} - \sqrt{3} \cdot \sqrt{2} - \sqrt{3} \cdot \sqrt{5}$$
$$= 5\sqrt{2} + 5\sqrt{5} - \sqrt{6} - \sqrt{15}$$

Now Try:

2.   Multiply, using the FOIL method.

a. $\left(\sqrt{14} - \sqrt{2}\right)\left(\sqrt{14} + \sqrt{2}\right)$

____

b. $\left(3 - \sqrt{2}\right)\left(2 + \sqrt{7}\right)$

____

c. $\left(\sqrt{11}-6\right)^2$ **c.** $\left(3-\sqrt{2}\right)^2$

$$\left(\sqrt{11}-6\right)^2 = \left(\sqrt{11}-6\right)\left(\sqrt{11}-6\right)$$
$$= \sqrt{11}\cdot\sqrt{11}-6\sqrt{11}-6\sqrt{11}+6\cdot 6$$
$$= 11-12\sqrt{11}+36$$
$$= 47-12\sqrt{11}$$

Objective 1 Practice Exercises

For extra help, see Examples 1–2 on pages 732–733 of your text.

Multiply each product, then simplify. Assume that variables represent positive real numbers.

1. $\left(\sqrt{5}+\sqrt{6}\right)\left(\sqrt{2}-4\right)$ 1. _______________

2. $\left(\sqrt{2}-\sqrt{12}\right)^2$ 2. _______________

3. $\left(2+\sqrt[3]{5}\right)\left(2-\sqrt[3]{5}\right)$ 3. _______________

Objective 2   Rationalize denominators with one radical term.

| **Review these examples for Objective 2:** | **Now Try:** |

Review these examples for Objective 2:

3. Rationalize the denominator.

$$\frac{4}{\sqrt{3}}$$

$$\frac{4}{\sqrt{3}} = \frac{4 \cdot \sqrt{3}}{\sqrt{3} \cdot \sqrt{3}} = \frac{4\sqrt{3}}{3}$$

4. Simplify the radical.

$$-\sqrt{\frac{27}{98}}$$

$$-\sqrt{\frac{27}{98}} = -\frac{\sqrt{27}}{\sqrt{98}}$$

$$= -\frac{\sqrt{9 \cdot 3}}{\sqrt{49 \cdot 2}}$$

$$= -\frac{3\sqrt{3}}{7\sqrt{2}}$$

$$= -\frac{3\sqrt{3} \cdot \sqrt{2}}{7\sqrt{2} \cdot \sqrt{2}}$$

$$= -\frac{3\sqrt{6}}{7 \cdot 2}$$

$$= -\frac{3\sqrt{6}}{14}$$

5. Simplify.

$$\sqrt[3]{\frac{16}{9}}$$

$$\sqrt[3]{\frac{16}{9}} = \frac{\sqrt[3]{8 \cdot 2}}{\sqrt[3]{9}} = \frac{2\sqrt[3]{2}}{\sqrt[3]{9}}$$

$$= \frac{2\sqrt[3]{2} \cdot \sqrt[3]{3}}{\sqrt[3]{9} \cdot \sqrt[3]{3}}$$

$$= \frac{2\sqrt[3]{6}}{\sqrt[3]{27}}$$

$$= \frac{2\sqrt[3]{6}}{3}$$

Now Try:

3. Rationalize the denominator.

$$\frac{2}{\sqrt{15}}$$

____

4. Simplify the radical.

$$-\sqrt{\frac{45}{32}}$$

____

5. Simplify.

$$\sqrt[3]{\frac{8}{100}}$$

____

Objective 2 Practice Exercises

For extra help, see Examples 3–5 on pages 734–736 of your text.

Simplify. Assume that variables represent positive real numbers.

4. $\sqrt{\dfrac{5a^2b^3}{6}}$ 4. ___________________

5. $\sqrt{\dfrac{7y^2}{12b}}$ 5. ___________________

6. $\sqrt[3]{\dfrac{5}{49x}}$ 6. ___________________

            Copyright © 2025 Pearson Education, Inc.

Objective 3   Rationalize denominators with binomials involving radicals.

Review this example for Objective 3: | **Now Try:**

6.   Rationalize the denominator.

$$\frac{5}{5+\sqrt{2}}$$

$$\frac{5}{5+\sqrt{2}} = \frac{5\left(5-\sqrt{2}\right)}{\left(5+\sqrt{2}\right)\left(5-\sqrt{2}\right)}$$

$$= \frac{5\left(5-\sqrt{2}\right)}{25-2}$$

$$= \frac{5\left(5-\sqrt{2}\right)}{23}$$

6.   Rationalize the denominator.

$$\frac{2}{\sqrt{3}-2}$$

____

Objective 3 Practice Exercises

For extra help, see Example 6 on pages 737–738 of your text.

Rationalize each denominator. Write quotients in lowest terms. Assume that variables represent positive real numbers.

7. $\dfrac{4}{\sqrt{3}+2}$

7. _______________

8. $\dfrac{5}{\sqrt{3}-\sqrt{10}}$

8. _______________

9. $\dfrac{\sqrt{6}+2}{\sqrt{2}-4}$

9. _______________

Objective 4   Write radical quotients in lowest terms.

Review this example for Objective 4:

7.   Write the quotient in lowest terms.

$$\frac{72\sqrt{2}-16\sqrt{7}}{24}$$

$$\frac{72\sqrt{2}-16\sqrt{7}}{24}=\frac{8\left(9\sqrt{2}-2\sqrt{7}\right)}{24}$$

$$=\frac{9\sqrt{2}-2\sqrt{7}}{3}$$

Now Try:

7.   Write the quotient in lowest terms.

$$\frac{9+6\sqrt{15}}{12}$$

____

Objective 4 Practice Exercises

For extra help, see Example 7 on page 738 of your text.

Write each quotient in lowest terms. Assume that variables represent positive real numbers.

10. $\dfrac{7-\sqrt{98}}{14}$

10. ________________

11. $\dfrac{16-12\sqrt{72}}{24}$

11. ________________

12. $\dfrac{2x-\sqrt{8x^2}}{4x}$

12. ________________

Chapter 10 ROOTS, RADICALS, AND ROOT FUNCTIONS

10.6   Solving Equations with Radicals

Learning Objectives
1   Solve radical equations using the power rule.
2   Solve radical equations that require additional steps.
3   Solve radical equations with indexes greater than 2.
4   Use the power rule to solve a formula for a specified variable.

Key Terms

Use the vocabulary terms listed below to complete each statement in exercises 1–2.

radical equation   **extraneous solution**

1.   A(n) _________________________ is a proposed solution to an equation that does not satisfy the equation.

2.   An equation with a variable in the radicand is a(n) _________________________.

Objective 1   Solve radical equations using the power rule.

Review these examples for Objective 1:

1.   Solve $\sqrt{3w+4} = 7$.

$$\left(\sqrt{3w+4}\right)^2 = 7^2$$
$$3w+4 = 49$$
$$3w = 45$$
$$w = 15$$

Check $\sqrt{3w+4} = 7$

$$\sqrt{3(15)+4} \overset{?}{=} 7$$
$$\sqrt{49} \overset{?}{=} 7$$
$$7 = 7 \quad \text{True}$$

Since 15 satisfies the original equation, the solution set is $\{15\}$.

2.   Solve $\sqrt{12p+1} + 7 = 0$.

Step 1 $\sqrt{12p+1} = -7$

Step 2 $\left(\sqrt{12p+1}\right)^2 = (-7)^2$

Now Try:

1.   Solve $\sqrt{7x-6} = 8$.

____

2.   Solve $\sqrt{4x-19} + 5 = 0$.

____

Step 3 $12p+1=49$

$12p=48$

$p=4$

Step 4 Check $\sqrt{12p+1}+7=0$

$\sqrt{12(4)+1}+7 \overset{?}{=} 0$

$\sqrt{49}+7 \overset{?}{=} 0$

$14=0$ False

The false result shows that the proposed solution 4 is not a solution of the original equation. It is extraneous. The solution set is $\varnothing$.

Objective 1 Practice Exercises

For extra help, see Examples 1–2 on pages 743–744 of your text.

Solve each equation.

1. $\sqrt{4x-19}=5$ 1. _______________

2. $\sqrt{12p+1}+7=0$ 2. _______________

3. $\sqrt{4x-3}=7$ 3. _______________

Objective 2   Solve radical equations that require additional steps.

Review these examples for Objective 2:	**Now Try:**
3. Solve $\sqrt{x+3} = x-3$.	**3.** Solve $\sqrt{x+11} = x-1$.

Step 1 The radical is isolated on the left side of the equation.

Step 2 Square each side.
$$\left(\sqrt{x+3}\right)^2 = (x-3)^2$$
$$x+3 = x^2 - 6x + 9$$

Step 3 Write the equation in standard form and solve.
$$0 = x^2 - 7x + 6$$
$$0 = (x-1)(x-6)$$
$$x-1 = 0 \quad \text{or} \quad x-6 = 0$$
$$x = 1 \quad \text{or} \quad x = 6$$

Step 4 Check each proposed solution in the original equation.

$$\sqrt{x+3} = x-3 \qquad\qquad \sqrt{x+3} = x-3$$
$$\sqrt{1+3} \overset{?}{=} 1-3 \qquad\qquad \sqrt{6+3} \overset{?}{=} 6-3$$
$$\sqrt{4} \overset{?}{=} -2 \qquad\qquad \sqrt{9} \overset{?}{=} 3$$
$$2 = -2 \quad \text{False} \qquad\qquad 3 = 3 \quad \text{True}$$

The solution set is $\{6\}$. The other proposed solution, 1, is extraneous.

4. Solve $\sqrt{x^2+7x-14} = x+2$.	**4.** Solve $\sqrt{x^2+8x-17} = x+1$.

$$\left(\sqrt{x^2+7x-14}\right)^2 = (x+2)^2$$
$$x^2 + 7x - 14 = x^2 + 4x + 4$$
$$3x = 18$$
$$x = 6$$

Check $\sqrt{x^2+7x-14} = x+2$
$$\sqrt{6^2 + 7(6) - 14} \overset{?}{=} 6+2$$
$$\sqrt{64} \overset{?}{=} 8$$
$$8 = 8 \quad \text{True}$$

The solution set is $\{6\}$.

5. Solve $\sqrt{3x} - 4 = \sqrt{x-2}$.

$$\left(\sqrt{3x} - 4\right)^2 = \left(\sqrt{x-2}\right)^2$$

$$3x - 8\sqrt{3x} + 16 = x - 2$$

$$-8\sqrt{3x} = -2x - 18$$

$$\left(-8\sqrt{3x}\right)^2 = (-2x - 18)^2$$

$$192x = 4x^2 + 72x + 324$$

$$0 = 4x^2 - 120x + 324$$

$$0 = 4\left(x^2 - 30x + 81\right)$$

$$0 = 4(x-3)(x-27)$$

$$x - 3 = 0 \quad \text{or} \quad x - 27 = 0$$

$$x = 3 \quad \text{or} \qquad x = 27$$

Check

$$\sqrt{3x} - 4 = \sqrt{x-2} \qquad\qquad \sqrt{3x} - 4 = \sqrt{x-2}$$

$$\sqrt{3(3)} - 4 \overset{?}{=} \sqrt{3-2} \qquad \sqrt{3(27)} - 4 \overset{?}{=} \sqrt{27-2}$$

$$3 - 4 \overset{?}{=} \sqrt{1} \qquad\qquad 9 - 4 \overset{?}{=} \sqrt{25}$$

$$-1 = 1 \quad \text{False} \qquad\qquad 5 = 5 \quad \text{True}$$

The proposed solution, 27, is valid, but 3 is extraneous and must be rejected. The solution set is {27}.

5. Solve $\sqrt{3x+4} = \sqrt{9x} - 2$.

____

Objective 2 Practice Exercises

For extra help, see Examples 3–5 on pages 745–746 of your text.

Solve each equation.

4. $\sqrt{k+10} + \sqrt{2k+19} = 2$

4. __________

5. $\sqrt{x-5} = x - 5$

5. __________

6. $\sqrt{x^2 - 3x - 15} = x - 3$

6. ___________________

Objective 3   Solve radical equations with indexes greater than 2.

Review this example for Objective 3:

6. Solve $\sqrt[3]{5r - 6} = \sqrt[3]{3r + 4}$.

$$\left(\sqrt[3]{5r - 6}\right)^3 = \left(\sqrt[3]{3r + 4}\right)^3$$
$$5r - 6 = 3r + 4$$
$$2r = 10$$
$$r = 5$$

Check $\sqrt[3]{5r - 6} = \sqrt[3]{3r + 4}$

$$\sqrt[3]{5(5) - 6} \overset{?}{=} \sqrt[3]{3(5) + 4}$$
$$\sqrt[3]{19} = \sqrt[3]{19} \quad \text{True}$$

The solution set is $\{5\}$.

Now Try:

6. Solve $\sqrt[4]{8x + 5} = \sqrt[4]{7x + 7}$.

____

Objective 3 Practice Exercises

For extra help, see Example 6 on page 747 of your text.

Solve each equation.

7. $\sqrt[3]{2a - 63} + 5 = 0$

7. ___________________

8. $\sqrt[5]{5a + 1} - \sqrt[5]{2a - 11} = 0$

8. ___________________

9. $\sqrt[4]{8x + 5} = \sqrt[4]{7x + 7}$

9. ___________________

Objective 4   Use the power rule to solve a formula for a specified variable.

Review this example for Objective 4:

7.   Solve the formula $d = \sqrt{\dfrac{H}{1.6n}}$ for n.

$$d^2 = \left(\sqrt{\dfrac{H}{1.6n}} \right)^2$$

$$d^2 = \dfrac{H}{1.6n}$$

$$1.6d^2 n = H$$

$$n = \dfrac{H}{1.6d^2}$$

Now Try:

7.   Solve the formula $r = \sqrt{\dfrac{3v}{\pi h}}$

for h.

____

Objective 4 Practice Exercises

For extra help, see Example 7 on page 747 of your text.

Solve each equation for the indicated variable.

10. $Z = \sqrt{\dfrac{L}{C}}$, for L

10. __________

11. $f = \dfrac{1}{2\pi\sqrt{LC}}$, for C

11. __________

12. $N = \dfrac{1}{2\pi}\sqrt{\dfrac{a}{r}}$, for r

12. __________

Chapter 10 ROOTS, RADICALS, AND ROOT FUNCTIONS

10.7   Complex Numbers

Learning Objectives
1   Simplify numbers of the form $\sqrt{-b}$, where $b > 0$.
2   Identify subsets of the complex numbers.
3   Add and subtract complex numbers.
4   Multiply complex numbers.
5   Divide complex numbers.
6   Simplify powers of i.

Key Terms

Use the vocabulary terms listed below to complete each statement in exercises 1−7.

complex number   **real part**   **imaginary part**

pure imaginary number   **standard form (of a complex number)**

nonreal complex number   **complex conjugate**

1.   A _________________________________ is a number that can be written in the form $a + bi$, where a and b are real numbers.

2.   The _________________________________ of $a + bi$ is $a - bi$.

3.   The _________________________________ of $a + bi$ is bi.

4.   The _________________________________ of $a + bi$ is a.

5.   A complex number is in _________________________________ if it is written in the form $a + bi$.

6.   A complex number $a + bi$ with $a = 0$ and $b \neq 0$ is called a

     _________________________________.

7.   A complex number $a + bi$ with $b \neq 0$ is called a _________________________________.

Objective 1   Simplify numbers of the form $\sqrt{-b}$, where $b > 0$.

Review these examples for Objective 1:

1. Write each number as a product of a real number and i.

   a. $\sqrt{-36}$

$$\sqrt{-36} = \sqrt{-1 \cdot 36} = \sqrt{-1} \cdot \sqrt{36} = i\sqrt{36} = 6i$$

Now Try:

1. Write each number as a product of a real number and i.

   a. $\sqrt{-16}$

     ____

b. $-\sqrt{-121}$

$-\sqrt{-121} = -i\sqrt{121} = -11i$

c. $\sqrt{-3}$

$\sqrt{-3} = i\sqrt{3}$

d. $\sqrt{-75}$

$\sqrt{-75} = i\sqrt{25 \cdot 3} = 5i\sqrt{3}$

2. Multiply.

$\sqrt{-6} \cdot \sqrt{-7}$

$$\sqrt{-6} \cdot \sqrt{-7} = i\sqrt{6} \cdot i\sqrt{7}$$
$$= i^2\sqrt{6 \cdot 7}$$
$$= (-1)\sqrt{42}$$
$$= -\sqrt{42}$$

3. Divide.

a. $\dfrac{\sqrt{-125}}{\sqrt{-5}}$

$$\frac{\sqrt{-125}}{\sqrt{-5}} = \frac{i\sqrt{125}}{i\sqrt{5}}$$
$$= \sqrt{\frac{125}{5}}$$
$$= \sqrt{25}$$
$$= 5$$

b. $\dfrac{\sqrt{-28}}{\sqrt{7}}$

$$\frac{\sqrt{-28}}{\sqrt{7}} = \frac{i\sqrt{28}}{\sqrt{7}}$$
$$= i\sqrt{\frac{28}{7}}$$
$$= i\sqrt{4}$$
$$= 2i$$

b. $-\sqrt{-144}$

____

c. $\sqrt{-11}$

____

d. $\sqrt{-128}$

____

2. Multiply.

$\sqrt{-5} \cdot \sqrt{-6}$

____

3. Divide.

a. $\dfrac{\sqrt{-200}}{\sqrt{-8}}$

____

b. $\dfrac{\sqrt{-80}}{\sqrt{5}}$

____

Objective 1 Practice Exercises

For extra help, see Examples 1–3 on pages 750–752 of your text.

Write the number as a product of a real number and i. Simplify all radical expressions.

1. $-\sqrt{-162}$ 1. ________________

Multiply or divide as indicated

2. $\sqrt{-5}\cdot\sqrt{-3}\cdot\sqrt{-7}$ 2. ________________

3. $\dfrac{\sqrt{-42}\cdot\sqrt{-6}}{\sqrt{-7}}$ 3. ________________

Objective 2   Identify subsets of the complex numbers.

For extra help, see pages 752–753 of your text.

Classify each of the following complex numbers as real *or* imaginary.

4. $\sqrt{5}$ 4. ________________

5. $\sqrt{3}-i\sqrt{5}$ 5. ________________

6. $i\sqrt{7}$ 6. ________________

Objective 3   Add and subtract complex numbers.

Review these examples for Objective 3:	**Now Try:**
4. Add.	**4.** Add.
$(2+9i)+(10-3i)$	$(4-7i)+(6-2i)$
$(2+9i)+(10-3i)=(2+10)+(9-3)i$	
$=12+6i$	___________

5. Subtract.

$(7-9i)-(-5-6i)$

$$(7-9i)-(-5-6i)=\left[7-(-5)\right]+\left[-9-(-6)\right]i$$
$$=(7+5)+(-9+6)i$$
$$=12-3i$$

5. Subtract.

$(12+2i)-(-12-2i)$

____

Objective 3 Practice Exercises

For extra help, see Examples 4–5 on pages 753–754 of your text.

Add or subtract as indicated. Write answers in standard form.

7. $(-7-2i)-(-3-3i)$

7. ____________

8. $4i-(9+5i)+(2+3i)$

8. ____________

9. $(7-9i)-(5-6i)$

9. ____________

Objective 4   Multiply complex numbers.

Review these examples for Objective 4:

6. Multiply.

  a. $6i(2-7i)$

$$6i(2-7i)=6i(2)+6i(-7i)$$
$$=12i-42i^2$$
$$=12i-42(-1)$$
$$=42+12i$$

  b. $(3+2i)(5-i)$

$$(3+2i)(5-i)=3(5)+3(-i)+2i(5)+2i(-i)$$
$$=15-3i+10i-2i^2$$
$$=15+7i-2(-1)$$
$$=15+7i+2$$
$$=17+7i$$

  c. $(1+6i)(2+5i)$

$$(1+6i)(2+5i)=2(1)+1(5i)+6i(2)+(6i)(5i)$$
$$=2+5i+12i+30i^2$$
$$=2+17i+30(-1)$$
$$=-28+17i$$

Now Try:

6. Multiply.

  a. $2i(4+7i)$

____

  b. $(12+5i)(1-i)$

____

  c. $(1+3i)(2-5i)$

____

Objective 4 Practice Exercises

For extra help, see Example 6 on page 754 of your text.

Multiply.

10. $(2-5i)(2+5i)$ **10.** ______________

11. $(1+3i)^2$ **11.** ______________

12. $(12+2i)(-1+i)$ **12.** ______________

Objective 5   Divide complex numbers.

Review this example for Objective 5:

7.   Find the quotient.

$$\frac{6-i}{2-3i}$$

Multiply the numerator and denominator by $2 + 3i$, the conjugate of the denominator.

$$\frac{6-i}{2-3i} = \frac{(6-i)(2+3i)}{(2-3i)(2+3i)}$$

$$= \frac{12+18i-2i-3i^2}{2^2+3^2}$$

$$= \frac{12+16i-3(-1)}{4+9}$$

$$= \frac{15+16i}{13}, \text{ or } \frac{15}{13}+\frac{16}{13}i$$

Now Try:

7.   Find the quotient.

$$\frac{4+i}{5-2i}$$

____

Objective 5 Practice Exercises

For extra help, see Example 7 on page 755 of your text.

Write each quotient in the form a + bi.

13. $\dfrac{3-2i}{2+i}$ **13.** ______________

14. $\dfrac{5+2i}{9-4i}$ 14. _______________

15. $\dfrac{6-i}{2-3i}$ 15. _______________

Objective 6   Simplify powers of *i*.

Review these examples for Objective 6:

8. Find each power of *i*.

   a. i^{100}

$$i^{100} = (i^4)^{25} = 1^{25} = 1$$

   b. i^{27}

$$i^{27} = i^{24} \cdot i^3 = (i^4)^6 \cdot (-i) = 1^6 \cdot (-i) = -i$$

   c. i^{-3}

$$i^{-3} = \frac{1}{i^3} = \frac{1}{i^3} \cdot \frac{i}{i} = \frac{i}{i^4} = \frac{i}{1} = i$$

   d. i^{49}

$$i^{49} = i^{48} \cdot i = (i^4)^{12} \cdot i = 1^{12} \cdot i = i$$

Now Try:

8. Find each power of *i*.

   a. i^{48}

    ____

   b. i^{77}

    ____

   c. i^{-5}

    ____

   d. i^{55}

    ____

Objective 6 Practice Exercises

For extra help, see Example 8 on page 756 of your text.

Find each power of i.

16. i^{14} 16. _______________

17. i^{113} 17. _______________

18. i^{-21} 18. _______________

Chapter 11 QUADRATIC EQUATIONS, INEQUALITIES, AND FUNCTIONS

11.1 Solving Quadratic Equations by the Square Root Property

Learning Objectives	
1	Solve quadratic equations using the zero-factor property. (Review)
2	Solve equations of the form $x^2 = k$, where $k > 0$.
3	Solve equations of the form $(ax + b)^2 = k$, where $k > 0$.
4	Solve quadratic equations with solutions that are not real numbers.

Key Terms

Use the vocabulary terms listed below to complete each statement in exercises 1−2.

quadratic equation   **zero-factor property**

1. An equation that can be written in the form $ax^2 + bx + c = 0$ is a

    ___________________________.

2. The ________________________________ states that if a product equals 0, then at least one of the factors of the product also equals zero.

Objective 1   Solve quadratic equations using the zero-factor property. (Review)

Review these examples for Objective 1:

1. Solve each equation by the zero-factor property.

a. $x^2 + 5x + 4 = 0$

$x^2 + 5x + 4 = 0$

$(x + 4)(x + 1) = 0$

$x + 4 = 0$ or $x + 1 = 0$

$x = -4$ or $x = -1$

The solution set is $\{-4, -1\}$.

b. $x^2 = 64$

$x^2 = 64$

$x^2 - 64 = 0$

$(x + 8)(x - 8) = 0$

$x + 8 = 0$ or $x - 8 = 0$

$x = -8$ or $x = 8$

The solution set is $\{-8, 8\}$.

Now Try:

1. Solve each equation by the zero-factor property.

a. $x^2 + 8x + 7 = 0$

    ____

b. $x^2 = 100$

    ____

Objective 1 Practice Exercises

For extra help, see Example 1 on page 772 of your text.

Solve each equation by using the zero-factor property.

1. $x^2 + 6x + 8 = 0$

1. ________________

2. $x^2 = 121$

2. ________________

3. $x^2 + 2x - 35 = 0$

3. ________________

Objective 2   Solve equations of the form $x^2 = k$, where $k > 0$.

Review these examples for Objective 2:

2. Solve the equation.

a. $x^2 = 121$

$$x^2 = 121$$
$$x = \sqrt{121} \quad \text{or} \quad x = -\sqrt{121}$$
$$x = 11 \quad\quad \text{or} \quad x = -11$$

The solution set is $\{11, -11\}$, or $\{\pm 11\}$.

b. $x^2 = 7$

$$x^2 = 7$$
$$x = \sqrt{7} \quad \text{or} \quad x = -\sqrt{7}$$

The solution set is $\{\sqrt{7}, -\sqrt{7}\}$, or $\{\pm\sqrt{7}\}$.

c. $5p^2 - 100 = 0$

$$5p^2 - 100 = 0$$
$$5p^2 = 100$$
$$p^2 = 20$$
$$p = \sqrt{20} \quad \text{or} \quad p = -\sqrt{20}$$
$$p = 2\sqrt{5} \quad \text{or} \quad p = -2\sqrt{5}$$

The solution set is $\{2\sqrt{5}, -2\sqrt{5}\}$, or $\{\pm 2\sqrt{5}\}$.

Now Try:

2. Solve the equation.

a. $r^2 = 169$

____

b. $r^2 = 13$

____

c. $3x^2 - 54 = 0$

____

3. Use Galileo's formula to determine how long it will take a penny dropped from the 86th floor Observatory deck of the Empire State Building to reach the ground. The deck is 1050 feet above the ground. Round your answer to the nearest tenth.

Galileo's formula is $d = 16t^2$, where d is the distance in feet that an object falls, and t is the time in seconds.

$$d = 16t^2$$

$$1050 = 16t^2$$

$$65.625 = t^2$$

$$t = \sqrt{65.625} \quad \text{or} \quad t = -\sqrt{65.625}$$

Time cannot be negative, so we discard $t = -\sqrt{65.625}$. Using a calculator, $\sqrt{65.625} \approx 8.1$, so $t \approx 8.1$. The penny would fall to the ground in about 8.1 seconds.

3. A child dropped a ball from a hotel balcony that is 113 ft above the ground. Use Galileo's formula to determine how long it takes for the ball to reach the ground. Round your answer to the nearest tenth.

____

Objective 2 Practice Exercises

For extra help, see Examples 2–3 on pages 773–774 of your text.

Solve each equation by using the square root property. Express all radicals in simplest form.

4. $r^2 = 30$

4. _____________

5. $x^2 - 98 = 0$

5. _____________

6. $3d^2 - 750 = 0$

6. _____________

Objective 3   Solve equations of the form $(ax+b)^2 = k$, where $k > 0$.

Review this example for Objective 3: | **Now Try:**

5. Solve $(5r-3)^2 = 12$.

$$5r-3 = \sqrt{12} \quad \text{or} \quad 5r-3 = -\sqrt{12}$$
$$5r-3 = 2\sqrt{3} \quad \text{or} \quad 5r-3 = -2\sqrt{3}$$
$$5r = 3+2\sqrt{3} \quad \text{or} \quad 5r = 3-2\sqrt{3}$$
$$r = \frac{3+2\sqrt{3}}{5} \quad \text{or} \quad r = \frac{3-2\sqrt{3}}{5}$$

Check
$$(5r-3)^2 = 12$$
$$\left[5 \cdot \frac{3+2\sqrt{3}}{5} - 3\right]^2 \overset{?}{=} 12$$
$$\left(3+2\sqrt{3}-3\right)^2 \overset{?}{=} 12$$
$$\left(2\sqrt{3}\right)^2 \overset{?}{=} 12$$
$$12 = 12 \quad \text{True}$$

The check of the other solution is similar. The solution set is

The solution set is $\left\{\dfrac{3+2\sqrt{3}}{5},\ \dfrac{3-2\sqrt{3}}{5}\right\}$.

Now Try:

5. Solve $(7r-3)^2 = 32$.

____

Objective 3 Practice Exercises

For extra help, see Examples 4–5 on pages 774–775 of your text.

Solve each equation by using the square root property. Express all radicals in simplest form.

7. $(y+2)^2 = 16$ **7.** ________________

8. $(7p-4)^2 = 289$ **8.** ________________

9. $(10m-5)^2 - 9 = 0$ **9.** ________________

Objective 4   Solve quadratic equations with solutions that are not real numbers.

Review these examples for Objective 4:

6.   Solve each equation.

a. $x^2 = -48$

$$x^2 = -48$$
$$x = \sqrt{-48} \quad \text{or} \quad x = -\sqrt{-48}$$
$$x = 4i\sqrt{3} \quad \text{or} \quad x = -4i\sqrt{3}$$

The solution set is $\left\{ -4i\sqrt{3},\ 4i\sqrt{3} \right\}$.

b. $(x-3)^2 = -25$

$$(x-3)^2 = -25$$
$$x - 3 = \sqrt{-25} \quad \text{or} \quad x - 3 = -\sqrt{-25}$$
$$x - 3 = 5i \quad \text{or} \quad x - 3 = -5i$$
$$x = 3 + 5i \quad \text{or} \quad x = 3 - 5i$$

The solution set is $\{3 - 5i,\ 3 + 5i\}$.

Now Try:

6.   Solve each equation.

a. $y^2 = -32$

____

b. $(x+2)^2 = -49$

____

Objective 4 Practice Exercises

For extra help, see Example 6 on page 776 of your text.

Find the complex solutions of each equation.

10. $(10m - 5)^2 + 9 = 0$

10. ________________

11. $(m + 1)^2 = -36$

11. ________________

12. $(x - 1)^2 + 2 = 0$

12. ________________

Chapter 11 QUADRATIC EQUATIONS, INEQUALITIES, AND FUNCTIONS

11.2   Solving Quadratic Equations by Completing the Square

Learning Objectives
1   Solve quadratic equations by completing the square when the coefficient of the second-degree term is 1.
2   Solve quadratic equations by completing the square when the coefficient of the second-degree term is not 1.
3   Simplify the terms of an equation before solving.

Key Terms

Use the vocabulary terms listed below to complete each statement in exercises 1−3.

completing the square   **perfect square trinomial**   **square root property**

1.   A ___________________________ can be written in the form $x^2 + 2kx + k^2$ or $x^2 - 2kx + k^2$

2.   The ___________________________ says that, if k is positive and $a^2 = k$, then $a = \pm\sqrt{k}$.

3.   Use the process called ___________________________ in order to rewrite an equation so it can be solved using the square root property.

Objective 1   **Solve quadratic equations by completing the square when the coefficient of the second-degree term is 1.**

Review these examples for Objective 1:	Now Try:
1. Solve $x^2 - 12x + 24 = 0$.	1. Solve $x^2 - 6x + 1 = 0$.

Review these examples for Objective 1:

1. Solve $x^2 - 12x + 24 = 0$.

$$x^2 - 12x + 24 = 0$$
$$x^2 - 12x = -24$$

Take half of the coefficient of the first-degree term, $-12x$, and square the result.

$$\left[\frac{1}{2}(-12)\right]^2 = (-6)^2 = 36$$

Add 36 to each side.

$$x^2 - 12x + 36 = -24 + 36$$
$$(x - 6)^2 = 12$$

Now Try:

1. Solve $x^2 - 6x + 1 = 0$.

____

$$x - 6 = \sqrt{12} \quad \text{or} \quad x - 6 = -\sqrt{12}$$
$$x = 6 + 2\sqrt{3} \quad \text{or} \quad x = 6 - 2\sqrt{3}$$

A check indicates the solution set is

$$\left\{ 6 - 2\sqrt{3},\ 6 + 2\sqrt{3} \right\}.$$

3. Solve $x^2 + 5x + 2 = 0$.

Since the coefficient of the second-degree term is 1, begin with Step 2.

Step 2 $x^2 + 5x = -2$

Step 3   Take half the coefficient of the first-degree term and square the result.

$$\left[\frac{1}{2}(5) \right]^2 = \left(\frac{5}{2} \right)^2 = \frac{25}{4}$$

$$x^2 + 5x + \frac{25}{4} = -2 + \frac{25}{4}$$

$$\left(x + \frac{5}{2} \right)^2 = \frac{17}{4}$$

Step 4

$$x + \frac{5}{2} = \sqrt{\frac{17}{4}} \quad \text{or} \quad x + \frac{5}{2} = -\sqrt{\frac{17}{4}}$$

$$x + \frac{5}{2} = \frac{\sqrt{17}}{2} \quad \text{or} \quad x + \frac{5}{2} = -\frac{\sqrt{17}}{2}$$

$$x = -\frac{5}{2} + \frac{\sqrt{17}}{2} \quad \text{or} \quad x = -\frac{5}{2} - \frac{\sqrt{17}}{2}$$

A check shows the solution set is

$$\left\{ -\frac{5}{2} - \frac{\sqrt{17}}{2},\ -\frac{5}{2} + \frac{\sqrt{17}}{2} \right\}.$$

3. Solve $x^2 - 11x + 8 = 0$.

____

Objective 1 Practice Exercises

For extra help, see Examples 1–3 on pages 780–781 of your text.

Solve each equation by completing the square.

1. $r^2 + 8r = -4$

1. ___________________

2. $x^2 - 4x = 2$

2. ___________________

      429

3. $x^2 + 2x = 63$ **3.** _______________

Objective 2   **Solve quadratic equations by completing the square when the coefficient of the second-degree term is not 1.**

Review these examples for Objective 2:

4. Solve $9x^2 + 18x = 7$.

Step 1 Before completing the square, the coefficient of x^2 must be 1, not 9. Divide each side by 9.

$$x^2 + 2x = \frac{7}{9}$$

Step 2 The equation is already in the correct form, with the variable terms on one side and the constant on the other.

Step 3 Complete the square. Take half the coefficient of x, and square it.

$$\frac{1}{2}(2) = 1 \quad \text{and} \quad 1^2 = 1$$

$$x^2 + 2x + 1 = \frac{7}{9} + 1$$

$$(x+1)^2 = \frac{16}{9}$$

Step 4 Solve the equation by using the square root property.

$$x + 1 = \sqrt{\frac{16}{9}} \quad \text{or} \quad x + 1 = -\sqrt{\frac{16}{9}}$$

$$x + 1 = \frac{4}{3} \qquad x + 1 = -\frac{4}{3}$$

$$x = \frac{1}{3} \quad \text{or} \quad x = -\frac{7}{3}$$

The two solutions $-\frac{7}{3}$ and $\frac{1}{3}$ check, so the solution set is $\left\{-\frac{7}{3},\ \frac{1}{3}\right\}$.

5. Solve $3x^2 - 6x - 2 = 0$.

$$x^2 - 2x - \frac{2}{3} = 0 \qquad \text{Step 1}$$

$$x^2 - 2x = \frac{2}{3} \qquad \text{Step 2}$$

$$\left[\frac{1}{2}(-2)\right]^2 = (-1)^2 = 1 \qquad \text{Step 3}$$

Now Try:

4. Solve $16x^2 - 64x = 55$.

____

5. Solve $2p^2 + 6p - 1 = 0$.

____

$$x^2 - 2x + 1 = \frac{2}{3} + 1$$

$$(x-1)^2 = \frac{5}{3}$$

$$x - 1 = \sqrt{\frac{5}{3}} \quad \text{or} \quad x - 1 = -\sqrt{\frac{5}{3}} \qquad \text{Step 4}$$

$$x = 1 + \sqrt{\frac{5}{3}} \quad \text{or} \quad x = 1 - \sqrt{\frac{5}{3}}$$

$$x = 1 + \frac{\sqrt{15}}{3} \quad \text{or} \quad x = 1 - \frac{\sqrt{15}}{3}$$

$$x = \frac{3 + \sqrt{15}}{3} \quad \text{or} \quad x = \frac{3 - \sqrt{15}}{3}$$

The solution set is $\left\{ \dfrac{3 - \sqrt{15}}{3}, \ \dfrac{3 + \sqrt{15}}{3} \right\}$.

Objective 2 Practice Exercises

For extra help, see Examples 4–6 on pages 782–784 of your text.

Solve each equation by completing the square.

4. $6x^2 - x = 15$

4. _________________

5. $6q^2 + 4q = 1$

5. _________________

6. $3t^2 + t - 2 = 0$

6. _________________

Objective 3   Simplify the terms of an equation before solving.

Review this example for Objective 3:

8. Solve $x(x+5)=7$.

$$x(x+5)=7$$

$$x^2+5x=7$$

$$x^2+5x+\frac{25}{4}=7+\frac{25}{4}$$

$$\left(x+\frac{5}{2}\right)^2=\frac{53}{4}$$

$$x+\frac{5}{2}=\sqrt{\frac{53}{4}} \quad \text{or} \quad x+\frac{5}{2}=-\sqrt{\frac{53}{4}}$$

$$x=-\frac{5}{2}+\frac{\sqrt{53}}{2} \quad \text{or} \quad x=-\frac{5}{2}-\frac{\sqrt{53}}{2}$$

$$x=\frac{-5+\sqrt{53}}{2} \quad \text{or} \quad x=\frac{-5-\sqrt{53}}{2}$$

A check confirms the solution set is

$$\left\{\frac{-5\pm\sqrt{53}}{2}\right\}.$$

Now Try:

8. Complete the square to solve $x(x+3)=6$.

____

Objective 3 Practice Exercises

For extra help, see Examples 7–8 on page 784 of your text.

Simplify each of the following equations and then solve by completing the square.

7. $6y^2+3y=4y^2+y+6$

7. _________________

8. $(b-1)(b+7)=9$

8. _________________

9. $(s+3)(s+1)=1$

9. _________________

Chapter 11 QUADRATIC EQUATIONS, INEQUALITIES, AND FUNCTIONS

11.3 Solving Quadratic Equations by the Quadratic Formula

Learning Objectives
1   Derive the quadratic formula.
2   Solve quadratic equations using the quadratic formula.
3   Use the discriminant to determine the number and type of solutions.

Key Terms

Use the vocabulary terms listed below to complete each statement in exercises 1–2.

quadratic formula   **discriminant**

1.   The expression under the radical in the quadratic formula is called the

_______________________________________.

2.   The formula $x = \dfrac{-b \pm \sqrt{b^2 - 4ac}}{2a}$ is called the _______________________________.

Objective 1   Derive the quadratic formula.

For extra help, see pages 787–788 of your text.

Objective 2   Solve quadratic equations using the quadratic formula.

Review these examples for Objective 2:

1.   Solve $5x^2 - 13x - 6 = 0$.

Use the quadratic formula with $a = 5$, $b = -13$, and $c = -6$.

$$x = \frac{-b \pm \sqrt{b^2 - 4ac}}{2a}$$

$$x = \frac{-(-13) \pm \sqrt{(-13)^2 - 4(5)(-6)}}{2(5)}$$

$$x = \frac{13 \pm \sqrt{169 + 120}}{10}$$

$$x = \frac{13 \pm \sqrt{289}}{10}$$

$$x = \frac{13 \pm 17}{10}$$

There are two solutions.

$$x = \frac{13 + 17}{10} = 3 \quad \text{or} \quad x = \frac{13 - 17}{10} = \frac{-4}{10} = -\frac{2}{5}$$

The solution set is $\left\{ -\dfrac{2}{5},\ 3 \right\}$.

Now Try:

1.   Solve $6x^2 - 17x + 12 = 0$.

____

2. Solve $16x^2 + 8x + 1 = 0$.

Use the quadratic formula with
$a = 16$, $b = 8$, $c = 1$

$$x = \frac{-b \pm \sqrt{b^2 - 4ac}}{2a}$$

$$x = \frac{-8 \pm \sqrt{(8)^2 - 4(16)(1)}}{2(16)}$$

$$x = \frac{-8 \pm \sqrt{64 - 64}}{32} = \frac{-8 \pm 0}{32}$$

$$x = -\frac{1}{4}$$

There is one *distinct* solution. The solution set is
$\left\{ -\frac{1}{4} \right\}$.

3. Solve $4x^2 = -4x + 1$.

First write the equation in standard
form as $4x^2 + 4x - 1 = 0$.
$a = 4$, $b = 4$, $c = -1$

$$x = \frac{-b \pm \sqrt{b^2 - 4ac}}{2a}$$

$$x = \frac{-4 \pm \sqrt{(4)^2 - 4(4)(-1)}}{2(4)}$$

$$x = \frac{-4 \pm \sqrt{16 + 16}}{8} = \frac{-4 \pm \sqrt{32}}{8}$$

$$x = \frac{-4 \pm 4\sqrt{2}}{8}$$

$$x = \frac{4(-1 \pm \sqrt{2})}{4(2)}$$

$$x = \frac{-1 \pm \sqrt{2}}{2}$$

The solution set is $\left\{ \frac{-1 - \sqrt{2}}{2}, \ \frac{-1 + \sqrt{2}}{2} \right\}$.

4. Solve $(5x - 2)(x + 2) = -9$.

$$(5x - 2)(x + 2) = -9$$
$$5x^2 + 8x - 4 = -9$$
$$5x^2 + 8x + 5 = 0$$

From the standard form, we identify
$a = 5$, $b = 8$, and $c = 5$.

2. Solve $25x^2 + 30x + 9 = 0$.

____

3. Solve $2x^2 = 2x + 3$.

____

4. Solve $(2x - 6)(x + 1) = -16$.

____

$$x = \frac{-b \pm \sqrt{b^2 - 4ac}}{2a}$$

$$x = \frac{-8 \pm \sqrt{(8)^2 - 4(5)(5)}}{2(5)}$$

$$x = \frac{-8 \pm \sqrt{-36}}{10} = \frac{-8 \pm 6i}{10}$$

$$x = \frac{2(-4 \pm 3i)}{2(5)}$$

$$x = \frac{-4 \pm 3i}{5} = -\frac{4}{5} \pm \frac{3}{5}i$$

The solution set is $\left\{ -\frac{4}{5} - \frac{3}{5}i, -\frac{4}{5} + \frac{3}{5}i \right\}$, or

$\left\{ -\frac{4}{5} \pm \frac{3}{5}i \right\}$.

Objective 2 Practice Exercises

For extra help, see Examples 1–4 on pages 788–790 of your text.

Use the quadratic formula to solve each equation. (All solutions for these equations are real numbers.)

1. $(z+2)^2 = 2(5z - 2)$

1. ___________________

2. $5k^2 + 4k - 2 = 0$

2. ___________________

3. $34 - 10x = -x^2$

3. ___________________

Objective 3   Use the discriminant to determine the number and type of solutions.

Review these examples for Objective 3:

5. Find the discriminant. Use it to predict the number and type of solutions for each equation. Then tell whether the equation can be solved by factoring or whether the quadratic formula should be used.

a. $3x^2 + x - 2 = 0$

First identify the values of a, b, and c.
   $a = 3$, $b = 1$, and $c = -2$.
Then find the discriminant.
$$b^2 - 4ac = 1^2 - 4(3)(-2)$$
$$= 1 + 24$$
$$= 25, \text{ or } 5^2$$

Since a, b, and c are integers and the discriminant 25 is a perfect square, there will be two rational solutions. The equation can be solved by factoring.

b. $6x^2 - 5x = 9$

Write in standard form: $6x^2 - 5x - 9 = 0$.
$a = 6$, $b = -5$, and $c = -9$
$$b^2 - 4ac = (-5)^2 - 4(6)(-9)$$
$$= 25 + 216$$
$$= 241$$
Because the discriminant, 241, is positive but *not* the square of an integer, this quadratic equation will have two irrational solutions. The equation can be solved using the quadratic formula.

c. $5y^2 - 5y + 2 = 0$

$a = 5$, $b = -5$, and $c = 2$
$$b^2 - 4ac = (-5)^2 - 4(5)(2)$$
$$= 25 - 40$$
$$= -15$$
Because the discriminant is negative and a, b, and c are integers, this quadratic equation will have two nonreal complex solutions. The quadratic equation should be used to solve it.

Now Try:

5. Find the discriminant. Use it to predict the number and type of solutions for each equation. Then tell whether the equation can be solved by factoring or whether the quadratic formula should be used.

a. $10x^2 + 21x + 9 = 0$

____

b. $5x^2 - 6x = 3$

____

c. $2y^2 + 4y + 8 = 0$

____

d. $16x^2 + 25 = 40x$ | **d.** $25x^2 + 9 = 30x$

Write in standard form: $16x^2 - 40x + 25 = 0$.
$a = 16$, $b = -40$, and $c = 25$

$$b^2 - 4ac = (-40)^2 - 4(16)(25)$$
$$= 1600 - 1600$$
$$= 0$$

Because the discriminant is 0, this quadratic equation will have one distinct rational solution. The equation can be solved by factoring.

____

Objective 3 Practice Exercises

For extra help, see Example 5 on pages 791–792 of your text.

Use the discriminant to determine whether the solutions for each equation are

    A. two rational numbers   *B. one rational number,*
    C. two irrational numbers   *D. two imaginary numbers.*

Do not actually solve.

4. $m^2 - 4m + 4 = 0$ **4.** _______________

5. $z^2 + 6z + 3 = 0$ **5.** _______________

6. $16x^2 - 12x + 9 = 0$ **6.** _______________

Chapter 11 QUADRATIC EQUATIONS, INEQUALITIES, AND FUNCTIONS

11.4   Equations That Lead to Quadratic Methods

Learning Objectives
1   Solve rational equations that lead to quadratic equations.
2   Solve applied problems involving quadratic equations.
3   Solve radical equations that lead to quadratic equations.
4   Solve equations that are quadratic in form.

Key Terms

Use the vocabulary terms listed below to complete each statement in exercises 1−2.

quadratic in form   **standard form**

1.   A quadratic equation written in the form $ax^2 + bx + c = 0$, $a \neq 0$ is written in

    ____.

2.   A nonquadratic equation that can be written as a quadratic equation is called

    ____.

Objective 1   Solve rational equations that lead to quadratic equations.

Review this example for Objective 1:

1.   Solve $5 + \dfrac{6}{m+1} = \dfrac{14}{m}$.

Multiply each side by the least common denominator, $m(m + 1)$. The domain must be restricted to $m \neq 0$, $m \neq -1$.

$$5 + \frac{6}{m+1} = \frac{14}{m}$$

$$m(m+1)\left(5 + \frac{6}{m+1}\right) = m(m+1)\frac{14}{m}$$

$$m(m+1)(5) + m(m+1)\frac{6}{m+1} = m(m+1)\frac{14}{m}$$

$$5m^2 + 5m + 6m = 14m + 14$$

$$5m^2 + 11m = 14m + 14$$

$$5m^2 - 3m - 14 = 0$$

$$(5m+7)(m-2) = 0$$

$$5m + 7 = 0 \quad \text{or} \quad m - 2 = 0$$

$$m = -\frac{7}{5} \quad \text{or} \quad m = 2$$

The solution set is $\left\{-\dfrac{7}{5}, 2\right\}$.

Now Try:

1.   Solve $4 - \dfrac{8}{x-1} = -\dfrac{35}{x}$.

    ____

Objective 1 Practice Exercises

For extra help, see Example 1 on page 794 of your text.

Solve each equation. Check your solutions.

1. $\dfrac{5}{x}+\dfrac{1}{2x+7}=-\dfrac{2}{3}$

1. ______________________

2. $\dfrac{2m}{m-5}+\dfrac{7}{m+1}=0$

2. ______________________

3. $1+\dfrac{49}{2x}=\dfrac{15}{x+1}$

3. ______________________

Objective 2   Solve applied problems involving quadratic equations.

Review these examples for Objective 2:

2. Amy rows her boat 6 miles upstream and then returns in $2\dfrac{6}{7}$ hours. The speed of the current is 2 miles per hour. How fast can she row?

Step 1 Read the problem carefully.

Step 2 Assign a variable. Let x = the rate that Amy rows in still water. The current slows Amy when she is going upstream, so Amy's rate going upstream is her rate in still water less the rate of the current, or $x-2$. Similarly, the current makes Amy row faster when she is going downstream,

Now Try:

2. Mike can row 3 miles per hour in still water. It takes him 3 hours and 36 minutes to row 3 miles upstream and return. Find the speed of the current.

____

so her downstream rate is $x + 2$.

Complete a table. Recall that $d = rt$.

	d	r	t
Upstream	6	$x - 2$	$\dfrac{6}{x-2}$
Downstream	6	$x + 2$	$\dfrac{6}{x+2}$

Step 3 Write an equation. The time upstream plus the time downstream equals the total time, $2\dfrac{6}{7}$ hours, or $\dfrac{20}{7}$ hours.

$$\frac{6}{x-2} + \frac{6}{x+2} = \frac{20}{7}$$

Step 4 Solve the equation. The LCD is $7(x-2)(x+2)$.

$$7(x-2)(x+2)\left(\frac{6}{x-2} + \frac{6}{x+2}\right)$$
$$= 7(x-2)(x+2)\left(\frac{20}{7}\right)$$
$$7(x+2)6 + 7(x-2)6 = (x-2)(x+2)20$$
$$42(x+2) + 42(x-2) = 20(x^2 - 4)$$
$$42x + 84 + 42x - 84 = 20x^2 - 80$$
$$20x^2 - 84x - 80 = 0$$
$$4(5x^2 - 21x - 20) = 0$$
$$4(5x + 4)(x - 5) = 0$$
$$5x + 4 = 0 \quad \text{or} \quad x - 5 = 0$$
$$x = -\frac{4}{5} \quad \text{or} \quad x = 5$$

Step 5 State the answer. The rate cannot be $-\dfrac{4}{5}$ mph, so the answer is Amy rows at 5 mph.

Step 6 Check that this value satisfies the original equation.

3. It takes two painters working together $4\dfrac{4}{5}$ hours to paint a house. If each worked alone, one of them could do the job in 4 hours less than the other. How long would it take each painter to complete the job alone?

Step 1 Read the problem carefully. There will be two answers.

3. Working together, Tom and Huck painted a fence in 8 hours. If each worked alone, Tom could paint the fence 12 hours faster than Huck could. How long would it take each to paint the fence alone?

Step 2 Assign a variable. Let x = the number of hours for the slower painter to complete the job alone. Then the faster painter could do the entire job in $x - 4$ hours. The faster painter's rate is $\dfrac{1}{x-4}$ and the slower painter's rate is $\dfrac{1}{x}$.

Together, they do the job in $4\dfrac{4}{5}$ hours.

Complete a table.

	Rate	Time working together	Fractional part of the job done
Slower painter	$\dfrac{1}{x}$	$\dfrac{24}{5}$	$\dfrac{1}{x}\left(\dfrac{24}{5}\right)$
Faster painter	$\dfrac{1}{x-4}$	$\dfrac{24}{5}$	$\dfrac{1}{x-4}\left(\dfrac{24}{5}\right)$

Step 3 Write an equation. The sum of the two fractional parts is 1.

$$\frac{24}{5x}+\frac{24}{5(x-4)}=1$$

Step 4 Solve the equation. The LCD is $5x(x-4)$.

$$5x(x-4)\left(\frac{24}{5x}+\frac{24}{5(x-4)}\right)=5x(x-4)(1)$$

$$24(x-4)+24x=5x(x-4)$$

$$24x-96+24x=5x^2-20x$$

$$48x-96=5x^2-20x$$

$$5x^2-68x+96=0$$

$$(5x-8)(x-12)=0$$

$$5x-8=0 \quad \text{or} \quad x-12=0$$

$$x=\frac{8}{5} \quad \text{or} \quad x=12$$

Step 5 State the answer.

If the slower painter can do the job in $\dfrac{8}{5}$ hours (or 1.6 hours), then the faster painter's time to do the job is $1.6 - 4 = -2.4$ hr, which cannot represent the time for the slower painter.
If the slower painter can do the job in 12 hours, then the faster painter's time is $12 - 4 = 8$ hours.

Step 6 Check that these results satisfy the original problem.

Objective 2 Practice Exercises

For extra help, see Examples 2–3 on pages 795–797 of your text.

Solve each problem. Round answers to the nearest tenth, if necessary.

4. Two pipes together can fill a large tank in 10 hours. One of the pipes, used alone, takes 15 hours longer than the other to fill the tank. How long would each pipe used alone take to fill the tank?

4. pipe 1 _____________

   pipe 2 _____________

5. A jet plane traveling at a constant speed goes 1200 miles with the wind, then turns around and travels for 1000 miles against the wind. If the speed of the wind is 50 miles per hour and the total flight takes 4 hours, find the speed of the plane.

5. _____________

6. A person rode a bicycle for 12 miles and then hiked an additional 8 miles. The total time for the trip was 5 hours. If the rate when the person was riding the bicycle was 10 miles per hour faster than the rate walking, what was each rate?

6. bike_____________

   hike_____________

Objective 3   Solve radical equations that lead to quadratic equations.

Review these examples for Objective 3:	**Now Try:**

4. Solve each equation.

   a. $y = \sqrt{y+42}$

Start by squaring each side.
$$y = \sqrt{y+42}$$
$$y^2 = \left(\sqrt{y+42}\right)^2$$
$$y^2 = y+42$$
$$y^2 - y - 42 = 0$$
$$(y+6)(y-7) = 0$$
$$y+6 = 0 \quad \text{or} \quad y-7 = 0$$
$$y = -6 \quad \text{or} \quad y = 7$$

We must check all proposed solutions in the original equation because squaring each side of an equation can introduce extraneous solutions.
Check

$y = \sqrt{y+42}$	$y = \sqrt{y+42}$
$-6 \overset{?}{=} \sqrt{-6+42}$	$7 \overset{?}{=} \sqrt{7+42}$
$-6 \overset{?}{=} \sqrt{36}$	$7 \overset{?}{=} \sqrt{49}$
$-6 = 6$ False	$7 = 7$ True

The solution set is $\{7\}$.

   b. $\sqrt{x}+2 = x$

$$\sqrt{x} = x-2$$
$$\left(\sqrt{x}\right)^2 = (x-2)^2$$
$$x = x^2 - 4x + 4$$
$$0 = x^2 - 5x + 4$$
$$0 = (x-1)(x-4)$$
$$x-1 = 0 \quad \text{or} \quad x-4 = 0$$
$$x = 1 \quad \text{or} \quad x = 4$$

Check each proposed solution in the original equation.

4. Solve each equation.

   a. $x = \sqrt{x+2}$

   ____

   b. $\sqrt{2x}+4 = x$

   ____

$$\sqrt{x}+2=x \qquad\qquad \sqrt{x}+2=x$$

$$\sqrt{1}+2\overset{?}{=}1 \qquad\qquad \sqrt{4}+2\overset{?}{=}4$$

$$1+2\overset{?}{=}1 \qquad\qquad 2+2\overset{?}{=}4$$

$$3=1 \quad \text{False} \qquad\qquad 4=4 \quad \text{True}$$

The solution set is $\{4\}$.

Objective 3 Practice Exercises

For extra help, see Example 4 on page 797 of your text.

Solve each equation. Check your solutions.

7. $\sqrt{7y-10}=y$ 7. _________________

8. $x=\sqrt{\dfrac{x+3}{2}}$ 8. _________________

9. $\sqrt{4x}+3=x$ 9. _________________

Objective 4   Solve equations that are quadratic in form.

Review these examples for Objective 4:	**Now Try:**

6. Solve the equation.

$$c^4 - 20c^2 + 64 = 0$$

Write this equation in quadratic form by substituting u for c^2.

$$u^2 - 20u + 64 = 0$$

$$(u-4)(u-16) = 0$$

$$u - 4 = 0 \quad \text{or} \quad u - 16 = 0$$

$$u = 4 \quad \text{or} \quad u = 16$$

$$c^2 = 4 \quad \text{or} \quad c^2 = 16$$

$$c = \pm 2 \quad \text{or} \quad c = \pm 4$$

The solution set is $\{-4, -2, 2, 4\}$.

6. Solve the equation.

$$x^4 - 5x^2 + 4 = 0$$

____

---

7. Solve each equation.

a. $(m+5)^2 + 6(m+5) + 8 = 0$

Step 1 Substitute u for $m + 5$.

$$(m+5)^2 + 6(m+5) + 8 = 0$$

$$u^2 + 6u + 8 = 0$$

Step 2 $\quad (u+4)(u+2) = 0$

$$u + 4 = 0 \quad \text{or} \quad u + 2 = 0$$

$$u = -4 \quad \text{or} \quad u = -2$$

Step 3 $\quad m + 5 = -4 \quad \text{or} \quad m + 5 = -2$

Step 4 $\quad m = -9 \quad \text{or} \quad m = -7$

Step 5 Check that the solution set of the original equation is $\{-9, -7\}$.

b. $x^{4/3} - 20x^{2/3} + 36 = 0$

Step 1 Substitute u for $x^{2/3}$.

$$x^{4/3} - 20x^{2/3} + 36 = 0$$

$$u^2 - 20u + 36 = 0$$

Step 2 $\quad (u-2)(u-18) = 0$

$$u - 2 = 0 \quad \text{or} \quad u - 18 = 0$$

$$u = 2 \quad \text{or} \quad u = 18$$

7. Solve each equation.

a. $(x-5)^2 + 2(x-5) - 35 = 0$

____

b. $x^{2/3} - 2x^{1/3} = 3$

____

Step 3 $\qquad x^{2/3} = 2 \qquad$ or $\qquad x^{2/3} = 18$

Step 4 $\ \left(x^{2/3}\right)^{3/2} = (2)^{3/2}$ or $\left(x^{2/3}\right)^{3/2} = (18)^{3/2}$

$\qquad\qquad\qquad x = 2\sqrt{2} \quad$ or $\qquad\quad x = 54\sqrt{2}$

Step 5   Check that the solution set of the original equation is $\left\{2\sqrt{2},\ 54\sqrt{2}\right\}$.

Objective 4 Practice Exercises

For extra help, see Examples 5–7 on pages 798–800 of your text.

Solve each equation. Check your solutions.

10. $\quad 4t^4 = 21t^2 - 5$ **10.** _______________

11. $\quad p^{4/3} - 12p^{2/3} + 27 = 0$ **11.** _______________

12. $\quad \left(t^2 - 3t\right)^2 = 14\left(t^2 - 3t\right) - 40$ **12.** _______________

Chapter 11 QUADRATIC EQUATIONS, INEQUALITIES, AND FUNCTIONS

11.5   Formulas and Further Applications

Learning Objectives
1   Solve formulas involving squares and square roots for specified variables.
2   Solve applied problems using the Pythagorean theorem.
3   Solve applied problems using area formulas.
4   Solve applied problems using quadratic functions as models.

Key Terms

Use the vocabulary terms listed below to complete each statement in exercises 1−2.

quadratic function   **Pythagorean theorem**

1.   A function defined by $f(x) = ax^2 + bx + c$, for real numbers a, b, and c, with $a \neq 0$, is a ________________________________.

2.   The ________________________________ states that the sum of the squares of the lengths of the legs of a right triangle equals the square of the length of the hypotenuse.

Objective 1   Solve formulas involving squares and square roots for specified variables.

Review these examples for Objective 1:

1.   Solve each formula for the given variable.

a. $y = \dfrac{1}{2}gt^2$ for t

The goal is to isolate t on one side.

$$y = \frac{1}{2}gt^2$$

$$2y = gt^2$$

$$\frac{2y}{g} = t^2$$

$$t = \pm\sqrt{\frac{2y}{g}}$$

$$t = \pm\frac{\sqrt{2y}}{\sqrt{g}} \cdot \frac{\sqrt{g}}{\sqrt{g}}$$

$$t = \pm\frac{\sqrt{2yg}}{g}$$

Now Try:

1.   Solve each formula for the given variable.

a. $F = \dfrac{mx}{t^2}$ for t

____

b. $D = \sqrt{kh}$ for k

The goal is to isolate k on one side.

$$D = \sqrt{kh}$$

$$D^2 = kh$$

$$\frac{D^2}{h} = k$$

b. $p = \dfrac{y}{\sqrt{6z}}$ for z

____

2. Solve $rk^2 - 3k = -s$ for k.

Write the equation in standard form and then use the quadratic formula to solve for k.

$$rk^2 - 3k = -s$$

$$rk^2 - 3k + s = 0$$

Let $a = r$, $b = -3$, and $c = s$.

$$k = \frac{-(-3) \pm \sqrt{(-3)^2 - 4(r)(s)}}{2r}$$

$$k = \frac{3 \pm \sqrt{9 - 4rs}}{2r}$$

The solutions are $\dfrac{3 + \sqrt{9 - 4rs}}{2r}$ and

$\dfrac{3 - \sqrt{9 - 4rs}}{2r}$.

2. Solve $p^2 q^2 + pkq = k^2$ for q.

____

Objective 1 Practice Exercises

For extra help, see Examples 1–2 on pages 806–807 of your text.

Solve each equation for the indicated variable. (Leave ± in your answers.)

1. $F = \dfrac{kl}{\sqrt{d}}$ for d

1. _________________

2. $p = \sqrt{\dfrac{kl}{g}}$ for k

2. _________________

3. $b^2 a^2 + 2bca = c^2$ for a

3. _________________

Objective 2   Solve applied problems using the Pythagorean theorem.

Review this example for Objective 2:

3. A 13-foot ladder is leaning against a building. The distance from the bottom of the ladder to the building is 2 feet more than twice the distance from the top of the ladder to the ground. How far is the bottom of the ladder from the building?

Step 1 Read the problem carefully.

Step 2 Assign a variable. Let x = the distance from the top of the ladder to the ground. Then $2x + 2$ = the distance from the bottom of the ladder to the building. Draw a picture to represent the problem.

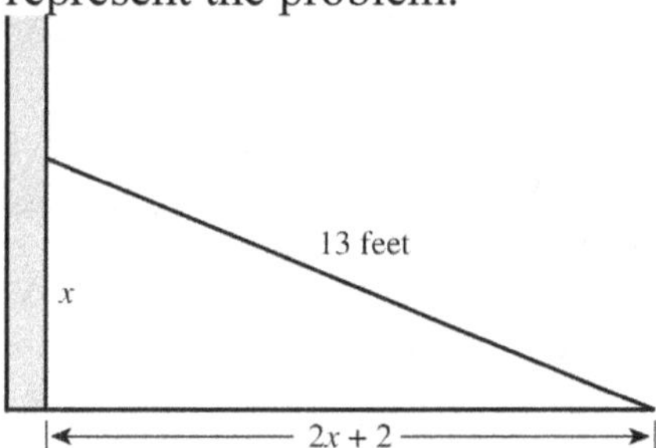

Step 3 Write an equation. Use the Pythagorean theorem.

$$a^2 + b^2 = c^2$$

$$x^2 + (2x+2)^2 = 13^2$$

Step 4 Solve.

$$x^2 + 4x^2 + 8x + 4 = 169$$

$$5x^2 + 8x - 165 = 0$$

$$(5x+33)(x-5) = 0$$

$$5x + 33 = 0 \quad \text{or} \quad x - 5 = 0$$

$$x = -\frac{33}{5} \quad \text{or} \quad x = 5$$

Step 5 State the answer. Length cannot be negative, so discard the negative solution. The distance from the top of the ladder to the ground is 5 feet. However, we are asked to find the distance from the bottom of the ladder to the building. This distance is $2(5) + 2 = 12$ feet.

Step 6 Check. Since $5^2 + 12^2 = 13^2$, the answer is correct.

Now Try:

3. Two cars left an intersection at the same time, one heading south, the other heading east. Sometime later, the car traveling south had gone 18 miles farther than the car headed east. At that time they were 90 miles apart. How far had each car traveled?

south   ____________

east   ____________

Objective 2 Practice Exercises

For extra help, see Example 3 on page 807 of your text.

Solve each problem.

4. A child flying a kite has let out 45 feet of string to the kite. The distance from the kite to the ground is 9 feet more than the distance from the child to a point directly below the kite. How high up is the kite?

4. _________________

5. A ladder is leaning against a building so that the top is 8 feet above the ground. The length of the ladder is 2 feet less than twice the distance of the bottom of the ladder from the building. Find the length of the ladder.

5. _________________

6. Two cars left an intersection at the same time, one heading north, the other heading west. Later they were exactly 95 miles apart. The car headed west had gone 38 miles less than twice as far as the car headed north. How far had each car traveled?

6. north _____________

    west _____________

Objective 3   Solve applied problems using area formulas.

Review this example for Objective 3:

4. A fish pond is 3 feet by 4 feet. How wide a strip of concrete can be poured around the pond if there is enough concrete for 44 square feet?

Step 1 Read the problem carefully.

Step 2 Assign a variable. Let x = the width of the strip of concrete. Then $2x + 3$ = the width of the fish pond with the two strips of concrete and $2x + 4$ = the length of the fish pond with the two strips of concrete.

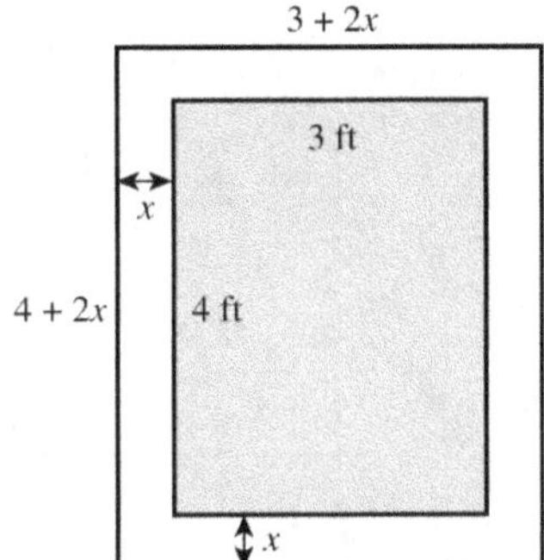

Step 3 Write an equation. The area of the strip is 44 sq ft and the area of the fish pond is $3(4) = 12$ sq ft, so the total area of the outer rectangle is $44 + 12 = 56$ sq ft.

$$(3+2x)(4+2x)=56$$

Step 4 Solve.

$$(3+2x)(4+2x)=56$$
$$12+14x+4x^2=56$$
$$4x^2+14x-44=0$$
$$2(2x^2+7x-22)=0$$
$$2(2x+11)(x-2)=0$$
$$2x+11=0 \qquad \text{or} \quad x-2=0$$
$$x=-\frac{11}{2} \quad \text{or} \qquad x=2$$

Step 5 State the answer. Since length cannot be negative, we disregard the negative solution. The concrete strip should be 2 ft wide.

Step 6 Check. If $x = 2$, then the area of the large rectangle is $(3+2\cdot2)(4+2\cdot2)=7(8)=56$ sq ft. The area of the fish pond is $3(4) = 12$ sq ft, so the area of the concrete strip is $56 - 12 = 44$ sq ft.

Now Try:

4. A picture 9 inches by 12 inches is to be mounted on a piece of mat board so that there is an even width of mat all around the picture. How wide will the matted border be if the area of the mounted picture is 238 square inches?

____

Objective 3 Practice Exercises

For extra help, see Example 4 on page 808 of your text.

Solve each problem.

7. A rug is to fit in a room so that a border of even
    width is left on all four sides. If the room is
    16 feet by 20 feet and the area of the rug is 165
    square feet, how wide to the nearest tenth of a foot
    will the border be?

7. ______________________

8. A rectangular garden has an area of 12 feet by 5 feet.
    A gravel path of equal width is to be built around the
    garden. How wide can the path be if there is enough
    gravel for 138 square feet?

8. ______________________

9. A doghouse 2 feet by 4 feet is to be built with a
    cement path around it of equal width on all sides.
    The area available for the doghouse and path is 120
    square feet. How wide will the path be?

9. ______________________

Name: Date:
Instructor: Section:

Objective 4   Solve applied problems using quadratic functions as models.

Review this example for Objective 4:

5. A certain projectile is located at a distance of $d(t) = 3t^2 - 6t + 1$ feet from its starting point after t seconds. How many seconds will it take the projectile to travel 10 feet?

Let $d = 10$ in the formula and solve for t.

$$d = 3t^2 - 6t + 1$$
$$10 = 3t^2 - 6t + 1$$
$$9 = 3t^2 - 6t$$
$$3 = t^2 - 2t$$
$$3 + 1 = t^2 - 2t + 1$$
$$4 = (t - 1)^2$$
$$\sqrt{4} = t - 1 \quad \text{or} \quad -\sqrt{4} = t - 1$$
$$2 = t - 1 \quad \text{or} \quad -2 = t - 1$$
$$3 = t \quad \text{or} \quad -1 = t$$

Since t represents time, we reject the negative solution. It will take 3 seconds for the projectile to travel 10 feet.

Now Try:

5. A baseball is thrown upward from a building 20 m high with a velocity of 15 m/sec. Its distance from the ground after t seconds is modeled by the function

$$f(t) = -4.9t^2 + 15t + 20.$$

When will the ball hit the ground? Round your answer to the nearest tenth.

____

For extra help, see Examples 5–6 on pages 809–810 of your text.

Solve each problem. Round answers to the nearest tenth.

10.   A population of microorganisms grows according to the function $p(x) = 100 + 0.2x + 0.5x^2$, where x is given in hours. How many hours does it take to reach a population of 250 microorganisms?

10. ________________

11. An object is thrown downward from a tower 280 feet high. The distance the object has fallen at time t in seconds is given by $s(t) = 16t^2 + 68t$. How long will it take the object to fall 100 feet?

11. _______________

12. A widget manufacturer estimates that the monthly revenue can be modeled by the function $R(x) = -0.006x^2 + 32x - 10{,}000$. What is the minimum number of items that must be sold for the revenue to equal \$30,000?

12. _______________

Chapter 11 QUADRATIC EQUATIONS, INEQUALITIES, AND FUNCTIONS

11.6   Graphs of Quadratic Functions

Learning Objectives
1   Graph a quadratic function.
2   Graph parabolas with horizontal and vertical shifts.
3   Use the coefficient of x^2 to predict the shape and direction in which a parabola opens.
4   Find a quadratic function to model data.

Key Terms

Use the vocabulary terms listed below to complete each statement in exercises 1–4.

       parabola   **vertex**   **axis**   **quadratic function**

1.   The vertical (or horizontal) line through the vertex of a vertical (or horizontal) parabola is its _____________________________.

2.   The point on a parabola that has the least y-value (if the parabola opens up) or the greatest y-value (if the parabola opens down) is called the __________________ of the parabola.

3.   A function defined by $f(x) = ax^2 + bx + c$, for real numbers a, b, and c, with $a \neq 0$, is a _____________________________.

4.   The graph of a quadratic function is a _____________________________.

Objective 1   Graph a quadratic function.

For extra help, see pages 816–817 of your text.

Objective 2   Graph parabolas with horizontal and vertical shifts.

Review these examples for Objective 2:	Now Try:
1. Graph $g(x) = x^2 + 2$. Give the vertex, axis, domain, and range. The graph of $g(x)$ has the same shape as that of $f(x) = x^2$ but shifted 2 units up with vertex $(0, 2)$. Every function value is 2 more than the corresponding function value of $f(x) = x^2$.	1. Graph $f(x) = x^2 - 1$. Give the vertex, axis, domain, and range. Vertex __________ Axis __________ Domain __________ Range __________

x	$f(x)=x^2$	$g(x)=x^2+2$
-2	4	6
-1	1	3
0	0	2
1	1	3
2	4	6

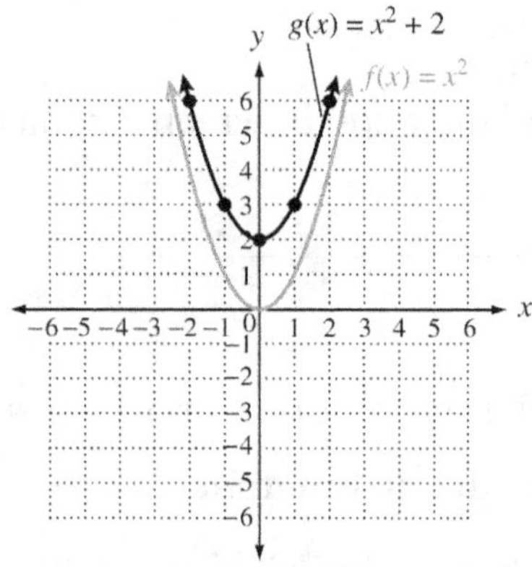

The vertex is $(0, 2)$. The axis is $x = 0$. The domain is $(-\infty, \infty)$. The range is $[2, \infty)$.

2. Graph $g(x)=(x-1)^2$. Give the vertex, axis, domain, and range.

The graph of $g(x)$ has the same shape as that of $f(x)=x^2$ but shifted 1 unit right with vertex $(1, 0)$.

x	$f(x)=x^2$	$g(x)=(x-1)^2$
-2	4	9
-1	1	4
0	0	1
1	1	0
2	4	1
3	9	4

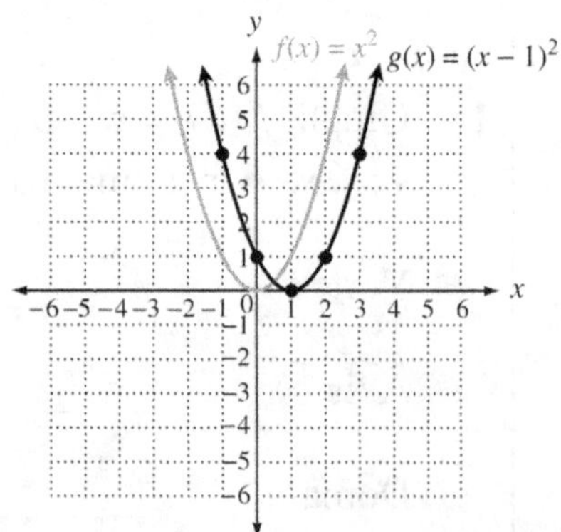

The vertex is $(1, 0)$. The axis is $x = 1$. The domain is $(-\infty, \infty)$. The range is $[0, \infty)$.

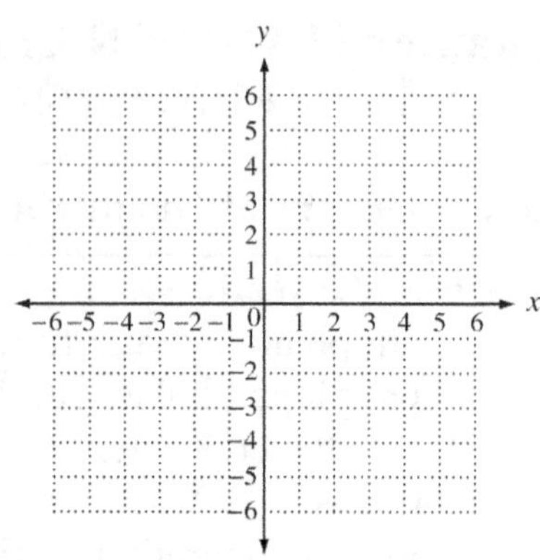

2. Graph $f(x)=(x-3)^2$. Give the vertex, axis, domain, and range.

Vertex   ____________

Axis   ____________

Domain   ____________

Range   ____________

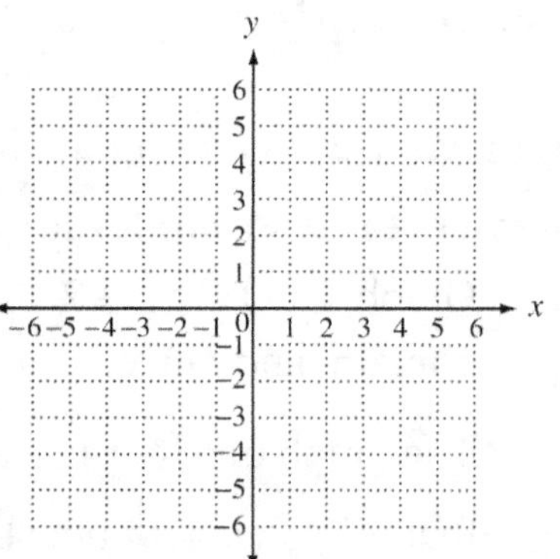

          Copyright © 2025 Pearson Education, Inc.

3. Graph $g(x) = (x-1)^2 - 2$. Give the vertex, axis, domain, and range.

The graph of $g(x)$ has the same shape as that of $f(x) = x^2$ but shifted 1 unit right (since $x - 1 = 0$ if $x = 1$) and 2 units down (because of the -2).

x	$g(x) = (x-1)^2 - 2$
-2	7
-1	2
0	-1
1	-2
2	-1
3	2
4	7

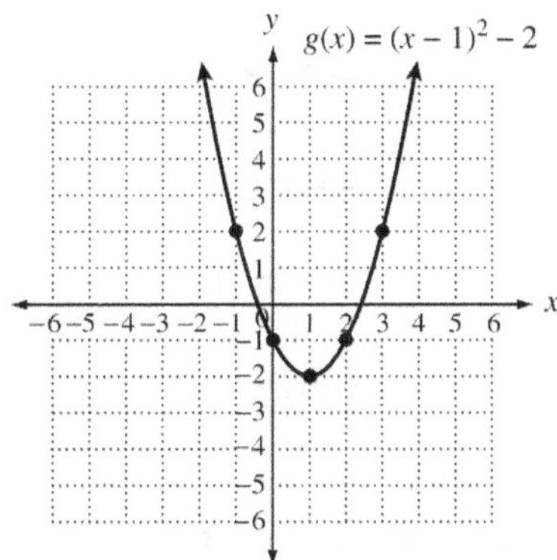

The vertex is $(1, -2)$. The axis is $x = 1$. The domain is $(-\infty, \infty)$. The range is $[-2, \infty)$.

3. Graph $f(x) = (x+2)^2 - 1$. Give the vertex, axis, domain, and range.

Vertex   ____________

Axis   ____________

Domain   ____________

Range   ____________

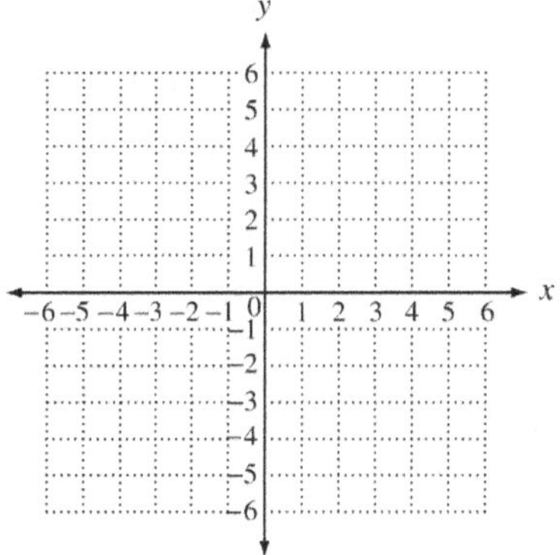

Objective 2 Practice Exercises

For extra help, see Examples 1–3 on pages 817–819 of your text.

Sketch the graph of each parabola. Give the vertex, axis, domain, and range.

1. $f(x) = x^2 - 4$

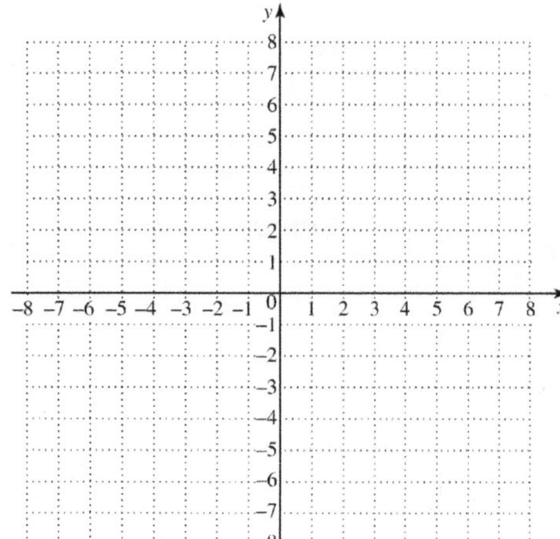

1. vertex ______________

    axis ______________

    domain ______________

    range ______________

2. $f(x) = (x - 3)^2$

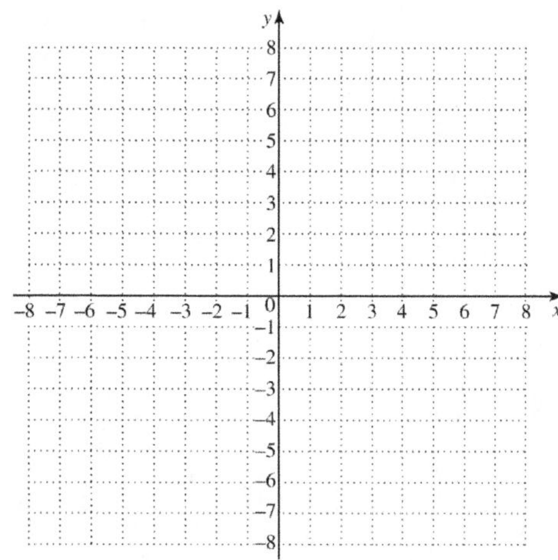

2. vertex ______________

    axis ______________

    domain ______________

    range ______________

3. $f(x) = (x-3)^2 - 1$

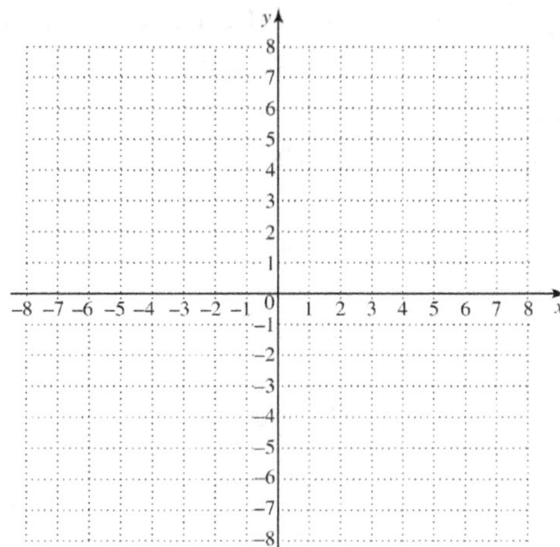

3. vertex _____________

     axis ______________

     domain ____________

     range _____________

Objective 3   **Use the coefficient of x^2 to predict the shape and direction in which a parabola opens.**

Review these examples for Objective 3:

4. Graph $g(x) = -2x^2$. Give the vertex, axis, domain, and range.

The graph of $g(x)$ has the same shape as that of $f(x) = x^2$ but is narrower and opens downward.

x	$g(x) = -2x^2$
-2	-8
-1	-2
0	0
1	-2
2	-8

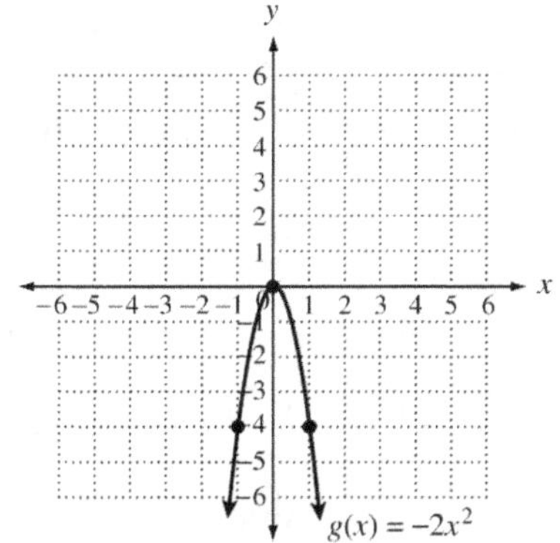

The vertex is $(0, 0)$. The axis is $x = 0$. The domain is $(-\infty, \infty)$. The range is $(-\infty, 0]$.

Now Try:

4. Graph $g(x) = -\dfrac{1}{4}x^2$. Give the vertex, axis, domain, and range.

Vertex   ___________

Axis   ___________

Domain   ___________

Range   ___________

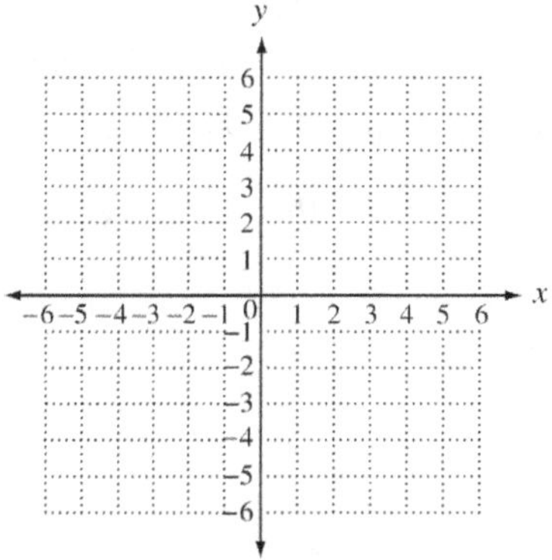

5. Graph $g(x) = -\dfrac{1}{2}(x+1)^2 - 2$. Give the vertex, axis, domain, and range.

The parabola opens down because $a < 0$ and is wider than the graph of $f(x) = x^2$. The parabola has vertex $(-1,-2)$.

x	$g(x) = -\dfrac{1}{2}(x+1)^2 - 2$
-3	-4
-2	-2.5
-1	-2
0	-2.5
1	-4

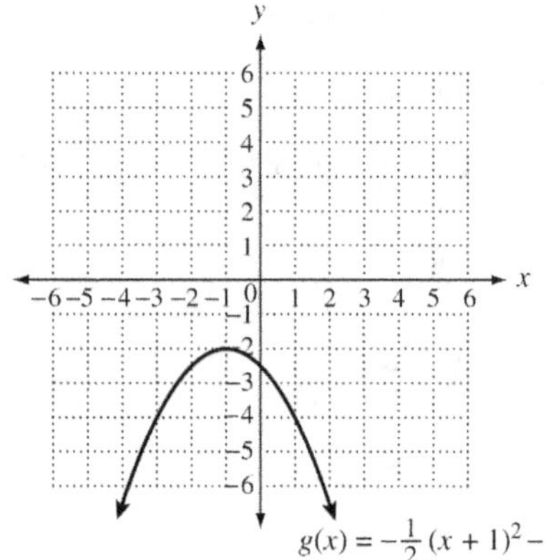

The vertex is $(-1,-2)$. The axis is $x = -1$. The domain is $(-\infty, \infty)$. The range is $(-\infty, -2]$.

5. Graph $f(x) = 3(x-1)^2 + 1$. Give the vertex, axis, domain, and range.

Vertex   ________________

Axis   ________________

Domain   ________________

Range   ________________

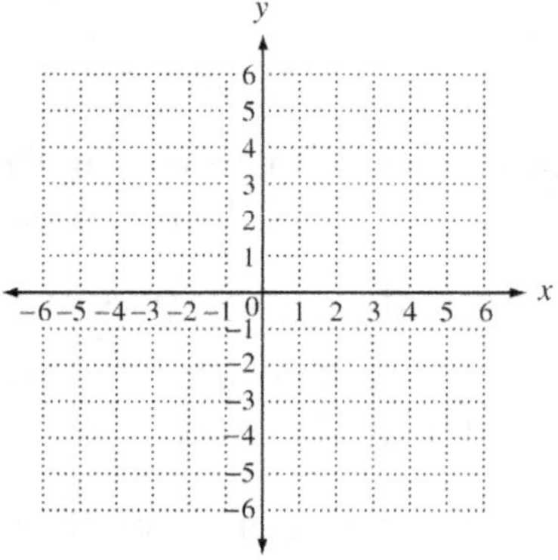

Objective 3 Practice Exercises

For extra help, see Examples 4–5 on pages 819–820 of your text.

For each quadratic function, tell whether the graph opens up or down and whether the graph is wider, narrower, or the same shape as the graph of $f(x) = x^2$. Then give the vertex, domain, and range.

4. $f(x) = -\dfrac{4}{3}x^2 - 1$

4. ________________

vertex ____________

domain ____________

range ____________

5. $f(x) = -2(x+1)^2$

5. _______________

vertex _______________

domain _______________

range _______________

6. $f(x) = \frac{5}{4}(x-1)^2 + 7$

6. _______________

vertex _______________

domain _______________

range _______________

Objective 4   Find a quadratic function to model data.

Review this example for Objective 4:

6.   The number of ice cream cones sold by an ice cream parlor from 2017–2023 is shown in the following table.

Year	Years since 2017, x	Number of cones sold
2017	0	1775
2018	1	4194
2019	2	5063
2020	3	5161
2021	4	4663
2022	5	4639
2023	6	3710

Use the ordered pairs (x, number of cones sold) to make a scatter diagram of the data. Determine a quadratic function that models these data by using a system of equations. Use the ordered pairs (0, 1775), (3, 5161), and (6, 3710). Round the values of a, b, and c in your model to the nearest tenth, as necessary.

Now Try:

6.   The table lists the average price of a Major League Baseball ticket.

Year	Years since 1990, x	Price
1991	1	$9.14
1994	4	$10.60
1997	7	$12.49
2000	10	$16.81
2004	14	$19.82
2010	20	$26.74

Use the ordered pairs (x, price) to make a scatter diagram of the data. Determine a quadratic function that models these data by using a system of equations. Use the ordered pairs (1, 9.14), (10, 16.81), and (20, 26.74). Round the values of a, b, and c in your model to the nearest hundredth, as necessary.

____

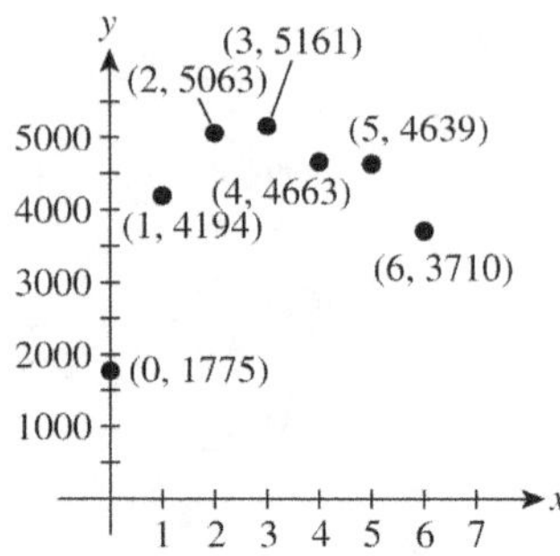
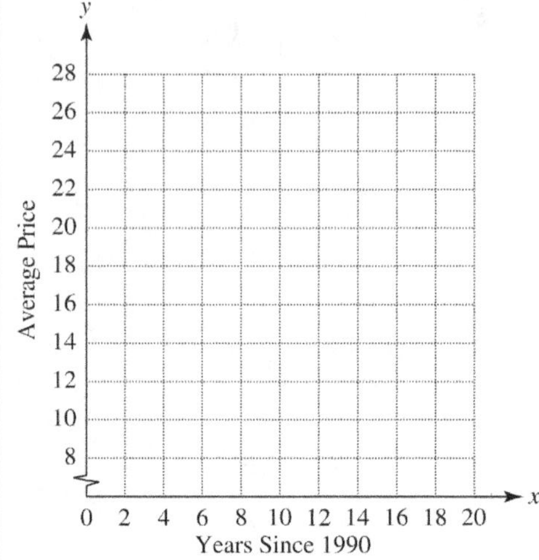

It appears that the parabola opens down, so the coefficient a is negative.

Using the chosen ordered pairs, we substitute the x- and y-values into the quadratic form

$y = ax^2 + bx + c$ to obtain the three equations.

$$a(0)^2 + b(0) + c = 1775 \quad (1)$$

$$a(3)^2 + b(3) + c = 5161 \quad (2)$$

$$a(6)^2 + b(6) + c = 3710 \quad (3)$$

Equation (1) simplifies to $c = 1775$, so substitute 1775 for c in equation (2) and (3).

$$9a + 3b + 1775 = 5161 \quad (2)$$

$$36a + 6b + 1775 = 3710 \quad (3)$$

Subtract 1775 from each side of both equations.

$$9a + 3b = 3386 \quad (2)$$

$$36a + 6b = 1935 \quad (3)$$

Solve by elimination. Multiply equation (2) by –2 and add to equation (3).

$$-18a - 6b = -6772 \quad \text{Multiply (2) by } -2.$$

$$\underline{36a + 6b = 1935 \quad (3)}$$

$$18a = -4837$$

$$a \approx -268.7 \quad \text{Round to the tenth.}$$

Substitute this value for a into equation (2) and solve for b.

$$9a + 3b = 3386 \quad (2)$$

$$9(-268.7) + 3b = 3386$$

$$-2418.3 + 3b = 3386$$

$$3b = 5804.3$$

$$b \approx 1934.8$$

Therefore, the model

is $y = -268.7x^2 + 1934.8x + 1775$.

Objective 4 Practice Exercises

For extra help, see Example 6 on pages 820–821 of your text.

Tell whether a linear or quadratic function would be a more appropriate model for each set of graphed data. If linear, tell whether the slope should be positive or negative. If quadratic, tell whether the coefficient a of x^2 should be positive or negative.

7. 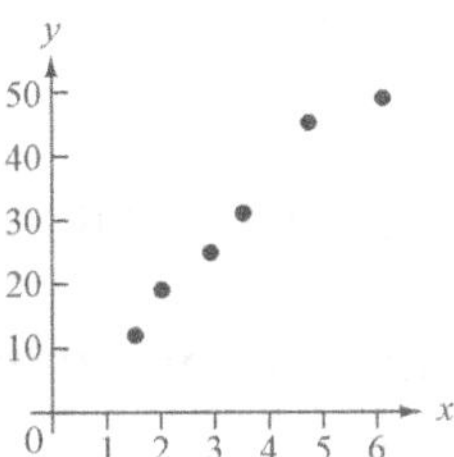

7. ______________________

8. 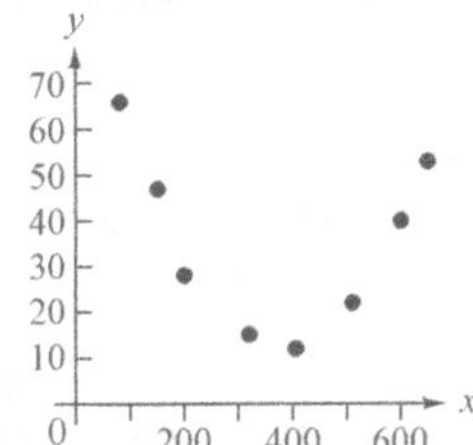

8. ______________________

Solve the problem.

9. The number of publicly traded companies filing for bankruptcy for selected years between 1990 and 2000 are shown in the table, with 0 representing 1990, 2 representing 1992, etc.

Year	Number of Bankruptcies
0	115
2	91
4	70
6	84
8	120
10	176

Use the ordered pairs to make a scatter diagram of the data.

Use the ordered pairs (0, 115), (4, 70), and (8, 120) to find a function that models the data. Round the values of *a*, *b*, and *c* to three decimal places, if necessary.

Source: Lial, Margaret L., John Hornsby, Terry McGinnis, *Intermediate Algebra* Eighth Edition. Boston: Pearson Education, 2006.

9.

____

Chapter 11 QUADRATIC EQUATIONS, INEQUALITIES, AND FUNCTIONS

11.7   More about Parabolas and Their Applications

Learning Objectives	
1	Find the vertex of a vertical parabola.
2	Graph a quadratic function.
3	Use the discriminant to find the number of x-intercepts of a parabola with a vertical axis.
4	Use quadratic functions to solve problems involving maximum or minimum value.
5	Graph parabolas with horizontal axes.

Key Terms

Use the vocabulary terms listed below to complete each statement in exercises 1–2.

        discriminant   **vertex**

1.   The_____________________________ of a quadratic function is found by using the formula $b^2 - 4ac$.

2.   The maximum or minimum value of a quadratic function occurs at the ______________________________ of its graph.

Objective 1   Find the vertex of a vertical parabola.

Review these examples for Objective 1:

1.   Find the vertex of the graph of
$$f(x) = x^2 + 6x + 10.$$

We can express $x^2 + 6x + 10$ in the form $(x - h)^2 + k$ by completing the square on $x^2 + 6x$. Because we want to keep $f(x)$ alone on one side of the equation, we add and subtract the appropriate number on just one side.

$$f(x) = x^2 + 6x + 10 \qquad \left[\frac{1}{2}(6)\right]^2 = 9$$

$$f(x) = (x^2 + 6x + 9 - 9) + 10$$

$$f(x) = (x^2 + 6x + 9) + 10 - 9$$

$$f(x) = (x + 3)^2 + 1$$

The vertex of the parabola is (–3, 1).

Now Try:

1.   Find the vertex of the graph of
$$f(x) = x^2 - 6x + 4.$$

____

2. Find the vertex of the graph of
$f(x) = 3x^2 + 6x + 10$.

Because the x^2-term has a coefficient other than 1, we factor that coefficient out of the first two terms before completing the square.

$$f(x) = 3x^2 + 6x + 10$$

$$f(x) = 3(x^2 + 2x) + 10 \qquad \left[\tfrac{1}{2}(2)\right]^2 = 1$$

$$f(x) = 3(x^2 + 2x + 1 - 1) + 10$$

$$f(x) = 3(x^2 + 2x + 1) + 3(-1) + 10$$

$$f(x) = 3(x+1)^2 + 7$$

The vertex is (−1, 7).

3. Use the vertex formula to find the vertex and axis of symmetry of the graph of
$f(x) = -4x^2 + 5x + 3$.

The x-coordinate of the vertex of the parabola is given by $\dfrac{-b}{2a}$.

$$\frac{-b}{2a} = \frac{-5}{2(-4)} = \frac{5}{8}$$

The y-coordinate is $f\left(\dfrac{-b}{2a}\right) = f\left(\dfrac{5}{8}\right)$.

$$f\left(\frac{5}{8}\right) = -4\left(\frac{5}{8}\right)^2 + 5\left(\frac{5}{8}\right) + 3 = \frac{73}{16}$$

The vertex is $\left(\dfrac{5}{8}, \dfrac{73}{16}\right)$ and $x = \dfrac{5}{8}$ is the axis of symmetry.

2. Find the vertex of the graph of
$f(x) = -2x^2 + 4x - 1$.

____

3. Use the vertex formula to find the vertex and axis of symmetry of the graph of
$f(x) = 2x^2 - 6x + 5$.

____

Objective 1 Practice Exercises

For extra help, see Examples 1–3 on pages 825–827 of your text.

Find the vertex of each parabola.

1. $f(x) = x^2 - 2x + 4$

2. $f(x) = 5x^2 - 4x + 1$

3. $f(x) = -\dfrac{1}{4}x^2 - 3x - 9$

1. _______________

2. _______________

3. _______________

Objective 2 Graph a quadratic function.

Review this example for Objective 2:

4. Graph the quadratic function defined by $f(x) = x^2 - 3x + 2$. Give the vertex, axis, domain, and range.

Step 1 From the equation, $a = 1$, so the graph opens up.

Step 2 The x-coordinate of the vertex is $\frac{3}{2}$. The y-coordinate of the vertex is

$f\left(\frac{3}{2}\right) = \left(\frac{3}{2}\right)^2 - 3\left(\frac{3}{2}\right) + 2 = -\frac{1}{4}$. The vertex is $\left(\frac{3}{2}, -\frac{1}{4}\right)$.

Step 3 Find any intercepts. Since the vertex is in quadrant IV and the graph opens up, there will be two x-intercepts. Let $f(x) = 0$ and solve.

$$x^2 - 3x + 2 = 0$$
$$(x-1)(x-2) = 0$$
$$x - 1 = 0 \quad \text{or} \quad x - 2 = 0$$
$$x = 1 \quad \text{or} \quad x = 2$$

The x-intercepts are $(1, 0)$ and $(2, 0)$. Find the y-intercept by evaluating $f(0)$.

$$f(0) = 0^2 - 3(0) + 2$$

The y-intercept is $(0, 2)$.

Step 4 Plot the vertex, intercepts, and additional points as needed using symmetry about the axis, $x = \frac{3}{2}$.

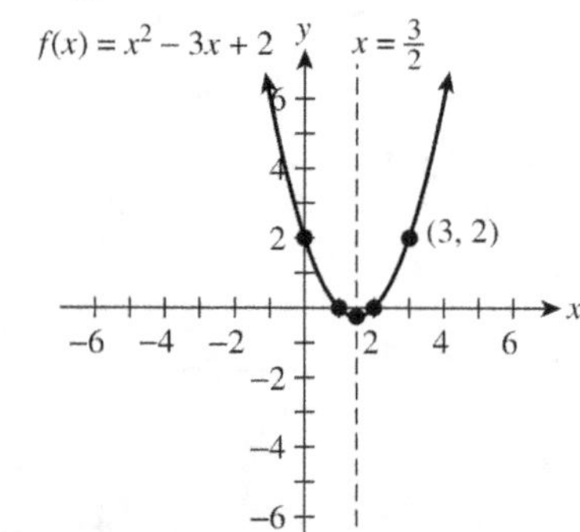

The domain is $(-\infty, \infty)$. The range is $\left[-\frac{1}{4}, \infty\right)$.

Now Try:

4. Graph the quadratic function defined by $f(x) = x^2 + 4x + 5$. Give the vertex, axis, domain, and range.

Vertex _______________________

Axis _______________________

Domain _______________________

Range _______________________

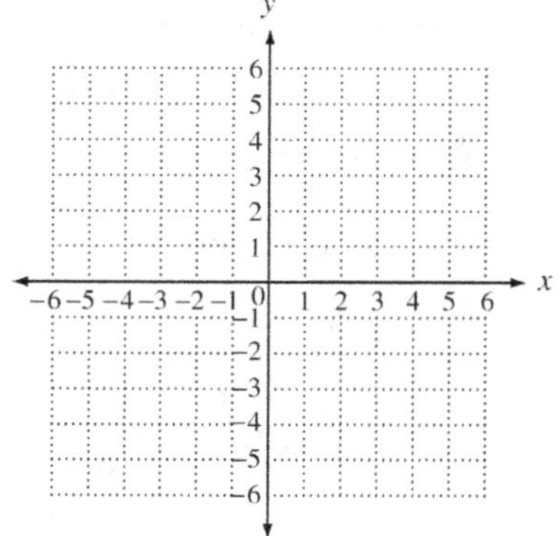

Name: ___________________ Date: ___________________

Instructor: ___________________ Section: ___________________

Objective 2 Practice Exercises

For extra help, see Example 4 on page 828 of your text.

Sketch the graph of each parabola. Give the vertex, axis, domain, and range.

4. $f(x) = -x^2 + 8x - 10$

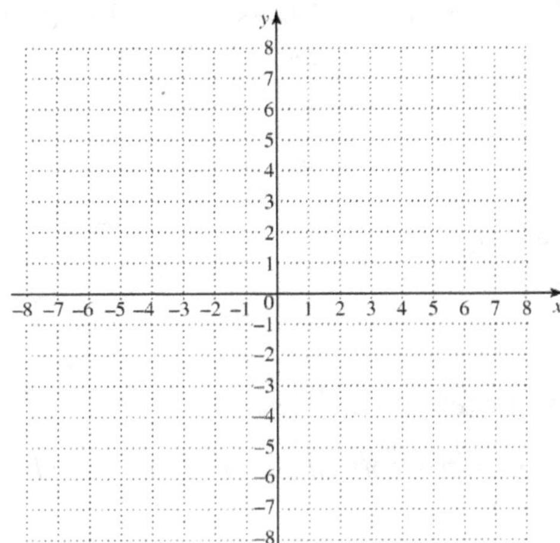

4. vertex ___________

axis ___________

domain ___________

range ___________

5. $f(x) = 3x^2 + 6x + 2$

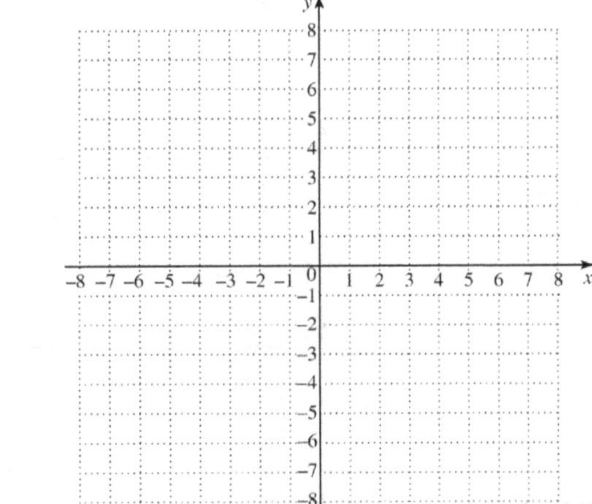

5. vertex ___________

axis ___________

domain ___________

range ___________

6. $f(x) = -2x^2 + 4x + 1$

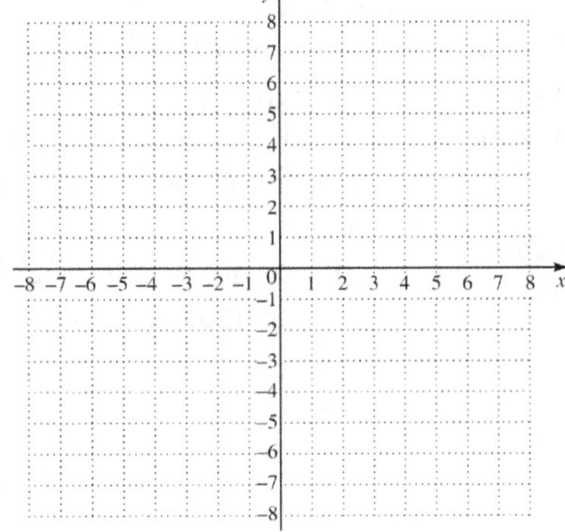

6. vertex ___________

axis ___________

domain ___________

range ___________

Objective 3   **Use the discriminant to find the number of x-intercepts of a parabola with a vertical axis.**

Review this example for Objective 3:	**Now Try:**

5. Use the discriminant to determine the number of x-intercepts of the graph of the quadratic function.

$$f(x) = 4x^2 + 12x + 9$$

$$b^2 - 4ac = 12^2 - 4(4)(9) = 0$$

Since the discriminant is zero, the graph has only one x-intercept, its vertex.

5. Use the discriminant to determine the number of x-intercepts of the graph of the quadratic function.

$$f(x) = 9x^2 - 24x + 16$$

____

Objective 3 Practice Exercises

For extra help, see Example 5 on page 829 of your text.

Use the discriminant to determine the number of x-intercepts of the graph of each function.

7. $f(x) = 2x^2 - 3x + 2$ **7.** ______________

8. $f(x) = -3x^2 - x + 5$ **8.** ______________

9. $f(x) = 3x^2 - 6x + 3$ **9.** ______________

Objective 4   **Use quadratic functions to solve problems involving maximum or minimum value.**

Review these examples for Objective 4:	**Now Try:**

6. A farmer has 1000 yards of fencing to enclose a rectangular field. What is the largest area that the farmer can enclose? What are the dimensions of the field when the area is maximized?

If the length of the field is represented by l and the width of the field is represented by w, the perimeter is given by $2l + 2w = 1000$ or $l + w = 500$ or $w = 500 - l$.

The area is given by $A = lw$ or

$$A(l) = l(500 - l) = 500l - l^2 = -l^2 + 500l.$$

6. A farmer has 1000 yards of fencing to enclose a rectangular field next to a building. What is the largest area that the farmer can enclose? What are the dimensions of the field when the area is maximized?

____

This is a quadratic equation, so its maximum occurs at the vertex of its graph. The x-coordinate is given by $\dfrac{-b}{2a} = \dfrac{-500}{2(-1)} = 250.$

The y-coordinate is

$A(250) = -250^2 + 500(250) = 62{,}500.$

If $l = 250$, then $w = 500 - 250 = 250$. Therefore, the maximum area is 62,500 sq yd when the length of the field is 250 yd and the width is 250 yd.

7. An object is launched directly upward at 64 feet per second from a platform 80 feet high. Its height above the ground is given by

$s(t) = -16t^2 + 64t + 80,$ where t is the number of seconds after launch. What will be the object's maximum height? When will it reach this height?

For this function, $a = -16$, $b = 64$, and $c = 80$.

The vertex formula gives $t = \dfrac{-b}{2a} = \dfrac{-64}{2(-16)} = 2.$

This indicates that the maximum height is reached at 2 seconds. Now calculate $s(2)$ to find the maximum height.

$s(2) = -16(2)^2 + 64(2) + 80 = 144$

Thus, it takes 2 seconds to reach the maximum height of 144 feet above the ground.

7. An object is launched directly upward at 48 feet per second from a platform 250 feet high. Its height above the ground is given by

$s(t) = -16t^2 + 48t + 250$

where t is the number of seconds after launch. What will be the object's maximum height? When will it reach this height?

____

Objective 4 Practice Exercises

For extra help, see Examples 6–7 on pages 830–831 of your text.

Solve each problem.

10. Jean sells ceramic pots. She has weekly costs of

$C(x) = x^2 - 100x + 2700,$ where x is the number of pots she sells each week. How many pots should she sell to minimize her costs? What is the minimum cost?

10. units _______________

cost _______________

11. The length and width of a rectangle have a sum of 48. What width will produce the maximum area?

11. _______________

12. A projectile is fired upward so that its distance (in feet) above the ground t seconds after firing is given by $s(t) = -16t^2 + 80t + 156$. Find the maximum height it reaches and the number of seconds it takes to reach that height.

12. height _______________

       time _______________

Objective 5   Graph parabolas with horizontal axes.

Review this example for Objective 5:

9. Graph $x = -y^2 + 6y - 9$. Give the vertex, axis, domain, and range.

We must complete the square in order to write the equation in $x = (y - k)^2 + h$ form.

$$x = -(y^2 - 6y) - 9$$

$$x = -(y^2 - 6y + 9 - 9) - 9$$

$$x = -(y^2 - 6y + 9) - 1(-9) - 9$$

$$x = -(y - 3)^2$$

The vertex is $(0, 3)$. The axis is $y = 3$.

x	y
-4	1
0	3
-4	5

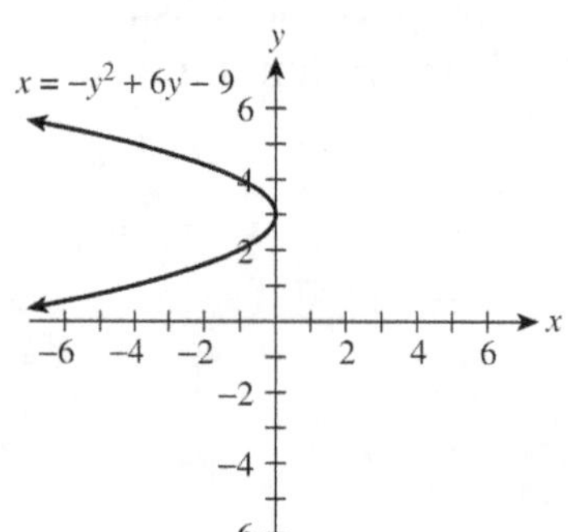

domain: $(-\infty, 0]$

range: $(-\infty, \infty)$

Now Try:

9. Graph $x = -y^2 + 4y - 4$. Give the vertex, axis, domain, and range.

Vertex _______________

Axis _______________

Domain _______________

Range _______________

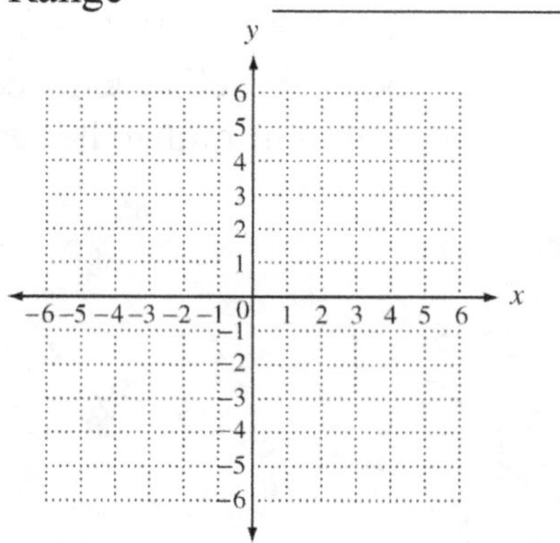

Objective 5 Practice Exercises

For extra help, see Examples 8–9 on pages 831–832 of your text.

Sketch the graph of each parabola. Give the vertex, axis, domain, and range.

13. $x = -y^2 + 2$

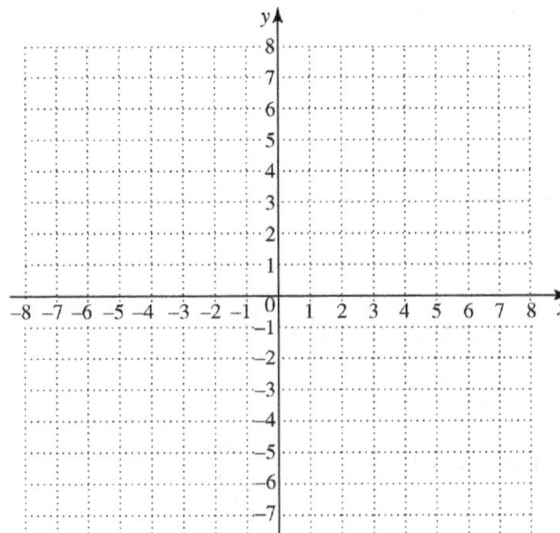

13. vertex _______________

axis _______________

domain _______________

range _______________

14. $x = y^2 - 3$

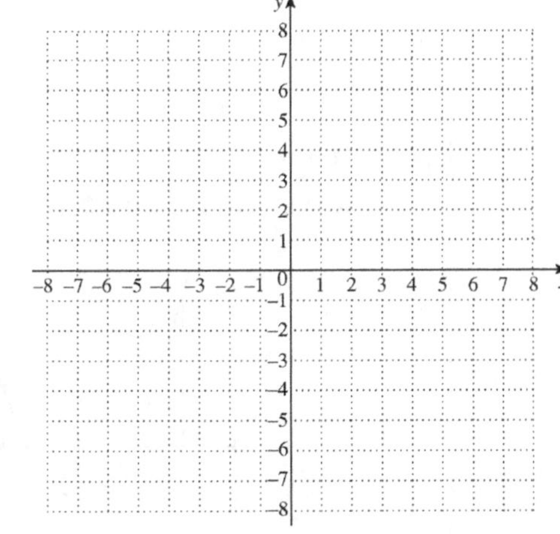

14. vertex _______________

axis _______________

domain _______________

range _______________

15. $x = -y^2 - 6y - 10$

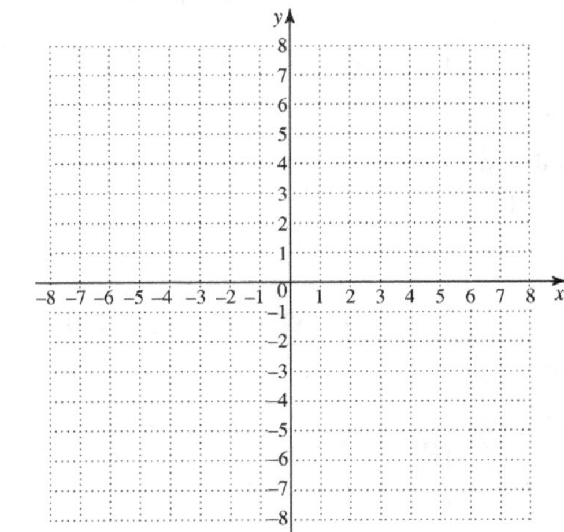

15. vertex _______________

axis _______________

domain _______________

range _______________

Chapter 11 QUADRATIC EQUATIONS, INEQUALITIES, AND FUNCTIONS

11.8   Polynomial and Rational Inequalities

Learning Objectives
1.   Solve quadratic inequalities.
2.   Solve polynomial inequalities of degree 3 or greater.
3.   Solve rational inequalities.

Key Terms

Use the vocabulary terms listed below to complete each statement in exercises 1−2.

      quadratic inequality   **rational inequality**

1.   An inequality that involves a rational expression is a ________________________.

2.   An inequality that can be written in the form $ax^2 + bx + c < 0$ or

     $ax^2 + bx + c > 0$, where a, b, and c are real numbers with $a \neq 0$ is called a

     ________________________________.

Objective 1   Solve quadratic inequalities.

Review these examples for Objective 1:

1.   Solve each inequality.

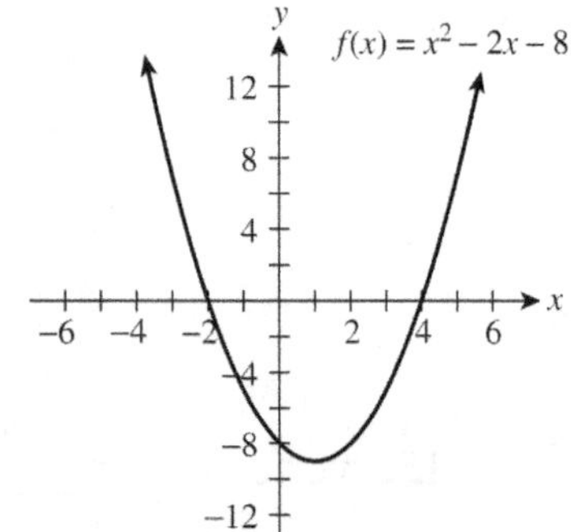

   a. $x^2 - 2x - 8 > 0$

From the graph, we see that the y-values are greater than 0 when the x-values are less than –2 or greater than 4. Therefore, the solution set of $x^2 - 2x - 8 > 0$ is $(-\infty, -2) \cup (4, \infty)$.

   b. $x^2 - 2x - 8 < 0$

From the graph, we see that the y-values are less than 0 when the x-values are greater than –2 and less than 4. Therefore, the solution set of $x^2 - 2x - 8 < 0$ is $(-2, 4)$.

Now Try:

1.   Solve each inequality.

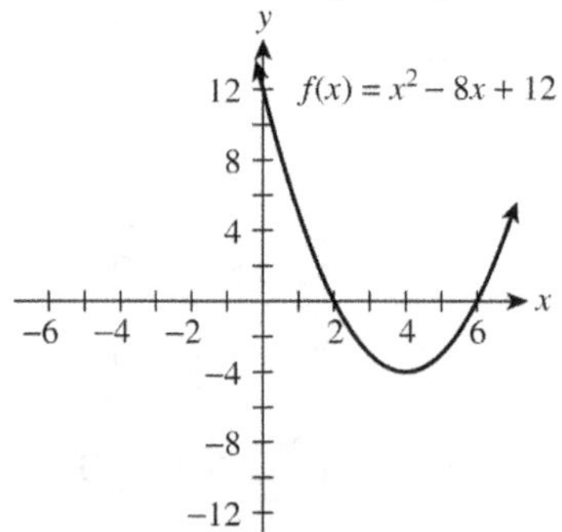

   a. $x^2 - 8x + 12 > 0$

                ____

   b. $x^2 - 8x + 12 < 0$

                ____

2. Solve and graph the solution set of $x^2 + 5x + 4 \geq 0$.

Solve the quadratic equation by factoring.

$$(x+1)(x+4) = 0$$
$$x+1 = 0 \quad \text{or} \quad x+4 = 0$$
$$x = -1 \quad \text{or} \quad x = -4$$

The numbers -4 and -1 divide a number line into intervals A, B, and C, as shown below.

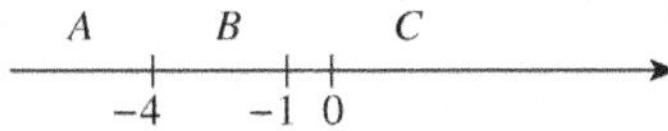

Since the numbers -4 and -1 are the only numbers that make the quadratic expression $x^2 + 5x + 4$ equal to 0, all other numbers make the expression either positive or negative. If one number in an interval satisfies the inequality, then all the numbers in that interval will satisfy the inequality.

Choose any number in interval A as a test number; we will choose -5.

$$x^2 + 5x + 4 \geq 0$$
$$(-5)^2 + 5(-5) + 4 \overset{?}{\geq} 0$$
$$4 \geq 0 \quad \text{True}$$

Because -5 satisfies the inequality, all numbers from interval A are solutions.

Now try -2 from interval B.

$$x^2 + 5x + 4 \geq 0$$
$$(-2)^2 + 5(-2) + 4 \overset{?}{\geq} 0$$
$$-2 \geq 0 \quad \text{False}$$

The numbers in interval B are not solutions.

Finally, try 0 from interval C.

$$x^2 + 5x + 4 \geq 0$$
$$0^2 + 5(0) + 4 \overset{?}{\geq} 0$$
$$4 \geq 0 \quad \text{True}$$

Because 0 satisfies the inequality, all numbers from interval C are solutions.

Because the inequality is greater than or equal to zero, we include the endpoints of the intervals in the solution set. Thus, the solution set is $(-\infty, -4] \cup [-1, \infty)$.

2. Solve and graph the solution set of $x^2 - x - 2 < 0$.

____

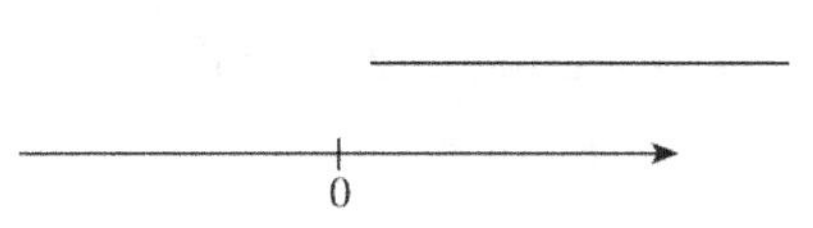

4. Solve each inequality.

   a. $(2k+5)^2 \geq -1$

Because $(2k+5)^2$ is never negative, it is always greater than -1. The solution set is $(-\infty, \infty)$.

   b. $(2k+5)^2 \leq -1$

Because $(2k+5)^2$ is never negative, there is no solution. The solution set is $\varnothing$.

4. Solve each inequality.

   a. $(4m+1)^2 \geq -3$

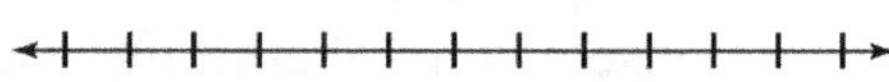

   b. $(4m+1)^2 \leq -3$

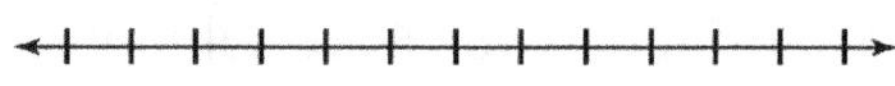

Objective 1 Practice Exercises

For extra help, see Examples 1–4 on pages 837–840 of your text.

Solve each inequality, and graph the solution set.

1. $a^2 - a - 2 \leq 0$

1. _______________________

2. $8k^2 + 10k > 3$

2. _______________________

3. $(3x-2)^2 < -1$

3. _______________________

                Copyright © 2025 Pearson Education, Inc.

Objective 2   Solve polynomial inequalities of degree 3 or greater.

Review this example for Objective 2:	**Now Try:**
5. Solve and graph the solution set of $(x+1)(x-2)(x+4) \leq 0$.	**5.** Solve and graph the solution set of $(2x-1)(2x+3)(3x+1) \leq 0$.

Set the factored polynomial equal to 0, then use the zero-factor property.

$$x+1=0 \quad \text{or} \quad x-2=0 \quad \text{or} \quad x+4=0$$
$$x=-1 \quad \text{or} \quad x=2 \quad \text{or} \quad x=-4$$

Locate -4, -1, and 2 on a number line to determine the intervals A, B, C, and D.

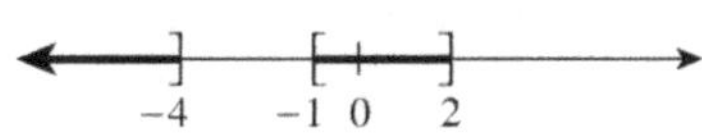

Substitute a test number from each interval in the original inequality to determine which intervals satisfy the inequality.

Interval	Test Number	Test of inequality	True or False?
A	-5	$-28 \leq 0$	T
B	-2	$8 \leq 0$	F
C	0	$-8 \leq 0$	T
D	5	$162 \leq 0$	F

The numbers in intervals A and C are in the solution set. The three endpoints are included in the solution set since the inequality symbol, $\leq$, includes equality. Thus, the solution set is $(-\infty, -4] \cup [-1,\ 2]$.

Objective 2 Practice Exercises

For extra help, see Example 5 on pages 841–842 of your text.

Solve each inequality, and graph the solution set.

4. $(y+2)(y-1)(y-2) < 0$ **4.** ______________________

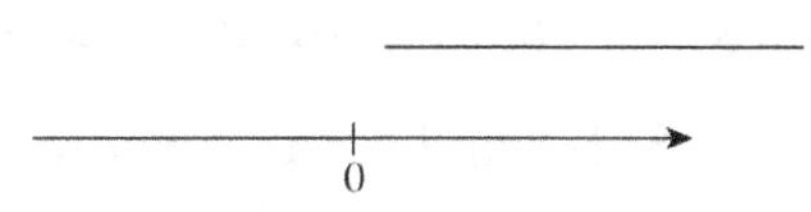

5. $(k+5)(k-1)(k+3) \leq 0$ **5.** ______________________

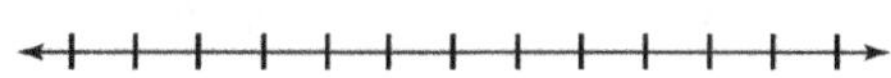

        475

6. $(x-1)(x-3)(x+2) \geq 0$

6. ________________________

Objective 3   Solve rational inequalities.

Review these examples for Objective 3:

6. Solve and graph the solution set of $\dfrac{7}{x-1} < 1$.

Write the inequality so that 0 is on one side.

$$\frac{7}{x-1} - 1 < 0$$

$$\frac{7}{x-1} - \frac{x-1}{x-1} < 0 \qquad \text{The LCD is } x-1.$$

$$\frac{7-x+1}{x-1} < 0$$

$$\frac{8-x}{x-1} < 0$$

The sign of $\dfrac{8-x}{x-1}$ will change from positive to negative or negative to positive only at those numbers that make the numerator or denominator 0. These two numbers, 1 and 8, divide a number line into three intervals.

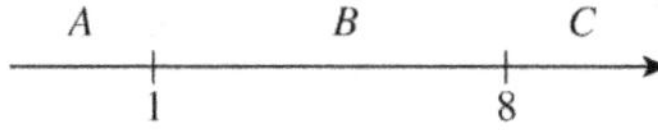

Test a number in each interval using the original inequality.

Interval	Test Number	Test of inequality	True or False?
A	0	$-7 < 1$	T
B	2	$7 < 1$	F
C	10	$\dfrac{7}{9} < 1$	T

The solution set is $(-\infty,\ 1) \cup (8,\ \infty)$. This interval does not include 1 because it would make the denominator of the original inequality 0. The number 8 is not included because the inequality symbol, $<$, does not include equality.

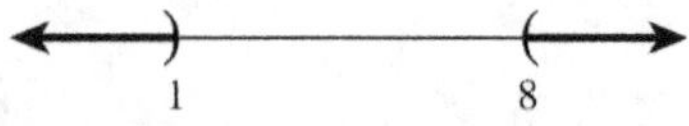

Now Try:

6. Solve and graph the solution set of $\dfrac{y}{y+1} > 3$.

____

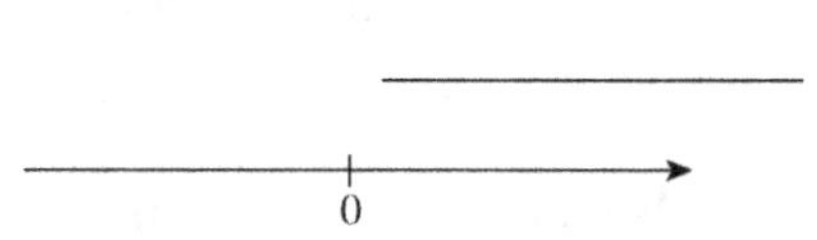

7. Solve and graph the solution set of $\dfrac{x+1}{x-5} \geq 3$.

Write the inequality so that 0 is on one side.

$$\frac{x+1}{x-5} - 3 \geq 0$$

$$\frac{x+1}{x-5} - \frac{3(x-5)}{x-5} \geq 0$$

$$\frac{x+1}{x-5} - \frac{3x-15}{x-5} \geq 0$$

$$\frac{x+1-3x+15}{x-5} \geq 0$$

$$\frac{-2x+16}{x-5} \geq 0$$

The sign of $\dfrac{-2x+16}{x-5}$ will change from positive to negative or negative to positive only at those numbers that make the numerator or denominator 0. These two numbers, 5 and 8, divide a number line into three intervals.

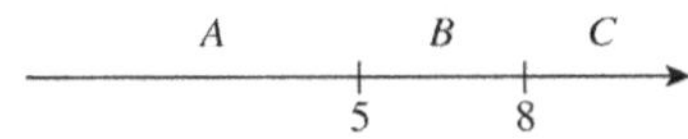

Test a number in each interval using original inequality.

Interval	Test Number	Test of inequality	True or False?
A	0	$-\dfrac{1}{5} \geq 3$	F
B	6	$7 \geq 3$	T
C	10	$\dfrac{11}{5} \geq 3$	F

The solution set is $(5,\ 8]$. This interval does not include 5 because it would make the denominator of the original inequality 0. The number 8 is included because the inequality symbol, $\geq$, does includes equality.

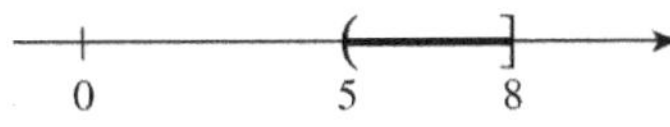

7. Solve and graph the solution set of $\dfrac{z+2}{z-3} \leq 2$.

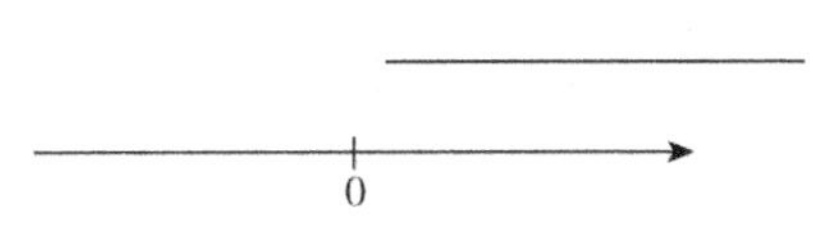

Objective 3 Practice Exercises

For extra help, see Examples 6–7 on pages 841–843 of your text.

Solve each inequality, and graph the solution set.

7. $\dfrac{7}{x-1} \le 1$

7. ________________________

8. $\dfrac{2p-1}{3p+1} \le 1$

8. ________________________

9. $\dfrac{5}{x-3} \le -1$

9. ________________________

Chapter 12 INVERSE, EXPONENTIAL, AND LOGARITHMIC FUNCTIONS

12.1   Inverse Functions

Learning Objectives
1 Decide whether a function is one-to-one and, if it is, find its inverse.
2 Use the horizontal line test to determine whether a function is one-to-one.
3 Find the equation of the inverse of a function.
4 Graph f^{-1} from the graph of f.

Key Terms

Use the vocabulary terms listed below to complete each statement in exercises 1−2.

one-to-one function   **inverse of a function f**

1.   A function in which each x-value corresponds to just one y-value and each y-value corresponds to just one x-value is a(n) ______________________________.

2.   If f is a one-to-one function, the ______________________________
     is the set of all ordered pairs of the form (y, x) where (x, y) belongs to f.

Objective 1   Decide whether a function is one-to-one and, if it is, find its inverse.

Review these examples for Objective 1:

1.   Find the inverse of each function that is one-to-one.

a. $G = \{(-3,-1), (-2, 0), (-1, 1), (0, 2)\}$

Every x-value in G corresponds to only one y-value, and every y-value corresponds to only one x-value, so G is a one-to-one function.
The inverse function is found by interchanging the x- and y-values in each ordered pair.

$$G^{-1} = \{(-1,-3), (0,-2), (1,-1), (2, 0)\}$$

b. $F = \{(2, 1), (-1, 1), (0, 0), (1, 1)\}$

Every x-value in F corresponds to only one y-value. However, the y-value 1 corresponds to two x-values, so F is not a one-to-one function.

Now Try:

1.   Find the inverse of each function that is one-to-one.

a. $G = \{(3, 2), (-3,-2), (2, 3), (-2,-3)\}$

____

b. $F = \{(2, 4), (-1, 1), (0, 0) (1, 1), (2, 6)\}$

____

c.

State	Number of National Parks
AK	8
AZ	3
CA	8
CO	4
FL	3
HI	2
UT	5

c.

State	Number of representatives
AK	1
AZ	8
CA	53
FL	25
NY	29
DE	1

____

Let N be the function defined in the table, with the states forming the domain and the number of national parks forming the range. Then, N is not one-to-one, because two different states have the same number of national parks.

Objective 1 Practice Exercises

For extra help, see Example 1 on page 859 of your text.

If the function is one-to-one, find its inverse.

   1. $\{(-3,-1), (-2, 2), (-1, 3), (0, 4)\}$ **1.** _______________

   2. $\{(1, 0), (2, 0), (3, 5), (4, 1)\}$ **2.** _______________

   3. $\{(0, 0), (1, 1), (-1,-1), (2, 2), (-2,-2)\}$ **3.** _______________

Objective 2   Use the horizontal line test to determine whether a function is one-to-one.

Review these examples for Objective 2:	**Now Try:**
2. Use the horizontal line test to determine whether each graph is the graph of a one-to-one function.	**2.** Use the horizontal line test to determine whether each graph is the graph of a one-to-one function.

a.

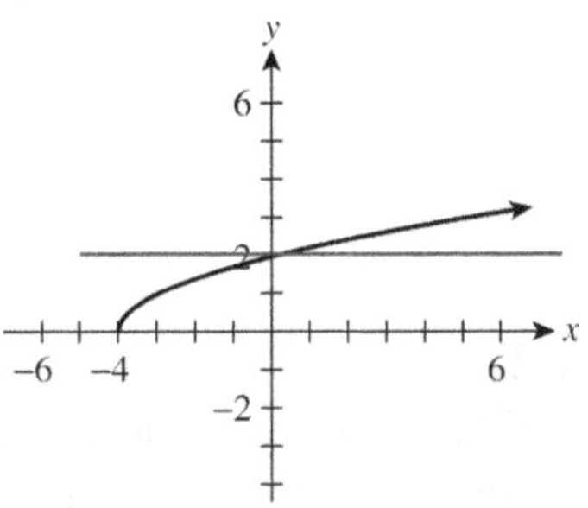

a.

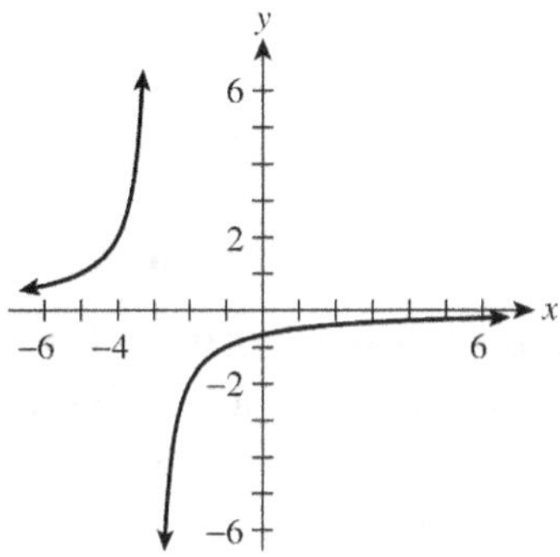

Every horizontal line will intersect the graph in exactly one point. The function is one-to-one.

____

b.

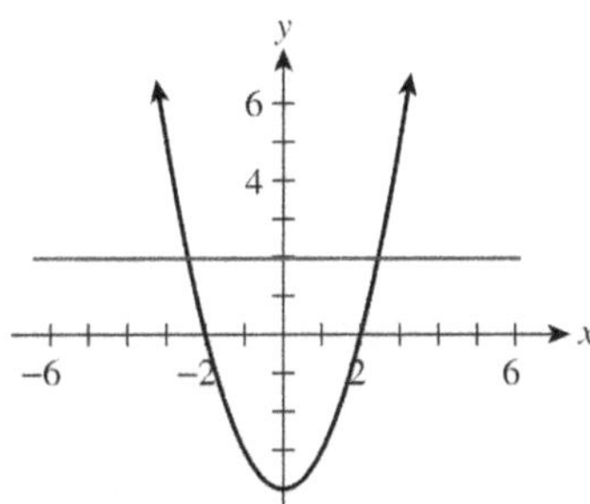

b.

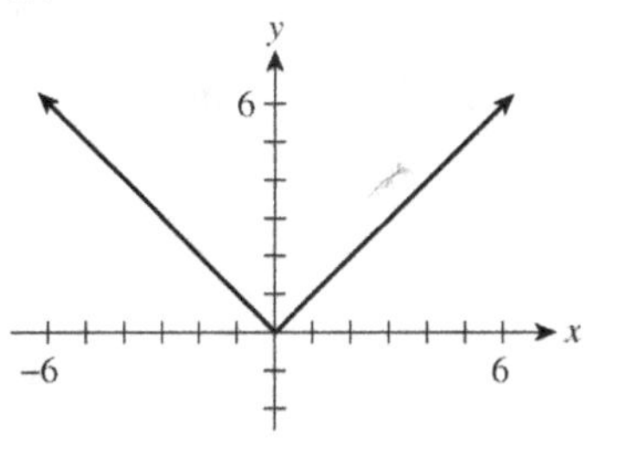

Because a horizontal line intersects the graph in more than one point, the function is not one-to-one.

____

Objective 2 Practice Exercises

For extra help, see Example 2 on page 860 of your text.

Use the horizontal line test to determine whether each function is one-to-one.

4.

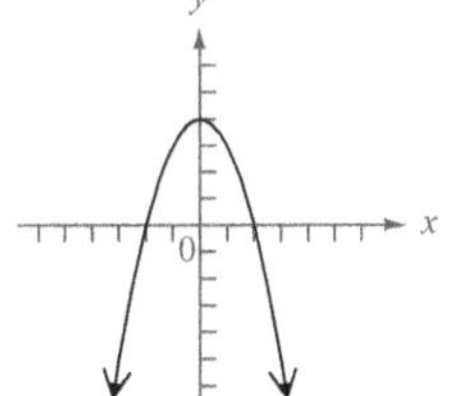

4. _________________________

5.

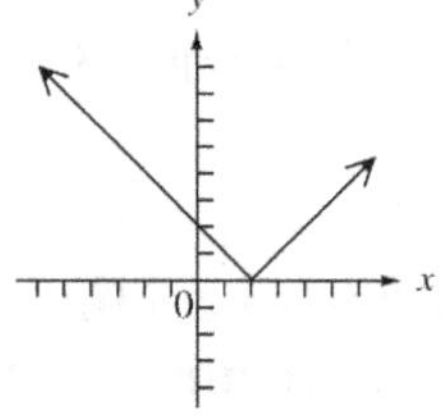

5. ________________

6.

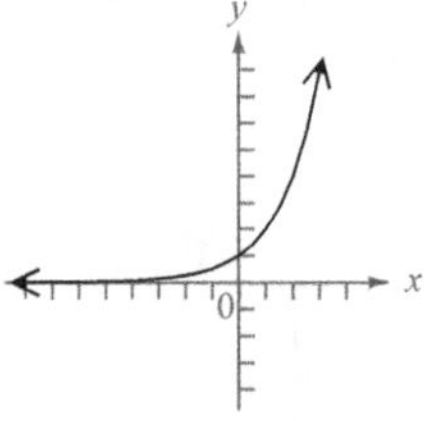

6. ________________

Objective 3   Find the equation of the inverse of a function.

Review these examples for Objective 3:	**Now Try:**

3. Decide whether each equation represents a one-to-one function. If so, find the equation for the inverse.

a. $f(x) = 3x - 5$

The graph of $y = 3x - 5$ is a nonvertical line, so by the horizontal line test, f is a one-to-one function. To find the inverse, let $y = f(x)$, interchange x and y, then solve for y.

$$y = 3x - 5$$
$$x = 3y - 5 \quad \text{Interchange } x \text{ and } y.$$
$$x + 5 = 3y$$
$$\frac{x + 5}{3} = y$$
$$f^{-1}(x) = \frac{x + 5}{3} = \frac{x}{3} + \frac{5}{3}$$
$$f^{-1}(x) = \frac{1}{3}x + \frac{5}{3}$$

b. $f(x) = 2x^2 + 3$

The graph of $y = 2x^2 + 3$ is a vertical parabola, so by the horizontal line test, f is not a one-to-one function and does not have an inverse.

Now Try:

3. Decide whether each equation represents a one-to-one function. If so, find the equation for the inverse.

a. $f(x) = 4x - 1$

____

b. $f(x) = -\frac{3}{2}x^2$

____

c. $f(x) = x^3 + 1$

The graph of $y = x^3 + 1$ is a cubing function.
The function is one-to-one and has an inverse.

$$y = x^3 + 1$$
$$x = y^3 + 1 \quad \text{Interchange } x \text{ and } y.$$
$$x - 1 = y^3$$
$$\sqrt[3]{x-1} = y$$
$$f^{-1}(x) = \sqrt[3]{x-1}$$

5. Find $f^{-1}(x)$ for $f(x) = \sqrt{x-2}$, $x \geq 2$.

$$f(x) = \sqrt{x-2}, \quad x \geq 2$$
$$y = \sqrt{x-2}, \quad x \geq 2$$
$$x = \sqrt{y-2}, \quad y \geq 2$$
$$x^2 = \left(\sqrt{y-2}\right)^2$$
$$x^2 = y - 2$$
$$y = x^2 + 2$$

If domain of f is $[2, \infty)$ and its range is $[0, \infty)$,

then the domain of $f^{-1}(x)$ is $[0, \infty)$ and its

range is $[2, \infty)$. Thus, $f^{-1}(x) = x^2 + 2$, $x \geq 0$.

c. $f(x) = 2x^3 - 3$

____

5. Find $f^{-1}(x)$ for
$f(x) = \sqrt{x+7}$, $x \geq -7$.

____

Objective 3 Practice Exercises

For extra help, see Examples 3–5 on pages 860–862 of your text.

If the function is one-to-one, find its inverse.

7. $f(x) = 2x - 5$

7. __________

8. $f(x) = x^3 - 1$

8. __________

9. $f(x) = x^2 - 1$

9. __________

Objective 4   Graph f^{-1} from the graph of f.

Review this example for Objective 4:

6.   Use the given graph to graph the inverse of f.

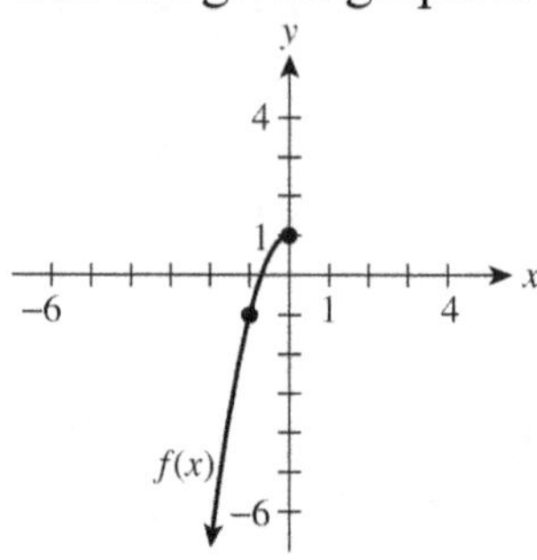

We can find the graph of f^{-1} from the graph of f by locating the mirror image of each point in f with respect to the line $y = x$.

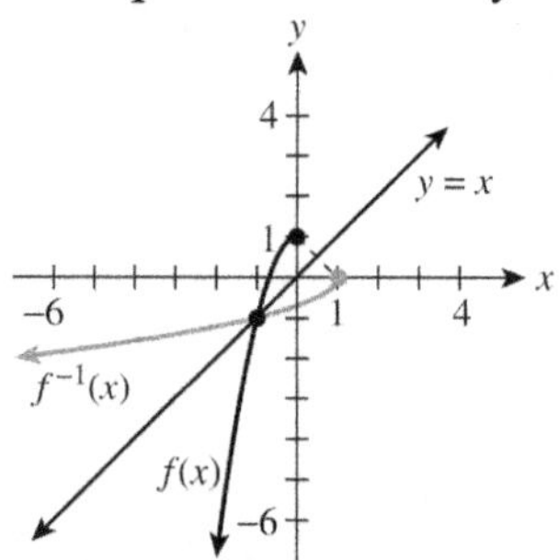

Now Try:

6.   Use the given graph to graph the inverse of f.

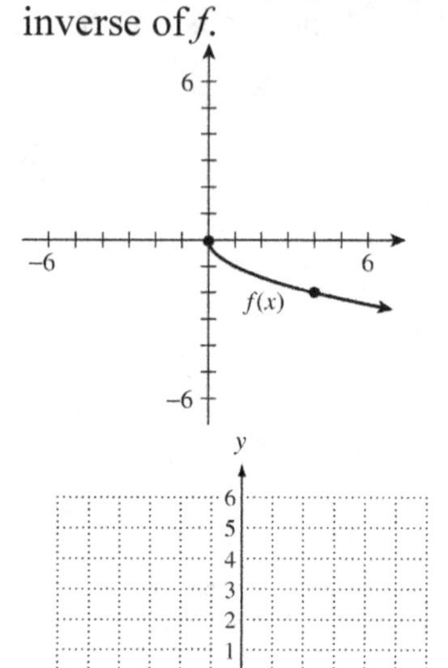

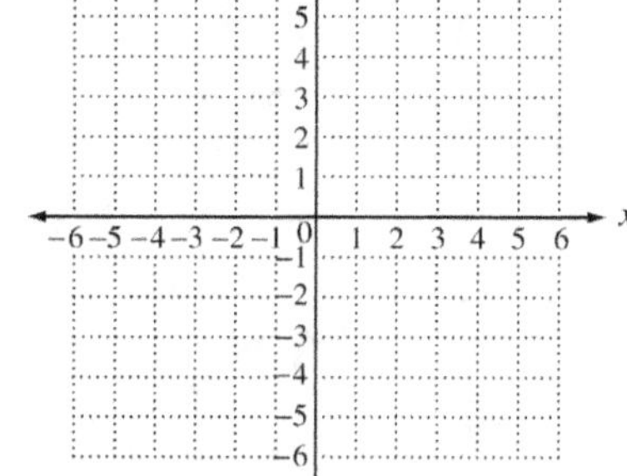

Objective 4 Practice Exercises

For extra help, see Example 6 on pages 863–864 of your text.

If the function is one-to-one, graph its inverse.

10.

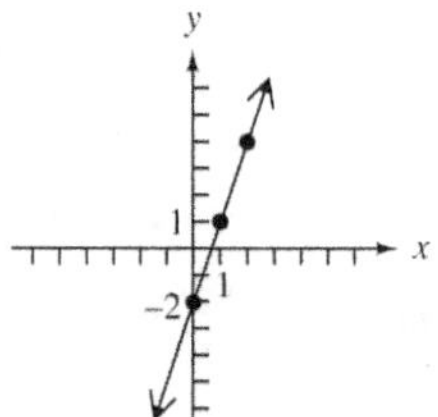

10. _______________

11.

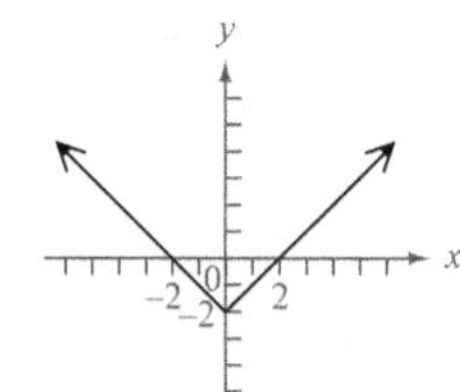

11. _______________

12.

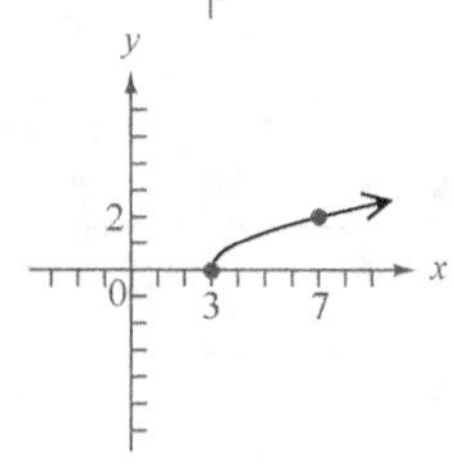

12. _______________

Chapter 12 INVERSE, EXPONENTIAL, AND LOGARITHMIC FUNCTIONS

12.2   Exponential Functions

Learning Objectives
1   Evaluate exponential expressions using a calculator.
2   Define and graph exponential functions.
3   Solve exponential equations of the form $a^x = a^k$ for x.
4   Use exponential functions in applications involving growth or decay.

Key Terms

Use the vocabulary terms listed below to complete each statement in exercises 1−3.

exponential equation   **asymptote**   **inverse**

1.   If f is a one-to-one function, then the ________________________ of f is the set of all ordered pairs formed by interchanging the coordinates of the ordered pairs of f.

2.   An equation that has a variable as an exponent, is an ________________________.

3.   An exponential function of the form $f(x) = a^x$ approaches but does not touch the x-axis, at an ________________________.

Objective 1   Evaluate exponential expressions using a calculator.

Review these examples for Objective 1:

1. Use a calculator to approximate each exponential expression to three decimal places.

    a. $3^{1.8}$

    $3^{1.8} \approx 7.225$

    b. $3^{-1.4}$

    $3^{-1.4} \approx 0.215$

    c. $3^{1/4}$

    $3^{1/4} \approx 1.316$

Now Try:

1. Use a calculator to approximate each exponential expression to three decimal places.

    a. $3^{1.9}$

    ____

    b. $3^{-1.6}$

    ____

    c. $3^{1/5}$

    ____

       485

Objective 1 Practice Exercises

For extra help, see Example 1 on page 868 of your text.

Use a calculator to find an approximation to three decimal places for each exponential expression.

1. $3^{1.2}$

2. $3^{-1.2}$

3. $3^{1/3}$

1. _________________

2. _________________

3. _________________

Objective 2   Define and graph exponential functions.

Review these examples for Objective 2:

2. Graph $f(x)=6^x$.

Create a table of values, then plot the points and draw a smooth curve through them.

x	$f(x)=6^x$
-2	$\dfrac{1}{36}$
-1	$\dfrac{1}{6}$
0	1
1	6
2	36
3	216

Now Try:

2. Graph $f(x)=3^x$.

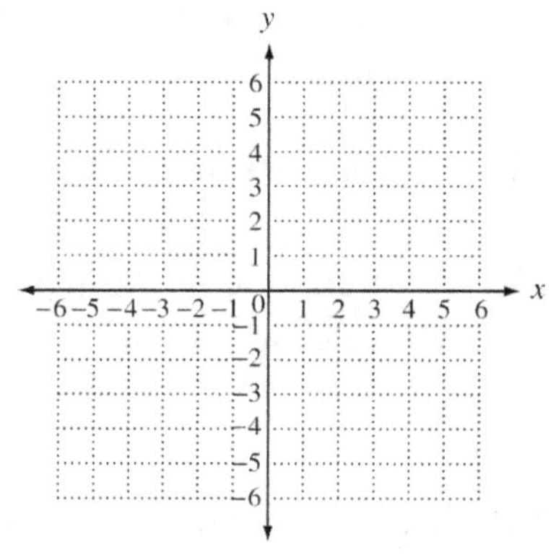

3. Graph $f(x)=\left(\dfrac{1}{6}\right)^x$.

Create a table of values, then plot the points and draw a smooth curve through them.

x	$f(x)=\left(\dfrac{1}{6}\right)^x$
-3	216
-2	36
-1	6
0	1
1	$\dfrac{1}{6}$
2	$\dfrac{1}{36}$

3. Graph $f(x)=\left(\dfrac{1}{3}\right)^x$.

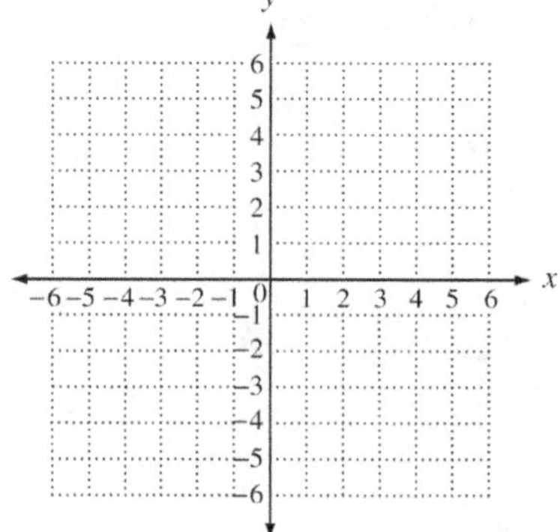

4. Graph $f(x) = 3^{2x-1}$.

Create a table of values, then plot the points and draw a smooth curve through them.

x	$2x-1$	$f(x) = 3^{2x-1}$
-1	-3	$\dfrac{1}{27}$
0	-1	$\dfrac{1}{3}$
1	1	3
2	3	27

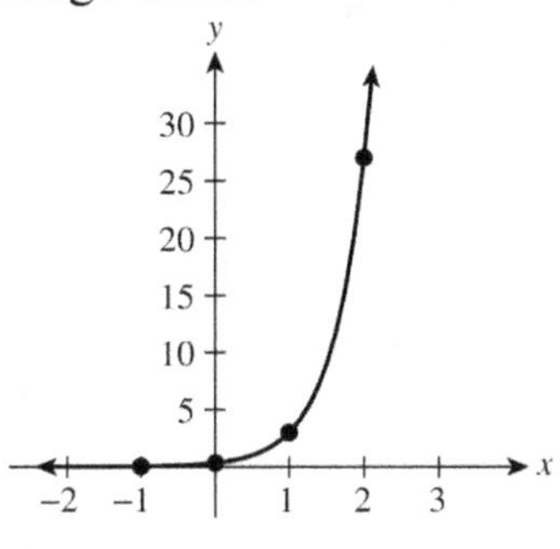

4. Graph $f(x) = 2^{1-x}$.

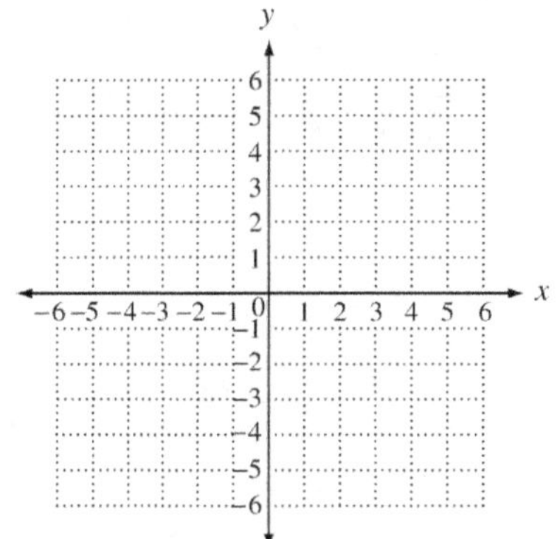

Objective 2 Practice Exercises

For extra help, see Examples 2–4 on pages 869–870 of your text.

Graph each exponential function.

4. $f(x) = 2^{-x}$

4.

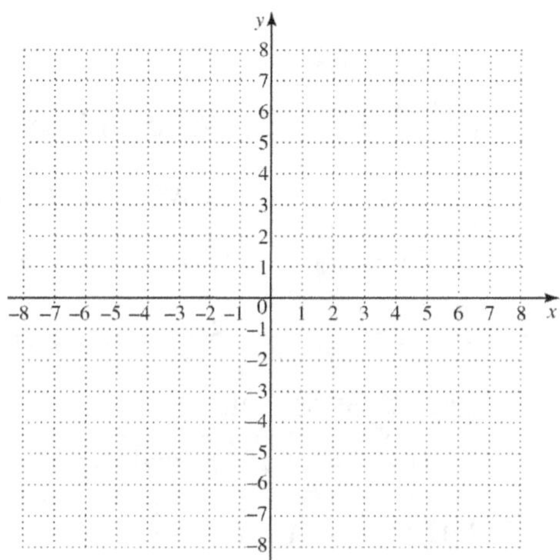

5. $f(x) = \left(\dfrac{1}{8}\right)^{x}$

5.

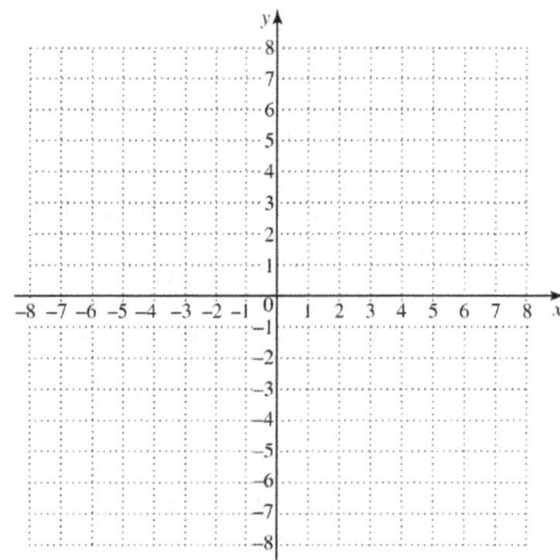

6. $f(x) = 4^{2x-3}$

6.

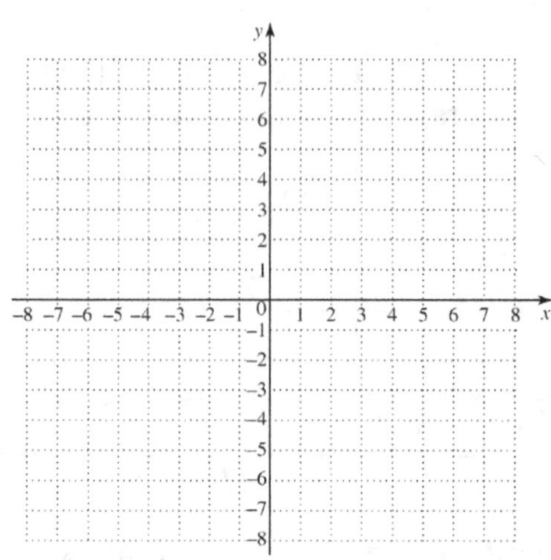

Objective 3   Solve exponential equations of the form $a^x = a^k$ for x.

Review these examples for Objective 3:

5. Solve the equation $16^x = 64$.

$$16^x = 64$$

$(2^4)^x = 2^6$ Write with the same base.

$2^{4x} = 2^6$ Power rule for exponents

$4x = 6$ If $a^x = a^y$, then $x = y$.

$x = \dfrac{6}{4} = \dfrac{3}{2}$ Solve for x; simplify.

Check: Substitute 3/2 for x.

$16^{3/2} = (16^{1/2})^3 = 4^3 = 64$

The solution set is $\left\{\dfrac{3}{2}\right\}$.

6. Solve each equation.

a. $16^{x-2} = 64^x$

$$16^{x-2} = 64^x$$

$(2^4)^{x-2} = (2^6)^x$ Write with the same base.

$2^{4x-8} = 2^{6x}$ Power rule for exponents

$4x - 8 = 6x$ If $a^x = a^y$, then $x = y$.

$-8 = 2x$ Solve for x.

$-4 = x$

The solution set is $\{-4\}$.

Now Try:

5. Solve the equation $25^x = 125$.

____

6. Solve each equation.

a. $4^{x-1} = 8^x$

____

b. $4^x = \dfrac{1}{64}$ $\qquad\qquad\qquad\qquad\qquad\qquad\qquad$ **b.** $3^x = \dfrac{1}{243}$

$4^x = \dfrac{1}{64}$

$\qquad\qquad\qquad\qquad\qquad\qquad\qquad\qquad$ _______________

$4^x = \dfrac{1}{4^3} \qquad 64 = 4^3$

$4^x = 4^{-3} \quad$ Write with the same base.

$\quad x = -3 \quad$ Set exponents equal.

The solution set is $\{-3\}$.

c. $\left(\dfrac{2}{5}\right)^x = \dfrac{125}{8}$ $\qquad\qquad\qquad\qquad\qquad$ **c.** $\left(\dfrac{3}{2}\right)^x = \dfrac{16}{81}$

$\left(\dfrac{2}{5}\right)^x = \dfrac{125}{8}$

$\qquad\qquad\qquad\qquad\qquad\qquad\qquad\qquad$ _______________

$\left(\dfrac{2}{5}\right)^x = \left(\dfrac{8}{125}\right)^{-1}$

$\left(\dfrac{2}{5}\right)^x = \left[\left(\dfrac{2}{5}\right)^3\right]^{-1} \quad$ Write with the same base.

$\left(\dfrac{2}{5}\right)^x = \left(\dfrac{2}{5}\right)^{-3} \qquad$ Power rule for exponents

$\qquad x = -3 \qquad$ Set exponents equal.

The solution set is $\{-3\}$.

Objective 3 Practice Exercises

For extra help, see Examples 5–6 on pages 871–872 of your text.

Solve each equation.

7. $\quad 25^{1-t} = 5$ $\qquad\qquad\qquad\qquad\qquad\qquad\qquad$ **7.** _______________

8. $\quad 8^{2x+1} = 4^{4x}$ $\qquad\qquad\qquad\qquad\qquad\qquad\quad$ **8.** _______________

9. $\quad \left(\dfrac{3}{4}\right)^x = \dfrac{16}{9}$ $\qquad\qquad\qquad\qquad\qquad\qquad$ **9.** _______________

 $\qquad\qquad$ **489**

Objective 4   Use exponential functions in applications involving growth or decay.

Review these examples for Objective 4:

7. Suppose the number of bacteria present in a certain culture after t minutes is given by the equation $Q(t) = 2500(2^{0.05t})$,

How many bacteria were present after 20 minutes?

Start with the given function. Replace t with 20.

$$Q(t) = 2500(2^{0.05t})$$

$$Q(20) = 2500(2^{0.05 \times 20})$$

$$Q(20) = 2500(2^1)$$

$$Q(20) = 5000$$

There were 5000 bacteria present after 20 minutes.

8. The amount of radioactive material in a sample is given by the function $A(t) = 90\left(\dfrac{1}{2}\right)^{t/18}$, where $A(t)$ is the amount present, in grams, t days after the initial measurement.

How many grams will be present after 3 days? Round to the nearest hundredth.

Start with the given function. Replace t with 3.

$$A(t) = 90\left(\frac{1}{2}\right)^{t/18}$$

$$A(3) = 90\left(\frac{1}{2}\right)^{3/18}$$

$$A(3) \approx 80.18$$

After 3 days, there were about 80.18 grams in the sample.

Now Try:

7. The population of Evergreen Park is now 16,000. The population t years from now is given by the formula $P = 16{,}000(2^{t/10})$. Using the model, what will be the population 40 years from now?

____

8. An industrial city in Ohio has found that its population is declining according to the equation $y = 70{,}000(2)^{-0.01x}$, where x is the time in years from 1910. According to the model, what will the city's population be in the year 2020?

____

Objective 4 Practice Exercises

For extra help, see Examples 7–8 on pages 873–874 of your text.

Solve each problem.

10. The population of Canadian geese that spend the
     summer at Gemini Lake each year has been growing
     according to the function $f(x) = 56(2)^{0.2x}$, where x
     is the time in years from 2010. Find the number of
     geese in 2030.

10. 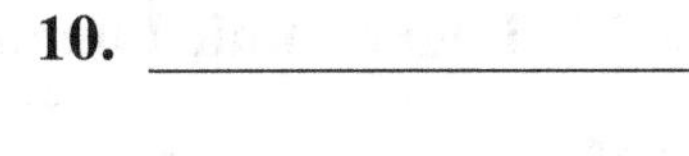

11. A sample of a radioactive substance with mass in
     grams decays according to the function
     $f(x) = 100(10)^{-0.2x}$, where x is the time in hours
     after the original measurement. Find the mass of the
     substance after 10 hours.

11. 

12. A culture of a certain kind of bacteria grows
     according to $f(x) = 7750(x)^{0.75x}$, where x is the
     number of hours after 12 noon. Find the number of
     bacteria in the culture at 12 noon.

12. ________________

Chapter 12 INVERSE, EXPONENTIAL, AND LOGARITHMIC FUNCTIONS

12.3   Logarithmic Functions

Learning Objectives	
1	Define a logarithm.
2	Convert between exponential and logarithmic forms, and evaluate logarithms.
3	Use the definition of logarithm to simplify logarithmic expressions.
4	Solve logarithmic equations of the form $\log_a b = k$ for a, b, or k.
5	Define and graph logarithmic functions.
6	Use logarithmic functions in applications involving growth or decay.

Key Terms

Use the vocabulary terms listed below to complete each statement in exercises 1–2.

logarithm   **logarithmic equation**

1.   The ______________________ of a positive number is the exponent indicating the power to which it is necessary to raise a given number (the base) to give the original number.

2.   An equation with a logarithm in at least one term is a ______________________.

Objective 1   Define a logarithm.

For extra help, see pages 877–878 of your text.

Objective 2   Convert between exponential and logarithmic forms, and evaluate logarithms.

Review these examples for Objective 2:
1.

   a. Write $5^3 = 125$ in logarithmic form.

   $\log_5 125 = 3$

   b. Write $\log_{16} 4 = \dfrac{1}{2}$ in exponential form.

   $16^{1/2} = 4$

Now Try:
1.

   a. Write $8^2 = 64$ in logarithmic form.

   ____

   b. Write $\log_{16} \dfrac{1}{4} = -\dfrac{1}{2}$ in exponential form.

   ____

          Copyright © 2025 Pearson Education, Inc.

2. Evaluate

   a. $\log_2 64$

   $\log_2 64 = 6$ because $2^6 = 64$.

   b. $\log_{10} 0.001$

   $\log_{10} 0.001 = -3$ because $10^{-3} = 0.001$.

   c. $\log_{49} 7$

   $\log_{49} 7 = \dfrac{1}{2}$ because $49^{1/2} = \sqrt{49} = 7$.

   d. $\log_4 \dfrac{1}{64}$

   $\log_4 \dfrac{1}{64} = -3$ because $4^{-3} = \dfrac{1}{4^3} = \dfrac{1}{64}$.

2. Evaluate

   a. $\log_2 256$

   ____

   b. $\log_{10} 0.00001$

   ____

   c. $\log_{121} 11$

   ____

   d. $\log_3 81$

   ____

Objective 2 Practice Exercises

For extra help, see Examples 1–2 on page 878 of your text.

Write in exponential form.

   1. $\log_{10} 0.001 = -3$

   1. _______________

Write in logarithmic form.

   2. $2^{-7} = \dfrac{1}{128}$

   2. _______________

Evaluate.

   3. $\log_{16} 2$

   3. _______________

Objective 3   Use the definition of logarithm to simplify logarithmic expressions.

Review these examples for Objective 3:

3. Use special properties to evaluate each expression.

   a. $\log_8 8$

   $\log_8 8 = 1$

   c. $\log_{64} 1$

   $\log_{64} 1 = 0$

   d. $\log_{0.2} 1$

   $\log_{0.2} 1 = 0$

Now Try:

3. Use special properties to evaluate each expression.

   a. $\log_4 4$

   ____

   c. $\log_{100} 1$

   ____

   d. $\log_{1/3} 1$

   ____

    e. $\log_4 4^{11}$ **e.** $\log_6 6^3$

    $\log_4 4^{11} = 11$

    g. $8^{\log_8 5}$ **g.** $6^{\log_6 9}$

    $8^{\log_8 5} = 5$

Objective 3 Practice Exercises

For extra help, see Example 3 on page 879 of your text.

Use the special properties to evaluate each expression.

4. $\log_{3.4} 1$ **4.** _______________

5. $\log_8 8^3$ **5.** _______________

6. $2^{\log_2 5}$ **6.** _______________

Objective 4   **Solve logarithmic equations of the form $\log_a b = k$ for a, b, or k.**

Review these examples for Objective 4:

4.   Solve each equation.

    a. $\log_{3/2} x = -2$

    By definition, $\log_{3/2} x = -2$ is equivalent to

    $x = \left(\dfrac{3}{2}\right)^{-2}$, and $\left(\dfrac{3}{2}\right)^{-2} = \left(\dfrac{2}{3}\right)^{2} = \dfrac{4}{9}$. The solution

    set is $\left\{\dfrac{4}{9}\right\}$.

    b. $\log_5(3x+1) = 2$

    $\log_5(3x+1) = 2$

        $3x+1 = 5^2$ Write in exponential form.

           $3x = 24$ Apply the exponent; subtract 1.

            $x = 8$ Divide by 3.

    The solution set is $\{8\}$.

Now Try:

4.   Solve each equation.

    a. $\log_4 x = -3$

    ____

    b. $\log_9(2x+1) = 2$

    ____

c. $\log_x 6 = 2$

$$\log_x 6 = 2$$
$$x^2 = 6 \qquad \text{Write in exponential form.}$$
$$x = \pm\sqrt{6} \qquad \text{Take square root.}$$

Only the principal square root satisfies the equation since the base must be a positive number. The solution set is $\{\sqrt{6}\}$.

d. $\log_{64} \sqrt[4]{8} = x$

$$\log_{64} \sqrt[4]{8} = x$$
$$64^x = \sqrt[4]{8} \qquad \text{Write in exponential form.}$$
$$\left(8^2\right)^x = 8^{1/4} \qquad \text{Write with the same base.}$$
$$8^{2x} = 8^{1/4} \qquad \text{Power rule for exponents}$$
$$2x = \frac{1}{4}$$
$$x = \frac{1}{8}$$

The solution set is $\left\{\frac{1}{8}\right\}$.

c. $\log_x 12 = 2$

____

d. $\log_{81} \sqrt[3]{9} = x$

____

Objective 4 Practice Exercises

For extra help, see Example 4 on pages 879–881 of your text.

Solve each equation.

7. $\quad x = \log_{32} 8$

7. _______________

8. $\quad \log_{1/3} r = -4$

8. _______________

9. $\quad \log_a 4 = \frac{1}{2}$

9. _______________

Objective 5   Define and graph logarithmic functions.

Review these examples for Objective 5:

5. Graph $f(x) = \log_5 x$.

Begin by writing $y = \log_5 x$ in exponential form as $x = 5^y$. Then, create a table of values, plot the points and draw a smooth curve through them.

$x = 5^y$	y
$\dfrac{1}{5}$	-1
1	0
5	1
25	2

6. Graph $f(x) = \log_{1/3} x$.

Begin by writing $y = \log_{1/3} x$. in exponential form. Then, create a table of values, plot the points and draw a smooth curve through them.

$x = \left(\dfrac{1}{3}\right)^y$	y
$\dfrac{1}{3}$	1
1	0
3	-1
9	-2

Now Try:

5. Graph $f(x) = \log_3 x$.

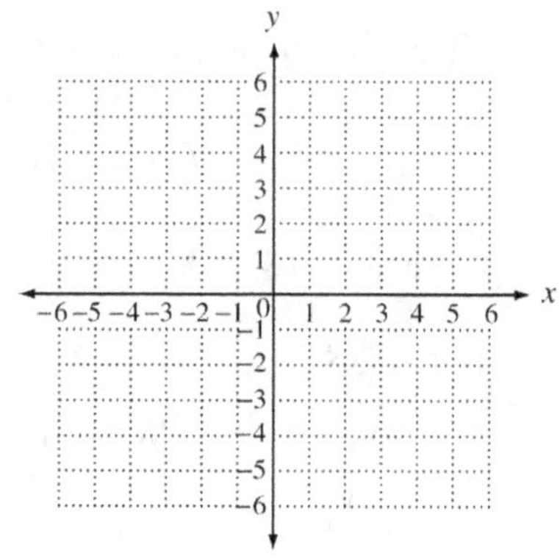

6. Graph $f(x) = \log_{1/4} x$.

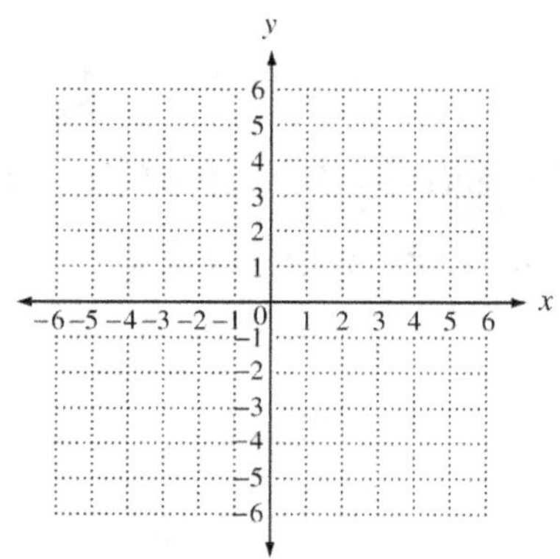

Copyright © 2025 Pearson Education, Inc.

Objective 5 Practice Exercises

For extra help, see Examples 5–6 on pages 881–882 of your text.

Graph each logarithmic function.

10. $y = \log_9 x$

10.

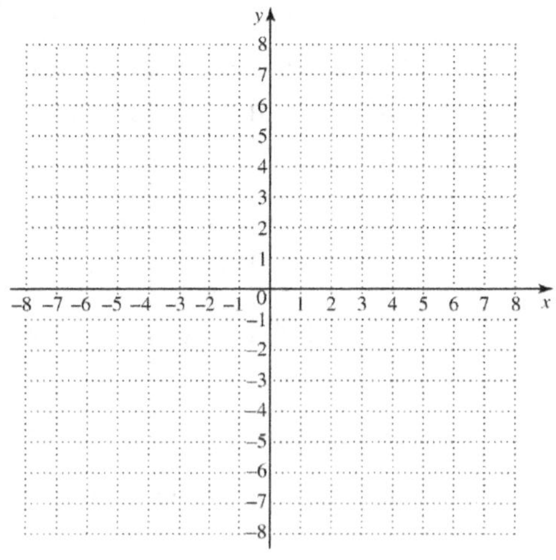

11. $y = \log_{1/4} x$

11.

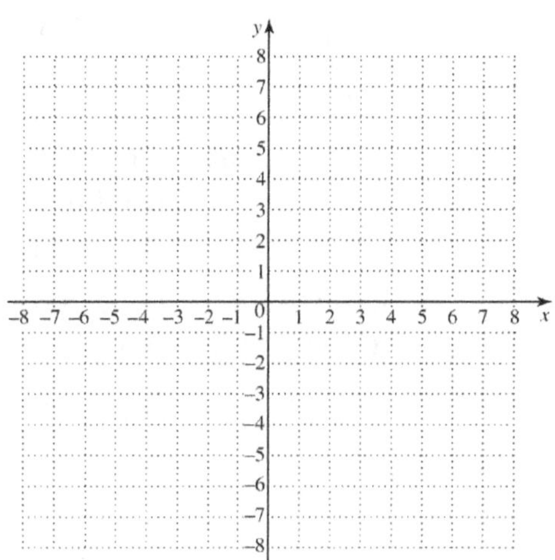

Objective 6   Use logarithmic functions in applications involving growth or decay.

Review this example for Objective 6:

Now Try:

7. A company analyst has found that total sales in thousands of dollars after a major advertising campaign are given by $S(x) = 100\log_2(x+2)$, where x is time in weeks after the campaign was introduced. Find the amount of sales two weeks after the campaign was introduced.

Two weeks after the campaign, $x = 2$, so we have

$$S(2) = 100\log_2(2+2)$$

$$S(2) = 100\log_2(2)^2$$

$$S(2) = 100 \cdot 2$$

$$S(2) = 200$$

Two weeks after the campaign, sales were $200,000.

7. The number of fish in an aquarium is given by the function $f(t) = 8\log_5(2t+5)$, where t is time in months. Find the number of fish present after 10 months.

____

Objective 6 Practice Exercises

For extra help, see Example 7 on page 882 of your text.

Solve each problem.

12. The population of foxes in an area t months after the foxes were introduced there is approximated by the function $F(t) = 500\log_{10}(2t+10)$. Find the number of foxes in the area when the foxes were first introduced into the area.

12. ___________________

13. A population of mites in a laboratory is growing according to the function
$p = 50\log_3(20t+7) - 25\log_9(80t+1)$, where t is the number of days after a study is begun. Find the number of mites present 1 day after the beginning of the study.

13. ___________________

14. Sales (in thousands) of a new product are
approximated by
$S = 125 + 20\log_2(30t + 4) + 30\log_4(35t - 6)$, where
t is the number of years after the product is
introduced. Find the total sales 2 years after the
product is introduced.

14. ______________

Chapter 12 INVERSE, EXPONENTIAL, AND LOGARITHMIC FUNCTIONS

12.4   Properties of Logarithms

Learning Objectives
1   Use the product rule for logarithms.
2   Use the quotient rule for logarithms.
3   Use the power rule for logarithms.
4   Use properties to write alternative forms of logarithmic expressions.

Key Terms

Use the vocabulary terms listed below to complete each statement in exercises 1−4.

   product rule for logarithms   **quotient rule for logarithms**

   power rule for logarithms   **special properties**

1.   The equations $b^{\log_b x} = x$, $x > 0$ and $\log_b b^x = x$ are referred to as

   ____________________ of logarithms.

2.   The equation $\log_b \dfrac{x}{y} = \log_b x - \log_b y$ is referred to as the ________________.

3.   The equation $\log_b xy = \log_b x + \log_b y$ is referred to as the ________________.

4.   The equation $\log_b x^r = r \log_b x$ is referred to as the ________________.

Objective 1   Use the product rule for logarithms.

Review these examples for Objective 1:

1.   Use the product rule to rewrite each logarithm. Assume $x > 0$.

   a. $\log_4 (6 \cdot 11)$

   Use the product rule.
   $\log_4 (6 \cdot 11) = \log_4 6 + \log_4 11$

   b. $\log_3 8 + \log_3 2$

   Use the product rule.
   $\log_3 8 + \log_3 2 = \log_3 (8 \cdot 2) = \log_3 16$

   c. $\log_2 (2x)$

   $\log_2 (2x) = \log_2 2 + \log_2 x$ Product rule
   $\qquad\quad = 1 + \log_2 x$ $\log_2 2 = 1$

Now Try:

1.   Use the product rule to rewrite each logarithm. Assume $x > 0$.

   a. $\log_6 (5 \cdot 3)$

   ____

   b. $\log_5 7 + \log_5 3$

   ____

   c. $\log_4 (4x)$

   ____

d. $\log_3 x^4$ **d.** $\log_5 x^4$

$$\log_3 x^4 = \log_3 (x \cdot x \cdot x \cdot x)$$
$$= \log_3 x + \log_3 x + \log_3 x + \log_3 x$$
$$= 4 \log_3 x$$

____

Objective 1 Practice Exercises

For extra help, see Example 1 on page 887 of your text.

Use the product rule to express each logarithm as a sum of logarithms.

1. $\log_7 5m$ 1. _____________________

2. $\log_2 6xy$ 2. _____________________

Use the product rule to express the sum as a single logarithm.

3. $\log_4 7 + \log_4 3$ 3. _____________________

Objective 2   Use the quotient rule for logarithms.

Review these examples for Objective 2: | **Now Try:**

2. Use the quotient rule to rewrite each logarithm. Assume $x > 0$.

     2. Use the quotient rule to rewrite each logarithm. Assume $x > 0$.

a. $\log_4 \dfrac{8}{7}$ **a.** $\log_5 \dfrac{4}{9}$

$$\log_4 \frac{8}{7} = \log_4 8 - \log_4 7$$

____

b. $\log_3 8 - \log_3 x$ **b.** $\log_6 x - \log_6 3$

$$\log_3 8 - \log_3 x = \log_3 \frac{8}{x}$$

____

c. $\log_2 \dfrac{16}{11}$ **c.** $\log_4 \dfrac{16}{11}$

$$\log_2 \frac{16}{11} = \log_2 16 - \log_2 11 = 4 - \log_2 11$$

____

d. $\log_5 54 - \log_5 6$ **d.** $\log_7 30 - \log_7 5$

$$\log_5 54 - \log_5 6 = \log_5 \frac{54}{6} = \log_5 9$$

____

Objective 2 Practice Exercises

For extra help, see Example 2 on page 887 of your text.

Use the quotient rule for logarithms to express each logarithm as a difference of logarithms, or as a single number if possible.

4. $\log_2 \dfrac{5}{m}$

4. _______________

5. $\log_6 \dfrac{k}{3}$

5. _______________

Use the quotient rule for logarithms to express the difference as a single logarithm.

6. $\log_2 7q^4 - \log_2 5q^2$

6. _______________

Objective 3   Use the power rule for logarithms.

Review these examples for Objective 3:

3. Use the power rule to rewrite each logarithm. Assume that $b > 0$, $x > 0$, and $b \neq 1$.

a. $\log_4 3^5$

$\log_4 3^5 = 5\log_4 3$

c. $\log_b \sqrt{11}$

$\log_b \sqrt{11} = \log_b 11^{1/2} = \dfrac{1}{2}\log_b 11$

d. $\log_2 \sqrt[5]{x^4}$

$\log_2 \sqrt[5]{x^4} = \log_2 x^{4/5} = \dfrac{4}{5}\log_2 x$

e. $\log_3 \dfrac{1}{x^5}$

$\log_3 \dfrac{1}{x^5} = \log_3 x^{-5} = -5\log_3 x$

Now Try:

3. Use the power rule to rewrite each logarithm. Assume that $b > 0$, $x > 0$, and $b \neq 1$.

a. $\log_6 4^3$

____

c. $\log_b \sqrt{13}$

____

d. $\log_3 \sqrt[4]{x^3}$

____

e. $\log_3 \dfrac{1}{x^7}$

____

Objective 3 Practice Exercises

For extra help, see Example 3 on page 889 of your text.

Use the power rule for logarithms to rewrite each logarithm or as a single number if possible.

7. $\log_m 2^7$ 7. _______________

8. $\log_3 \sqrt[3]{5}$ 8. _______________

9. $3^{\log_3 \sqrt[3]{7}}$ 9. _______________

Objective 4   Use properties to write alternative forms of logarithmic expressions.

Review these examples for Objective 4:

4. Use the properties of logarithms to rewrite each expression if possible. Assume that all variables represent positive real numbers.

a. $\log_5 6x^3$

$$\log_5 6x^3 = \log_5 6 + \log_5 x^3$$
$$= \log_5 6 + 3\log_5 x$$

b. $\log_b \sqrt{\dfrac{5}{x}}$

$$\log_b \sqrt{\dfrac{5}{x}} = \log_b \left(\dfrac{5}{x}\right)^{1/2}$$
$$= \dfrac{1}{2}\log_b \dfrac{5}{x}$$
$$= \dfrac{1}{2}(\log_b 5 - \log_b x)$$

d. $3\log_b x - \left(2\log_b y + \dfrac{1}{2}\log_b z\right)$

$$3\log_b x - \left(2\log_b y + \dfrac{1}{2}\log_b z\right)$$
$$= \log_b x^3 - (\log_b y^2 + \log_b z^{1/2})$$
$$= \log_b x^3 - \log_b y^2 \sqrt{z}$$
$$= \log_b \dfrac{x^3}{y^2 \sqrt{z}}$$

Now Try:

4. Use the properties of logarithms to rewrite each expression if possible. Assume that all variables represent positive real numbers.

a. $\log_6 36x^3$

____

b. $\log_b \sqrt{\dfrac{x}{3}}$

____

d. $\log_b x + 4\log_b y - \log_b z$

____

e. $2\log_3 x + \log_3(x-1) - \frac{1}{2}\log_3(x+1)$

$2\log_3 x + \log_3(x-1) - \frac{1}{2}\log_3(x+1)$

$\quad = \log_3 x^2 + \log_3(x-1) - \log_3(x+1)^{1/2}$

$\quad = \log_3\left(x^2(x-1)\right) - \log_3\sqrt{x+1}$

$\quad = \log_3 \dfrac{x^3 - x^2}{\sqrt{x+1}}$

f. $\log_2(2x+3y)$

$\log_2(2x+3y)$ cannot be rewritten using the properties of logarithms.

5. Given that $\log_2 9 \approx 3.1699$ and $\log_2 11 \approx 3.4594,$ use properties of logarithms to evaluate each expression.

a. $\log_2 99$

$\log_2 99 = \log_2(9\cdot 11)$

$\quad\quad = \log_2 9 + \log_2 11$

$\quad\quad = 3.1699 + 3.4594$

$\quad\quad = 6.6293$

b. $\log_2 \dfrac{1}{9}$

$\log_2 \dfrac{1}{9} = \log_2 1 - \log_2 9$

$\quad\quad = 0 - 3.1699$

$\quad\quad = -3.1699$

c. $\log_2 121$

$\log_2 121 = \log_2 11^2$

$\quad\quad = 2\log_2 11$

$\quad\quad = 2(3.4594)$

$\quad\quad = 6.9188$

e.

$2\log_2 x + \log_2(x-1) - \frac{1}{3}\log_2(x^2+1)$

____

f. $\log_3(3x-y)$

____

5. Given that $\log_2 9 \approx 3.1699$ and $\log_2 11 \approx 3.4594,$ use properties of logarithms to evaluate each experssion.

a. $\log_2 198$

____

b. $\log_2 \dfrac{1}{11}$

____

c. $\log_2 729$

____

6. Decide whether each statement is *true* or *false*.

   a. $\log_2 32 - \log_2 16 = \log_2 16$

Evaluate each side.
Left side:
$$\log_2 32 - \log_2 16 = \log_2 2^5 - \log_2 2^4$$
$$= 5 - 4\log_2 2$$
$$= 1$$
Right side:
$$\log_2 16 = \log_2 2^4 = 4$$
The statement is false because $1 \neq 4$.

   b. $\log_3(\log_4 64) = \dfrac{\log_{12} 144}{\log_6 36}$

Evaluate each side.
Left side:
$$\log_3(\log_4 64) = \log_3(\log_4 4^3)$$
$$= \log_3 3$$
$$= 1$$
Right side:
$$\frac{\log_{12} 144}{\log_6 36} = \frac{\log_{12} 12^2}{\log_6 6^2} = \frac{2}{2} = 1$$
The statement is true because $1 = 1$.

6. Decide whether each statement is *true* or *false*.

   a. $\log_3 27 + \log_3 9 = \log_5 5$

____

   b. $(\log_2 8)(\log_2 4) = \log_2 32$

____

For extra help, see Examples 4–6 on pages 889–891 of your text.

Use the properties of logarithms to express the sum or difference of logarithms as a single logarithm, or as a single number if possible.

10. $\log_4 10y + \log_4 3y - \log_4 6y^3$

10. _______________

Given that $\log_2 6 \approx 2.5850$ *and* $\log_2 12 \approx 3.5850$, *use properties of logarithms to evaluate the expression.*

11. $\log_2 72$

11. _______________

Decide whether the statement is true *or* false.

12. $\log_2 4p^3 = 6 + 3\log_2 p$

12. _______________

Chapter 12 INVERSE, EXPONENTIAL, AND LOGARITHMIC FUNCTIONS

12.5   Common and Natural Logarithms

Learning Objectives
1   Evaluate common logarithms using a calculator.
2   Use common logarithms in applications.
3   Evaluate natural logarithms using a calculator.
4   Use natural logarithms in applications.
5   Use the change-of-base rule.

Key Terms

Use the vocabulary terms listed below to complete each statement in exercises 1–2.

common logarithm   **natural logarithm**

1.   A logarithm to the base e is a _______________________________________.

2.   A logarithm to the base 10 is a _______________________________________.

Objective 1   Evaluate common logarithms using a calculator.

Review this example for Objective 1:	**Now Try:**
1. Using a calculator, evaluate the logarithm to four decimal places. $\log 436.2$ $\log 436.2 \approx 2.6397$	1. Using a calculator, evaluate the logarithm to four decimal places. $\log 983.5$ _______________

Objective 1 Practice Exercises

For extra help, see Example 1 on pages 893–894 of your text.

Use a calculator to find each logarithm. Give an approximation to four decimal places.

1. $\log 57.23$ 1. _______________

2. $\log 0.0914$ 2. _______________

3. $\log 87{,}123$ 3. _______________

          Copyright © 2025 Pearson Education, Inc.

Objective 2   Use common logarithms in applications.

Review these examples for Objective 2:

Now Try:

2. Wetlands are classified as bogs, fens, marshes, and swamps, on the basis of pH values. A pH value between 6.0 and 7.5 indicates that the wetland is a "rich fen." When the pH is between 3.0 and 6.0, the wetland is a "poor fen," and if the pH falls to 3.0 or less, it is a "bog." Suppose that the hydronium ion concentration of a sample of water from a wetland is 5.4×10^{-4}. Find the pH value for the water and determine how the wetland should be classified.

$$\text{pH} = -\log\left(5.4 \times 10^{-4}\right) \quad \text{Definition of pH}$$

$$= -\left(\log 5.4 + \log 10^{-4}\right) \quad \text{Product rule}$$

$$= -(0.7324 - 4) \quad \begin{array}{l}\text{Use a calculator to}\\\text{find } \log 5.4.\end{array}$$

$$= 3.2676$$

Since the pH is between 3.0 and 6.0, the wetland is a poor fen.

2. Suppose that the hydronium ion concentration of a sample of water from a wetland is 6.2×10^{-8}. Find the pH value for the water and determine how the wetland should be classified.

____

3. Find the hydronium ion concentration of a solution with pH 5.4.

$$\text{pH} = -\log\left[H_3O^+\right] \quad \text{Definition of pH}$$

$$5.4 = -\log\left[H_3O^+\right]$$

$$\log\left[H_3O^+\right] = -5.4 \quad \text{Multiply by } -1.$$

$$H_3O^+ = 10^{-5.4} \quad \begin{array}{l}\text{Write in}\\\text{exponential form.}\end{array}$$

$$\approx 4.0 \times 10^{-6} \quad \text{Use a calculator.}$$

3. Find the hydronium ion concentration of a solution with pH 3.6.

____

4. Find the decibel level to the nearest whole number of the sound with intensity I of $5.012 \times 10^{10} I_0$.

$$D = 10\log\left(\frac{I}{I_0}\right) = 10\log\left(\frac{5.012 \times 10^{10} I_0}{I_0}\right)$$

$$= 10\log\left(5.012 \times 10^{10}\right)$$

$$\approx 107 \text{ db}$$

4. Find the decibel level to the nearest whole number of the sound with intensity I of $3.16 \times 10^8 I_0$.

____

Objective 2 Practice Exercises

For extra help, see Examples 2–4 on pages 894–895 of your text.

Solve each problem.

4. Find the pH of a solution with the given hydronium ion concentration. Round the answer to the nearest tenth.

   a. 4.3×10^{-9} **b.** 2.8×10^{-6}

4. a._______________

   b._______________

5. Find the decibel level to the nearest whole number of the sound with intensity I of $2.5 \times 10^{13} I_0$.

5. _______________

6. Find the hydronium ion concentration of a solution with the given pH value.

   a. 5.2   **b.** 1.3

6. a._______________

   b._______________

Objective 3   Evaluate natural logarithms using a calculator.

Review this example for Objective 3:

5. Using a calculator, evaluate the logarithm to four decimal places.

   $\ln 436.2$

   $\ln 436.2 \approx 6.0781$

Now Try:

5. Using a calculator, evaluate the logarithm to four decimal places.

   $\ln 98$

   ____

Objective 3 Practice Exercises

For extra help, see Example 5 on page 896 of your text.

Find each natural logarithm. Give an approximation to four decimal places.

7. $\ln 76.3$

7. _______________

8. $\ln 0.102$

8. _______________

9. $\ln 50$

9. _______________

Objective 4   Use natural logarithms in applications.

Review this example for Objective 4:

6. The time t in years for an investment increasing at a rate of r percent (in decimal form) to double is given by

$$t = \frac{\ln 2}{\ln(1+r)}.$$

This is called the doubling time. Find the doubling time to the nearest tenth for an investment at 4%.

$4\% = 0.04$, so $t = \dfrac{\ln 2}{\ln(1+0.04)} = \dfrac{\ln 2}{\ln 1.04} \approx 17.7$

The doubling time for the investment is about 17.7 years.

Now Try:

6. Use the formula at the left to find the doubling time to the nearest tenth for an investment at 6%.

____

Objective 4 Practice Exercises

For extra help, see Example 6 on page 897 of your text.

The time t in years for an amount increasing at a rate of r (in decimal form) to double (the doubling time) is given by $t = \dfrac{\ln 2}{\ln(1+r)}$. *Find the doubling time for an investment at the interest rate. Round to the nearest whole number.*

10.   3%

10. _______________

The half-life of a radioactive substance is the time it takes for half of the material to decay. The amount A in pounds of substance remaining after t years is given by $\ln \dfrac{A}{C} = -\dfrac{t}{h} \ln 2$, *where C is the initial amount in pounds, and h is its half-life in years. Use the formula to solve the following problems. Round to the nearest whole number.*

11.   The half-life of radium-226 is 1620 years. How long, to the nearest year, will it take for 100 pounds to decay to 25 pounds?

11. _______________

Newton's Law of Cooling describes the cooling of a warmer object to the cooler temperature of the surrounding environment. The formula can be given as

$$t = \frac{1}{k} \ln \frac{T_s - T_1}{T_s - T_2},$$ *where t is the elapsed time, T_1 is the initial temperature measurement of the object, T_2 is the second temperature measurement of the object, and T_s is the temperature of the surrounding environment. Use this formula to solve the problem. Round to the nearest tenth.*

12.   A corpse was discovered in a motel room at midnight and its temperature was 80°F. The temperature in the room was 60°F. Assuming that the person's temperature at the time of death was 98.6° F and using $k = 0.1438$, determine t and the time of death.

12. t _________________

time _________________

Objective 5   Use the change-of-base rule.

Review this example for Objective 5:

7.   Use the change-of-base rule to approximate $\log_7 28$ to four decimal places.

$$\log_7 28 = \frac{\log 28}{\log 7} = 1.7124$$

Now Try:

7.   Use the change-of-base rule to approximate $\log_5 180$ to four decimal places.

____

Objective 5 Practice Exercises

For extra help, see Example 7 on page 898 of your text.

Use the change-of-base rule to find each logarithm. Give approximations to four decimal places.

13. $\log_{16} 27$

13. _________________

14. $\log_6 0.25$

14. _________________

15. $\log_{1/2} 5$

15. _________________

Chapter 12 INVERSE, EXPONENTIAL, AND LOGARITHMIC FUNCTIONS

12.6   Exponential and Logarithmic Equations; Further Applications

Learning Objectives
1   Solve equations involving variables in the exponents.
2   Solve equations involving logarithms.
3   Solve applications of compound interest.
4   Solve applications involving base e exponential growth and decay.

Key Terms

Use the vocabulary terms listed below to complete each statement in exercises 1–2.

compound interest   **continuous compounding**

1.   The formula for ___________________________ is $A = Pe^{rt}$.

2.   The formula for ___________________________ is $A = P\left(1+\dfrac{r}{n}\right)^{nt}$

Objective 1   Solve equations involving variables in the exponents.

Review these examples for Objective 1:

1.   Solve $4^x = 30$. Approximate the solution to three decimal places.

$$4^x = 30$$

$$\log 4^x = \log 30 \quad \begin{array}{l}\text{If } x = y, \text{ and } x > 0,\ y > 0, \\ \text{then } \log_b x = \log_b y.\end{array}$$

$$x \log 4 = \log 30 \quad \text{Power rule}$$

$$x = \frac{\log 30}{\log 4} \quad \text{Divide by log 4.}$$

$$x \approx 2.453 \quad \text{Use a calculator.}$$

Check

$$4^x = 4^{2.453} \approx 30$$

The solution set is $\{2.453\}$.

Now Try:

1.   Solve $3^x = 15$. Approximate the solution to three decimal places.

____

2. Solve $e^{0.005x} = 9$. Approximate the solution to three decimal places.

$$e^{0.005x} = 9$$

$$\ln e^{0.005x} = \ln 9 \qquad \text{If } x = y, \text{ and } x > 0,$$
$$y > 0, \text{ then } \ln x = \ln y.$$

$$0.005x \ln e = \ln 9 \qquad \text{Power rule}$$

$$0.005x = \ln 9 \qquad \ln e = 1$$

$$x = \frac{\ln 9}{0.005} \qquad \text{Divide by 0.005.}$$

$$x \approx 439.445 \quad \text{Use a calculator.}$$

The solution set is $\{439.445\}$.

2. Solve $e^{0.4x} = 15$. Approximate the solution to three decimal places.

____

Objective 1 Practice Exercises

For extra help, see Examples 1–3 on pages 901–902 of your text.

Solve each equation. Approximate solutions to three decimal places.

1. $25^{x+2} = 125^{3-x}$

1. ______________

2. $4^{x-1} = 3^{2x}$

2. ______________

3. $e^{-0.45x} = 7$

3. ______________

Name:

Instructor:

Date:

Section:

Objective 2   Solve equations involving logarithms.

Review these examples for Objective 2:

Now Try:

5.   Solve $\log_3 (x-1)^3 = 5$. Give the exact solution.

$$\log_3 (x-1)^3 = 5$$

$$(x-1)^3 = 3^5 \qquad \text{Write in exponential form.}$$

$$(x-1)^3 = 243$$

$$x-1 = \sqrt[3]{243} \qquad \text{Take the cube root of each side.}$$

$$x-1 = 3\sqrt[3]{9} \qquad \text{Simplify the cube root.}$$

$$x = 1 + 3\sqrt[3]{9} \quad \text{Add 1.}$$

Check:

$$\log_3 (x-1)^3 = 5$$

$$\log_3 \left(1 + 3\sqrt[3]{9} - 1\right)^3 \overset{?}{=} 5$$

$$\log_3 \left(\sqrt[3]{243}\right)^3 \overset{?}{=} 5$$

$$\log_3 (243) \overset{?}{=} 5$$

$$\log_3 3^5 \overset{?}{=} 5$$

$$5 = 5$$

The solution set is $\left\{1 + 3\sqrt[3]{9}\right\}$.

5.   Solve $\log_6 (x+1)^3 = 2$. Give the exact solution.

____

6.   Solve $\log_3 (5x+42) - \log_3 x = \log_3 26$.

$$\log_3 (5x+42) - \log_3 x = \log_3 26.$$

$$\log_3 \frac{5x+42}{x} = \log_3 26$$

$$\frac{5x+42}{x} = 26 \qquad \text{If } \log_b x = \log_b y \text{ then } x = y.$$

$$5x + 42 = 26x \quad \text{Multiply by } x.$$

$$42 = 21x \quad \text{Subtract } 5x.$$

$$2 = x \quad \text{Divide by 21.}$$

6.   Solve
$$\log_6 (2x+7) - \log_6 x = \log_6 16.$$

____

Check:

$$\log_3 (5x+42) - \log_3 x = \log_3 26$$

$$\log_3 (5\cdot 2+42) - \log_3 2 \overset{?}{=} \log_3 26$$

$$\log_3 52 - \log_3 2 \overset{?}{=} \log_3 26$$

$$\log_3 \tfrac{52}{2} \overset{?}{=} \log_3 26$$

$$\log_3 26 = \log_3 26$$

The solution set is $\{2\}$.

7. Solve $\log_2 (x+7) + \log_2 (x+3) = \log_2 77$.

$$\log_2 (x+7) + \log_2 (x+3) = \log_2 77$$

$$\log_2 [(x+7)(x+3)] = \log_2 77$$

Product rule

$$(x+7)(x+3) = 77$$

If $\log_b x = \log_b y$
then $x = y$.

$$x^2 + 10x + 21 = 77 \quad \text{Multiply.}$$

$$x^2 + 10x - 56 = 0 \quad \text{Subtract 77.}$$

$$(x-4)(x+14) = 0 \quad \text{Factor.}$$

$$x - 4 = 0 \quad \text{or} \quad x + 14 = 0$$

$$x = 4 \qquad\qquad x = -14$$

The value -14 must be rejected since it leads to the logarithm of a negative number in the original equation.
A check shows that the only solution is 4.
The solution set is $\{4\}$.

7. Solve.
$$\log_4 (4x-3) + \log_4 x = \log_4 (2x-1)$$

____

Objective 2 Practice Exercises

For extra help, see Examples 4–7 on pages 902–904 of your text.

Solve each equation. Give exact solution.

4. $\log(-a) + \log 4 = \log(2a+5)$

4. ______________

5. $\log_3(x^2 - 10) - \log_3 x = 1$

5. _______________

6. $\ln(x+4) + \ln(x-2) = \ln 7$

6. _______________

Objective 3   Solve applications of compound interest.

Review these examples for Objective 3:

8. How much money will be in an account at the end of 5 years if $5000 is deposited at 4% compounded monthly?

Because interest is compounded monthly, $n = 12$. The other given values are $P = 5000$, $r = 0.04$, and $t = 5$.

$$A = P\left(1 + \frac{r}{n}\right)^{nt}$$

$$A = 5000\left(1 + \frac{0.04}{12}\right)^{12 \cdot 5}$$

$$A = 5000(1.0033)^{60}$$

$$A = 6104.98$$

There will be $6104.98 in the account at the end of 5 years.

9. Approximate the time it would take for money deposited in an account paying 5% interest compounded quarterly to double. Round to the nearest hundredth.

We want the number of years t for P dollars to grow to $2P$ dollars at a rate of 5% per year. In the compound interest formula, we substitute $2P$ for A, and let $r = 0.05$ and $n = 4$.

Now Try:

8. How much money will be in an account at the end of 5 years if $10,000 is deposited at 4% compounded quarterly?

____

9. Approximate the time it would take for money deposited in an account paying 5% interest compounded monthly to double. Round to the nearest hundredth.

____

$$2P = P\left(1 + \frac{0.05}{4}\right)^{4t}$$

$$2 = 1.0125^{4t}$$

$$\log 2 = \log 1.025^{4t}$$

$$\log 2 = 4t \log 1.0125$$

$$t = \frac{\log 2}{4 \log 1.0125}$$

$$t \approx 13.95$$

It will take about 13.95 years for the investment to double.

10. Suppose that $5000 is invested at 4% interest for 3 years.

a. How much will the investment be worth if it is compounded continuously?

$$A = Pe^{rt}$$

$$A = 5000e^{0.04 \cdot 3}$$

$$A = 5000e^{0.12}$$

$$A = 5637.48$$

The investment will be worth $5637.48.

b. Approximate the amount of time it would take for the investment to double. Round to the nearest tenth.

Find the value of t that will cause A to be $2(\$5000) = \$10,000$.

$$A = Pe^{rt}$$

$$10,000 = 5000e^{0.04t}$$

$$2 = e^{0.04t} \qquad \text{Divide by 5000.}$$

$$\ln 2 = \ln e^{0.04t} \qquad \begin{array}{l}\text{If } x = y, \\ \text{then } \ln x = \ln y.\end{array}$$

$$\ln 2 = 0.04t \qquad \ln e^k = k$$

$$\frac{\ln 2}{0.04} = t \qquad \text{Divide by 0.04.}$$

$$t \approx 17.3$$

It will take about 17.3 years for the amount to double.

10. Suppose that $5000 is invested at 2% interest for 3 years.

a. How much will the investment be worth if it is compounded continuously?

____

b. Approximate the amount of time it would take for the investment to double. Round to the nearest tenth.

____

Objective 3 Practice Exercises

For extra help, see Examples 8–10 on pages 905–906 of your text.

Solve each problem.

7. How much will be in an account after 10 years if $25,000 is invested at 8% compounded quarterly? Round to the nearest cent.

7. ___________________

8. How much will be in an account after 5 years if $10,000 is invested at 4.5% compounded continuously? Round to the nearest cent.

8. ___________________

9. How long will it take an investment to double if it is placed in an account paying 9% interest compounded continuously? Round to the nearest tenth.

9. ___________________

Objective 4   Solve applications involving base e exponential growth and decay.

Review these examples for Objective 4:

11. A sample of 500 g of lead-210 decays according to the function $y = y_0 e^{-0.032t}$, where t is the time in years, y is the amount of the sample at time t, and y_0 is the initial amount present at $t = 0$.

a. How much lead will be left in the sample after 20 years? Round to the nearest tenth of a gram.

Let $t = 20$ and $y_0 = 500$.

$$y = 500e^{-0.032 \cdot 20} \approx 263.6$$

There will be about 263.6 grams after 20 years.

Now Try:

11. Cesium-137, a radioactive isotope used in radiation therapy, decays according to the function $y = y_0 e^{-0.0231t}$, where t is the time in years and y_0 is the initial amount present at $t = 0$.

a. If an initial sample contains 36 mg of cesium-137, how much cesium-137 will be left in the sample after 50 years? Round to the nearest tenth.

____

b. Approximate the half-life of lead-210 to the nearest tenth.

Let $y = \frac{1}{2}(500) = 250.$

$$250 = 500e^{-0.032t}$$

$$0.5 = e^{-0.032t}$$

$$\ln 0.5 = \ln e^{-0.032t}$$

$$\ln 0.5 = -0.032t$$

$$t = \frac{\ln 0.5}{-0.032} \approx 21.7$$

The half-life of lead-210 is about 21.7 years.

b. Approximate the half-life of cesium-137 to the nearest tenth.

____

Objective 4 Practice Exercises

For extra help, see Example 11 on page 907 of your text.

Solve each problem.

10. Radioactive strontium decays according to the function $y = y_0 e^{-0.0239t}$, where t is the time in years. If an initial sample contains $y_0 = 15$ g of radioactive strontium, how many grams will be present after 25 years? Round to the nearest hundredth of a gram.

10. _________________

11. How long will it take the initial sample of strontium in exercise 19 to decay to half of its original amount?

11. _________________

12. The concentration of a drug in a person's system decreases according to the function $C(t) = 2e^{-0.2t}$, where $C(t)$ is given in mg and t is in hours. How much of the drug will be in the person's system after one hour? Approximate answer to the nearest hundredth.

12. _________________

Chapter 13 NONLINEAR FUNCTIONS, CONIC SECTIONS, AND NONLINEAR SYSTEMS

13.1   Additional Graphs of Functions

Learning Objectives
1   Graph absolute value, reciprocal, and square root functions.
2   Graph step functions.

Key Terms

Use the vocabulary terms listed below to complete each statement in exercises 1–6.

absolute value function	**reciprocal function**	**asymptotes**
square root function	**greatest integer function**	**step function**

1.   A _________________________________ is a function that looks like a series of steps.

2.   The function defined by $f(x) = [\![x]\!]$ is called the _____________________________.

3.   The function defined by $f(x) = \sqrt{x}$ is called the _____________________________.

4.   The function defined by $f(x) = \dfrac{1}{x}$ is called the _____________________________.

5.   Lines that a graph approaches without actually touching are called

   _____________________.

6.   The function defined by $f(x) = |x|$ is called the _____________________________.

Objective 1   Graph absolute value, reciprocal, and square root functions.

Review these examples for Objective 1:

1.   Graph $f(x) = |x + 4|$. Give the domain and range.

The graph of $f(x) = |x + 4|$ is found by shifting the graph of $y = |x|$ four units to the left.

x	y
-6	2
-5	1
-4	0
-3	1
-2	2

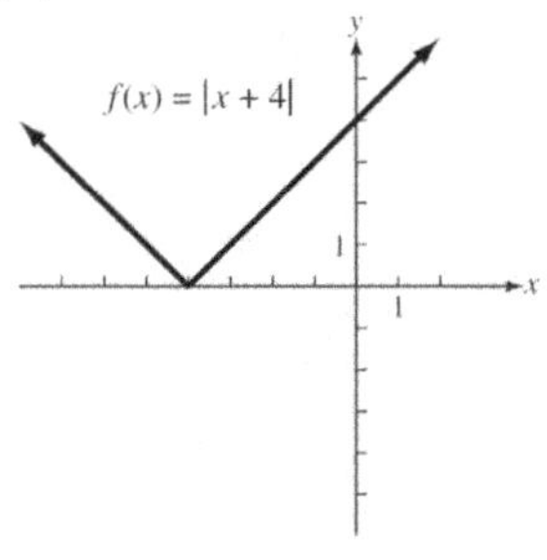

The domain is $(-\infty,\ \infty)$. The range is $[0,\ \infty)$.

Now Try:

1.   Graph $f(x) = \sqrt{x + 3}$. Give the domain and range.

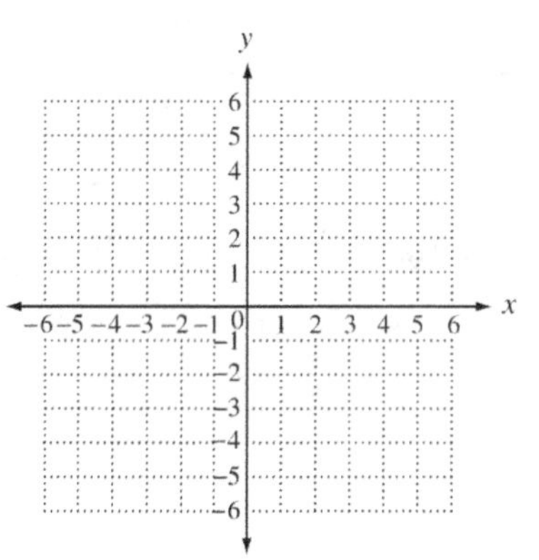

2. Graph $f(x)=\sqrt{x}-2$. Give the domain and range.

The graph of $y=\sqrt{x}-2$ is obtained by shifting the graph of $y=\sqrt{x}$ two units down.

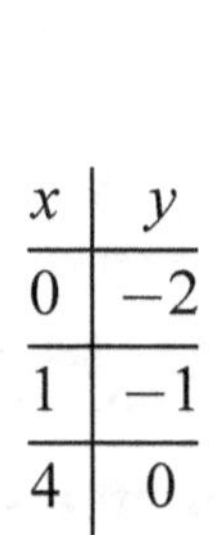

x	y
0	-2
1	-1
4	0

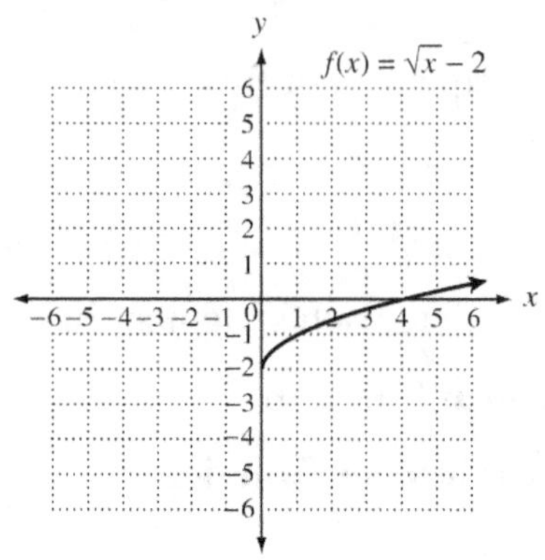

The domain is $[0,\infty)$. The range is $[-2,\infty)$.

3. Graph $f(x)=\dfrac{1}{x+2}-1$. Give the domain and range.

The graph of $y=\dfrac{1}{x+2}-1$ is obtained by shifting the graph of $y=\dfrac{1}{x}$ two units to the left and then one unit down.

x	y	x	y
-6	$-\dfrac{5}{4}$	$-\dfrac{3}{2}$	1
-5	$-\dfrac{4}{3}$	-1	0
-4	$-\dfrac{3}{2}$	0	$-\dfrac{1}{2}$
-3	-2	1	$-\dfrac{2}{3}$
$-\dfrac{5}{2}$	-3	2	$-\dfrac{3}{4}$

The domain is $(-\infty,-2)\cup(-2,\infty)$.
The range is $(-\infty,-1)\cup(-1,\infty)$.

2. Graph $f(x)=|x|-1$. Give the domain and range.

____

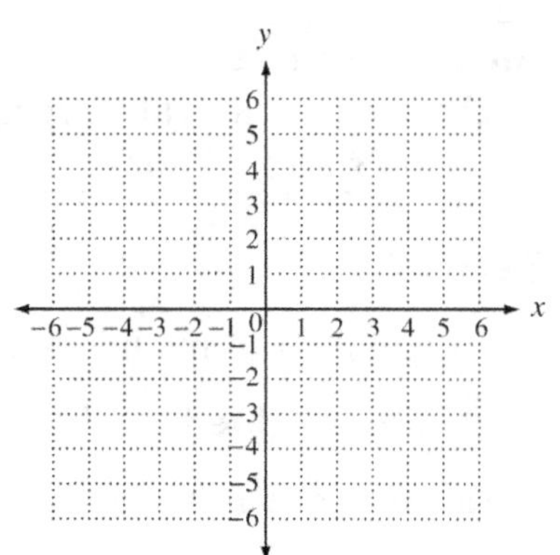

3. Graph $f(x)=\dfrac{1}{x-1}-2$. Give the domain and range.

____

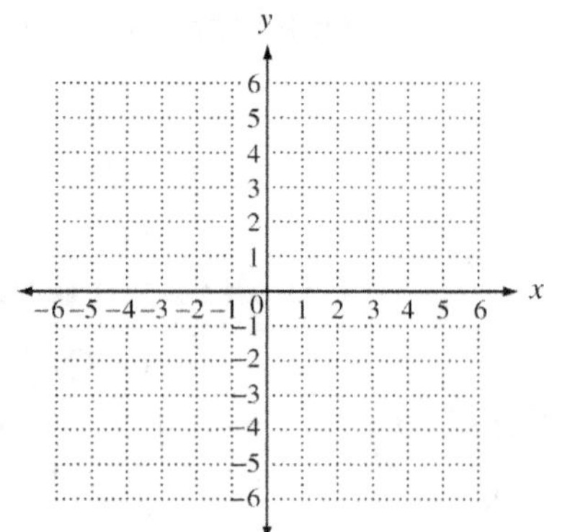

Objective 1 Practice Exercises

For extra help, see Examples 1–3 on pages 925–926 of your text.

Graph each function. Give the domain and range.

1. $f(x) = |x-2| + 3$

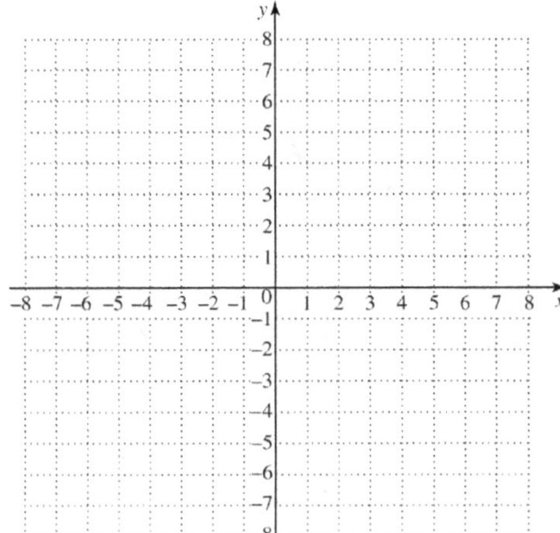

    1. domain _______________

       range _______________

2. $f(x) = \sqrt{x} + 3$

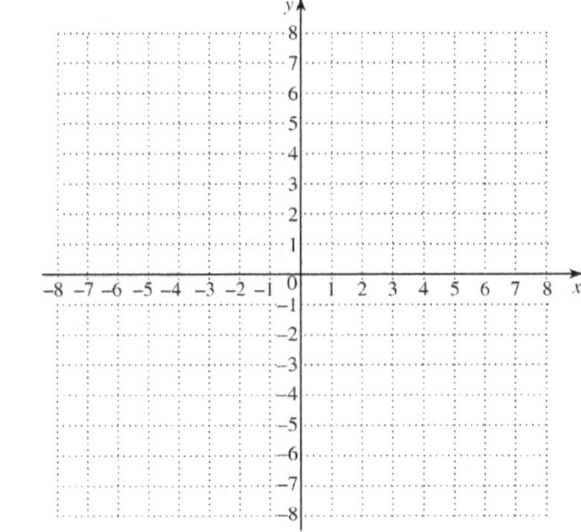

    2. domain _______________

       range _______________

3. $f(x) = \dfrac{1}{x-3}$

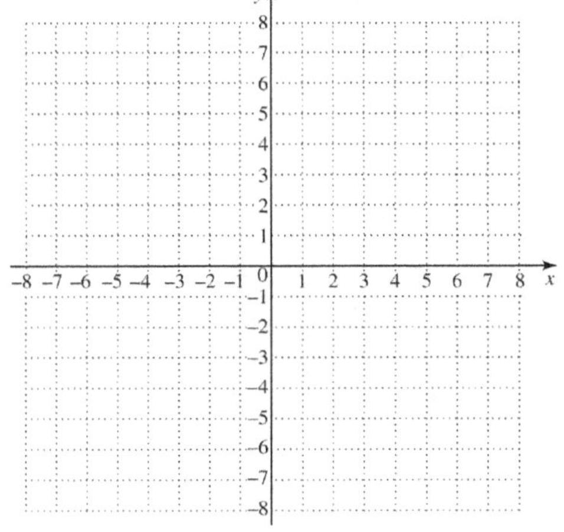

    3. domain _______________

       range _______________

Objective 2   Graph step functions.

Review these examples for Objective 2:

4. Evaluate each expression.

    d. $\left[\!\left[\dfrac{3}{5}\right]\!\right]$

$$\left[\!\left[\dfrac{3}{5}\right]\!\right]=0$$

    e. $[\![9.7]\!]$

$$[\![9.7]\!]=9$$

    f. $[\![-3.5]\!]$

$$[\![-3.5]\!]=-4$$

5. Graph $f(x)=[\![x]\!]+2.$ Give the domain and range.

The graph of $y=[\![x]\!]+2$ is obtained by shifting the graph of $y=[\![x]\!]$ two units up.

If $-3\le x<-2$, then $[\![x]\!]+2=-3+2=-1.$
If $-2\le x<-1$, then $[\![x]\!]+2=-2+2=0.$
If $-1\le x<0$, then $[\![x]\!]+2=-1+2=1.$
If $0\le x<1$, then $[\![x]\!]+2=0+2=2$, etc.

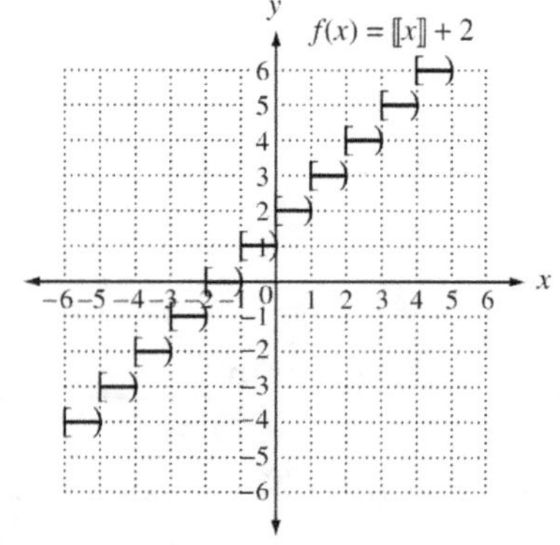

The domain is $(-\infty,\ \infty)$. The range is $\{...,-2,-1,\ 0,\ 1,\ 2,...\}$ (the set of integers)

Now Try:

4. Evaluate each expression.

    d. $\left[\!\left[\dfrac{5}{9}\right]\!\right]$

____

    e. $[\![1.6]\!]$

____

    f. $[\![-10.1]\!]$

____

5. Graph $f(x)=[\![x-2]\!].$ Give the domain and range.

____

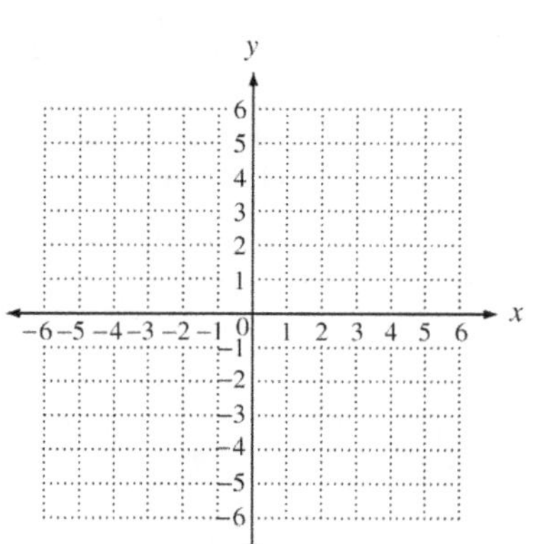

                     Copyright © 2025 Pearson Education, Inc.

6. In 2011, U.S. first class letter postage is 46¢ for the first ounce and 18¢ for each additional ounce. Let $p(x)$ = the cost of sending a letter that weighs x ounces. Graph the function over the interval (0, 6].

For x in the interval (0, 1], $y = \$0.46$.
For x in the interval (1, 2],
   $y = \$0.46 + \$0.18 = \$0.64$.
For x in the interval (2, 3],
   $y = \$0.64 + \$0.18 = \$0.82$.
For x in the interval (3, 4],
   $y = \$0.82 + \$0.18 = \$1.00$.
For x in the interval (4, 5],
   $y = \$1.00 + \$0.18 = \$1.18$.
For x in the interval (5, 6],
   $y = \$1.18 + \$0.18 = \$1.36$.

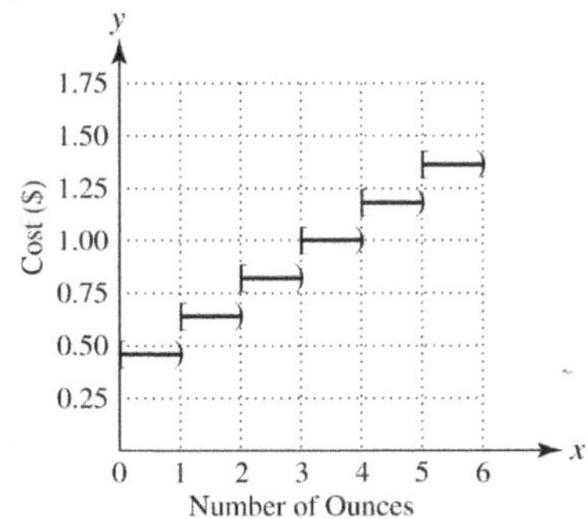

6. At the end of December, Mary's dad lent her $500 to buy an iPad. She agreed to repay the loan at $50 per month on the first of each month, starting in January. If $f(x)$ represents the amount to be repaid in month x, graph the function over its entire domain.

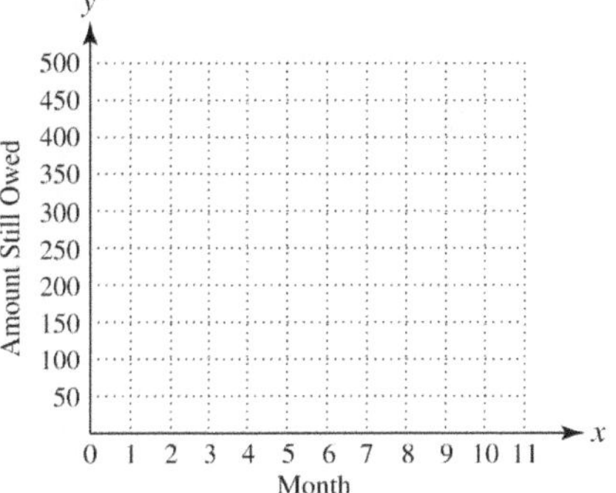

Objective 2 Practice Exercises

For extra help, see Examples 4–6 on pages 926–928 of your text.

Graph each function.

4. $f(x) = [\![x + 2]\!] - 3$

4.

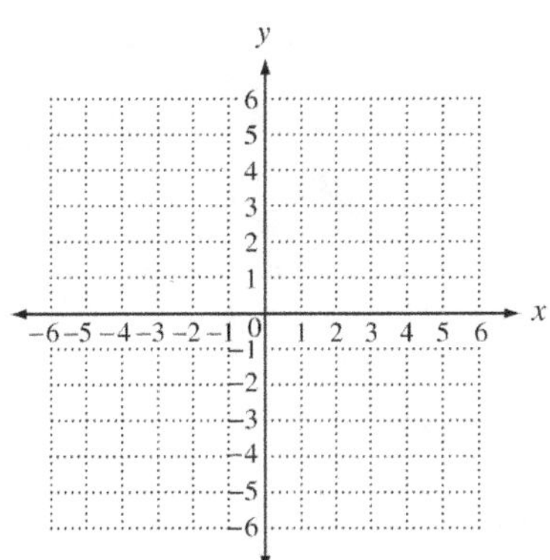

5. The cost of parking a car in an hourly parking lot is \$6.00 for the first hour and \$4.00 for each additional hour or fraction of an hour. Graph the function f that models the cost of parking a car for x hours over the interval $(0, 6]$.

5.

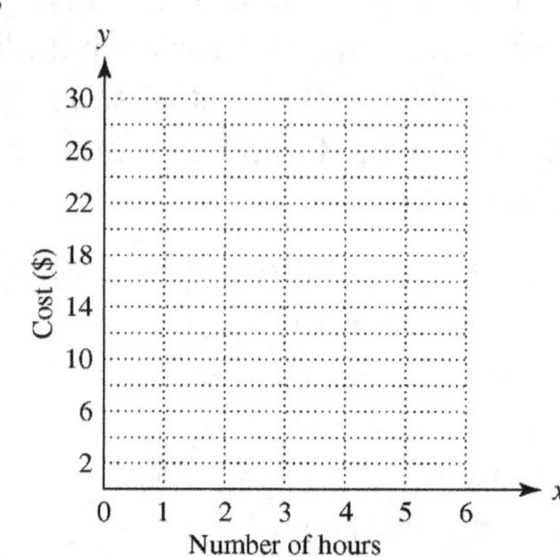

Chapter 13 NONLINEAR FUNCTIONS, CONIC SECTIONS, AND NONLINEAR SYSTEMS

13.2   Circles Revisited and Ellipses

Learning Objectives
1   Graph circles.
2   Write an equation of a circle given its center and radius.
3   Determine the center and radius of a circle given its equation.
4   Recognize the equation of an ellipse.
5   Graph ellipses.

Key Terms

Use the vocabulary terms listed below to complete each statement in exercises 1–8.

conic sections	**circle**	**center (of a circle)**	**radius**
center-radius form	**ellipse**	**foci**	**center (of an ellipse)**

1. A(n) ___________________________ is the set of all points in a plane that lie a fixed distance from a fixed point.

2. A(n) ___________________________ is the set of all points in a plane the sum of whose distances from two fixed points is constant.

3. Figures that result from the intersection of an infinite cone with a plane are called

    ___________________________________.

4. A fixed point such that every point on a circle is a fixed distance from it is the

    ___________________________.

5. The distance from the center of a circle to a point on the circle is called the

    ___________________________.

6. The ___________________ of the equation of a circle with center (h, k) and radius r is $(x-h)^2 + (y-k)^2 = r^2$.

7. ___________________ are fixed points used to determine the points that form a parabola, and ellipse, or a hyperbola.

8. The ___________________ of the ellipse defined by $\dfrac{(x-3)^2}{9} + \dfrac{(y+2)^2}{16} = 1$ is $(3, -2)$.

Name: _______________________   Date: _______________________

Instructor: _______________________   Section: _______________________

Objective 1   Graph circles.

Review these examples for Objective 1:

1. Find the center and radius of each circle. Then graph the circle.

   a. $x^2 + y^2 = 4$

   $(x-0)^2 + (y-0)^2 = 2^2$
   Here, $h = 0$, and $k = 0$.
   The center is (0, 0) and the radius is 2.

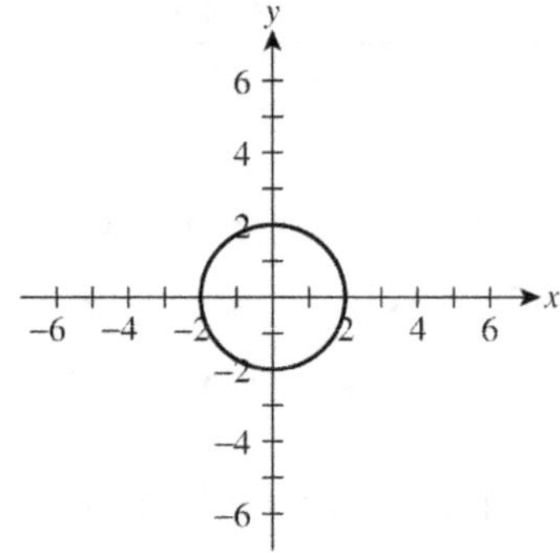

   b. $(x+3)^2 + (y-2)^2 = 9$

   $[x-(-3)]^2 + (y-2)^2 = 3^2$
   The center (h, k) is (−3, 2) and the radius is 3.

   To graph the circle, plot the center (−3, 2), then move three units right, left, up, and down from the center, plotting the points (0, 2), (−3, 5), (−6, 2), and (−3, −1). Draw a smooth curve through the points.

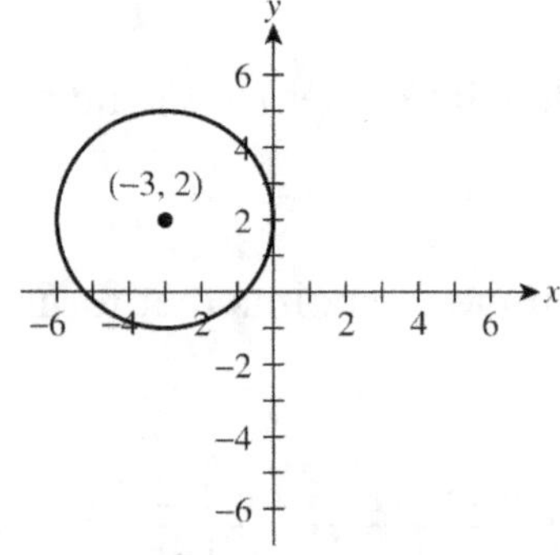

Now Try:

1. Find the center and radius of each circle. Then graph the circle.

   a. $x^2 + y^2 = 25$

   Center: _______________

   Radius: _______________

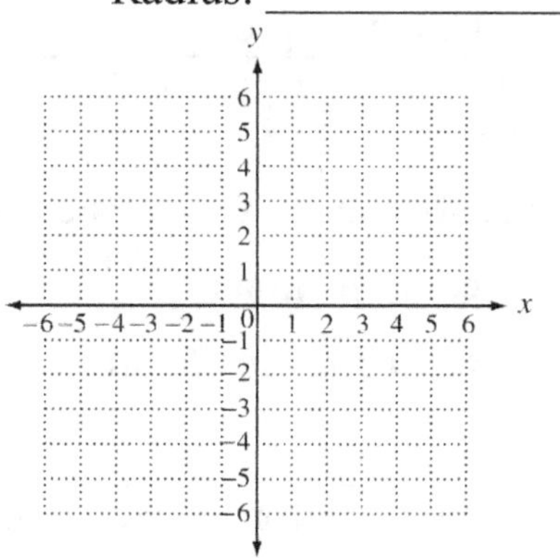

   b. $(x+5)^2 + (y-4)^2 = 16$

   Center: _______________

   Radius: _______________

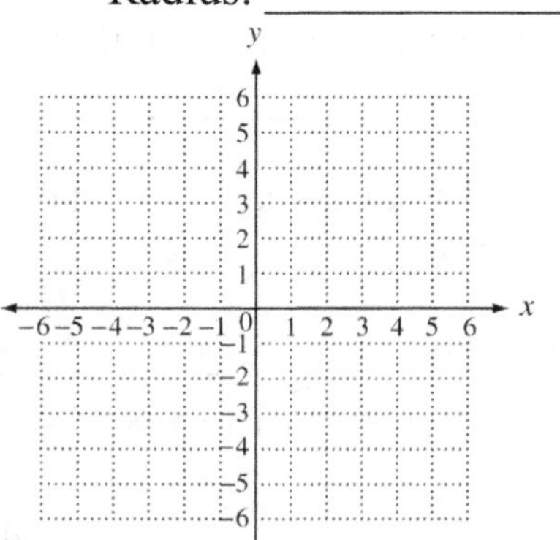

Name: _______________________ Date: _______________________

Instructor: _______________________ Section: _______________________

Objective 1 Practice Exercises

For extra help, see Example 1 on page 932 of your text.

Find the center and radius of each circle. Then graph the circle.

1. $(x-1)^2 + (y-4)^2 = 4$

1. Center: _______________________

Radius: _______________________

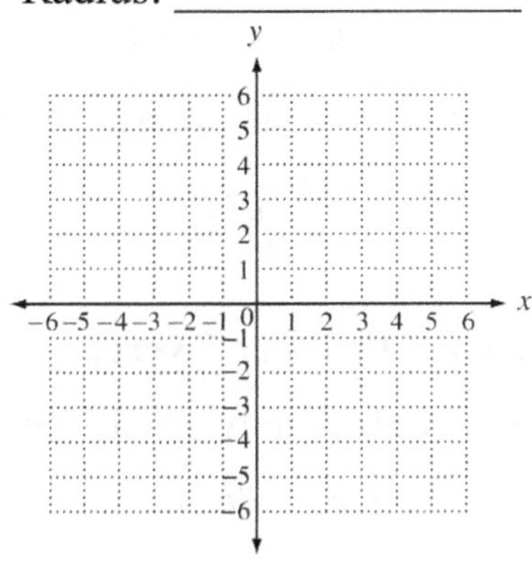

2. $(x-2)^2 + (y+3)^2 = 25$

2. Center: _______________________

Radius: _______________________

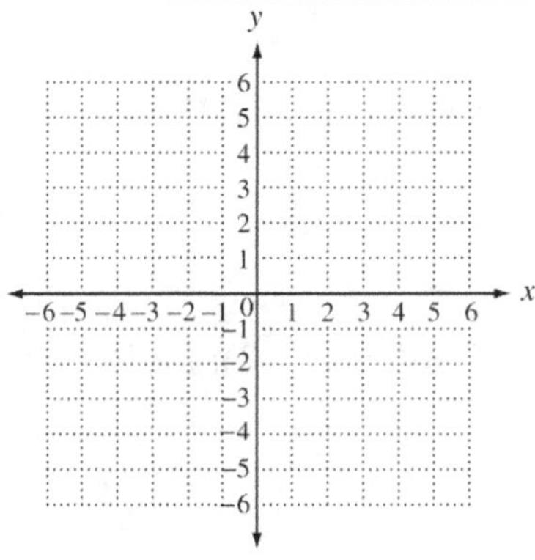

3. $x^2 + (y-5)^2 = 9$

3. Center: _______________________

Radius: _______________________

Objective 2   Write an equation of a circle given its center and radius.

Review this example for Objective 2:

2. Find an equation of the circle with center $(-2,-4)$ and radius $2\sqrt{5}$.

Here $(h, k) = (-2, -4)$ and $r = 2\sqrt{5}$.
$$(x-h)^2 + (y-k)^2 = r^2$$
$$[x-(-2)]^2 + [y-(-4)]^2 = (2\sqrt{5})^2$$
$$(x+2)^2 + (y+4)^2 = 20$$

Now Try:

2. Find an equation of the circle with center $(0, 3)$ and radius $\sqrt{2}$.

____

Objective 2 Practice Exercises

For extra help, see Example 2 on page 933 of your text.

Write the center-radius form of each circle described.

4.   Center: $(-1, 3)$; radius: $\sqrt{7}$

4. ________________

5.   Center: $(4, 0)$; radius: $\sqrt{11}$

5. ________________

6.   Center: $(5,-2)$; radius: $\sqrt{18}$

6. ________________

Objective 3   Determine the center and radius of a circle given its equation.

Review this example for Objective 3:

3. Find the center and radius of the circle.

$$x^2 + y^2 + 8x + 4y - 29 = 0$$

To find the center and radius, complete the squares on x and y.

$$x^2 + y^2 + 8x + 4y - 29 = 0$$

$$x^2 + y^2 + 8x + 4y = 29$$

$$\left(x^2 + 8x \quad\right) + \left(y^2 + 4y \quad\right) = 29$$

$$\left[\tfrac{1}{2}(8)\right]^2 = 16 \quad \left[\tfrac{1}{2}(4)\right]^2 = 4$$

$$(x^2 + 8x + 16) + (y^2 + 4y + 4) = 29 + 16 + 4$$

$$(x+4)^2 + (y+2)^2 = 49$$

$$\left[x - (-4)\right]^2 + \left[y - (-2)\right]^2 = 7^2$$

The center is $(-4, -2)$ and the radius is 7.

Now Try:

3. Find the center and radius of the circle.

$$x^2 + y^2 - 8x - 2y + 15 = 0$$

Center: _______________

Radius: _______________

Objective 3 Practice Exercises

For extra help, see Example 3 on page 934 of your text.

Find the center and radius of each circle.

7. $x^2 + y^2 - 4x + 8y + 11 = 0$

7.

Center: _______________

Radius: _______________

8. $x^2 + y^2 - 2x - 6y - 15 = 0$

8.

Center: _______________

Radius: _______________

9. $3x^2 + 3y^2 + 12y + 30x = 21$

9.

Center: _______________

Radius: _______________

Objective 4   Recognize the equation of an ellipse.

For extra help, see pages 934–935 of your text.

Objective 5   Graph ellipses.

Review these examples for Objective 5:

4.   Graph $\dfrac{x^2}{25}+\dfrac{y^2}{9}=1$.

$a^2=25$, so $a=5$, and
   the x-intercepts are $(5, 0)$ and $(-5, 0)$.
$b^2=9$, so $b=3$, and
   the y-intercepts are $(0, 3)$ and $(0, -3)$.

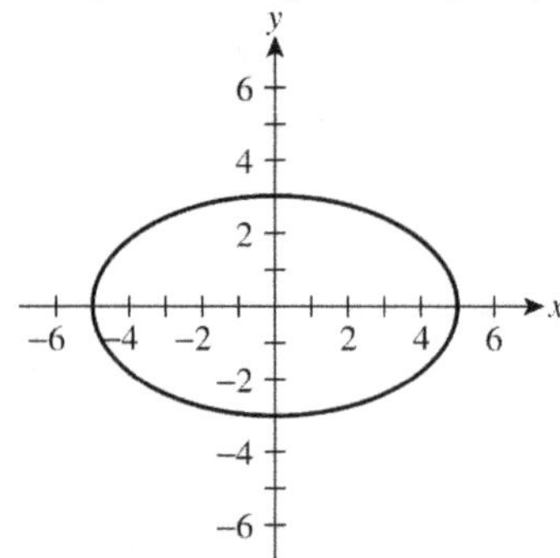

6.   Graph $\dfrac{(x-1)^2}{9}+\dfrac{(y+2)^2}{4}=1$.

The center is $(1, -2)$, $a=3$, and $b=2$. The
ellipse passes through the points $(4, -2)$,
$(1, -4)$, $(-2, -2)$, and $(1, 0)$.

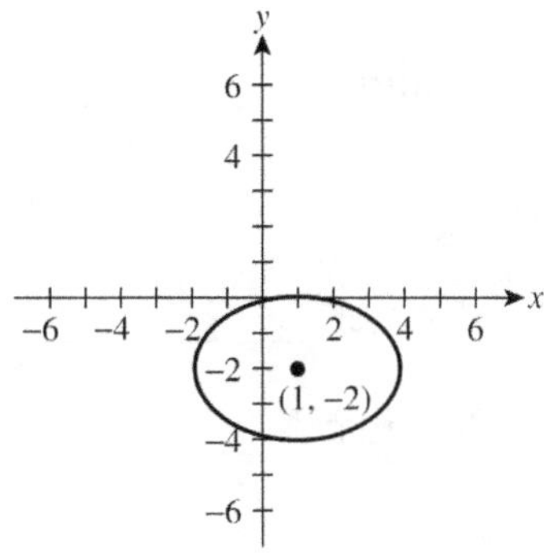

Now Try:

4.   Graph $\dfrac{x^2}{16}+\dfrac{y^2}{9}=1$.

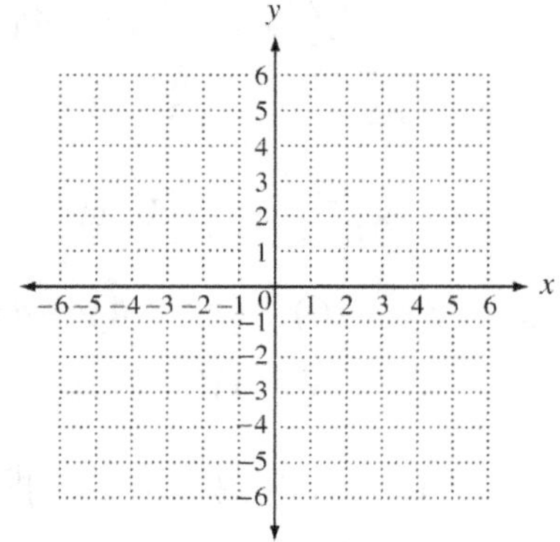

6.   Graph $\dfrac{(x+1)^2}{4}+\dfrac{(y-2)^2}{9}=1$.

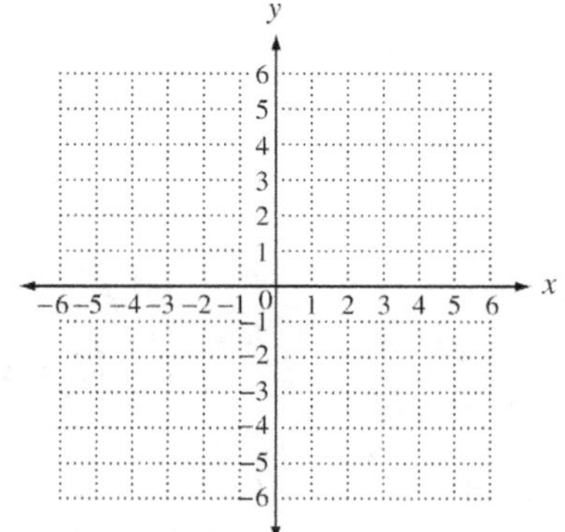

Objective 5 Practice Exercises

For extra help, see Examples 4–6 on pages 935–936 of your text.

Graph each ellipse.

10. $\dfrac{x^2}{36} + \dfrac{y^2}{9} = 1$

10.

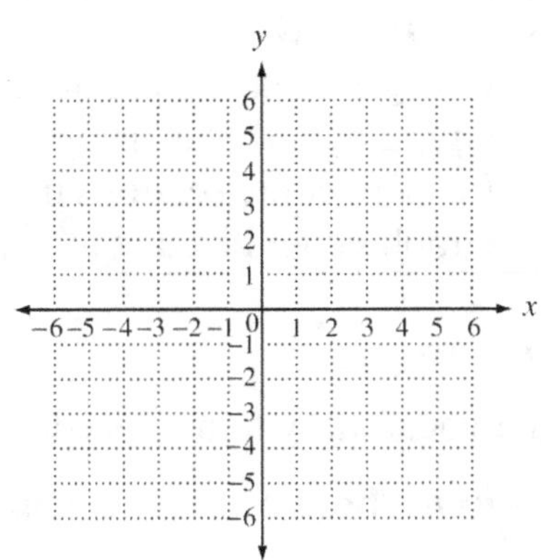

11. $\dfrac{(x+2)^2}{4} + \dfrac{(y-3)^2}{25} = 1$

11.

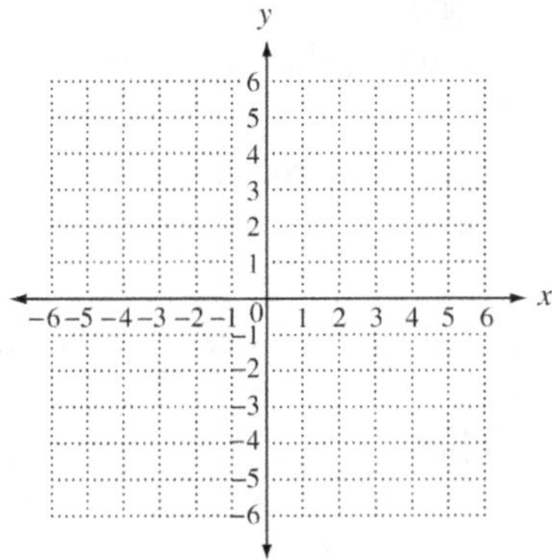

12. $\dfrac{(x+3)^2}{9} + \dfrac{(y+3)^2}{4} = 1$

12.

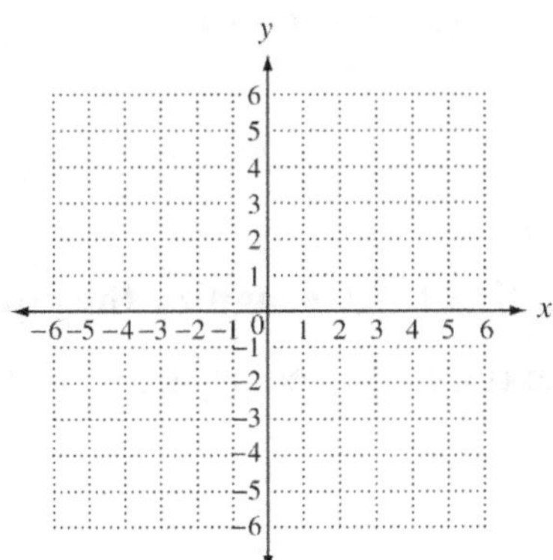

Chapter 13 NONLINEAR FUNCTIONS, CONIC SECTIONS, AND NONLINEAR SYSTEMS

13.3   Hyperbolas and Functions Defined by Radicals

Learning Objectives	
1	Recognize the equation of a hyperbola.
2	Graph hyperbolas using asymptotes.
3	Identify conic sections using their equations.
4	Graph generalized square root functions.

Key Terms

Use the vocabulary terms listed below to complete each statement in exercises 1−5.

hyperbola   **transverse axis**   **asymptotes**

fundamental rectangle   **generalized square root function**

1.   A(n) __________________ is the set of all points in a plane such that the absolute value of the difference of the distances from two fixed points is constant.

2.   Two intersecting lines that the branches of a hyperbola approach, but never reach, are its __________________.

3.   The asymptotes of a hyperbola are the extended diagonals of its __________________.

4.   The __________________ of a hyperbola with x-intercepts $(a, 0)$ and $(-a, 0)$ lies on the x-axis.

5.   For an algebraic expression in x defined by u, with $u \geq 0$, a function of the form $f(x) = \sqrt{u}$, is a __________________.

Objective 1   Recognize the equation of a hyperbola.

For extra help, see page 941 of your text.

          Copyright © 2025 Pearson Education, Inc.

Objective 2   Graph hyperbolas using asymptotes.

Review this example for Objective 2:

1. Graph $\dfrac{x^2}{9} - \dfrac{y^2}{16} = 1$.

   Step 1: $a = 3$, $b = 4$. The x-intercepts are $(3, 0)$ and $(-3, 0)$.

   Step 2: The four points $(3, 4)$, $(3, -4)$, $(-3, -4)$, and $(-3, 4)$ are the vertices of the fundamental rectangle.

   Step 3: The equations of the asymptotes are $y = \pm\frac{4}{3}x$.

   Step 4: Sketch the graph.

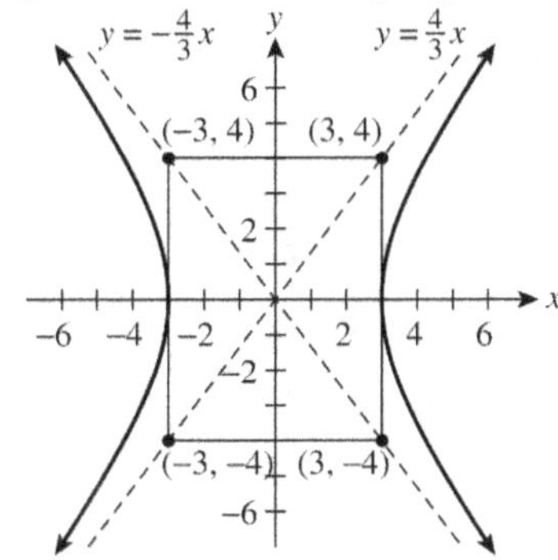

Now Try:

1. Graph $\dfrac{x^2}{16} - \dfrac{y^2}{4} = 1$.

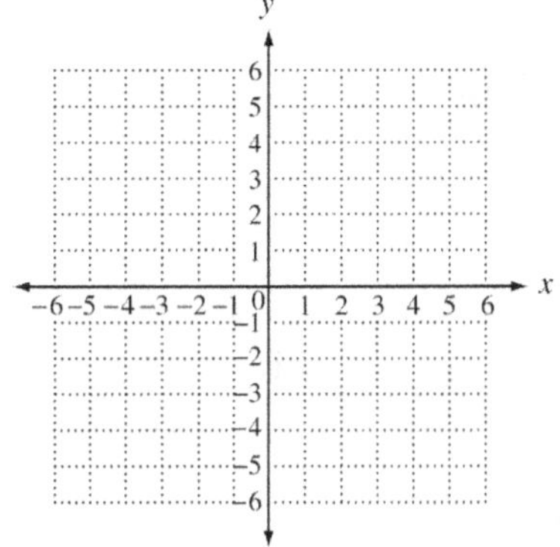

Objective 2 Practice Exercises

For extra help, see Examples 1–2 on pages 942–943 of your text.

Graph each hyperbola.

1. $\dfrac{x^2}{36} - \dfrac{y^2}{49} = 1$

1.

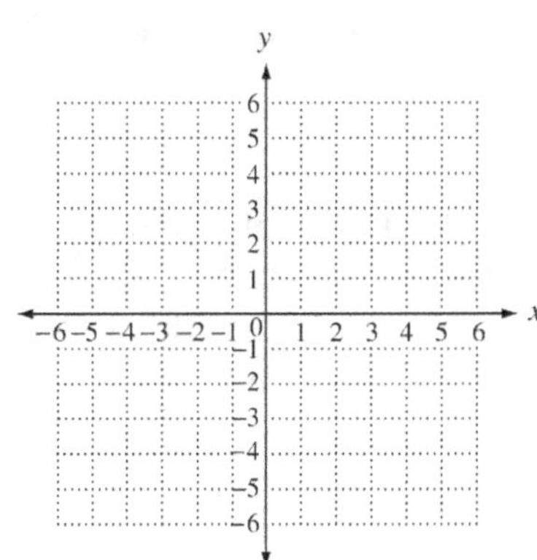

                                      533

2. $\dfrac{y^2}{4} - \dfrac{x^2}{25} = 1$

2.

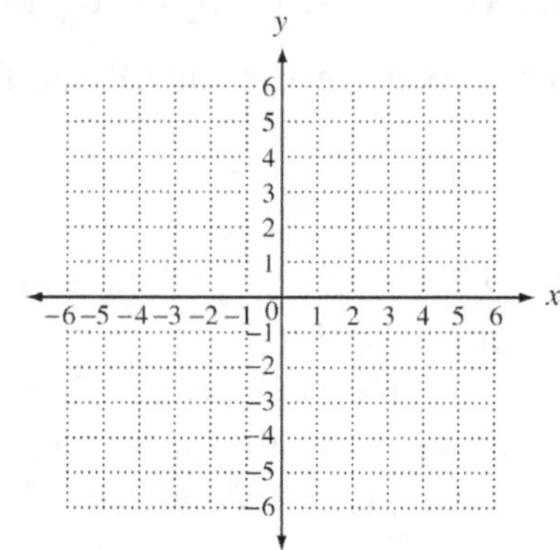

3. $\dfrac{x^2}{16} - y^2 = 1$

3.

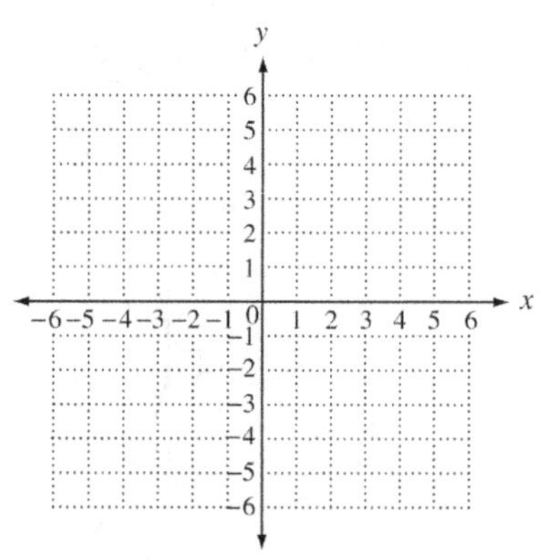

Objective 3   Identify conic sections using their equations.

Review these examples for Objective 3:

3. Identify the graph of each equation.

    a. $3x^2 + y = 36$

Only one of the variables is squared, so this is the vertical parabola $y = -3x^2 + 36$.

    b. $4x^2 = 36 - 4y^2$

Both variables are squared, so the graph is either an ellipse or a hyperbola. Note that a circle is a special case of an ellipse. Rewrite the equation so that the x^2- and y^2-terms are on one side of the equation.

$$4x^2 + 4y^2 = 36$$
$$x^2 + y^2 = 9$$

The graph of this equation is a circle with center $(0, 0)$ and radius 3.

Now Try:

3. Identify the graph of each equation.

    a. $5x^2 - 6y^2 = 30$

____

    b. $4y^2 = 12 - 3x^2$

____

c. $4x^2 = 36 + 4y^2$

Again, both variables are squared, so the graph is either an ellipse or a hyperbola. Rewrite the equation so that the x^2- and y^2-terms are on one side of the equation.

$$4x^2 - 4y^2 = 36$$

$$\frac{x^2}{9} - \frac{y^2}{9} = 1$$

The graph of this equation is a hyperbola.

c. $2x - 9 = y^2$

____

Objective 3 Practice Exercises

For extra help, see Example 3 on page 943 of your text.

Identify the graph of each equation as a parabola, circle, ellipse, or hyperbola.

4. $2x^2 + y^2 = 16$

4. __________

5. $25y^2 + 100 = 4x^2$

5. __________

6. $5x^2 = 25 - 5y^2$

6. __________

Objective 4   Graph generalized square root functions.

| **Review these examples for Objective 4:** | **Now Try:** |

Review these examples for Objective 4:

4.   Graph $f(x) = \sqrt{36 - x^2}$. Give the domain and range.

$$f(x) = \sqrt{36 - x^2}$$
$$y = \sqrt{36 - x^2}$$
$$y^2 = 36 - x^2$$
$$x^2 + y^2 = 36$$

This is the graph of a circle with center at $(0, 0)$ and radius 6. Since $f(x)$ represents a principal square root in the original equation, $f(x)$ must be nonnegative. This restricts the graph to the upper half of the circle.

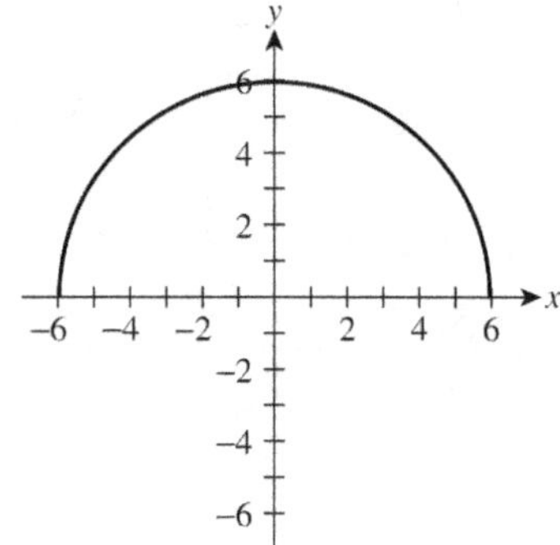

Domain: $[-6, 6]$; range: $[0, 6]$.

Now Try:

4.   Graph $f(x) = \sqrt{9 - x^2}$. Give the domain and range.

Domain: _______________

Range: _______________

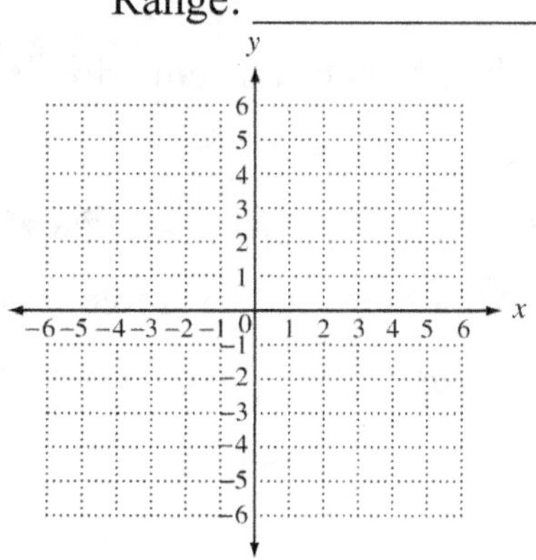

5. Graph $f(x) = -3\sqrt{1 - \dfrac{x^2}{4}}$. Give the domain and range.

$$f(x) = -3\sqrt{1 - \frac{x^2}{4}}$$

$$y^2 = 9\left(1 - \frac{x^2}{4}\right)$$

$$\frac{y^2}{9} = 1 - \frac{x^2}{4}$$

$$\frac{x^2}{4} + \frac{y^2}{9} = 1$$

This is the equation of an ellipse with x-intercepts $(2, 0)$ and $(-2, 0)$ and y-intercepts $(0, 3)$ and $(0, -3)$.

Since the original equation has a negative square root, y must be nonpositive, restricting the graph to the lower half of the ellipse.

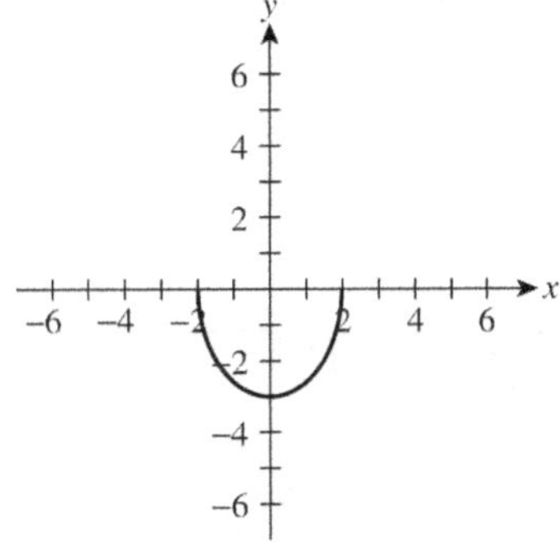

Domain: $[-2, 2]$; range: $[0, -3]$.

5. Graph $f(x) = \sqrt{9 - 9x^2}$. Give the domain and range.

Domain: ______________

Range: ______________

Name: Date:
Instructor: Section:

Objective 4 Practice Exercises

For extra help, see Examples 4–5 on pages 945–946 of your text.

Graph each generalized square root function. Give the domain and range.

7. $f(x) = -\sqrt{16 - x^2}$

7. Domain: ______________

Range: ______________

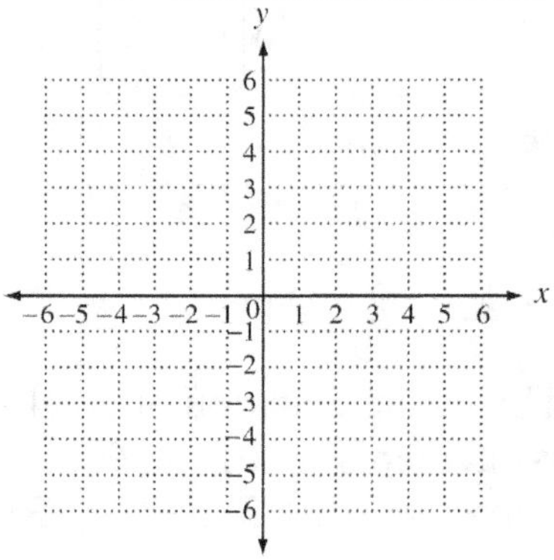

8. $f(x) = -\sqrt{1 + x}$

8. Domain: ______________

Range: ______________

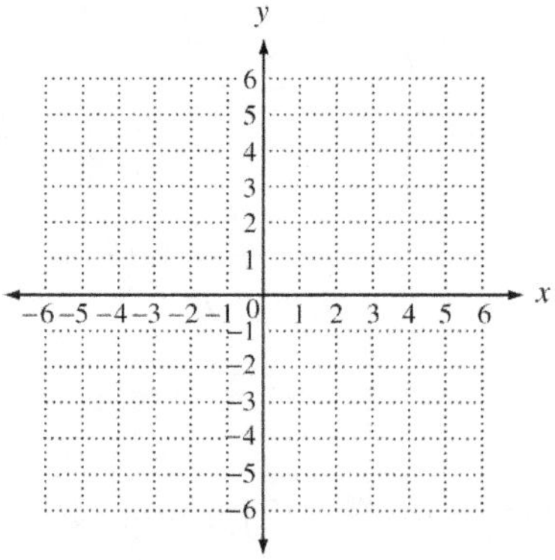

9. $f(x) = -3\sqrt{1 + \frac{x^2}{25}}$

9. Domain: ______________

Range: ______________

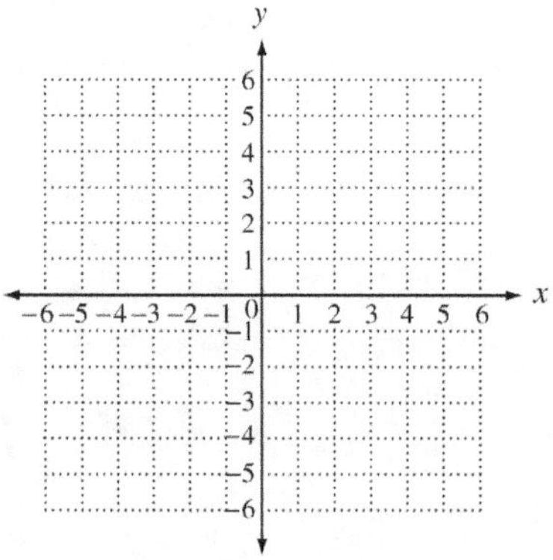

Chapter 13 NONLINEAR FUNCTIONS, CONIC SECTIONS, AND NONLINEAR SYSTEMS

13.4   Nonlinear Systems of Equations

Learning Objectives
1   Solve a nonlinear system using substitution.
2   Solve a nonlinear system with two second-degree equations using elimination.
3   Solve a nonlinear system that requires a combination of methods.

Key Terms

Use the vocabulary terms listed below to complete each statement in exercises 1−2.

nonlinear equation   **nonlinear system of equations**

1.   An equation in which some terms have more than one variable or a variable of degree 2 or greater is a ________________________________.

2.   A system with at least one nonlinear equation is a ____________________.

Objective 1   Solve a nonlinear system using substitution.

Review these examples for Objective 1:

1.   Solve the system.

$$x^2 + 3y^2 = 3 \qquad (1)$$
$$x + y = -1 \qquad (2)$$

The graph of (1) is an ellipse and the graph of (2) is a line, so the graphs could intersect in zero, one, or two points.

Solve (2) for x, then substitute that expression into (1) and solve for y.

$$x + y = -1 \quad (2)$$
$$x = -y - 1$$
$$x^2 + 3y^2 = 3 \quad (1)$$
$$(-y - 1)^2 + 3y^2 = 3$$
$$y^2 + 2y + 1 + 3y^2 = 3$$
$$4y^2 + 2y - 2 = 0$$
$$2(2y - 1)(y + 1) = 0$$
$$(2y - 1)(y + 1) = 0$$
$$2y - 1 = 0 \quad \text{or} \quad y + 1 = 0$$
$$y = \tfrac{1}{2} \qquad\qquad y = -1$$

Now Try:

1.   Solve the system.

$$x^2 + y^2 = 17$$
$$x + y = -3$$

____

To solve for x, substitute $y = \frac{1}{2}$ in (2): $x = -\frac{3}{2}$.

To solve for x, substitute $y = -1$ in (1): $x = 0$.

The solution set is $\left\{ \left(-\frac{3}{2}, \frac{1}{2}\right), (0, -1) \right\}$.

2. Solve the system.
$$xy = -10 \quad (1)$$
$$2x - y = 9 \quad (2)$$

The graph of (1) is a hyperbola and the graph of (2) is a line. There may be zero, one, or two points of intersection. Since neither equation has a squared term, solve either equation for one of the variables and then substitute the result into the other equation.

Solving (1) for y gives $y = -\frac{10}{x}$. Substituting into (2) gives

$$2x - \left(-\frac{10}{x}\right) = 9$$
$$2x + \frac{10}{x} = 9$$
$$x\left[2x + \frac{10}{x}\right] = 9x$$
$$2x^2 + 10 = 9x$$
$$2x^2 - 9x + 10 = 0$$
$$(2x - 5)(x - 2) = 0$$
$$2x - 5 = 0 \quad \text{or} \quad x - 2 = 0$$
$$x = \frac{5}{2} \qquad\qquad x = 2$$

To solve for y, substitute $x = \frac{5}{2}$ in (1): $y = -4$.

To solve for y, substitute $x = 2$ in (1): $y = -5$.

The solution set is $\left\{ (2, -5), \left(\frac{5}{2}, -4\right) \right\}$.

2. Solve the system.
$$xy = 1$$
$$x + y = 2$$

____

Objective 1 Practice Exercises

For extra help, see Examples 1–2 on pages 949–951 of your text.

Solve each system by the substitution method.

1. $2x^2 - y^2 = -1$
$\quad\quad 2x + y = 7$

1. _______________

2. $xy = -6$ **2.** ______________

             $x + y = 1$

3. $xy = 24$ **3.** ______________

            $y = 2x + 2$

Objective 2   **Solve a nonlinear system with two second-degree equations using elimination.**

Review this example for Objective 2:

3. Solve the system.

$$3x^2 + y^2 = 35 \quad (1)$$
$$2x^2 - y^2 = 15 \quad (2)$$

The graph of (1) is an ellipse and the graph of (2) is a hyperbola. There may be zero, one, or two points of intersection. Adding the two equations will eliminate y.

$$3x^2 + y^2 = 35 \quad (1)$$
$$\underline{2x^2 - y^2 = 15} \quad (2)$$
$$5x^2 = 50$$
$$x^2 = 10$$
$$x = \pm\sqrt{10}$$

Substitute the values for x in (1) and solve for y.

For $x = \sqrt{10}$, $3(\sqrt{10})^2 + y^2 = 35$
$$30 + y^2 = 35$$
$$y^2 = 5$$
$$y = \pm\sqrt{5}$$

For $x = -\sqrt{10}$, $3(-\sqrt{10})^2 + y^2 = 35$
$$30 + y^2 = 35$$
$$y^2 = 5$$
$$y = \pm\sqrt{5}$$

The solution set is $\{(\sqrt{10}, -\sqrt{5}), (\sqrt{10}, \sqrt{5}), (-\sqrt{10}, \sqrt{5}), (-\sqrt{10}, -\sqrt{5})\}$.

Now Try:

3. Solve the system.

$$x^2 - 2y = 8$$
$$x^2 + y^2 = 16$$

____

Name: Date:
Instructor: Section:

Objective 2 Practice Exercises

For extra help, see Example 3 on pages 951–952 of your text.

Solve each system by the elimination method.

4. $5x^2 - y^2 = 55$

    $2x^2 + y^2 = 57$ 4. ______________

5. $x^2 + 2y^2 = 11$

    $2x^2 - y^2 = 17$ 5. ______________

6. $3x^2 + 2y^2 = 30$

    $2x^2 + y^2 = 17$ 6. ______________

Objective 3   Solve a nonlinear system that requires a combination of methods.

Review this example for Objective 3:

4. Solve the system.

    $x^2 + 5xy - y^2 = 13$ (1)

    $x^2 - \quad y^2 = 3$ (2)

    We will use the elimination method in
    combination with the substitution method.
    Multiply Eq (2) by -1 and add to Eq (1).

    $x^2 + 5xy - y^2 = 13$ (1)

    $\underline{-x^2 + \quad y^2 = -3}$ (2)

    $\qquad\qquad 5xy = 10$

    $\qquad\qquad\quad y = \frac{2}{x}$ (3)

Now Try:

4. Solve the system.

    $5x^2 - xy + 5y^2 = 89$

    $x^2 + \quad y^2 = 17$

    ____

                Copyright © 2025 Pearson Education, Inc.

Now substitute $\frac{2}{x}$ for y in (2) and solve for x.

$$x^2 - \left(\frac{2}{x}\right)^2 = 3$$

$$x^2 - \frac{4}{x^2} = 3$$

$$x^4 - 4 = 3x^2$$

$$x^4 - 3x^2 - 4 = 0$$

$$(x-2)(x+2)(x^2+1) = 0$$

Using the zero-factor property, we have
$x = 2$, $x = -2$, $x = i$, $x = -i$.
Substitute each of these values into (3) and solve
for y.
If $x = 2$, then $y = 1$.
If $x = -2$, then $y = -1$.
If $x = i$, then $y = \frac{2}{i} = \frac{2}{i} \cdot \frac{-i}{-i} = \frac{-2i}{-i^2} = -2i$.

If $x = -i$, then $y = \frac{2}{-i} = \frac{2}{-i} \cdot \frac{i}{i} = \frac{2i}{-i^2} = 2i$.

It is important to check all answers in the
original equations because it is possible to
obtain extraneous solutions. The solution
set is $\{(2,\ 1),\ (-2,\ -1),\ (i,\ -2i),\ (-i,\ 2i)\}$.

Objective 3 Practice Exercises

For extra help, see Example 4 on pages 952–953 of your text.

Solve each system.

7. $4x^2 - 2xy + 4y^2 = 64$
    $x^2 + y^2 = 13$

7. ______________

8. $x^2 + 3xy + 2y^2 = 12$
    $-x^2 + 8xy - 2y^2 = 10$

8. ______________

9. $x^2 + 5xy - y^2 = 20$
    $x^2 - 2xy - y^2 = -8$

9. ______________

Chapter 13 NONLINEAR FUNCTIONS, CONIC SECTIONS, AND NONLINEAR SYSTEMS

13.5   Second-Degree Inequalities and Systems of Inequalities

Learning Objectives
1   Graph second-degree inequalities.
2   Graph the solution set of a system of inequalities.

Key Terms

Use the vocabulary terms listed below to complete each statement in exercises 1–2.

second-degree inequality   **system of inequalities**

1.   A _______________________________ consists of two or more inequalities to be solved at the same time.

2.   A(n)____ is an inequality with at least one variable of degree 2 and no variable with degree greater than 2.

Objective 1   Graph second-degree inequalities.

Review these examples for Objective 1:

1.   Graph $y < -x^2 + 3$.

The boundary, $y = -x^2 + 3$, is a parabola that opens down with vertex $(0, 3)$.
Use $(0, 0)$ as a test point.

$$y < -x^2 + 3$$

$$0 \overset{?}{<} -0^2 + 3$$

$$0 < 3 \quad \text{True}$$

Because the final inequality is a true statement, the points in the region containing $(0, 0)$ satisfy the inequality. The parabola is drawn as a dashed curve since the points on the parabola itself do not satisfy the inequality and the region inside (or below) the parabola is shaded.

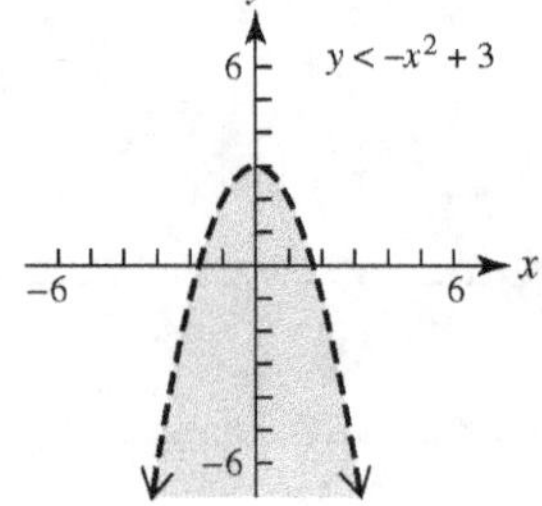

Now Try:

1.   Graph $y \geq x^2 - 4$.

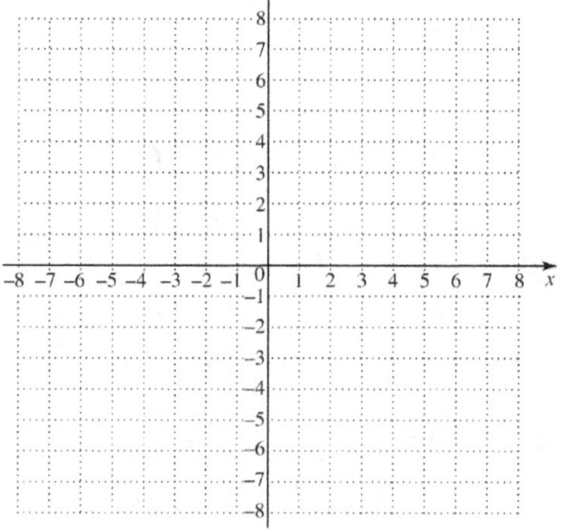

      Copyright © 2025 Pearson Education, Inc.

3. Graph $16x^2 < 9y^2 + 144$.

Rewrite the inequality.
$$16x^2 - 9y^2 < 144$$
$$\frac{x^2}{9} - \frac{y^2}{16} < 1$$

The boundary, drawn as a dashed curve, is the following hyperbola. $\dfrac{x^2}{9} - \dfrac{y^2}{16} = 1$

Since the graph is a horizontal hyperbola, the desired region will be either between the branches or the regions to the right of the right branch and to the left of the left branch. Using the test point $(0, 0)$ into the original inequality leads to $0 < 1$, a true statement. So the region between the branches containing $(0, 0)$ is shaded.

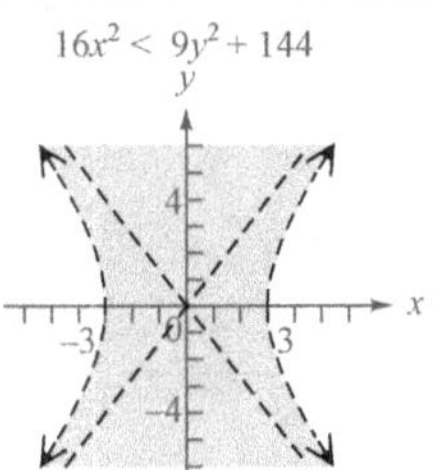

3. Graph $9y^2 - 36x^2 > 144$.

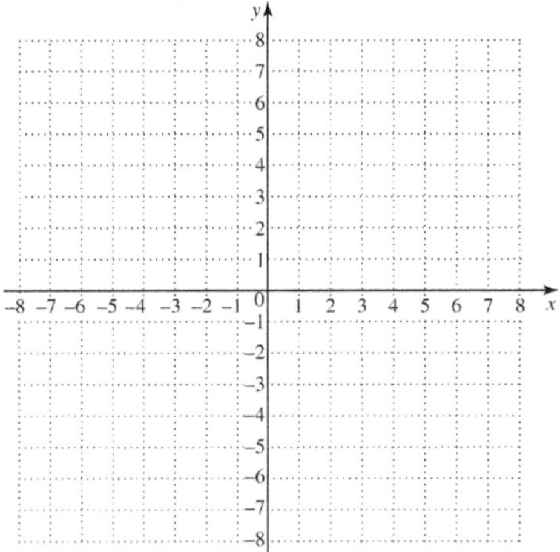

Objective 1 Practice Exercises

For extra help, see Examples 1–3 on pages 956–957 of your text.

Graph each inequality.

1. $x \le 2y^2 + 8y + 9$

1.

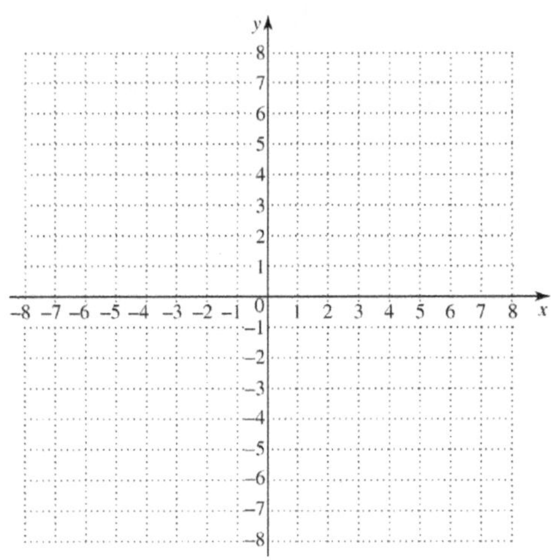

2. $x^2 + 9y^2 > 36$

2.

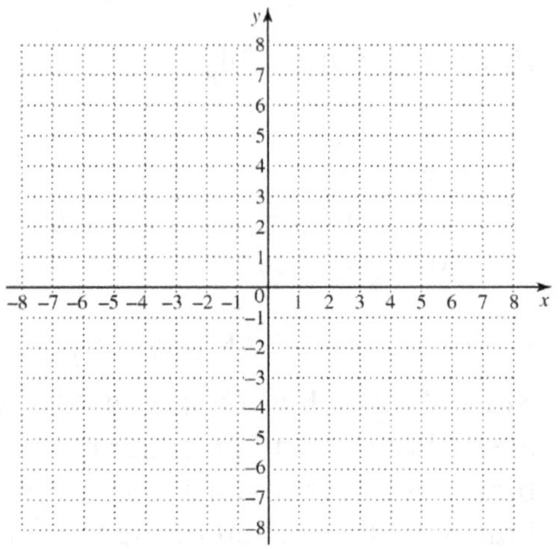

3. $9x^2 - y^2 < 36$

3.

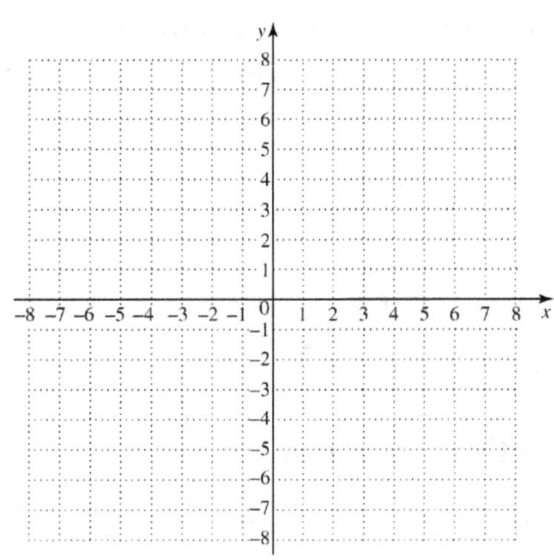

  Copyright © 2025 Pearson Education, Inc.

Objective 2   Graph the solution set of a system of inequalities.

Review these examples for Objective 2:

5. Graph the solution set of the system.

$$x^2 + y^2 \leq 16$$
$$y > x$$

Begin by graphing $x^2 + y^2 \leq 16$. The boundary line is a circle centered at the origin with radius 4. The test point $(0, 0)$ leads to a true statement, so we shade inside the circle.

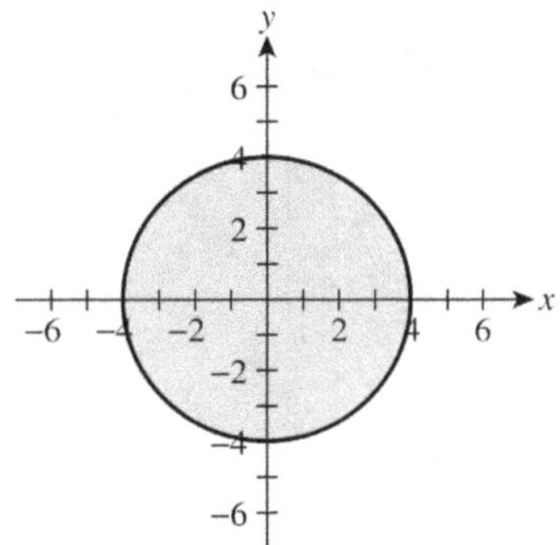

The boundary of the solution set of $y > x$ is a dashed line passing through $(0, 0)$. Using the test point $(0, 1)$ leads to a true statement, so shade above the line.

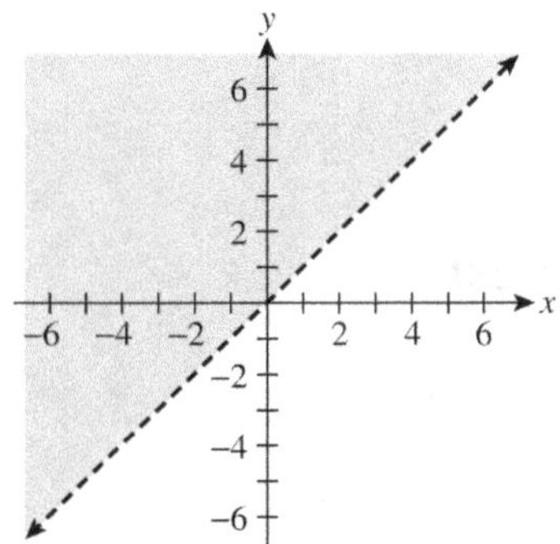

The graph of the solution set of the system is the intersection of the graphs of the two inequalities.

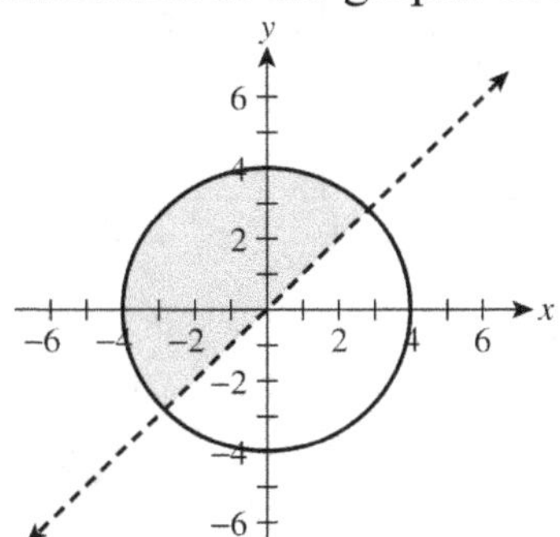

Now Try:

5. Graph the solution set of the system.

$$4y + x^2 < 0$$
$$x \geq 0$$

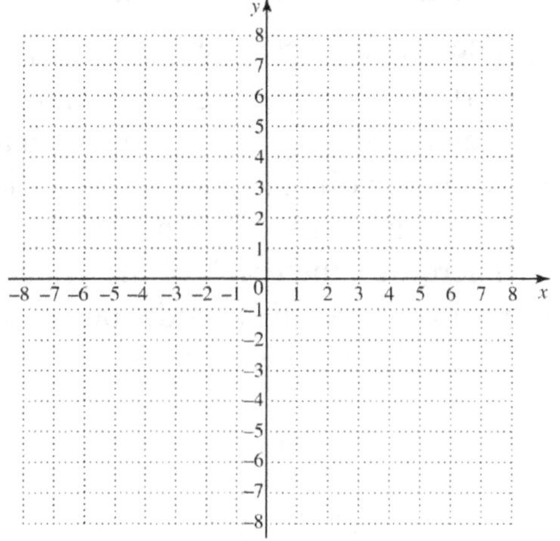

7. Graph the solution set of the system.

$$25x^2 + 9y^2 < 225$$
$$y \le -x^2 + 4$$
$$y < -x$$

The graph of $25x^2 + 9y^2 < 225$ is a dashed ellipse with $a = 3$ and $y = 5$. To satisfy the inequality, a point must lie inside the ellipse. The graph of $y \le -x^2 + 4$ is a parabola with vertex (0, 4) opening downward. The inequality includes the points on the boundary along with the points inside the parabola. The graph of $y < -x$ includes all points below the line $y = -x$. Therefore, the graph of the system is the shaded region, which lies inside the ellipse and the parabola, and below the line.

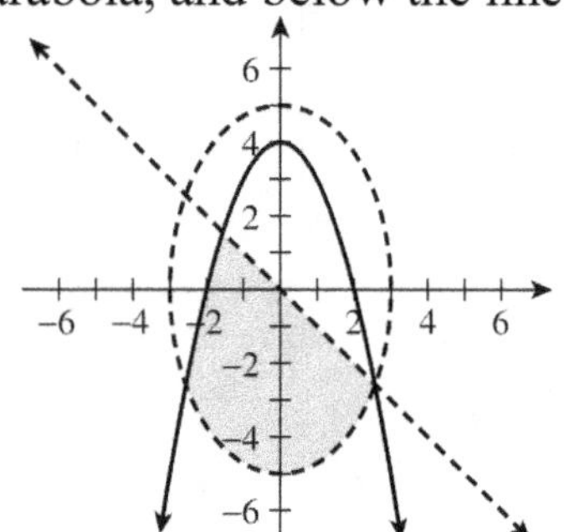

7. Graph the solution set of the system.

$$y \ge (x+2)^2 - 5$$
$$2x + y < -5$$
$$(x+3)^2 + (y-1)^2 < 9$$

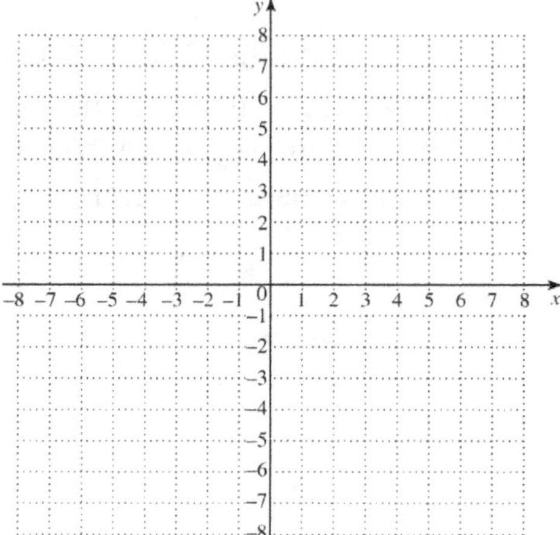

Objective 2 Practice Exercises

For extra help, see Examples 4–7 on pages 958–960 of your text.

Graph each system of inequalities.

4. $-x + y > 2$
     $3x + y > 6$

4.

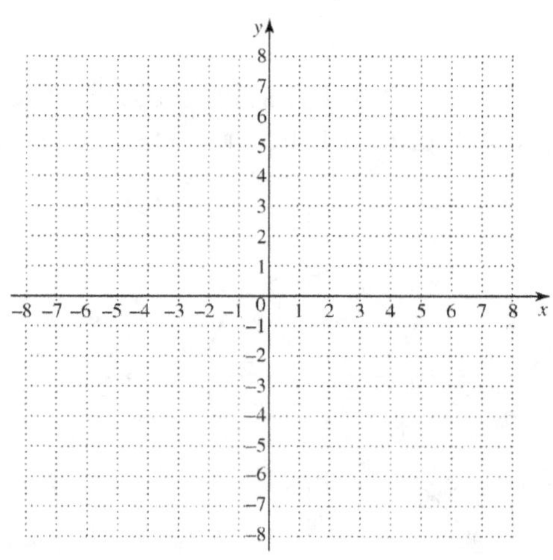

5. $\quad x^2 + y^2 \leq 25$
$\quad\quad 3x - 5y > -15$

5.

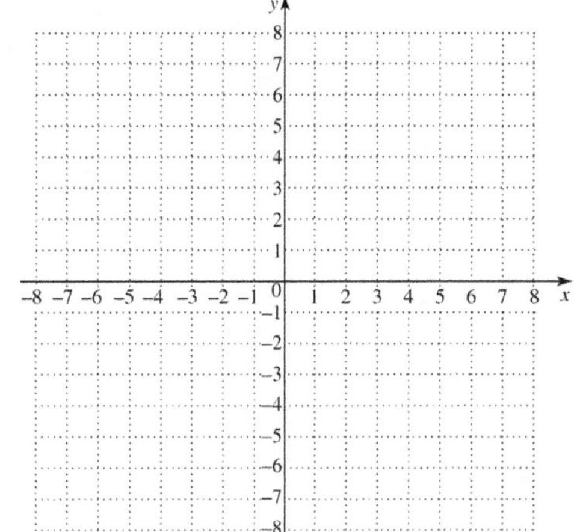

6. $\quad x^2 + 4y^2 \leq 36$
$\quad\quad -5 < x < 2$
$\quad\quad y \geq 0$

6.

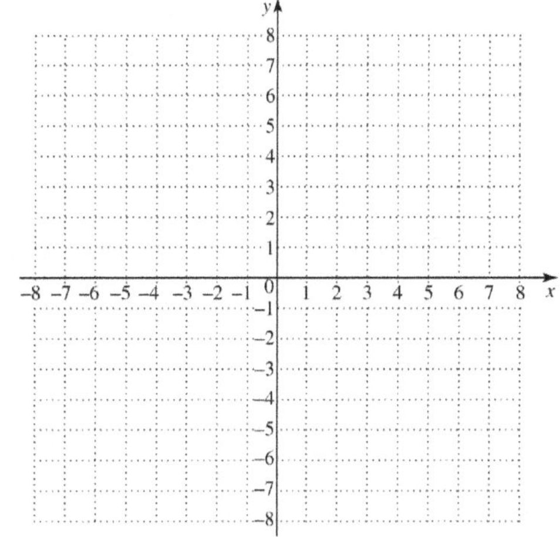

Chapter 14   FURTHER TOPICS IN ALGEBRA

14.1   Sequences and Series

Learning Objectives
1   Define infinite and finite sequences.
2   Find the terms of a sequence, given the general term.
3   Find the general term of a sequence.
4   Use sequences to solve applied problems.
5   Use summation notation to evaluate a series.
6   Write a series using summation notation.
7   Find the arithmetic mean (average) of a group of numbers.

Key Terms

Use the vocabulary terms listed below to complete each statement in exercises 1−8.

infinite sequence	**finite sequence**	**terms of a sequence**
general term	**series**	**summation notation**
index of summation	**arithmetic mean (average)**	

1. A(n) ____________________ is a function whose domain is the set of positive integers.

2. A ____________________ is the sum of the terms of a sequence.

3. ____________________ is a compact way of writing a series using the general term of the corresponding sequence.

4. When using summation notation $\displaystyle\sum_{i}^{n} f(i)$, the letter i is called the _____________.

5. The expression a_n which defines a sequence, is called the ___________________ of the sequence.

6. The __________________ of a group of numbers is given by the formula $\dfrac{\displaystyle\sum_{i=1}^{n} x_i}{n}$.

7. The function values $a_1, a_2, a_3, \ldots$, written in order, are the ___________________.

8. A(n) ____________________ is a function with domain of the form $\{1, 2, 3, \ldots, n\}$, where n is a positive integer.

Objective 1   Define infinite and finite sequences.

For extra help, see page 974 of your text.

Objective 2   Find the terms of a sequence, given the general term.

Review this example for Objective 1:	**Now Try:**

Review this example for Objective 1:

1. Write the first five terms of the sequence $a_n = 3n - 2$.

$$n = 1; \quad a_1 = 3(1) - 2 = 1$$
$$n = 2; \quad a_2 = 3(2) - 2 = 4$$
$$n = 3; \quad a_3 = 3(3) - 2 = 7$$
$$n = 4; \quad a_4 = 3(4) - 2 = 10$$
$$n = 5; \quad a_5 = 3(5) - 2 = 13$$

The first five terms are 1, 4, 7, 10, and 13.

Now Try:

1. Write the first five terms of the sequence $a_n = \dfrac{n+2}{5n}$.

____

Objective 2 Practice Exercises

For extra help, see Example 1 on pages 947–975 of your text.

Write out the first five terms of each sequence.

1. $a_n = (-1)^n$

1. _________________

2. $a_n = \dfrac{1+n}{n}$

2. _________________

Find the indicated term for the sequence.

3. $a_n = \dfrac{3n-2}{5n+2}; \quad a_8$

3. _________________

Objective 3   Find the general term of a sequence.

Review this example for Objective 3:

2. Find an expression for the general term a_n of the sequence $1, \dfrac{1}{2}, \dfrac{1}{3}, \dfrac{1}{4}, \dfrac{1}{5}, \ldots.$

Notice that the denominators are consecutive integers, so the general term is $a_n = \dfrac{1}{n}$.

Now Try:

2. Find an expression for the general term a_n of the sequence $2, -4, 8, -16, 32, \ldots.$

____

Objective 3 Practice Exercises

For extra help, see Example 2 on page 975 of your text.

Find a general term a_n for the given terms of each sequence.

4.   3, 5, 7, 9, 11, ...

4. _________________

5. $\sqrt{3}$, 3, $3\sqrt{3}$, 9, $9\sqrt{3}$, ...

5. _________________

6. $-\dfrac{1}{5}, \dfrac{1}{10}, -\dfrac{1}{15}, \dfrac{1}{20}, -\dfrac{1}{25}, \dots$

6. _________________

Objective 4   Use sequences to solve applied problems.

Review this example for Objective 4:

3.   Suppose a copier loses $\frac{1}{4}$ of its value each year; that is, at the end of any given year, the value is $\frac{3}{4}$ of its value at the beginning of the year. If a new copier costs \$4800, what is its value at the end of 4 years?

At the end of the first year, the value of the copier is $\frac{3}{4}(\$4800) = \$3600$.

At the end of the second year, the value of the copier is $\frac{3}{4}(\$3600) = \$2700$.

At the end of the third year, the value of the copier is $\frac{3}{4}(\$2700) = \$2025$.

At the end of the fourth year, the value of the copier is $\frac{3}{4}(\$2025) = \$1518.75$.

Now Try:

3.   Carla borrows \$6000 and agrees to pay \$500 monthly plus interest of 2% on the unpaid balance from the beginning of the first month. Find the payments for the first four months and the remaining debt at the end of that period.

Month 1 ______________

Month 2 ______________

Month 3 ______________

Month 4 ______________

Remaining balance _______

Objective 4 Practice Exercises

For extra help, see Example 3 on page 975 of your text.

Solve each applied problem by writing the first few terms of each sequence.

7. A colony of bacteria doubles in weight every hour. If   **7.** __________________
the colony weighs 2 grams at the beginning of an
experiment, find the weight after 3 hours.

8. Ms. Burley is offered a new job with a salary of   **8.** __________________
$35,000 per year and a 2% raise at the end of each
year. Write a sequence showing her salary for each
of the first five years. (Round to the nearest dollar.)

Objective 5   Use summation notation to evaluate a series.

Review this example for Objective 5:

4. Write out the series as a sum of terms and then find the sum.

$$\sum_{i=1}^{6}(2i-6)$$

$$\sum_{i=1}^{6}(2i-6)$$

$$=[2(1)-6]+[2(2)-6]+[2(3)-6]$$
$$+[2(4)-6]+[2(5)-6]+[2(6)-6]$$
$$=-4+(-2)+0+2+4+6$$
$$=6$$

Now Try:

4. Write out the series as a sum of terms and then find the sum.

$$\sum_{i=1}^{7}(4i-3)$$

____

Objective 5 Practice Exercises

For extra help, see Example 4 on pages 976–977 of your text.

Write out each series as a sum of terms and then find the sum.

9. $\displaystyle\sum_{i=1}^{4}(2i+3)$ **9.** __________________

10. $\displaystyle\sum_{i=1}^{4}\left(-\frac{1}{2i}\right)$ 10. _______________

11. $\displaystyle\sum_{i=1}^{6}(i^2+1)$ 11. _______________

Objective 6   Write a series using summation notation.

Review this example for Objective 6:

5. Write out the sum using summation notation.

$$3+9+27+81+243$$

Each term is a power of 3, so $a_n=3^n$. Thus, the

sum can be written as $\displaystyle\sum_{i=1}^{5}3^i$.

Now Try:

5. Write out the sum using summation notation.

$$4+8+12+16+20+24$$

____

Objective 6 Practice Exercises

For extra help, see Example 5 on page 977 of your text.

Write each series using summation notation.

12. $1+4+9+16+25$ 12. _______________

13. $\dfrac{1}{5}+\dfrac{1}{8}+\dfrac{1}{11}+\dfrac{1}{14}+\dfrac{1}{17}$ 13. _______________

14. $1+2x+3x^2+4x^3+5x^4$ 14. _______________

Objective 7   Find the arithmetic mean (average) of a group of numbers.

Review this example for Objective 7:

6. The amount of rain that fell in Philadelphia from March through August is given in the table.

Month	Amount of Precipitation
March	7.55 in.
April	3.85 in.
May	2.61 in.
June	2.15 in.
July	8.33 in.
August	1.08 in.

Source: Franklin Institute

What was the average amount of rain each month for this six-month period?

$$\overline{x} = \frac{\sum\limits_{i=1}^{6} x_i}{6}$$

$$= \frac{7.55 + 3.85 + 2.61 + 2.15 + 8.33 + 1.08}{6}$$

$$= \frac{25.57}{6} \approx 4.26$$

The average amount of rain was about 4.26 inches per month during the six-month period.

Now Try:

6. The amount of rain that fell in Smalltown during October from 2017 through 2023 is given in the table.

Month	Amount of Precipitation
2017	2.81 in.
2018	3.01 in.
2019	1.55 in.
2020	3.32 in.
2021	2.17 in.
2022	5.54 in.
2023	5.24 in.

What is the average amount of rain that fell during October for the seven-year period?

____

Objective 7 Practice Exercises

For extra help, see Example 6 on page 978 of your text.

Find the arithmetic mean for each collection of numbers.

15.   2, 8, 18, 32, –5, –7

15. _______________

16. $\dfrac{1}{3}, -\dfrac{1}{2}, \dfrac{1}{4}, \dfrac{7}{6}$

16. _______________

Chapter 14   FURTHER TOPICS IN ALGEBRA

14.2   Arithmetic Sequences

Learning Objectives
1   Find the common difference of an arithmetic sequence.
2   Find the general term of an arithmetic sequence.
3   Use an arithmetic sequence in an application.
4   Find any specified term or the number of terms of an arithmetic sequence.
5   Find the sum of a specified number of terms of an arithmetic sequence.

Key Terms

Use the vocabulary terms listed below to complete each statement in exercises 1–2.

arithmetic sequence (arithmetic progression)   **common difference**

1.   A(n) _____________________ is a sequence in which each term after the first differs from the preceding term by a constant amount.

2.   The _____________________ d is the difference between any two adjacent terms of an arithmetic sequence.

Objective 1   Find the common difference of an arithmetic sequence.

Review these examples for Objective 1:

1.   Determine the common difference d for the arithmetic sequence.
$$35, 32, 29, 26, \dots$$

Since the sequence is arithmetic, d is the difference between any two adjacent terms. We arbitrarily choose the terms 29 and 26.
$d = 26 - 29 = -3$

2.   Write the first five terms of the arithmetic sequence with first term 9 and common difference 5.

The second term is found by adding 5 to the first term 9, getting 14.
For the next term, add 5 to 14, and so on.
The first five terms are 9, 14, 19, 24, 29.

Now Try:

1.   Determine the common difference d for the arithmetic sequence.
$$-13, -11, -9, -7, \dots$$

____

2.   Write the first five terms of the arithmetic sequence with first term 9 and common difference -5.

____

Name: Date:

Instructor: Section:

Objective 1 Practice Exercises

For extra help, see Examples 1–2 on page 981 of your text.

Find the common difference d for the arithmetic sequence.

1. $-9, -5, -1, 3, \ldots$ 1. _______________

2. $\dfrac{1}{4}, -\dfrac{1}{4}, -\dfrac{3}{4}, -\dfrac{5}{4}, \ldots$ 2. _______________

3.   Write the first five terms of the arithmetic sequence   3. _______________
     with first term 6 and common difference –3.

Objective 2   Find the general term of an arithmetic sequence.

Review this example for Objective 2:

3. Determine an expression for the general term of the arithmetic sequence where $a_1 = -3$ and $d = -4$. Then find a_{15}.

   Now find a_n.
   $$a_n = a_1 + (n-1)d$$
   $$a_n = -3 + (n-1)(-4)$$
   $$a_n = -3 - 4n + 4$$
   $$a_n = -4n + 1$$
   Find a_{15}.
   $$a_{15} = -4(15) + 1 = -59$$

Now Try:

3. Determine an expression for the general term of the arithmetic sequence where $a_1 = 35$ and $d = 4$. Then find a_{15}.

   ____

Objective 2 Practice Exercises

For extra help, see Example 3 on page 982 of your text.

Use the formula for a_n to find the general term of each arithmetic sequence.

4. $a_1 = -7, \ d = -5$ 4. _______________

5. $\dfrac{1}{2}, \dfrac{5}{6}, \dfrac{7}{6}, \dfrac{3}{2}, \dfrac{11}{6}, \ldots$ 5. _______________

6. $0.8, 0.5, 0.2, \ldots, -29.8, \ldots$ 6. _______________

Objective 3   Use an arithmetic sequence in an application.

Review this example for Objective 3:	**Now Try:**
4. After knee surgery, your trainer tells you to return to your jogging program slowly. Your trainer suggests jogging for 12 minutes each day for the first week. Each week thereafter, your trainer suggests that you increase that time by 6 minutes per day. How long will you be jogging in the 6th week?	**4.** You accept a teaching job that pays \$32,530 the first year with a guarantee of a \$1,030 raise each year thereafter. What will your salary be in your 10th year on the job?

After n weeks, you will be jogging
$$a_n = 12 + 6(n-1) \text{ minutes.}$$
To find how long you will be jogging in the 6th week, find a_6.
$$a_6 = 12 + 6(6-1) = 42$$
You will be jogging 42 minutes in the 6th week.

____

Objective 3 Practice Exercises

For extra help, see Example 4 on page 982 of your text.

Solve each problem.

7. Ben's father has started a savings fund for Ben's college education. He makes an initial deposit of \$5000 and each month contributes an additional \$100. How much money will be in the account after 60 months? (Disregard any interest.)

7. ______________________

8. Liza is starting a swimming program. Liza plans to swim 50 laps per day the first week and add 5 laps per day each week until Liza is swimming 125 laps. In which week will this occur?

8. ______________________

9. The starting salary at a supermarket is $13.75 per hour. Every six months an employee receives a raise of $0.75 per hour. What will the employee's salary be after two years?

9. ______________

Objective 4   **Find any specified term or the number of terms of an arithmetic sequence.**

Review these examples for Objective 4:

5. Evaluate the indicated term for each arithmetic sequence.

   a. $a_1 = 28, \ d = 10, \ a_{12}$

$$a_n = a_1 + (n-1)d$$
$$a_{12} = 28 + (12-1)(10)$$
$$a_{12} = 138$$

   b. $a_4 = 24, \ a_{10} = -6, \ a_{20}$

We obtain a_{10} by adding the common difference to a_6 six times.

$$a_{10} = a_4 + 6d$$
$$-6 = 24 + 6d$$
$$-30 = 6d$$
$$-5 = d$$

To find a_{20}, we add the common difference, $d = -5$, a_{10} ten times.

$$a_{20} = a_{10} + 10d$$
$$a_{20} = -6 + 10(-5)$$
$$a_{20} = -6 - 50$$
$$a_{20} = -56$$

Now Try:

5. Evaluate the indicated term for each arithmetic sequence.

   a. $a_1 = 9, \ d = 6, \ a_{15}$

____

   b. $a_3 = 7, \ a_{10} = -14, \ a_{20}$

____

6. Find the number of terms in the arithmetic sequence 7, 10, 13, …, 55.

$$a_1 = 7, \quad a_n = 55, \quad d = 10 - 7 = 3$$
$$a_n = a_1 + (n-1)d$$
$$55 = 7 + (n-1)3$$
$$55 = 7 + 3n - 3$$
$$55 = 4 + 3n$$
$$51 = 3n$$
$$17 = n$$

There are 17 terms in the sequence.

6. Find the number of terms in the arithmetic sequence

$$3, \ \frac{9}{2}, \ 6, \ \frac{15}{2}, \ 9, \ ..., \ \frac{51}{2}.$$

____

Objective 4 Practice Exercises

For extra help, see Examples 5–6 on page 983 of your text.

Find the indicated term for each arithmetic sequence.

10. $a_1 = -5, \ d = 9; \ a_{12}$

10. ___________________

11. $a_1 = -\frac{1}{2}, \ d = \frac{3}{2}; \ a_{15}$

11. ___________________

12. Find the number of terms in the arithmetic sequence 19, 26, 33, …, 96.

12. ___________________

Objective 5   Find the sum of a specified number of terms of an arithmetic sequence.

Review these examples for Objective 5:

7. Find the sum of the first eight terms of the sequence in which $a_n = 7 - 2n$.

Begin by evaluating a_1 and a_8.
$$a_1 = 7 - 2(1) = 5$$
$$a_8 = 7 - 2(8) = -9$$

Now find the sum using $a_1 = 5, \ a_8 = -9,$ and $n = 8$.
$$S_n = \frac{n}{2}(a_1 + a_n)$$
$$S_8 = \frac{8}{2}(5 + (-9)) = 4(-4) = -16$$

Now Try:

7. Find the sum of the first six terms of the sequence in which $a_n = 3n - 4$.

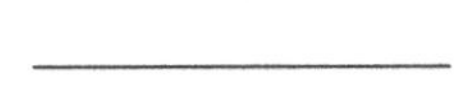

8. Find the sum of the first 15 terms of the arithmetic sequence having first term 5 and common difference -4.

$$a_1 = 5, \ d = -4, \ n = 15$$

$$S_n = \frac{n}{2}\left[2a_1 + (n-1)d\right]$$

$$S_{15} = \frac{15}{2}\left[2(5) + (15-1)(-4)\right]$$

$$= \frac{15}{2}(-46)$$

$$= -345$$

9. Evaluate $\displaystyle\sum_{i=1}^{15}(3i+2)$.

Begin by evaluating a_1 and a_{15}.

$$a_1 = 3(1) + 2 = 5$$

$$a_{15} = 3(15) + 2 = 47$$

Now find the sum using $a_1 = 5$, $a_{15} = 47$, and $n = 15$.

$$S_n = \frac{n}{2}\left(a_1 + a_n\right)$$

$$S_{15} = \frac{15}{2}[5 + 47] = \frac{15}{2}(52) = 390$$

8. Find the sum of the first 13 terms of the arithmetic sequence having first term 2 and common difference 10.

____

9. Evaluate $\displaystyle\sum_{i=1}^{10}(7i-2)$.

____

Objective 5 Practice Exercises

For extra help, see Examples 7–9 on pages 984–985 of your text.

Solve each problem.

13. Find S_{10} for the arithmetic sequence defined by $a_n = 2 - 3n$.

13. _______________

14. Find S_8 for the arithmetic sequence with $a_1 = 3$ and $d = 2$.

14. _______________

15. Evaluate $\displaystyle\sum_{i=1}^{10}(2i+3)$.

15. _______________

Chapter 14 FURTHER TOPICS IN ALGEBRA

14.3 Geometric Sequences

Learning Objectives
1 Find the common ratio of a geometric sequence.
2 Find the general term of a geometric sequence.
3 Find any specified term of a geometric sequence.
4 Find the sum of a specified number of terms of a geometric sequence.
5 Apply the formula for the future value of an ordinary annuity.
6 Find the sum of an infinite number of terms of a certain geometric sequence.

Key Terms

Use the vocabulary terms listed below to complete each statement in exercises 1−7.

geometric sequence (geometric progression)　　　　　**common ratio**

annuity　　　　　**ordinary annuity**　　　　　**payment period**

future value of an annuity　　　　　**term of an annuity**

1. A(n) _________________ r is the constant multiplier between adjacent terms in a geometric sequence.

2. A(n) _________________ is a sequence of equal payments made at equal periods of time.

3. A(n) _________________ is a sequence in which each term after the first is a constant multiple of the preceding term.

4. The _________________ is the sum of the compound amounts of all the payments, compounded to the end of the term.

5. If the payments for an annuity are made at the end of the period, and if the frequency of payments is the same as the frequency of compounding, the annuity is called a(n) _________________.

6. The time from the beginning of the first payment period to the end of the last is called the _________________.

7. The time between payments of an annuity is called the _________________.

Objective 1   Find the common ratio of a geometric sequence.

Review this example for Objective 1:

1. Determine the common ratio r for the geometric sequence.

    $4, -8, 16, -32, 64, \ldots$

    To find r, choose any two successive terms and divide the second one by the first. We choose the third and fourth terms of the sequence.

    $$r = \frac{a_4}{a_3} = \frac{-32}{16} = -2$$

Now Try:

1. Determine the common ratio r for the geometric sequence.

    $3, \dfrac{3}{4}, \dfrac{3}{16}, \dfrac{3}{64}, \ldots$

    ____

Objective 1 Practice Exercises

For extra help, see Example 1 on page 989 of your text.

Find the common ratio for the geometric sequence.

1. $10, \ 10\sqrt{2}, \ 20, \ 20\sqrt{2}, \ 40, \ \ldots$

1. _______________

2. $-\dfrac{1}{2}, \dfrac{1}{4}, -\dfrac{1}{8}, \dfrac{1}{16}, -\dfrac{1}{32}, \ldots$

2. _______________

Objective 2   Find the general term of a geometric sequence.

Review this example for Objective 2:

2. Determine an expression for the general term of the sequence.

    $-3, -15, -75, -375, \ldots$

    The first term is $a_1 = -3$ and the common ratio is $r = 5$.

    $$a_n = a_1 r^{n-1} = -3(5)^{n-1}$$

Now Try:

2. Determine an expression for the general term of the sequence.

    $\sqrt{3}, \ 3, \ 3\sqrt{3}, \ 9, \ 9\sqrt{3}, \ 27, \ \ldots$

    ____

Objective 2 Practice Exercises

For extra help, see Example 2 on page 989 of your text.

Determine an expression for the general term of the geometric sequence.

3.   6, 12, 24, 48, 96, …

3. _________________

4. $-\dfrac{2}{3}, \dfrac{2}{9}, -\dfrac{2}{27}, \dfrac{2}{81}, \cdots$

4. _________________

5. $-\dfrac{4}{5}, -\dfrac{4}{25}, -\dfrac{4}{125}, -\dfrac{4}{625}, \cdots$

5. _________________

Objective 3   Find any specified term of a geometric sequence.

Review these examples for Objective 3:

3.   Evaluate the indicated term for each geometric sequence.

   a. $a_1 = 64$, $r = -\dfrac{1}{2}$; a_6

$$a_n = a_1 r^{n-1}$$

$$a_6 = 64\left(-\dfrac{1}{2}\right)^{6-1} = -2$$

   b. 4, 12, 36, 108, …; a_9

$$a_n = a_1 r^{n-1}$$

$$a_9 = 4(3)^{9-1} = 26,244$$

4.   Write the first five terms of the geometric sequence whose first term is -3 and whose common ratio is $\dfrac{2}{3}$.

   Use $a_n = a_1 r^{n-1}$ with $a_1 = -3$, $r = \dfrac{2}{3}$, and $n = 1, 2, 3, 4, 5$.

Now Try:

3.   Evaluate the indicated term for each geometric sequence.

   a. $a_1 = -1$, $r = 3$; a_7

   ____

   b. $\dfrac{4}{3}, \dfrac{2}{3}, \dfrac{1}{3}, \dfrac{1}{6}, \cdots$; a_{10}

   ____

4.   Write the first five terms of the geometric sequence whose first term is 0.8 and whose common ratio is -5.

   ____

Name: _______________________ Date: _______________________

Instructor: _______________________ Section: _______________________

Objective 3 Practice Exercises

For extra help, see Examples 3–4 on page 990 of your text.

Evaluate the indicated term for each geometric sequence.

6. $a_1 = -4,\ r = 2;\ a_7$

6. _______________________

7. $a_1 = \dfrac{2}{3},\ r = -3;\ a_6$

7. _______________________

8. Write the first five terms of the geometric sequence whose first term is $\dfrac{1}{2}$ and common ratio 5.

8. _______________________

Objective 4   Find the sum of a specified number of terms of a geometric sequence.

Review these examples for Objective 4:

5. Evaluate the sum of the first seven terms of the geometric sequence with first term 36 and common ratio −6.

$$S_n = \frac{a_1\left(1-r^n\right)}{1-r}$$

$$S_7 = \frac{36\left(1-(-6)^7\right)}{1-(-6)}$$

$$= \frac{36\left(1-(-279{,}936)\right)}{7}$$

$$= \frac{36(279{,}937)}{7}$$

$$= 1{,}439{,}676$$

Now Try:

5. Evaluate the sum of the first six terms of the geometric sequence with first term $\frac{2}{3}$ and common ratio 3.

____

6. Evaluate $\displaystyle\sum_{i=1}^{8} 4(3)^i$.

Since the series is in the form $\displaystyle\sum_{i=1}^{n} a \cdot b^i$, it represents the sum of the first n terms of the geometric sequence with $a_1 = a \cdot b^1$ and $r = b$. $a_1 = 4 \cdot (3)^1 = 12$ and $r = 3$.

$$S_n = \frac{a_1\left(1 - r^n\right)}{1 - r}$$

$$S_8 = \frac{12\left(1 - 3^8\right)}{1 - 3}$$

$$= \frac{12(1 - 6561)}{-2}$$

$$= 39{,}360$$

6. Evaluate $\displaystyle\sum_{i=1}^{5} \frac{1}{2}(4^i)$.

____

Objective 4 Practice Exercises

For extra help, see Examples 5–6 on page 991 of your text.

9. Evaluate the sum of the first seven terms of the geometric sequence having first term 2 and common ratio 3.

9. ______________

Evaluate each sum.

10. $\displaystyle\sum_{i=1}^{12} 2^i$

10. ______________

11. $\displaystyle\sum_{i=1}^{7} 3(-2)^i$

11. ______________

Objective 5   Apply the formula for the future value of an ordinary annuity.

Review these examples for Objective 5:

7. Work each problem. Give answers to the nearest cent.

a. Matt decides to start a college fund for his child. He plans on depositing $1500 at the end of each year into an account paying 4.5% per year, compounded annually. How much will be in the account after 18 years?

The payments form an ordinary annuity with $R = 1500$, $n = 18$, and $i = 0.045$.

$$S = R\left[\frac{(1+i)^n - 1}{i}\right]$$

$$S = 1500\left[\frac{(1+0.045)^{18} - 1}{0.045}\right]$$

$$\approx 40,282.63$$

b. How much will be in the account after 18 years if Matt deposits $1500 at the end of each year into an account paying 3.5% per year, compounded quarterly?

The payments form an ordinary annuity with $R = 1500$, $n = 4(18)$, and $i = \dfrac{0.035}{4}$.

$$S = R\left[\frac{(1+i)^n - 1}{i}\right]$$

$$S = 1500\left[\frac{\left(1+\dfrac{0.035}{4}\right)^{4\cdot18} - 1}{\dfrac{0.035}{4}}\right]$$

$$\approx 149,566.71$$

Now Try:

7. Work each problem. Give answers to the nearest cent.

a. To save for retirement, Sara decides to deposit $2000 into an IRA at the end of each year. How much will be in the account at the end of 30 years if it is invested at 10% compounded annually?

____

b. To save for retirement, Sara decides to deposit $2000 into an IRA at the end of each year. How much will be in the account at the end of 30 years if it is invested at 5% compounded annually?

____

Objective 5 Practice Exercises

For extra help, see Example 7 on page 992 of your text.

Work each problem. Give answers to the nearest cent.

12. Julio is a professional wrestler who believes that his wrestling career will last ten years. To prepare for retirement, he deposits $14,000 at the end of each year for ten years into an account paying 7% compounded annually. How much will he have on deposit after ten years?

12. ___________________

13. Ginnie wants to buy a new car. If Ginnie puts $1,100 at the end of each year into an account paying 4.9% compounded annually for 3 years, how much will Ginnie have to spend on a car?

13. ___________________

Objective 6   Find the sum of an infinite number of terms of a certain geometric sequence.

Review these examples for Objective 6:

8. Evaluate the sum of the terms of the infinite geometric sequence with $a_1 = -3$ and $r = \frac{2}{3}$.

$$S = \frac{a_1}{1-r} \quad \text{Infinite sum formula}$$

$$= \frac{-3}{1-\frac{2}{3}}$$

$$= \frac{-3}{\frac{1}{3}}$$

$$= -9$$

Now Try:

8. Evaluate the sum of the terms of the infinite geometric sequence with $a_1 = 5$ and $r = -\frac{1}{5}$.

____

9. Evaluate $\displaystyle\sum_{i=1}^{\infty} 3\left(\tfrac{1}{4}\right)^{i}$.

This is the infinite geometric series with $a_1 = \dfrac{3}{4}$ and $r = \dfrac{1}{4}$. Since $|r| < 1$, find the sum using the formula $S = \dfrac{a_1}{1-r}$.

$$S = \frac{\frac{3}{4}}{1-\frac{1}{4}} = 1$$

9. Evaluate $\displaystyle\sum_{i=1}^{\infty} 3^{i}$.

____

Objective 6 Practice Exercises

For extra help, see Examples 8–9 on pages 994–995 of your text.

Evaluate the sum, if possible, of the infinite geometric sequence.

14. $a_1 = 2,\ \ r = -\dfrac{1}{3}$

14. _______________

15. $\displaystyle\sum_{i=1}^{\infty} -5\left(-\dfrac{5}{3}\right)^{i}$

15. _______________

Chapter 14   FURTHER TOPICS IN ALGEBRA

14.4   The Binomial Theorem

Learning Objectives
1   Expand a binomial raised to a power.
2   Find any specified term of the expansion of a binomial.

Key Terms

Use the vocabulary terms listed below to complete each statement in exercises 1–2.

Pascal's triangle   **binomial theorem (general binomial expansion)**

1. The _____________________ is a formula used to expand a binomial raised to a
    power.

2. Writing the coefficients of the terms of binomial expansions in a triangular pattern
    gives ______________________________________.

Objective 1   Expand a binomial raised to a power.

Review this example for Objective 1:

1. Evaluate 8!.

$$8! = 8 \cdot 7 \cdot 6 \cdot 5 \cdot 4 \cdot 3 \cdot 2 \cdot 1 = 40,320$$

2. Find the value of each expression.

    a. $\dfrac{6!}{4!2!}$

$$\frac{6!}{4!2!} = \frac{6 \cdot 5 \cdot 4 \cdot 3 \cdot 2 \cdot 1}{(4 \cdot 3 \cdot 2 \cdot 1)(2 \cdot 1)}$$

$$= \frac{6 \cdot 5}{2 \cdot 1}$$

$$= 15$$

    b. $\dfrac{8!}{4!4!}$

$$\frac{8!}{4!4!} = \frac{8 \cdot 7 \cdot 6 \cdot 5 \cdot 4 \cdot 3 \cdot 2 \cdot 1}{(4 \cdot 3 \cdot 2 \cdot 1)(4 \cdot 3 \cdot 2 \cdot 1)}$$

$$= \frac{8 \cdot 7 \cdot 6 \cdot 5}{4 \cdot 3 \cdot 2 \cdot 1}$$

$$= 70$$

Now Try:

1. Evaluate 10!.

    ____

2. Find the value of each
    expression.

    a. $\dfrac{9!}{9!0!}$

    ____

    b. $\dfrac{9!}{7!2!}$

    ____

c. $\dfrac{8!}{8!0!}$

$$\dfrac{8!}{8!0!} = \dfrac{8\cdot 7\cdot 6\cdot 5\cdot 4\cdot 3\cdot 2\cdot 1}{(8\cdot 7\cdot 6\cdot 5\cdot 4\cdot 3\cdot 2\cdot 1)(1)}$$

$$= 1$$

d. $\dfrac{8!}{7!1!}$

$$\dfrac{8!}{7!1!} = \dfrac{8\cdot 7\cdot 6\cdot 5\cdot 4\cdot 3\cdot 2\cdot 1}{(7\cdot 6\cdot 5\cdot 4\cdot 3\cdot 2\cdot 1)(1)}$$

$$= \dfrac{8}{1}$$

$$= 8$$

3. Evaluate $_9C_6$.

$$_nC_r = \dfrac{n!}{r!(n-r)!}$$

$$_9C_6 = \dfrac{9!}{6!(9-6)!}$$

$$= \dfrac{9!}{6!3!}$$

$$= \dfrac{9\cdot 8\cdot 7\cdot 6\cdot 5\cdot 4\cdot 3\cdot 2\cdot 1}{(6\cdot 5\cdot 4\cdot 3\cdot 2\cdot 1)(3\cdot 2\cdot 1)}$$

$$= 84$$

4. Expand $(2x+5y)^4$.

$$(2x+5y)^4$$

$$= (2x)^4 + \dfrac{4!}{1!3!}(2x)^3(5y) + \dfrac{4!}{2!2!}(2x)^2(5y)^2$$

$$+ \dfrac{4!}{3!1!}(2x)(5y)^3 + (5y)^4$$

$$= (2x)^4 + 4(2x)^3(5y) + 6(2x)^2(5y)^2$$

$$+ 4(2x)(5y)^3 + (5y)^4$$

$$= 16x^4 + 160x^3y + 600x^2y^2$$

$$+ 1000xy^3 + 625y^4$$

c. $\dfrac{9!}{8!1!}$

____

d. $\dfrac{9!}{5!4!}$

____

3. Evaluate $_{10}C_8$.

____

4. Expand $(3r-2s)^4$.

____

5. Expand $\left(\dfrac{1}{2}-a^2\right)^4$.

$\left(\dfrac{1}{2}-a^2\right)^4$

$=\left(\dfrac{1}{2}\right)^4+\dfrac{4!}{1!3!}\left(\dfrac{1}{2}\right)^3(-a)^2+\dfrac{4!}{2!2!}\left(\dfrac{1}{2}\right)^2\left(-a^2\right)^2$

$\qquad+\dfrac{4!}{3!1!}\left(\dfrac{1}{2}\right)\left(-a^2\right)^3+\left(-a^2\right)^4$

$=\left(\dfrac{1}{2}\right)^4+4\left(\dfrac{1}{2}\right)^3\left(-a^2\right)+6\left(\dfrac{1}{2}\right)^2\left(-a^2\right)^2$

$\qquad+4\left(\dfrac{1}{2}\right)\left(-a^2\right)^3+\left(-a^2\right)^4$

$=a^8-2a^6+\dfrac{3}{2}a^4-\dfrac{1}{2}a^2+\dfrac{1}{16}$

5. Expand $\left(2-\dfrac{x}{2}\right)^5$.

____

Objective 1 Practice Exercises

For extra help, see Examples 1–5 on pages 999–1002 of your text.

Evaluate each expression.

1. $\dfrac{7!}{3!4!}$

1. ___________________

2. $_{15}C_3$

2. ___________________

Use the binomial theorem to expand the expression.

3. $(2y+3z)^5$

3. ___________________

Objective 2   Find any specified term of the expansion of a binomial.

Review this example for Objective 2:

6. Find the sixth term of the expansion of $(a-2b)^9$.

Using the general form of the binomial expansion, we let $n = 9$, $x = a$, $y = -2b$, and $r = 6$. In the sixth term, a has an exponent of $9 - (6 - 1) = 4$ and $2b$ has an exponent of $6 - 1 = 5$. Thus, we have

$$\frac{9!}{5!4!}a^4(-2b)^5 = \frac{9\cdot 8\cdot 7\cdot 6}{4\cdot 3\cdot 2\cdot 1}a^4(-2b)^5$$

$$= 126a^4\left(-32b^5\right)$$

$$= -4032a^4b^5$$

Now Try:

6. Find the sixth term of the expansion of $\left(m^3 - 3r\right)^7$.

____

Objective 2 Practice Exercises

For extra help, see Example 6 on page 1003 of your text.

Find the indicated term of each binomial expansion.

4.   Third term of $(m-3)^5$

4. _________________

5.   Fourth term of $(4x-3y)^5$

5. _________________

6.   Eighth term of $\left(3k - p^2\right)^9$

6. _________________

INTEGRATED
REVIEW WORKSHEETS

Chapter 1 The Real Number System

Learning Objectives

Use the < and > symbols to compare integers.
Add integers.
Combine adding and subtracting of integers.
Multiply integers.
Use the order of operations to evaluate the expressions.
Identify variables, constants and coefficients.
Plot a point, given the coordinates, and find the coordinates, given a point.

Key Terms

Use the vocabulary terms listed below to complete each statement in exercises 1−7.

number line   **sum**   **factors**   **addends**

order of operations   **product**   **exponent**

1. A ___________________ is used to show how numbers relate to each other.

2. The answer to an addition problem is called the ___________________.

3. In addition, the numbers being added are called the ___________________.

4. In multiplication, the numbers being multiplied are called the

    ___________________.

5. The answer to a multiplication problem is called the ___________________.

6. For problems or expressions with more than one operation, the ______________
    tells what to do first, second, and so on, to obtain the correct answer.

7. An ___________________ tells how many times a number is used as a factor
    in repeated multiplication.

Objective 1:   Use the < and > symbols to compare integers.

Review these examples:

1. Write < or > between each pair of integers to make a true statement.

    a. 0 _____ 7

    0 is to the left of 7 on the number line, so 0 is less than 7. Write $0 < 7$.

    b. –2 _____ –5

    –2 is to the right of –5, so –2 is greater than –5. Write $-2 > -5$.

    c. –6 _____ 7

    –6 is to the left of 7, so –6 is less than 7. Write $-6 < 7$.

Now Try:

1. Write < or > between each pair of integers to make a true statement.

    a. 0 _______ –8

    b. –3 _____ –9

    c. 11 _____ –4

Practice Exercises

Write < or > in each blank to make a true statement.

1. -23 _____ -32

2.   –6 _____ 0

3. -5 _____ -3

1. ________________

2. ________________

3. ________________

Objective 2:   Add integers.

Review these examples:

2.   Use a number line to find $(-3)+(-2)$.

Start at 0. Move 3 places to the left. Then move 2 places to the left.

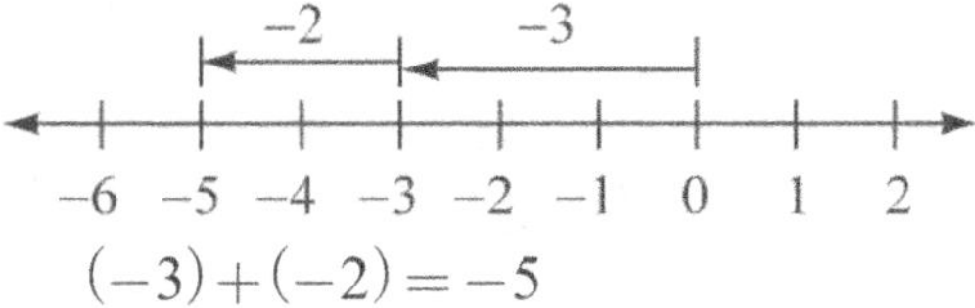

$(-3)+(-2)=-5$

3.   Add.

a. $-9+(-4)$

Step 1 Add the absolute values.
$|-9|=9$ and $|-4|=4$
Add $9+4$ to get 13.
Step 2 Use the common sign as the sign of the sum. Both numbers are negative, so the sum is negative.
$-9+(-4)=-13$

b. $7+8$

Both numbers are positive, therefore the sum is positive.
$7+8=15$

4.   Add.

a. $-7+4$

Step 1 $|-7|=7$ and $|4|=4$
    Subtract $7-4$ to get 3.
Step 2 -7 has the greater absolute value and is negative, so the sum is also negative.
$-7+4=-3$

b. $-6+15$

Step 1 $|-6|=6$ and $|15|=15$
    Subtract $15-6$ to get 9.
Step 2 15 has the greater absolute value and is positive, so the sum is also positive.
$-6+15=9$

Now Try:

2.   Use a number line to find $(-4)+(-1)$.

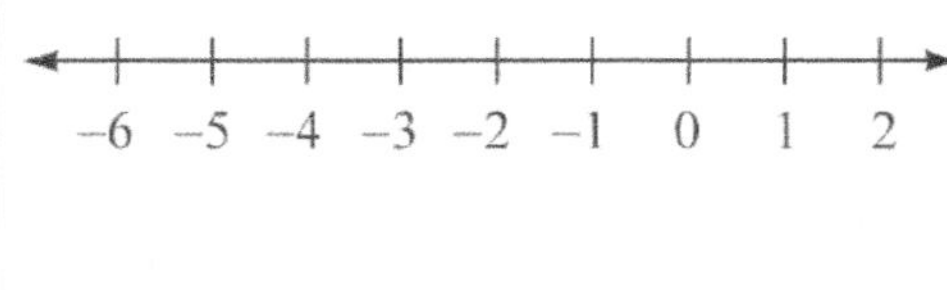

____

3.   Add.

a. $(-25)+(-11)$

____

b. $19+21$

____

4.   Add.

a. $-19+3$

____

b. $-4+25$

____

Practice Exercises

Add by using a number line.

4. $-8+5$

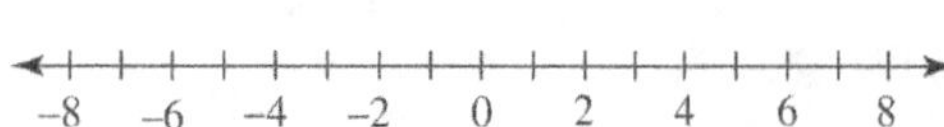

4. __________________

Add.

5. $7+(-22)$

5. __________________

Write an addition problem for the situation and find the sum.

6.   While playing a card game, Diane first gained 23 points, then lost 16 points, and finally gained 11 points. What was her final score?

6. __________________

Objective 3:   Combine adding and subtracting of integers.

Review this example:

5.   Simplify by completing all the calculations.
$$-6-15-18+3$$

Change all subtractions to adding the opposite. Change 15 to –15. Change 18 to –18. Then add from left to right.

$$-6 \;-\; 15 \;-\; 18 \;+\; 3$$

$$-6+(-15)+(-18)+3$$

$$-21 \;+\; (-18)+3$$

$$-39 \qquad +3$$

$$-36$$

Now Try:

5.   Simplify by completing all the calculations.
$$-9-14-21+6$$

____

Practice Exercises

Simplify.

7. $-2-16-(-18)$

8. $13-(-8)+(-7)-7$

9. $-14-(-3)-0+(-11)$

7. _______________

8. _______________

9. _______________

Objective 4: Multiply integers.

Review these examples:	**Now Try:**
Multiply.	Multiply.

6a. $-3 \cdot 9$

The factors have different signs, so the product is negative.
$$-3 \cdot 9 = -27$$

6a. $-6 \cdot 7$

____

b. $-11(-7)$

The factors have the same sign, so the product is positive.
$$-11(-7) = 77$$

b. $-4(-15)$

____

c. $-3 \cdot (-3) \cdot (-3)$

There is no work to do inside the parentheses, so multiply $-3 \cdot (-3)$ first. The factors have the same sign, so the product is positive.
$$-3 \cdot (-3) \cdot (-3)$$
$$9 \ \cdot \ (-3)$$
$$-27$$
Note that the factors 9 and –3 have different signs, so the product is negative.

c. $-4 \cdot (-4) \cdot (-4)$

____

d. $8(-12)$

The factors have different signs, so the product is negative.
$$8(-12) = -96$$

d. $9(-6)$

____

Practice Exercises

Multiply.

10. $-7(-1)(-4)$

10. _______________

11. $-2 \cdot 31$

11. _______________

12. $-5 \cdot (-2) \cdot (6)$

12. _______________

Objective 5:   Use the order of operations.

Review these examples:

7a. Simplify.

$$6^2 - (-7)^2$$

$$6^2 - (-7)^2$$
$$6^2 - 49$$
$$36 - 49$$
$$-13$$

b.   Simplify $8 + 5(24 - 6) \div 9$.

$$8 + 5(24 - 6) \div 9$$
$$8 + 5\ (18)\ \div 9$$
$$8 + 90 \div 9$$
$$8 + 10$$
$$18$$

c.   Simplify.

$$(-3)^4 - (5 - 7)^2 (-9)$$

$$(-3)^4 - (5 - 7)^2 (-9)$$
$$(-3)^4 - (-2)^2 (-9)$$
$$81 - 4(-9)$$
$$81 - (-36)$$
$$81 + 36$$
$$117$$

d.   Simplify.

$$-9 \div (7 - 4) - 13$$

$$-9 \div (7 - 4) - 13$$
$$-9 \div (3) - 13$$
$$-3 - 13$$
$$-16$$

Now Try:

7a. Simplify.

$$8^2 - (-9)^2$$

____

b.   Simplify $5 + 4(25 - 3) \div 11$.

____

c.   Simplify.

$$(-5)^3 - (6 - 9)^2 (-8)$$

____

d.   Simplify.

$$-10 \div (8 - 6) - 19$$

____

Practice Exercises

Simplify.

13. $3(-2+6)-(8-13)$

13. ______________________

14. $3-(-6)\cdot(-1)^8$

14. ______________________

15. $(-2)^3 \cdot (7-9)^2 \div 2$

15. ______________________

Objective 6:   Identify variables, constants, and expressions.

Review this example:

8. *Identify the parts of each expression. Choose from* **variable**, **constant**, *and* **coefficient**.

$$8x + 4$$

The coefficient of the variable x is 8 and 4 is the constant term.

Now Try:

8. *Identify the parts of each expression. Choose from* **variable**, **constant**, *and* **coefficient**.

$$h - 6$$

____

Practice Exercises

Identify the parts of each expression. Choose from **variable**, **constant**, *and* **coefficient**.

16. $-7 + h$

16. _______________

17. $-2w$

17. _______________

18. $9k + 1$

18. _______________

Objective 7:   Plot a point, given the coordinates, and find the coordinates, given a point.

Review these examples:	**Now Try:**

Review these examples:

9.   Use the grid below to plot each point.

   a. (2, 5)

Start at (0, 0). Move to the right along the horizontal axis until you reach 2. Then move up 5 units so that you are aligned with 5 on the vertical axis. Make a dot and label it (2, 5).

   b. (–5, 4)

Start at (0, 0). Move to the left along the horizontal axis until you reach –5. Then move up 4 units so that you are aligned with 4 on the vertical axis. Make a dot and label it (–5, 4).

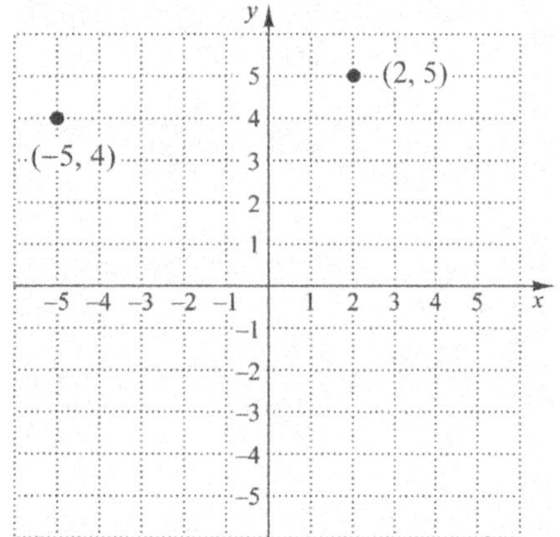

10. Use the grid below to plot each point.

   a. (1,–2)

Move right 1 unit. Move down 2 units. Make a dot and label it (1,–2).

   b. (–4, 0)

Move left 4 units. Make a dot and label it (–4, 0).

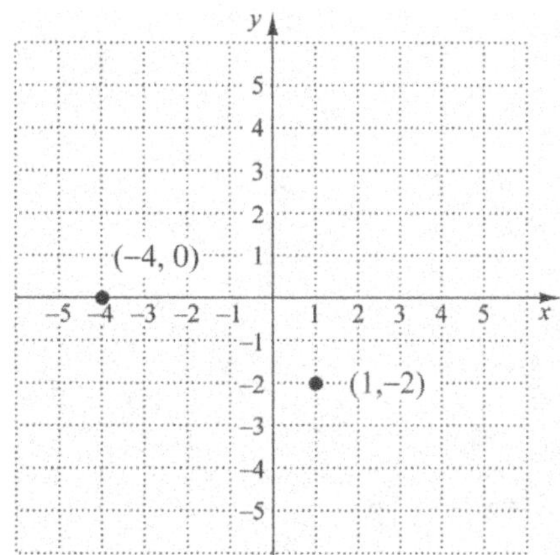

Now Try:

9.   Use the grid to plot each point.

   a. (1, 3)

   b. (4,–1)

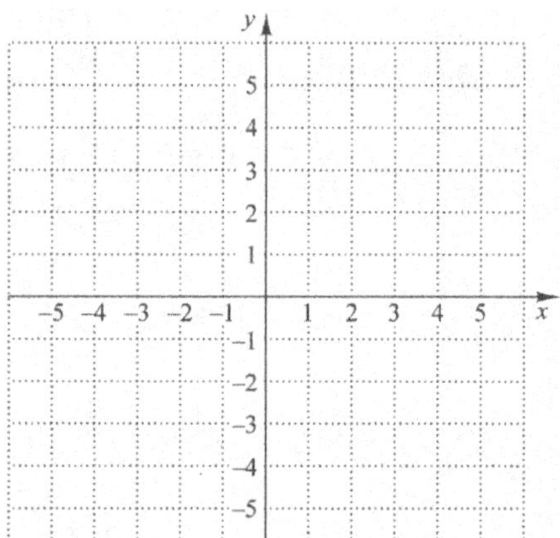

10. Use the grid to plot each point.

   a. (–2, 0)

   b. (0, 5)

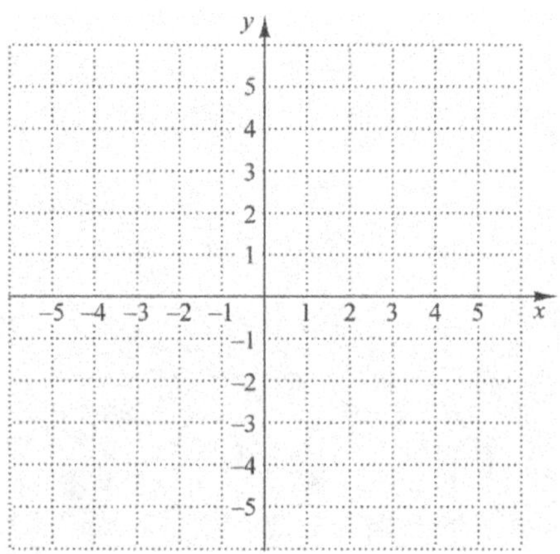

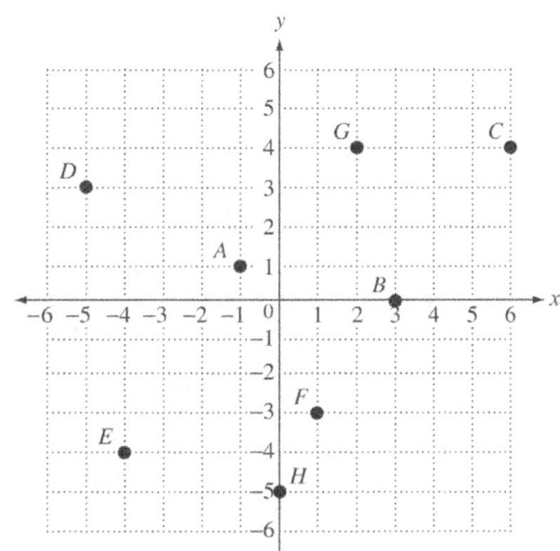

11. Give the coordinates of each point.

   a. *A*

To reach point *A* from the origin, move 1 unit to the left; then move up 1 unit. The coordinates are (–1, 1).

   b. *B*

To reach point *B* from the origin, move 3 units to the left; do not move up or down. The coordinates are (3, 0).

   c. *C*

To reach point *C* from the origin, move 6 units to the right; then move up 4 units. The coordinates are (6, 4).

   d. *D*

To reach point *D* from the origin, move 5 units to the left; then move up 3 units. The coordinates are (–5, 3).

11. Give the coordinates of each point.

   a. *E*

____

   b. *F*

____

   c. *G*

____

   d. *H*

____

Practice Exercises

19. *Plot each point on the rectangular coordinate system below. Label each point with its coordinates.*

a. $(-3, 4)$

b. $(2, -1)$

c. $(0, -5)$

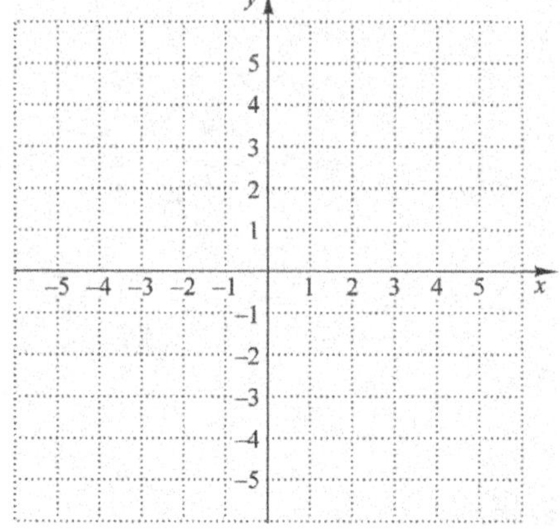

20. *Give the coordinates of each point.*

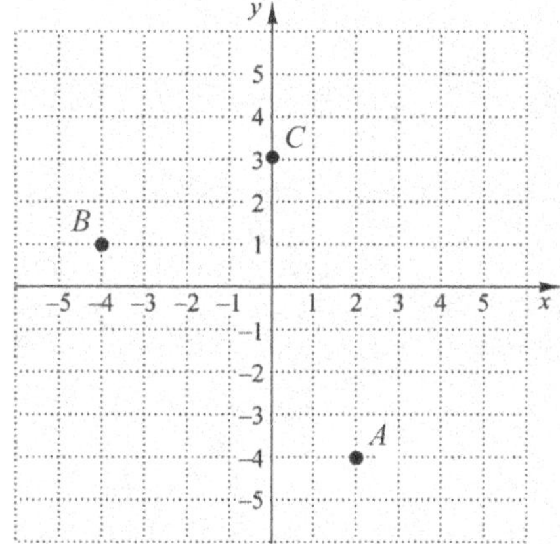

A ___________________

B ___________________

C ___________________

Chapter 2 Linear Equations and Inequalities in One Variable

Learning Objectives

Evaluate variable expressions for given replacement values.
Simplify expressions by using by combining like terms.
Write decimals as percents.
Simplify expressions by using the distributive property to multiply.
Write fractions as percents.
Simplify expressions using a combination of techniques.
Translate word phrases into algebraic expressions.

Key Terms

Use the vocabulary terms listed below to complete each statement in exercises 1−10.

simplify an expression	**term**	**constant term**
variable term	**like terms**	**evaluate the expression**
sum	**difference** **product**	**per**
quotient **increased by**	**less than**	**double**

1. _________________________________are terms with exactly the same variable parts.

2. Each addend in an expression is called a _________________.

3. To _________________________________, combine all the like terms.

4. A _____________________ has a coefficient multiplied by a variable part.

5. A _____________________ is a type of term that is just a number.

6. To _____________________________, replace each variable with specific values and then follow the order of operations.

7. _____________________ and _____________________ are words that mean addition.

8. _____________________ and _____________________ are words that mean multiplication.

9. _____________________ and _____________________ are words that mean division.

10. _____________________ and _____________________ are words that mean subtraction.

Objective 1:   Evaluate variable expressions for given replacement values.

Review these examples:

1. Use this rule for ordering lunches: Order the class limit minus 8. The expression is $c - 8$.

    a. Evaluate the expression when the class limit is 39.

    Replace c with 39 and follow the rule.
    $$c - 8$$
    $$39 - 8$$
    31   Order 31 lunches.

    b. Evaluate the expression when the class limit is 22.

    Replace c with 22 and follow the rule.
    $$c - 8$$
    $$22 - 8$$
    14   Order 14 lunches.

2. The expression (rule) for finding the perimeter of a square shape is $4s$. Evaluate the expression when the length of a side of a square garden is 20 feet.

    Replace s with 20 feet.

    The total distance around the square garden is 80 feet.

3.
    Find your average score if you bowl three games and your total score for all three games is 384.

    Use the expression (rule) for finding your average score. Replace t with your total score of 384, and replace g with 3, the number of games.
    $$\frac{t}{g} \rightarrow \frac{384}{3}$$
    Divide 384 by 3 to get 128. Your average score is 128.

Now Try:

1. Use this rule (from Now Try 1) for ordering lunches: Order the class limit plus 5. The expression is $c + 5$.

    a. Evaluate the expression when the class limit is 24.

    ____

    b. Evaluate the expression when the class limit is 45.

    ____

2. The expression (rule) for finding the perimeter of a square shape is $4s$. Evaluate the expression when the length of a side of a square garden is 8 yards.

    ____

3.
    Find your average score if you bowl four games and your total score for all four games is 384.

    ____

Practice Exercises

Evaluate each expression.

1. The expression (rule) for the weight of an object (in Newtons) is $10m$, where m is the object's mass in kilograms. Evaluate the expression when
   (a) the object has a mass of 14 kilograms.
   (b) the object has a mass of 92 kilograms.

   1. a. _______________
      b. _______________

2. The expression (rule) for the total number of dollars spent on a car, when there is a $2300 down payment and monthly payments of $215, is $2300 + 215t$, where t is the number of monthly payments. Evaluate the expression when
   (a) there are 36 monthly payments.
   (b) there are 48 monthly payments.

   2. a. _______________
      b. _______________

3. The expression (rule) for the amount of insurance paid out on a policy with a $2500 deductible is $\frac{8c}{10} - \$2500$, where c is total medical costs. Evaluate the expression when
   (a) total medical costs are $10,625.
   (b) total medical costs are $12,480.

   3. a. _______________
      b. _______________

Objective 2: Simplify expressions by combining like terms.

Review this example:	**Now Try:**
4. Combine like terms.	4. Combine like terms.

Review this example:

4. Combine like terms.

$$7x + x + 9x$$

$$7x + x + 9x$$

$$7x + 1x + 9x$$

$$(7 + 1 + 9)x$$

$$17x$$

Now Try:

4. Combine like terms.

$$12x + 7x + x$$

____

Practice Exercises

Identify the like terms in the expression. Then identify the coefficients of the like terms.

4. $x^2 + 2x + (-5x^3) + (-3x) + 8$

4. _______________

Simplify each expression.

5. $6d + d$

5. _______________

6. $c^3 z^4 - 11c^3 z^4 - 9c^3 z^4$

6. _______________

Objective 3:   Write decimals as percents.

Review these examples:	**Now Try:**
5. Write each decimal as a percent by moving the decimal point two places to the right.	**5.** Write each decimal as a percent by moving the decimal point two places to the right.

a.   0.31   **a.**   0.43

Decimal point is moved two places to the right and percent symbol is attached.
$0.31 = 31\%$

    1.6   **b.**   2.3

0 is attached so the decimal point can be moved two places to the right.
$1.6 = 1.60 = 160\%$

   b.   0.904   **c.**   0.751

$0.904 = 90.4\%$

Practice Exercises

Write each decimal as a percent.

7.   0.2   **7.** _______________

8.   0.564   **8.** _______________

9.   4.93   **9.** _______________

Objective 4: Simplify expressions by using the distributive property to multiply.

Review these examples:	**Now Try:**

Review these examples:

6. Simplify.

   a. $8(d-9)$

   $8(d-9)$ can be written as $8[d+(-9)]$
$$8 \cdot d + 8(-9)$$
$$8d + (-72)$$
$$8d - 72$$
So, $8(d-9)$ simplifies as $8d - 72$.

   b. $7(4x+3)$

   $7(4x+3)$ can be written as $7 \cdot 4x + 7 \cdot 3$
$$7 \cdot 4 \cdot x + 21$$
$$28 \cdot x + 21$$
$$28x + 21$$
So, $7(4x+3)$ simplifies as $28x + 21$.

   c. $-5(8b+4)$

   $-5(8b+4)$ can be written as $-5 \cdot 8b + (-5) \cdot 4$
$$-5 \cdot 8 \cdot b + (-20)$$
$$-40 \cdot b + (-20)$$
$$-40b + (-20)$$
So, $-5(8b+4)$ simplifies as $-40b - 20$.

7. Simplify: $9 + 4(x-5)$

Use the distributive property first.
$$9 + 4(x-5)$$
$$9 + 4 \cdot x - 4 \cdot 5$$
$$9 + 4x - 20$$
$$4x + 9 - 20$$
$$4x \quad -11$$
The simplified expression is $4x - 11$.

Now Try:

6. Simplify.

   a. $28(x-4)$

   ____

   b. $6(5x+2)$

   ____

   c. $-9(6x+4)$

   ____

7. Simplify: $12 + 6(x-3)$

   ____

Practice Exercises

Use the distributive property to simplify each expression.

10. $-5(t+4)$

10. _______________

11. $-8(3s-4)$

11. _______________

Simplify the expression.

12. $-q+7(3q-2)+11$

12. _______________

Objective 5: Write fractions as percents.

Review this example:
8. Write the fraction as a percent.

$$\frac{5}{16}$$

$$\frac{5}{16} = \left(\frac{5}{16}\right)(100\%) = \left(\frac{5}{16}\right)\left(\frac{100}{1}\%\right)$$

$$= \left(\frac{5}{4\cdot 4}\right)\left(\frac{4\cdot 25}{1}\%\right)$$

$$= \frac{125}{4}\% = 31\frac{1}{4}\%$$

Now Try:
8. Write the fraction as a percent.

$$\frac{15}{16}$$

____

Practice Exercises

Write each fraction or mixed number as a percent. If you're using a calculator, first work each one by hand. Then use your calculator and round to the nearest tenth of a percent, if necessary.

13. $\dfrac{47}{50}$ **13.** ___________________

14. $\dfrac{11}{40}$ **14.** ___________________

15. $\dfrac{64}{75}$ **15.** ___________________

Objective 6: Simplify expressions using a combination of techniques.

Review this example:	**Now Try:**
9. Simplify.	9. Simplify.

<table>
<tr><td>

$4(x - 3) + 7x - 2(5x)$

$= 4x - 12 + 7x - 10x$

$= x - 12$

</td><td>

$-2(x + 5) + 12 + 3(-2x)$

____

</td></tr>
</table>

Practice Exercises

Simplify the expressions.

16. $-3(2h - 4) + 21h$

16. __________________

17. $13 + 6(5 - 7m)$

17. __________________

18. $-3t + 5(t - 8) + 15$

18. __________________

Objective 7:   Translate word phrases into algebraic expressions.

Review these examples:

10. Write each phrase as an algebraic expression. Use x as the variable.

a. −66 added to a number

Algebraic expression: $-66 + x$ or $x + (-66)$

b. 39 minus a number

Algebraic expression: $39 - x$

11. Write the phrase as an algebraic expression. Use x as the variable.

17 subtracted from 8 times a number

Algebraic expression: $8x - 17$

Now Try:

10. Write each phrase as an algebraic expression. Use x as the variable.

a. −20 added to a number

____

b. 48 minus a number

____

11. Write the phrase as an algebraic expression. Use x as the variable.

81 subtracted from 6 times a number

____

Practice Exercises

Write an algebraic expression using x as the variable.

19. The product of −6 and a number

19. ____________

20. The quotient of a number and 10

20. ____________

21. One more than three times a number

21. ____________

Chapter 3 Linear Equations in Two Variables

Learning Objectives

Determine whether a given number is a solution of an equation.
Use the four steps for solving a linear equation.
Graph intervals on a number line.
Solve linear inequalities using both properties of inequality.

Key Terms

Use the vocabulary terms listed below to complete each statement in exercises 1−8.

number line	coordinate	variable
inequalities	interval	constant
linear inequality	algebraic expression	

1. A ________________________ is a symbol, usually a letter, used to represent an unknown number.

2. A ________________________ is a fixed, unchanging number.

3. A collection of numbers, variables, operation symbols, and grouping symbols is an________________________.

4. A ________________________ shows the ordering of the real numbers on a line.

5. Numbers that can be represented by points on the number line are

    ________________________.

6. A(n) ________________________ can be written in the form $Ax + B < C$, $Ax + B \leq C$, $Ax + B > C$, or $Ax + B \geq C$, where A, B, and C are real numbers with $A \neq 0$.

7. Algebraic expressions related by $<$, $\leq$, $>$, or $\geq$ are called ________________________.

8. The ________________________ for $a \leq x < b$ is $[a, b)$.

Objective 1:   Determine whether a given number is a solution of an equation.

Review this example:

1. Which of these numbers, 65, 45, or 85, is the solution of the equation $c - 15 = 70$?

    Replace c with each of the numbers. The one that makes the equation balance is the solution.

    $65 - 15 \neq 70$ Does not balance: $65 - 15 = 50$.

    $45 - 15 \neq 70$ Does not balance: $45 - 15 = 30$.

    $85 - 15 = 70$ Balances: $85 - 15 = 70$.

    The solution is 85 because when c is 85, the equation balances.

Now Try:

1. Which of these numbers, 25, 29, or 21, is the solution of the equation $25 = c - 4$?

    ____

Practice Exercises

In each list of numbers, find the one that is a solution of the given equation.

1. $r - 12 = 0$
    $-8, -2, \ 12$

    1. _______________

2. $p + 7 = 10$
    $-7, 3, 7$

    2. _______________

3. $6 - 2y = -8$
    $-3, 5, 7$

    3. _______________

Objective 2   Use the four steps for solving a linear equation.

Review these examples for Objective 2:	**Now Try:**

2. Solve $-5x + 8 = 23$.

2. Solve $-8x + 11 = 59$.

Step 1 There are no parentheses, fractions, or decimals in this equation, so this step is not necessary.

$$-5x + 8 = 23$$

Step 2 $\quad -5x + 8 - 8 = 23 - 8$

$$-5x = 15$$

Step 3 $\qquad \dfrac{-5x}{-5} = \dfrac{15}{-5}$

$$x = -3$$

Step 4 Check by substituting –3 for x in the original equation.

$$-5x + 8 = 23$$
$$-5(-3) + 8 \overset{?}{=} 23$$
$$15 + 8 \overset{?}{=} 23$$
$$23 = 23 \quad \text{True}$$

The solution, –3, checks, so the solution set is {–3}.

3. Solve $4x + 3 = 6x - 11$.

3. Solve $5x + 4 = 8x - 20$.

Step 1 There are no parentheses, fractions, or decimals in this equation, so begin with Step 2.

$$4x + 3 = 6x - 11$$

Step 2 $\quad 4x + 3 - 4x = 6x - 11 - 4x$

$$3 = 2x - 11$$
$$3 + 11 = 2x - 11 + 11$$
$$14 = 2x$$

Step 3 $\qquad \dfrac{14}{2} = \dfrac{2x}{2}$

$$7 = x$$

Step 4 Check by substituting 7 for x in the original equation.

$$4x + 3 = 6x - 11$$
$$4(7) + 3 \overset{?}{=} 6(7) - 11$$
$$28 + 3 \overset{?}{=} 42 - 11$$
$$31 = 31 \quad \text{True}$$

The solution, 7, checks, so the solution set is {7}.

4. Solve $9a - (4 + 3a) = 2a + 5$.

$$9a - (4 + 3a) = 2a + 5$$

Step 1 $\quad 9a - 4 - 3a = 2a + 5$

$$6a - 4 = 2a + 5$$

Step 2 $\quad 6a - 4 - 2a = 2a + 5 - 2a$

$$4a - 4 = 5$$

$$4a - 4 + 4 = 5 + 4$$

$$4a = 9$$

Step 3 $\quad \dfrac{4a}{4} = \dfrac{9}{4}$

$$a = \dfrac{9}{4}$$

Step 4 Check that the solution set is $\left\{\dfrac{9}{4}\right\}$.

4. Solve $10a - (11 + 3a) = 5a + 4$.

____

Objective 2 Practice Exercises

Solve each equation and check your solution.

4. $\quad 7t + 6 = 11t - 4$

4. ________________

5. $\quad 3a - 6a + 4(a - 4) = -2(a + 2)$

5. ________________

6. $\quad 3(t + 5) = 6 - 2(t - 4)$

6. ________________

Objective 3   Graph intervals on a number line.

Review this example for Objective 3:

5. Write the inequality in interval notation, and graph the interval.

$x > -3$

The statement $x > -3$ says that x can represent any value greater than –3, but cannot equal –3, written $(-3, \infty)$. We graph this interval by placing a parenthesis at –3 and drawing an arrow to the right. The parenthesis indicates that –3 is not part of the graph.

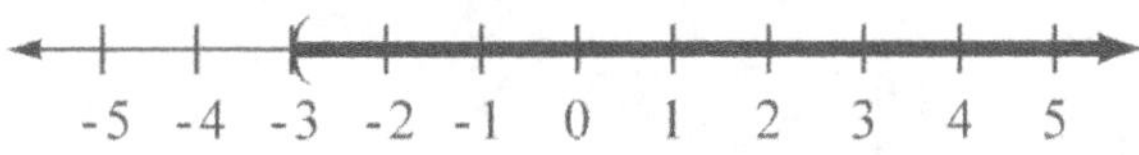

Now Try:

5. Write the inequality in interval notation, and graph the interval.

$x > -1$

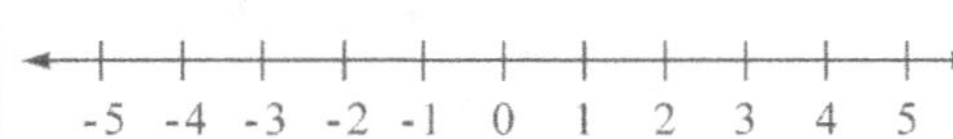

Objective 3 Practice Exercises

Write each inequality in interval notation and graph the interval.

7. $3 < a$

7. _______________________

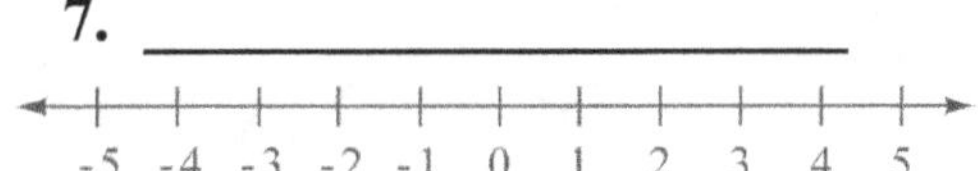

8. $y \geq -2$

8. _______________________

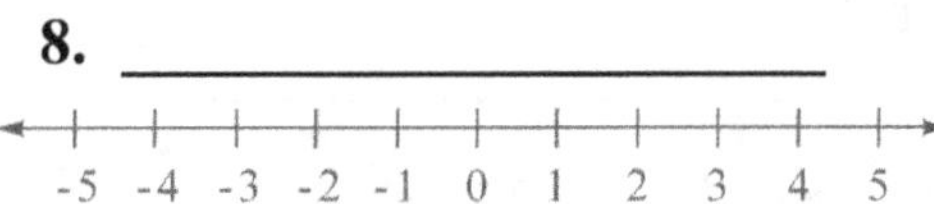

9. $x < -4$

9. _______________________

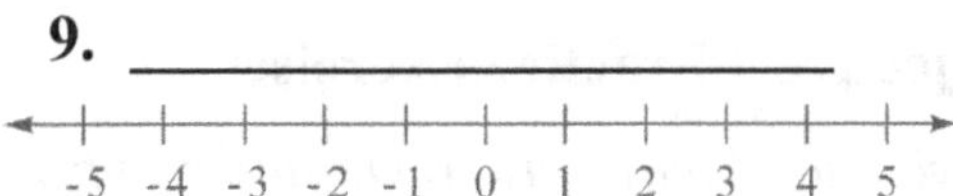

Objective 4   Solve linear inequalities using both properties of inequality.

| **Review this example for Objective 4:** | **Now Try:** |

Review this example for Objective 4:

6. Solve $4x + 3 - 7 > -2x + 8 + 3x$. Graph the solution set.

Step 1 Combine like terms and simplify.
$$4x + 3 - 7 > -2x + 8 + 3x$$
$$4x - 4 > x + 8$$

Step 2 Use the addition property of inequality.
$$4x - 4 - x > x + 8 - x$$
$$3x - 4 > 8$$
$$3x - 4 + 4 > 8 + 4$$
$$3x > 12$$

Step 3 Use the multiplication property of inequality.
$$\frac{3x}{3} > \frac{12}{3}$$
$$x > 4$$

The solution set is $(4, \infty)$. The graph is shown below.

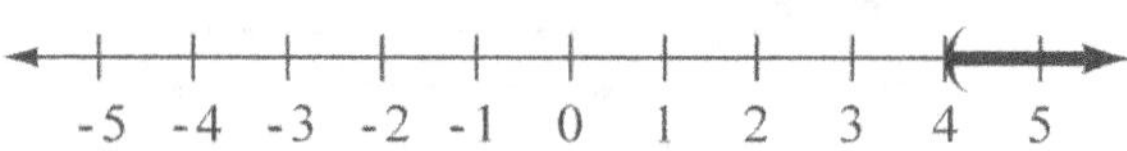

Now Try:

6. Solve $8x - 5 + 4 \geq 6x - 3x + 9$. Graph the solution set.

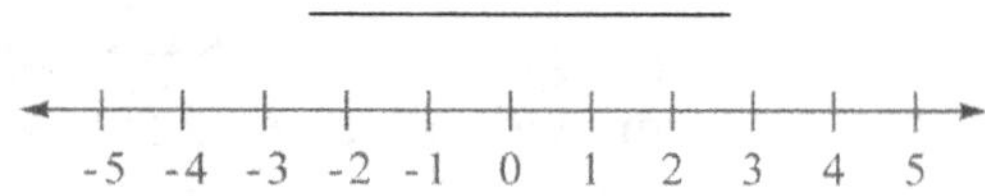

Objective 4 Practice Exercises

Solve each inequality. Write the solution set in interval notation and then graph it.

10. $4(y - 3) + 2 > 3(y - 2)$

10. _______________________

11. $-3(m + 2) + 3 \leq -4(m - 2) - 6$

11. _______________________

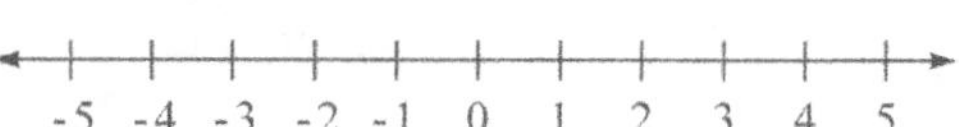

12. $7(2 - x) \leq -2(x - 3) - x$

12. _______________________

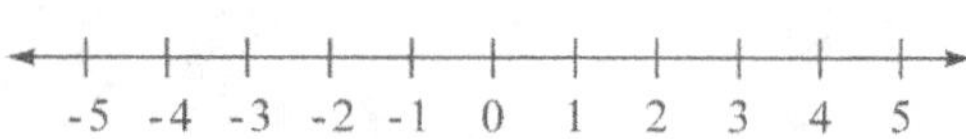
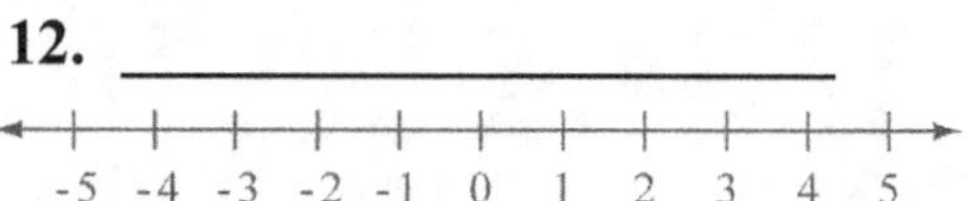

Chapter 4 Exponents and Polynomials

Learning Objectives

Review the use of exponents.
Use the product rule of exponents.
Use the power rules for exponents.
Simplify polynomials by combining like terms.

Key Terms

Use the vocabulary terms listed below to complete each statement in exercises 1−6.

exponential expression	**base**	**power**
numerical coefficient	**base**	**like terms**

1. In the term $4x^2$, "4" is the_______________________________.

2. A number, a variable, or a product or quotient of a number and one or more variables raised to powers is called a _______________________________.

3. Terms with exactly the same variables, including the same exponents, are called _______________________________.

4. 2^5 is read "2 to the fifth _______________________".

5. A number written with an exponent is called a(n) _______________________________.

6. The _______________ is the number being multiplied repeatedly.

Objective 1:   Review the use of exponents.

Review these examples.
1. Name the base and exponent of each expression. Then evaluate.

  a. 3^4

  Base: 3
  Exponent: 4
  Value: $3^4 = 3 \cdot 3 \cdot 3 \cdot 3 = 81$

  b. -3^4

  Base: 3
  Exponent: 4
  Value: $-3^4 = -1 \cdot (3 \cdot 3 \cdot 3 \cdot 3) = -81$

  c. $(-3)^4$

  Base: -3
  Exponent: 4
  Value: $(-3)^4 = (-3)(-3)(-3)(-3) = 81$

1. Name the base and exponent of each expression. Then evaluate.

  a. 2^6

  ____

  b. -2^6

  ____

  c. $(-2)^6$

  ____

Practice Exercises

Write the expression in exponential form and evaluate, if possible.

  1. $\left(\frac{1}{3}\right)\left(\frac{1}{3}\right)\left(\frac{1}{3}\right)\left(\frac{1}{3}\right)\left(\frac{1}{3}\right)$

  1. _________________

Evaluate each exponential expression. Name the base and the exponent.

  2. $(-4)^4$

  2. _________________

  base_____________

  exponent__________

  3. -3^8

  3. _________________

  base_____________

  exponent__________

Objective 2: Use the product rule for exponents.

Review this example:

2. Multiply $5x^4$ and $6x^9$.

$$5x^4 \cdot 6x^9 = (5 \cdot 6) \cdot \left(x^4 \cdot x^9\right)$$
$$= 30x^{4+9}$$
$$= 30x^{13}$$

Now Try:

2. Multiply $6x^5$ and $3x^6$.

____

Practice Exercises

Use the product rule to simplify each expression, if possible. Write each answer in exponential form.

4. $7^4 \cdot 7^3$

4. _______________

5. $\left(-2c^7\right)\left(-4c^8\right)$

5. _______________

6. $\left(3k^7\right)\left(-8k^2\right)\left(-2k^9\right)$

6. _______________

Objective 3:   Use the power rules for exponents.

Review this example:

3.   Use power rule (a) for exponents to simplify.

$$\left(m^3\right)^4$$

$$\left(m^3\right)^4 = m^{3\cdot 4}$$

$$= m^{12}$$

Now Try:

3.   Use power rule (a) for exponents to simplify.

$$\left(n^5\right)^6$$

____

Practice Exercises

Simplify each expression. Write all answers in exponential form.

7. $\left(7^3\right)^4$

7. __________

8. $-\left(v^4\right)^9$

8. __________

9. $\left[2^3\right]^7$

9. __________

Objective 4:   Simplify polynomials by combining like terms.

Review these examples for Objective 4:

4.   Simplify each expression by combining like terms.

a. $-7x^2 + 10x^2$

$$-7x^2 + 10x^2 = (-7 + 10)x^2$$
$$= 3x^2$$

b. $4x^5 - 15x^5 + x^5$

$$4x^5 - 15x^5 + x^5 = (4 - 15 + 1)x^5$$
$$= -10x^5$$

c. $19m^3 + 6m + 5m^3$

$$19m^3 + 6m + 5m^3 = (19 + 5)m^3 + 6m$$
$$= 24m^3 + 6m$$

d. $5rs^3 + rs^3 - 4rs^3$

$$5rs^3 + rs^3 - 4rs^3 = (5 + 1 - 4)rs^3$$
$$= 2rs^3$$

Now Try:

4.   Simplify each expression by combining like terms.

a. $-6x^4 + 11x^4$

____

b. $9x^7 - 18x^7 + x^7$

____

c. $22m^2 + 15m^3 + 7m^2$

____

d. $6p^2q - 3p^2q + 2p^2q$

____

Practice Exercises

In each polynomial, combine like terms whenever possible. Write the result with descending powers.

10. $7z^3 - 4z^3 + 5z^3 - 11z^3$

10. _______________

11. $-1.3z^7 + 0.4z^7 + 2.6z^8$

11. _______________

12. $6c^3 - 9c^2 - 2c^2 + 14 + 3c^2 - 6c - 8 + 2c^3$

12. _______________

Chapter 5 Factoring and Applications

Learning Objectives

Write a number as a product of prime factors.
Multiply a monomial and a polynomial.
Multiply two polynomials.

Key Terms

Use the vocabulary terms listed below to complete each statement in exercises 1−5.

factors	**composite number**	**prime number**
factorizations		**prime factorization**

1. The numbers that can be multiplied to give a specific number (product) are ______________________ of that number.

2. A ______________________________ has at least one factor other than itself and 1.

3. In a ______________________________________ every factor is a prime number.

4. The only factors of a ______________________________ are itself and 1.

5. Numbers that are multiplied to give a product are ______________________________.

Objective 1:   Write a number as a product of prime factors.

Review these examples:	**Now Try:**

Review these examples:

1. Find the prime factorization of 18.

Try to divide 18 by the first prime, 2.
$$18 \div 2 = 9.$$
so
$$18 = 2 \cdot 9.$$
Try to divide 9 by the prime, 3.
$$9 \div 3 = 3.$$
so
$$18 = 2 \cdot 3 \cdot 3.$$
Because all factors are prime, the prime factorization of 18 is $2 \cdot 3 \cdot 3$.

Now Try:

1. Find the prime factorization of 24.

____

2. Find the prime factorization of 72.

$3\overline{)3}^{\,1}$ Continue to divide until the quotient is 1.
Divide 3 by 3.

$3\overline{)9}$ Divide 9 by 3.

$2\overline{)18}$ Divide 18 by 2.

$2\overline{)36}$ Divide 36 by 2.

$2\overline{)72}$ Divide 72 by 2.

Because all factors (divisors) are prime, the prime factorization of 72 is
$$72 = 2 \cdot 2 \cdot 2 \cdot 3 \cdot 3 \text{ or } 2^3 \cdot 3^2.$$

2. Find the prime factorization of 56.

____

3. Find the prime factorization of 375.

$5\overline{)5}^{\,1}$ Continue to divide until the quotient is 1.
Divide 5 by 5.

$5\overline{)25}$ Divide 25 by 5.

$5\overline{)125}$ Divide 125 by 5.

$3\overline{)375}$ Divide 375 by 3.

Because all factors (divisors) are prime, the prime factorization of 375 is
$$375 = 3 \cdot 5 \cdot 5 \cdot 5 \text{ or } 3 \cdot 5^3.$$

3. Find the prime factorization of 150.

____

Practice Exercises

Find the prime factorization of each number. Write the answer with exponents when repeated factors appear.

1. 28

1. _______________

2. 72

2. _______________

3. 450

3. _______________

Objective 2:   Multiply a monomial and a polynomial.

Review this example:

4. Find the product.

$$5x^2(7x+3)$$

Use the distributive property.
$$5x^2(7x+3) = 5x^2(7x)+5x^2(3)$$
$$= 35x^3+15x^2$$

Now Try:

4. Find the product.

$$8x^3(4x+8)$$

____

Practice Exercises

Find each product.

4. $7z(5z^3+2)$

4. _______________

5. $2m(3+7m^2+3m^3)$

5. _______________

6. $-3y^2(2y^3+3y^2-4y+11)$

6. _______________

Objective 3:   Multiply two polynomials.

Review these examples:

5. Multiply $(x^2 + 6)(5x^3 - 4x^2 + 3x)$.

Multiply each term of the second polynomial by each term of the first.

$$(x^2 + 6)(5x^3 - 4x^2 + 3x)$$
$$= x^2(5x^3) + x^2(-4x^2) + x^2(3x)$$
$$\quad + 6(5x^3) + 6(-4x^2) + 6(3x)$$
$$= 5x^5 - 4x^4 + 3x^3 + 30x^3 - 24x^2 + 18x$$
$$= 5x^5 - 4x^4 + 33x^3 - 24x^2 + 18x$$

6. Multiply $(2x^3 + 7x^2 + 5x - 1)(4x + 6)$ vertically.

Write the polynomials vertically.
$$2x^3 + 7x^2 + 5x - 1$$
$$\underline{\qquad\qquad 4x + 6}$$

Begin by multiplying each term in the top row by 6.
$$2x^3 + 7x^2 + 5x - 1$$
$$\underline{\qquad\qquad 4x + 6}$$
$$12x^3 + 42x^2 + 30x - 6$$

Now multiply each term in the top row by $4x$. Then add like terms.
$$2x^3 + 7x^2 + 5x - 1$$
$$\underline{\qquad\qquad 4x + 6}$$
$$12x^3 + 42x^2 + 30x - 6$$
$$\underline{8x^4 + 28x^3 + 20x^2 - 4x \qquad}$$
$$8x^4 + 40x^3 + 62x^2 + 26x - 6$$

The product is $8x^4 + 40x^3 + 62x^2 + 26x - 6$.

Now Try:

5. Multiply
$$(x^3 + 9)(4x^4 - 2x^2 + x)$$

____

6. Multiply
$$(4x^3 - 3x^2 + 6x + 5)(7x - 3)$$
vertically.

____

7. Find the product of $-16m^3 + 12m^2 + 4$ and $\frac{1}{4}m^2 + \frac{3}{4}$.

Multiply each term of the second polynomial by each term of the first.

$$\left(-16m^3 + 12m^2 + 4\right)\left(\frac{1}{4}m^2 + \frac{3}{4}\right)$$

$$= -16m^3\left(\frac{1}{4}m^2\right) - 16m^3\left(\frac{3}{4}\right) + 12m^2\left(\frac{1}{4}m^2\right)$$

$$\qquad + 12m^2\left(\frac{3}{4}\right) + 4\left(\frac{1}{4}m^2\right) + 4\left(\frac{3}{4}\right)$$

$$= -12m^3 + 9m^2 + 3 - 5m^5 + 3m^4 + m^2$$

$$= -4m^5 + 3m^4 - 12m^3 + 10m^2 + 3$$

The product is $-4m^5 + 3m^4 - 12m^3 + 10m^2 + 3$.

7. Find the product of $12x^3 - 36x^2 + 6$ and $\frac{1}{6}x^2 + \frac{5}{6}$.

____

Practice Exercises

Find each product.

7. $(x+3)\left(x^2 - 3x + 9\right)$

7. _______________

8. $\left(2m^2 + 1\right)\left(3m^3 + 2m^2 - 4m\right)$

8. _______________

9. $\left(3x^2 + x\right)\left(2x^2 + 3x - 4\right)$

9. _______________

Chapter 6 Rational Expressions and Applications

<table>
<tr><td>

Learning Objectives

Write a fraction in lowest terms using common factors.
Write a fraction with variables in lowest terms.
Multiply signed fractions.
Multiply fractions that involve variables.
Use the multiplication property of equality to solve equations containing fractions.
Add and subtract unlike fractions.
Add and subtract fractions that contain variables.

</td></tr>
</table>

Key Terms

Use the vocabulary terms listed below to complete each statement in exercises 1−8.

equivalent fractions   **common factor**   **lowest terms**

like fractions   **unlike fractions**   **least common denominator**

mixed number   **improper fraction**

1. A fraction is written in _________________________________ when its numerator and denominator have no common factor other than 1.

2. A _________________________________ is a number that can be divided into two or more whole numbers.

3. Two fractions are _________________________________ when they represent the same portion of a whole.

4. A(n) _________________________________ includes a fraction and a whole number written together.

5. A mixed number can be rewritten as a(n) _________________________________.

6. Fractions with different denominators are called _________________________________.

7. Fractions with the same denominator are called _________________________________.

8. The _________________________________ of two whole numbers is the smallest whole number divisible by both of the numbers.

Objective 1:   Write a fraction in lowest terms using common factors.

Review these examples:

1. Write each fraction in lowest terms.

a. $\dfrac{32}{48}$

Divide both numerator and denominator by 16.

$$\dfrac{32}{48} = \dfrac{32 \div 16}{48 \div 16} = \dfrac{2}{3}$$

b. $\dfrac{28}{56}$

Suppose we thought that 4 was the greatest common factor of 28 and 56. Dividing by 4 would give

$$\dfrac{28}{56} = \dfrac{28 \div 4}{56 \div 4} = \dfrac{7}{14}$$

But $\dfrac{7}{14}$ is not in lowest terms, because 7 and 14 have a common factor of 7. So we divide by 7.

$$\dfrac{7}{14} = \dfrac{7 \div 7}{14 \div 7} = \dfrac{1}{2}$$

The fraction $\dfrac{28}{56}$ could have been written in lowest terms in one step by dividing by 28, the greatest common factor of 28 and 56.

$$\dfrac{28}{56} = \dfrac{28 \div 28}{56 \div 28} = \dfrac{1}{2}$$

Now Try:

1. Write each fraction in lowest terms.

a. $\dfrac{18}{27}$

____

b. $\dfrac{27}{45}$

____

Practice Exercises

Write each fraction in lowest terms.

1. $\dfrac{14}{49}$

1. _______________

2. $\dfrac{8}{36}$

2. _______________

3. $\dfrac{30}{42}$

3. _______________

Objective 2:   Write a fraction with variables in lowest terms.

Review these examples:

2.   Write each fraction in lowest terms.

a. $\dfrac{10}{15x}$

$$\frac{10}{15x} = \frac{2 \cdot \cancel{5}}{3 \cdot \cancel{5} \cdot x} = \frac{2}{3x}$$

b. $\dfrac{7xy}{28xy}$

$$\frac{7xy}{28xy} = \frac{\cancel{7} \cdot \cancel{x} \cdot \cancel{y}}{2 \cdot 2 \cdot \cancel{7} \cdot \cancel{x} \cdot \cancel{y}} = \frac{1}{4}$$

c. $\dfrac{6b^4}{12ab^2}$

$$\frac{6b^4}{12ab^2} = \frac{\cancel{2} \cdot \cancel{3} \cdot b \cdot b \cdot \cancel{b} \cdot \cancel{b}}{\cancel{2} \cdot 2 \cdot \cancel{3} \cdot a \cdot \cancel{b} \cdot \cancel{b}} = \frac{b^2}{2a}$$

Now Try:

2.   Write each fraction in lowest terms.

a. $\dfrac{25}{30x}$

____

b. $\dfrac{8ab}{24ab}$

____

c. $\dfrac{9b^5}{36ab^3}$

____

Practice Exercises

Write each fraction in lowest terms.

4. $\dfrac{12r^2 s}{4rs^3}$

4. _______________

5. $\dfrac{16b^2cd}{40b^2d}$

5. _______________

6. $\dfrac{8xy^2}{6x^2y^2}$

6. _______________

Objective 3:   Multiply signed fractions.

Review these examples:

3. Multiply. Write the product in lowest terms.

a. $-\dfrac{7}{9} \cdot -\dfrac{5}{11}$

Multiply the numerators and multiply the denominators.

$$-\dfrac{7}{9} \cdot -\dfrac{5}{11} = \dfrac{7 \cdot 5}{9 \cdot 11} = \dfrac{35}{99}$$

The answer is in lowest terms because 35 and 99 have no common factor other than 1.

b. $-\dfrac{9}{7}\left(\dfrac{7}{15}\right)$

Multiplying a negative number times a positive number gives a negative product.

$$-\dfrac{9}{7}\left(\dfrac{7}{15}\right) = -\dfrac{3 \cdot 3 \cdot 7}{7 \cdot 3 \cdot 5} = -\dfrac{\cancel{3} \cdot 3 \cdot \cancel{7}}{\cancel{7} \cdot \cancel{3} \cdot 5} = -\dfrac{3}{5}$$

c. Find $\dfrac{3}{8}$ of $\dfrac{4}{9}$.

Recall that "of" indicates multiplication.

$$\dfrac{3}{8} \cdot \dfrac{4}{9} = \dfrac{3 \cdot 2 \cdot 2}{2 \cdot 2 \cdot 2 \cdot 3 \cdot 3} = \dfrac{\cancel{3} \cdot \cancel{2} \cdot \cancel{2}}{\cancel{2} \cdot \cancel{2} \cdot 2 \cdot \cancel{3} \cdot 3} = \dfrac{1}{6}$$

Now Try:

3. Multiply. Write the product in lowest terms.

a. $-\dfrac{10}{11} \cdot -\dfrac{4}{13}$

____

b. $-\dfrac{11}{7}\left(\dfrac{14}{33}\right)$

____

c. Find $\dfrac{2}{7}$ of $\dfrac{21}{40}$.

____

Practice Exercises

Multiply. Write the products in lowest terms.

7. $-\dfrac{10}{42} \cdot \dfrac{3}{5}$

7. ________________

8. $\dfrac{6}{18} \cdot \dfrac{9}{2}$

8. ________________

9. $\dfrac{5}{9}$ of 81

9. ________________

Objective 4:   Multiply fractions that involve variables.

<table>
<tr><td>

Review these examples:
4.   Find each product.

a. $\dfrac{6x}{7} \cdot \dfrac{5}{18x}$

$$\frac{6x}{7} \cdot \frac{5}{18x} = \frac{2 \cdot 3 \cdot x \cdot 5}{7 \cdot 2 \cdot 3 \cdot 3 \cdot x} = \frac{\cancel{2} \cdot \cancel{3} \cdot \cancel{x} \cdot 5}{7 \cdot \cancel{2} \cdot \cancel{3} \cdot 3 \cdot \cancel{x}} = \frac{5}{21}$$

b. $\left(\dfrac{9a}{5b}\right)\left(\dfrac{10b^2}{3a}\right)$

$$\left(\frac{9a}{5b}\right)\left(\frac{10b^2}{3a}\right) = \frac{3 \cdot 3 \cdot a \cdot 2 \cdot 5 \cdot b \cdot b}{5 \cdot b \cdot 3 \cdot a}$$

$$= \frac{\cancel{3} \cdot 3 \cdot \cancel{a} \cdot 2 \cdot \cancel{5} \cdot \cancel{b} \cdot b}{\cancel{5} \cdot \cancel{b} \cdot \cancel{3} \cdot \cancel{a}}$$

$$= 6b$$

</td><td>

Now Try:
4.   Find each product.

a. $\dfrac{12x}{5} \cdot \dfrac{7}{18x}$

____

b. $\left(\dfrac{7x}{9c}\right)\dfrac{36c^2}{35x}$

____

</td></tr>
</table>

Practice Exercises

Use prime factorization to find these products.

10. $\dfrac{3c}{5} \cdot \dfrac{c}{9}$ 10. _______________

11. $\left(\dfrac{m}{10}\right)\dfrac{25}{m^2}$ 11. _______________

12. $\dfrac{5a}{7b^2} \cdot \dfrac{14b}{10a^2}$ 12. _______________

Objective 5: Use the multiplication property of equality to solve equations containing fractions.

Review these examples:	**Now Try:**

Review these examples:

5. Solve each equation and check each solution.

a. $\dfrac{1}{5}b = 12$

Multiply both sides by $\dfrac{5}{1}$.

$$\dfrac{1}{5}b = 12$$

$$\dfrac{5}{1}\left(\dfrac{1}{5}b\right) = \dfrac{5}{1}(12)$$

$$\left(\dfrac{5}{1}\cdot\dfrac{1}{5}\right)b = \dfrac{5}{1}\left(\dfrac{12}{1}\right)$$

$$1b = \dfrac{60}{1}$$

$$b = 60$$

Check $\qquad \dfrac{1}{5}b = 12$

$$\dfrac{1}{5}(60) = 12$$

$$\dfrac{1\cdot\overset{1}{\cancel{5}}\cdot 12}{\underset{1}{\cancel{5}}} = 12$$

$$12 = 12$$

The solution is 60.

b. $15 = -\dfrac{3}{5}x$

Multiply both sides by $-\dfrac{5}{3}$.

$$-\dfrac{5}{3}(15) = -\dfrac{5}{3}\left(-\dfrac{3}{5}x\right)$$

$$-25 = x$$

Check $\;\; 15 = -\dfrac{3}{5}x$

$$15 = -\dfrac{3}{5}(-25)$$

$$15 = \dfrac{3\cdot\overset{1}{\cancel{5}}\cdot 5}{\underset{1}{\cancel{5}}}$$

$$15 = 15$$

The solution is –25.

Now Try:

5. Solve each equation and check each solution.

a. $\dfrac{1}{7}b = 3$

____

b. $25 = -\dfrac{5}{6}x$

____

c. $-\dfrac{5}{6}n = -\dfrac{1}{4}$

$$-\dfrac{6}{5}\left(-\dfrac{5}{6}n\right) = -\dfrac{6}{5}\left(-\dfrac{1}{4}\right)$$

$$n = \dfrac{2\cdot 3}{5\cdot 2\cdot 2}$$

$$n = \dfrac{3}{10}$$

Check
$$-\dfrac{5}{6}n = -\dfrac{1}{4}$$

$$-\dfrac{5}{6}\left(\dfrac{3}{10}\right) = -\dfrac{1}{4}$$

$$-\dfrac{\overset{1}{\cancel{5}}\cdot\overset{1}{\cancel{3}}}{2\cdot\cancel{3}\cdot 2\cdot\cancel{5}} = -\dfrac{1}{4}$$

$$-\dfrac{1}{4} = -\dfrac{1}{4}$$

The solution is $\dfrac{3}{10}$.

c. $-\dfrac{7}{9}n = -\dfrac{14}{45}$

____

Practice Exercises

Solve each equation and check each solution.

13. $\quad -30 = \dfrac{5}{6}b$

13. _______________

14. $\quad \dfrac{2}{9}n = 18$

14. _______________

15. $\quad -\dfrac{8}{9}h = -\dfrac{1}{6}$

15. ________________

Objective 6:   Add and subtract unlike fractions.

Review these examples:	**Now Try:**

Review these examples:
6. Find each sum or difference.

a. $\dfrac{1}{2}+\dfrac{1}{6}$

Step 1 The larger denominator (6) is the LCD.

Step 2 $\dfrac{1}{2}=\dfrac{1\cdot 3}{2\cdot 3}=\dfrac{3}{6}$ and $\dfrac{1}{6}$ already has the LCD.

Step 3 Add the numerators.

$$\dfrac{1}{2}+\dfrac{1}{6}=\dfrac{3}{6}+\dfrac{1}{6}=\dfrac{3+1}{6}=\dfrac{4}{6}$$

Step 4 Write $\dfrac{4}{6}$ in lowest terms.

$$\dfrac{4}{6}=\dfrac{2\cdot\overset{1}{\cancel{2}}}{\underset{1}{\cancel{2}}\cdot 3}=\dfrac{2}{3}$$

b. $\dfrac{3}{8}-\dfrac{7}{12}$

Step 1 The LCD is 24.

Step 2 $\dfrac{3}{8}=\dfrac{3\cdot 3}{8\cdot 3}=\dfrac{9}{24}$ and $\dfrac{7}{12}=\dfrac{7\cdot 2}{12\cdot 2}=\dfrac{14}{24}$

Step 3 Subtract the numerators.

$$\dfrac{3}{8}-\dfrac{7}{12}=\dfrac{9}{24}-\dfrac{14}{24}=\dfrac{9-14}{24}=\dfrac{-5}{24}, \text{ or } -\dfrac{5}{24}$$

Step 4 $-\dfrac{5}{24}$ is in lowest terms.

c. $-\dfrac{7}{18}+\dfrac{5}{12}$

Step 1 Use prime factorization to find the LCD.

$$18 = 2\cdot 3\cdot 3$$
$$12 = 2\cdot 2\cdot 3$$
$$\text{LCD} = 2\cdot 2\cdot 3\cdot 3 = 36$$

Step 2

$$-\dfrac{7}{18}=-\dfrac{7\cdot 2}{18\cdot 2}=-\dfrac{14}{36} \text{ and } \dfrac{5}{12}=\dfrac{5\cdot 3}{12\cdot 3}=\dfrac{15}{36}$$

Step 3 Add the numerators.

$$-\dfrac{7}{18}+\dfrac{5}{12}=-\dfrac{14}{36}+\dfrac{15}{36}=\dfrac{-14+15}{36}=\dfrac{1}{36}$$

Step 4 $\dfrac{1}{36}$ is in lowest terms.

Now Try:
6. Find each sum or difference.

a. $\dfrac{5}{9}+\dfrac{7}{18}$

____

b. $\dfrac{8}{15}-\dfrac{7}{10}$

____

c. $-\dfrac{5}{24}+\dfrac{7}{9}$

____

Practice Exercises

Find each sum or difference. Write all answers in lowest terms.

16. $\dfrac{1}{6} + \dfrac{2}{15}$

16. _________________

17. $-\dfrac{1}{2} + \dfrac{7}{12}$

17. _________________

18. $\dfrac{33}{40} - \dfrac{7}{24}$

18. _________________

Objective 7:   Add and subtract unlike fractions that contain variables.

Review these examples:

7.   Find each sum or difference.

a. $\dfrac{1}{3} + \dfrac{x}{2}$

Step 1 The LCD is 6.

Step 2 $\dfrac{1}{3} = \dfrac{1 \cdot 2}{3 \cdot 2} = \dfrac{2}{6}$ and $\dfrac{x}{2} = \dfrac{x \cdot 3}{2 \cdot 3} = \dfrac{3x}{6}$

Step 3 $\dfrac{1}{3} + \dfrac{x}{2} = \dfrac{2}{6} + \dfrac{3x}{6} = \dfrac{2 + 3x}{6}$

Step 4 $\dfrac{2 + 3x}{6}$ is in lowest terms.

b. $\dfrac{5}{6} - \dfrac{7}{x}$

Step 1 The LCD is $6x$.

Step 2 $\dfrac{5}{6} = \dfrac{5 \cdot x}{6 \cdot x} = \dfrac{5x}{6x}$ and $\dfrac{7}{x} = \dfrac{7 \cdot 6}{x \cdot 6} = \dfrac{42}{6x}$

Step 3 $\dfrac{5}{6} - \dfrac{7}{x} = \dfrac{5x}{6x} - \dfrac{42}{6x} = \dfrac{5x - 42}{6x}$

Step 4 $\dfrac{5x - 42}{6x}$ is in lowest terms.

Now Try:

7.   Find each sum or difference.

a. $\dfrac{2}{3} + \dfrac{x}{5}$

____

b. $\dfrac{5}{8} - \dfrac{8}{x}$

____

Practice Exercises

Find each sum or difference. Write all answers in lowest terms.

19. $\dfrac{3}{n} + \dfrac{3}{5}$

19.   _______________

20. $\dfrac{1}{6} + \dfrac{x}{3}$

20.   _______________

21. $\dfrac{3}{a} - \dfrac{b}{4}$

21.   _______________

Chapter 7 Linear Equations, Graphs, and Systems

Learning Objectives
Graph linear equations in the form $Ax + By = C$
Determine if the slope of a line is positive, negative, zero, or undefined.
Use slopes to determine whether two lines are parallel, perpendicular, or neither.
Write an equation of a line using its slope and any point on the line.
Determine whether a given ordered pair is a solution to a system.
Solve linear systems by substitution.

Key Terms

Use the vocabulary terms listed below to complete each statement in exercises 1−6.

graph	**graphing**	**y-intercept**	**x-intercept**
perpendicular lines	**parallel lines**		

1. If a graph intersects the y-axis at k, then the _________________ is $(0, k)$.

2. If a graph intersects the x-axis at k, then the _________________ is $(k, 0)$.

3. Two lines in a plane that never intersect are called _________________.

4. The process of plotting the ordered pairs that satisfy a linear equation and drawing a line through them is called _________________.

5. The set of all points that correspond to the ordered pairs that satisfy the equation is called the _________________ of the equation.

6. Two lines that intersect in a 90° angle are called _________________.

Objective 1   Graph linear equations of the form $Ax + By = 0$.

Review this example for Objective 1:

1.   Graph $x + 5y = 0$.

To find the y-intercept, let $x = 0$.
To find the x-intercept, let $y = 0$.

$$0 + 5y = 0 \quad | \quad x + 5(0) = 0$$
$$5y = 0 \quad | \quad x + 0 = 0$$
$$y = 0 \quad | \quad x = 0$$

The x- and y-intercepts are the same point $(0, 0)$. We must select two other values for x or y to find two other points. We choose $y = 1$ and $y = -1$.

$$x + 5(1) = 0 \quad | \quad x + 5(-1) = 0$$
$$x + 5 = 0 \quad | \quad x - 5 = 0$$
$$x = -5 \quad | \quad x = 5$$

We use $(-5, 1)$, $(0, 0)$, and $(5, -1)$ to draw the graph.

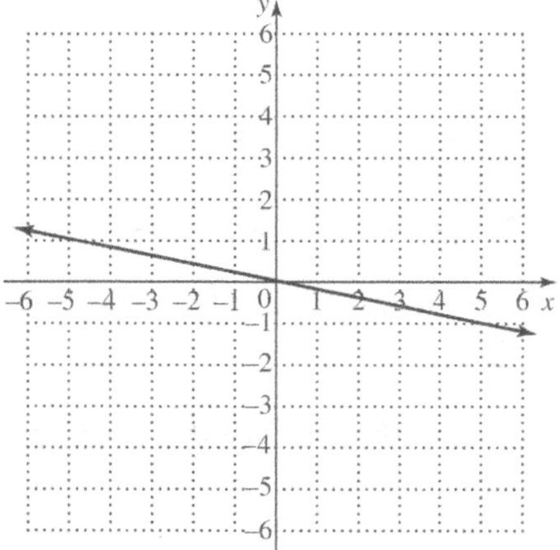

Now Try:

1.   Graph $3x - y = 0$.

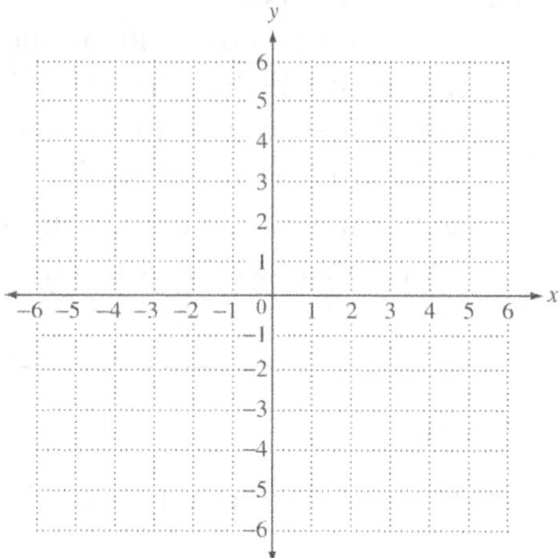

                Copyright © 2025 Pearson Education, Inc.

Objective 1 Practice Exercises

Graph each equation.

1. $-3x - 2y = 0$

1.

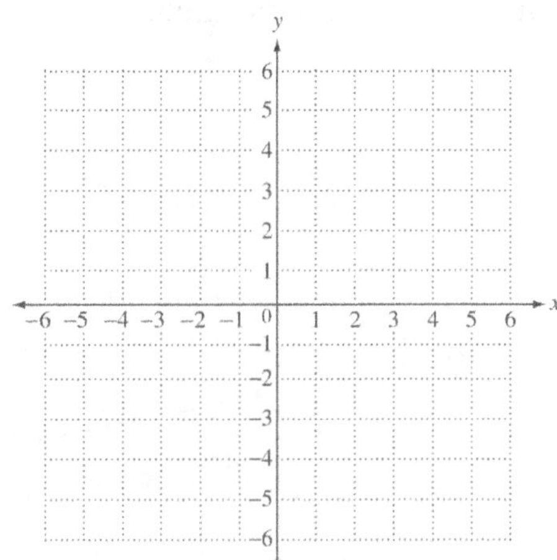

2. $x + y = 0$

2.

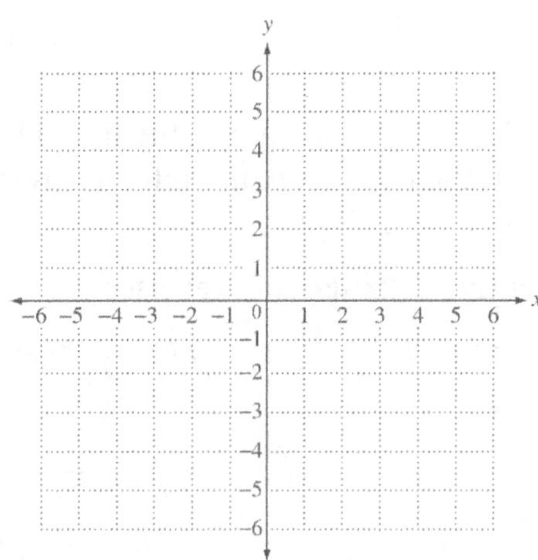

3. $y = 2x$

3.

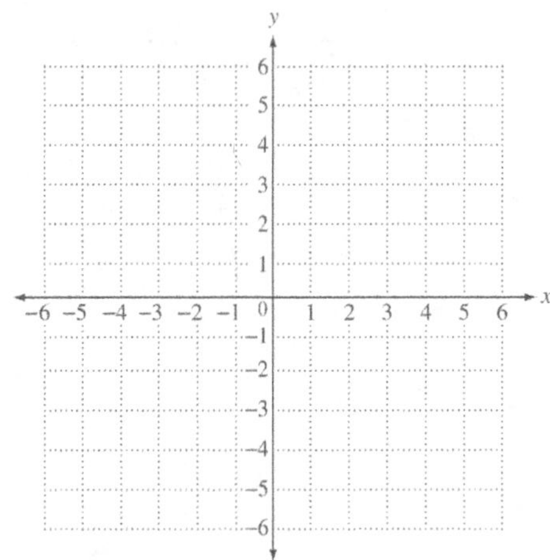

Objective 2   Determine if the slope of a line is positive, negative, zero, or undefined.

Review these examples for Objective 2:

2.   Determine if the slope of a line is positive, negative, zero, or undefined.

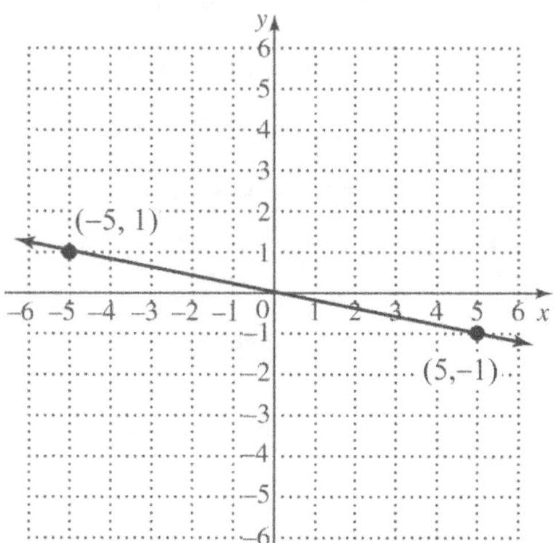

Looking at the graph the line falls as we move to the right from the left. Therefore it has a negative slope.

Now Try:

2.   Determine if the slope of a line is positive, negative, zero, or undefined.

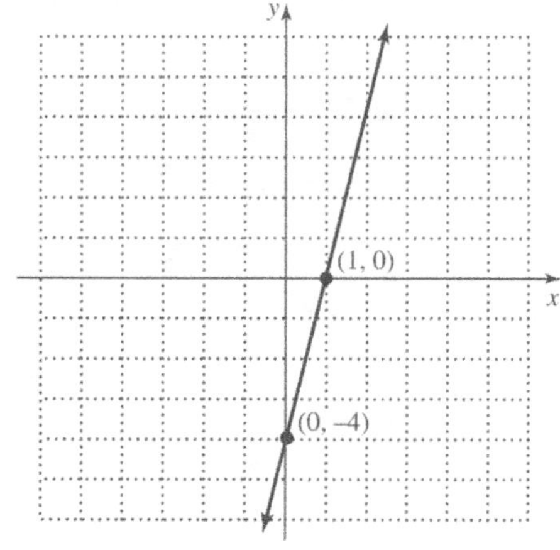

____

Objective 2 Practice Exercises

Determine if the slope of a line is positive, negative, zero, or undefined.

4.

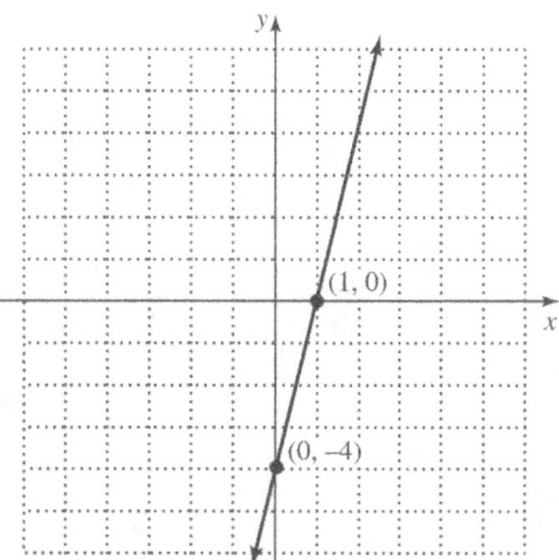

4. ____________________

     Copyright © 2025 Pearson Education, Inc.

5. 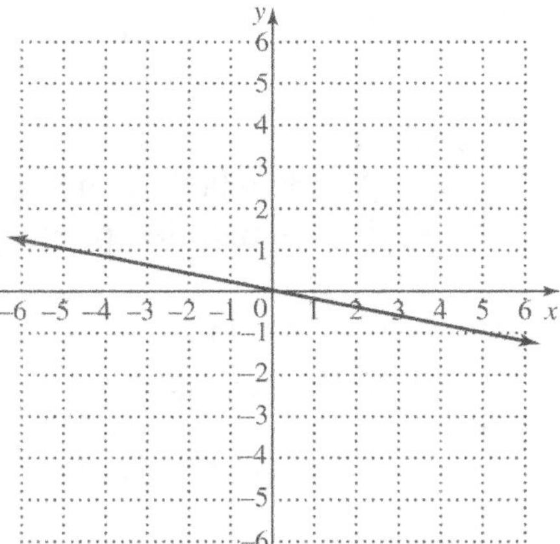

5. ______________________

6.

6. ______________________

Objective 3 Use slopes to determine whether two lines are parallel, perpendicular, or neither.

Review these examples for Objective 3:

3. Decide whether each pair of lines is parallel, perpendicular, or neither.

   a. $5x - y = 3$

   $15x - 3y = 12$

   Solve each equation for y.

   $y = 5x - 3$

   $y = 5x - 4$

   Both lines have slope 5, so the lines are parallel.

   b. $x + 3y = 8$ and $-3x + y = 5$

   Find the slope of each line by first solving each equation for y.

   $$3y = -x + 8 \qquad \bigg| \qquad y = 3x + 5$$

   $$y = -\frac{1}{3}x + \frac{8}{3}$$

   The slope is $-\dfrac{1}{3}$. $\bigg|$ The slope is 3.

   Because the slopes are not equal, the lines are not parallel.

   Check the product of the slopes: $-\dfrac{1}{3}(3) = -1$.

   The two lines are perpendicular because the product of their slopes is –1.

Now Try:

3. Decide whether each pair of lines is parallel, perpendicular, or neither.

   a. $2x - 4y = 7$

   $3x - 6y = 8$

   ____

   b. $9x - y = 7$ and $x + 9y = 11$

   ____

Objective 3 Practice Exercises

In each pair of equations, give the slope of each line, and then determine whether the two lines are **parallel,** **perpendicular,** *or* **neither.**

7. $-x + y = -7$

   $x - y = -3$

7. _________________

8. $4x + 2y = 8$

$x + 4y = -3$

8. _______________

9. $9x + 3y = 2$

$x - 3y = 5$

9. _______________

Objective 4 Write an equation of a line using its slope and any point on the line.

Review these examples for Objective 4:

4. Write an equation in slope-intercept form of the line passing through the given point and having the given slope.

a. $(0,\ 2),\ m = -3$

Because the point $(0, 2)$ is the y-intercept, $b = 2$. Substitute $b = 2$ and $m = -3$ directly in the slope-intercept form.

$$y = mx + b$$
$$y = -3x + 2$$

b. $(2,\ 9),\ m = 5$

Since the line passes through the point $(2, 9)$, we can substitute $x = 2$, $y = 9$, and slope $m = 5$ into $y = mx + b$ and solve for b.

$$y = mx + b$$
$$9 = 5(2) + b$$
$$-1 = b$$

Now substitute the values of m and b into slope-intercept form.

$$y = mx + b$$
$$y = 5x - 1$$

Now Try:

4. Write an equation in slope-intercept form of the line passing through the given point and having the given slope.

a. $(0, -5),\ m = \dfrac{3}{4}$

____

b. $(-1,\ 4),\ m = 6$

____

Objective 4 Practice Exercises

Write an equation in slope-intercept form of the line passing through the given point and having the given slope.

10. $(0, -4),\ m = \dfrac{2}{3}$

10. _______________

11. $(3,\ 6),\ m = -2$

11. _______________

12. $(-2,\ 0),\ m = 1$

12. _______________

Objective 5   Determine whether a given ordered pair is a solution of a system.

Review this example for Objective 5:

5. Determine whether the ordered pair $(5, -2)$ is a solution of the system.

$$4x + 5y = 10$$

$$3x + 8y = 6$$

Again, substitute 5 for x and -2 for y in each equation.

$$4x + 5y = 10 \qquad\qquad 3x + 8y = 6$$
$$4(5) + 5(-2) \overset{?}{=} 10 \qquad 3(5) + 8(-2) \overset{?}{=} 6$$
$$20 - 10 \overset{?}{=} 10 \qquad\qquad 15 - 16 \overset{?}{=} 6$$
$$10 = 10 \ \text{True} \qquad \text{False} \ -1 = 6$$

The ordered pair $(5, -2)$ is not a solution of this system because it does not satisfy the second equation.

Now Try:

5. Determine whether the ordered pair $(6, 5)$ is a solution of the system.

$$5x - 6y = 0$$

$$6x + 5y = 50$$

____

Objective 5 Practice Exercises

Decide whether the given ordered pair is a solution of the given system.

13. $(2, -4)$

$$2x + 3y = 6$$

$$3x - 2y = 14$$

13. __________________

14. $(-3, -1)$

$$5x - 3y = -12$$

$$2x + 3y = -9$$

14. __________________

15. $(4, \ 0)$

$$4x + 3y = 16$$

$$x - 4y = -4$$

15. __________________

Objective 6   Solve linear systems by substitution.

Review these examples for Objective 6:

6. Solve the system by the substitution method.
$$2x+5y=22 \quad (1)$$
$$y=4x \quad (2)$$

Equation (2) is already solved for y. We substitute $4x$ for y in equation (1).
$$2x+5y=22$$
$$2x+5(4x)=22$$
$$2x+20x=22$$
$$22x=22$$
$$x=1$$

Find the value of y by substituting 1 for x in either equation. We use equation (2).
$$y=4x$$
$$y=4(1)=4$$

We check the solution $(1, 4)$ by substituting 1 for x and 4 for y in both equations.

$$
\begin{array}{c|c}
2x+5y=22 & y=4x \\
2(1)+5(4)\overset{?}{=}22 & 4\overset{?}{=}4(1) \\
2+20\overset{?}{=}22 & \text{True } 4=4 \\
22=22 \ \text{True} &
\end{array}
$$

Since $(1, 4)$ satisfies both equations, the solution set of the system is $\{(1, 4)\}$.

7. Solve the system by the substitution method.
$$4x+5y=13 \qquad (1)$$
$$x=-y+2 \quad (2)$$

Equation (2) gives x in terms of y. We substitute $-y+2$ for x in equation (1).
$$4x+5y=13$$
$$4(-y+2)+5y=13$$
$$-4y+8+5y=13$$
$$y+8=13$$
$$y=5$$

Find the value of x by substituting 5 for y in either equation. We use equation (2).
$$x=-y+2$$
$$x=-5+2=-3$$

Now Try:

6. Solve the system by the substitution method.
$$x+y=7$$
$$y=6x$$

____

7. Solve the system by the substitution method.
$$2x+3y=6$$
$$x=5-y$$

____

We check the solution $(-3, 5)$ by substituting -3 for x and 5 for y in both equations.

$$4x + 5y = 13 \qquad\qquad x = -y + 2$$
$$4(-3) + 5(5) \overset{?}{=} 13 \qquad\qquad -3 \overset{?}{=} -5 + 2$$
$$-12 + 25 \overset{?}{=} 13 \qquad \text{True} \;\; -3 = -3$$
$$13 = 13 \;\; \text{True}$$

Both results are true, so the solution set of the system is $\{(-3, 5)\}$.

Objective 6 Practice Exercises

Solve each system by the substitution method. Check each solution.

16. $\quad 3x + 2y = 14$

$\qquad\quad y = x + 2$

16. _______________

17. $\quad x + y = 9$

$\qquad\quad 5x - 2y = -4$

17. _______________

18. $\quad 3x - 21 = y$

$\qquad\quad y + 2x = -1$

18. _______________

Chapter 8 Linear Equations, Inequalities, and Applications

Learning Objectives
> Simplify, and then use the addition property of equality to solve equations.
> Simplify, and then use the multiplication property of equality to solve equations.
> Solve linear inequalities using both properties of inequalities.
> Graph inequalities in two variables.

Key Terms

Use the vocabulary terms listed below to complete each statement in exercises 1−3.

linear equation   **solution set**   **equivalent equations**

1.   Equations that have exactly the same solutions sets are called

   _______________________________.

2.   An equation that can be written in the form $Ax + B = C$, where A, B, and C are real numbers and $A \neq 0$, is called a _______________________________.

3.   The set of all numbers that satisfy an equation is called its _______________.

Objective 1   Simplify, and then use the addition property of equality to solve equations.

Review these examples for Objective 1:	**Now Try:**

Review these examples for Objective 1:

1. Solve $5t - 16 + t + 4 = 9 + 5t + 6$.

$$5t - 16 + t + 4 = 9 + 5t + 6$$
$$6t - 12 = 15 + 5t$$
$$6t - 12 - 5t = 15 + 5t - 5t$$
$$t - 12 = 15$$
$$t - 12 + 12 = 15 + 12$$
$$t = 27$$

Check by substituting 27 in the original equation. The solution set is $\{27\}$.

2. Solve $4(3 + 6x) - (5 + 23x) = 19$.

$$4(3 + 6x) - (5 + 23x) = 19$$
$$4(3) + 4(6x) - 1(5) - 1(23x) = 19$$
$$12 + 24x - 5 - 23x = 19$$
$$x + 7 = 19$$
$$x + 7 - 7 = 19 - 7$$
$$x = 12$$

Check by substituting 12 in the original equation. The solution set is $\{12\}$.

Now Try:

1. Solve
$$8t - 9 + t + 7 = 12 + 8t + 15.$$

____

2. Solve
$$5(7 + 8x) - (29 + 39x) = 14.$$

____

Objective 1 Practice Exercises

Solve each equation. First simplify each side of the equation as much as possible. Check each solution.

1. $3(t + 3) - (2t + 7) = 9$

1. _______________

2. $-4(5g - 7) + 3(8g - 3) = 15 - 4 + 3g$

2. _______________

3. $3.6p + 4.8 + 4.0p = 8.6p - 3.1 + 0.7$

3. _______________

Objective 2   Simplify, and then use the multiplication property of equality to solve equations.

Review these examples for Objective 2:

3. Solve $9m + 4m = 39$.

$$9m + 4m = 39$$
$$13m = 39$$
$$\frac{13m}{13} = \frac{39}{13}$$
$$m = 3$$

Check by substituting 3 in the original equation. The solution set is {3}.

4. Solve $3(2x - 7) + 21 = -18$.

$$3(2x - 7) + 21 = -18$$
$$3(2x) + 3(-7) + 21 = -18$$
$$6x - 21 + 21 = -18$$
$$6x = -18$$
$$\frac{6x}{6} = \frac{-18}{6}$$
$$x = -3$$

Check by substituting –3 in the original equation. The solution set is {–3}.

Now Try:

3. Solve $12m + 8m = 80$.

____

4. Solve $4(2x - 5) + 20 = -32$.

____

Objective 2 Practice Exercises

Solve each equation and check your solution.

4. $-7b + 12b = 125$

4. _____________

5. $3w - 7w = 20$

5. _____________

6. $-11h - 6h + 14h = -21$

6. _____________

Objective 3   Solve linear inequalities using both properties of inequality.

Review this example for Objective 3:

5.   Solve $4x+3-7>-2x+8+3x$. Graph the solution set.

Step 1 Combine like terms and simplify.
$$4x+3-7>-2x+8+3x$$
$$4x-4>x+8$$

Step 2 Use the addition property of inequality.
$$4x-4-x>x+8-x$$
$$3x-4>8$$
$$3x-4+4>8+4$$
$$3x>12$$

Step 3 Use the multiplication property of inequality.
$$\frac{3x}{3}>\frac{12}{3}$$
$$x>4$$

The solution set is $(4,\ \infty)$. The graph is shown below.

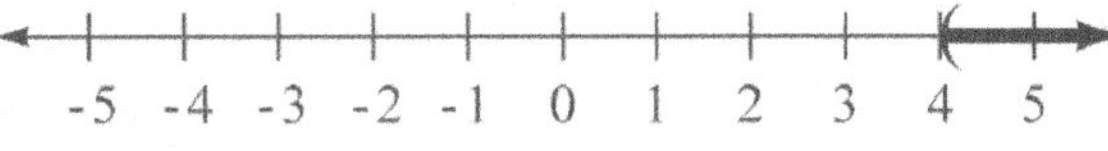

Now Try:

5.   Solve $8x-5+4\geq 6x-3x+9$. Graph the solution set.

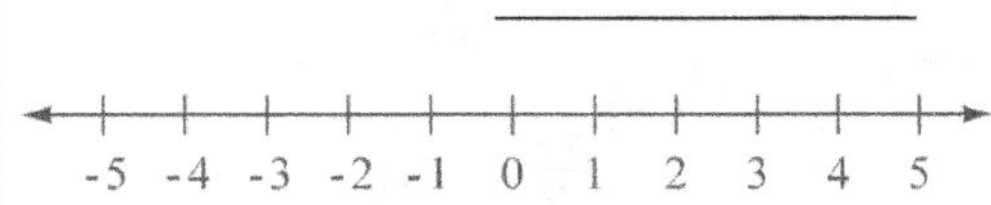

Objective 3 Practice Exercises

Solve each inequality. Write the solution set in interval notation and then graph it.

7. $4\ y-3)+2>3\ y-2)$

7.   __________________

8. $-3(m+2)+3\leq -4\ m-2)-6$

8.   __________________

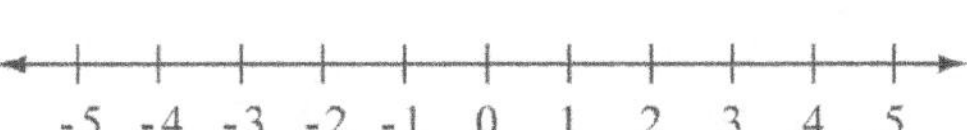

9. $7(2-x)\leq -2(x-3)-x$

9.   __________________

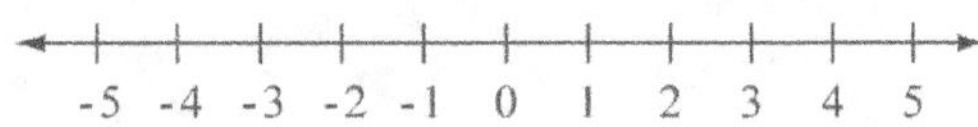

Objective 4   Graph linear inequalities in two variables.

Review these examples for Objective 4:

6. Graph $3x - 2y \leq 6$.

The inequality $3x - 2y \leq 6$ means that
$$3x - 2y < 6 \text{ or } 3x - 2y = 6.$$
We begin by graphing the line $3x - 2y = 6$ with intercepts $(0, -3)$ and $(2, 0)$. This boundary line divides the plane into two regions, one of which satisfies the inequality. We use the test point $(0, 0)$ to see whether the resulting statement is true or false, thereby determining whether the point is in the shaded region or not.

$$3x - 2y \leq 6$$
$$3(0) - 2(0) \overset{?}{\leq} 6$$
$$0 - 0 \overset{?}{\leq} 6$$
$$0 \leq 6 \quad \text{True}$$

Since the last statement is true, we shade the region that includes the test point $(0, 0)$. The shaded region, along with the boundary line, is the desired graph.

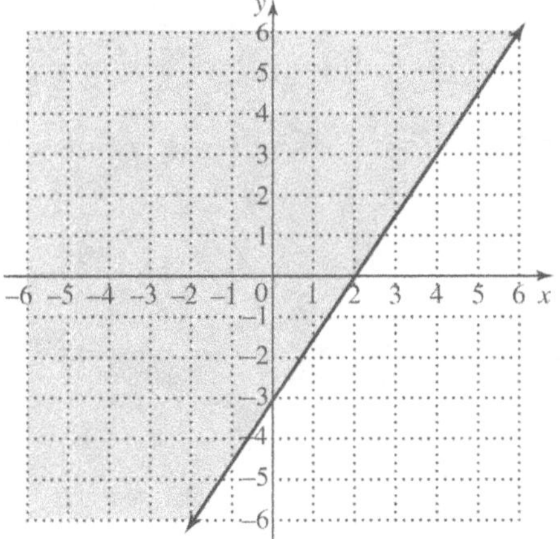

Now Try:

6. Graph $2x + 5y \leq -8$.

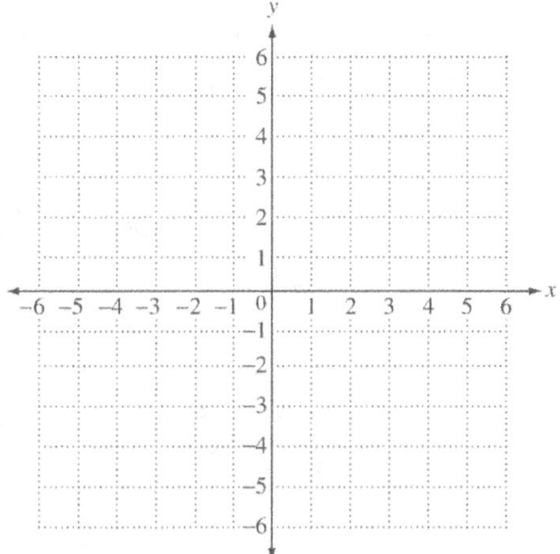

7. Graph $x - 4 \leq -1$.

First, solve the inequality for x.
$$x \leq 3$$
Now graph the line $x = 3$, a vertical line through the point (3, 0). Use a solid line, and choose (0, 0) as a test point.
$$0 \leq 3 \quad \text{True}$$
Because $0 \leq 3$ is true, we shade the region containing (0, 0).

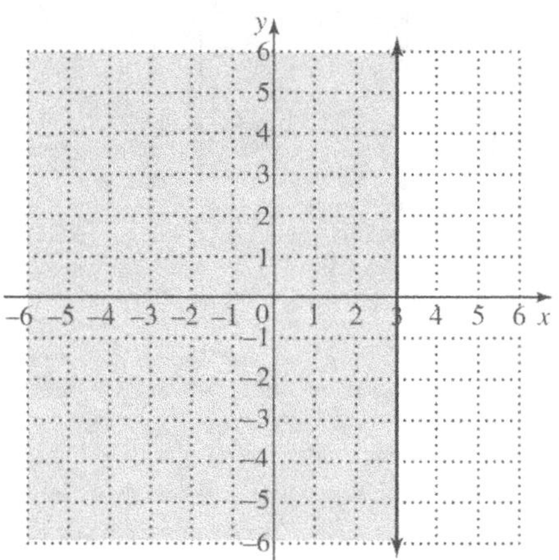

7. Graph $y \geq -1$.

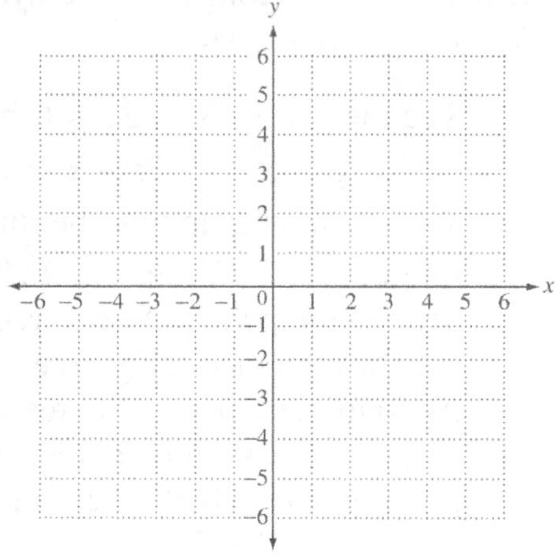

Objective 4 Practice Exercises

Graph each linear inequality.

10. $y \geq x - 1$

10.

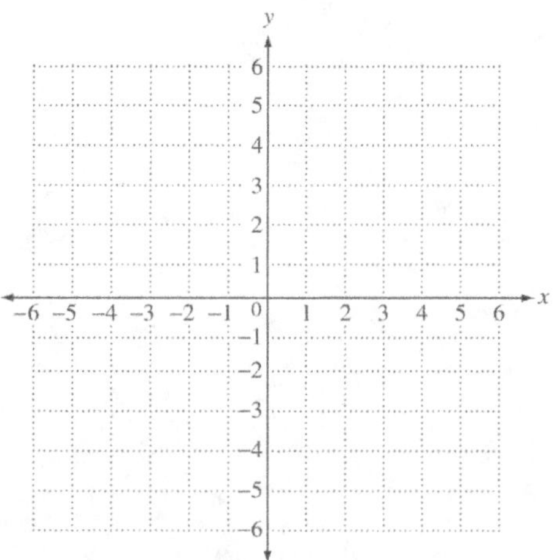

11. $y > -x + 2$

11.

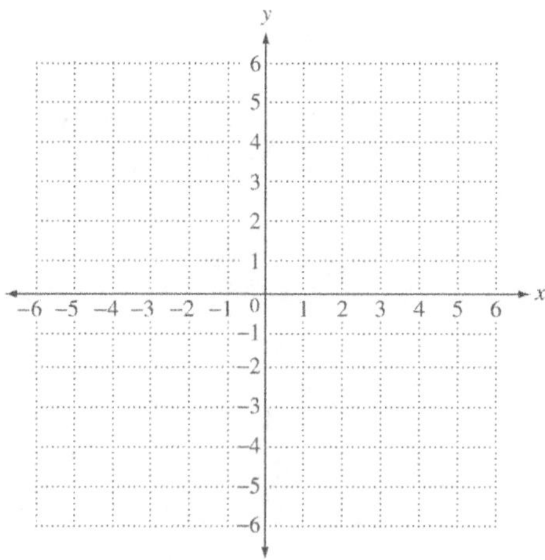

12. $3x - 4y - 12 > 0$

12.

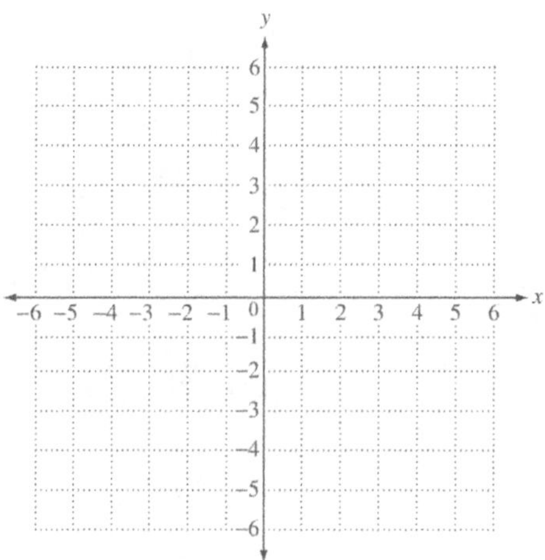

Chapter 9 Relations and Functions

Learning Objectives
Determine whether a graph or an equation represents a function.
Use function notation to evaluate a function at a specific value.
Find domain and range of given functions.
Determine whether situations and equations represent direct and inverse variations.
Solve direct variation problems.
Solve inverse variation problems.

Key Terms

Use the vocabulary terms listed below to complete each statement in exercises 1–8.

    components   relation   domain   range   function

    direct variation   constant of variation   inverse variation

1. In the equation $y = kx$, the number k is called the _______________________________.

2. Any set of ordered pairs is called a _______________________________.

3. The set of all second components in the ordered pairs of a relation is the
_______________________________ of the relation.

4. A _______________________________ is a set of ordered pairs in which each first
component corresponds to exactly one second component.

5. If two positive quantities x and y are in _______________________________ and
the constant of variation is positive, then as x increases, y also increases.

6. In an ordered pair (x, y), x and y are the _______________________________.

7. If two positive quantities x and y are in _______________________________ and
the constant of variation is positive, then as x increases, y decreases.

8. The set of all first components in the ordered pairs of a relation is the
_______________________________ of the relation.

Objective 1   Determine whether a graph or an equation represents a function.

Review these examples for Objective 1:	**Now Try:**

Review these examples for Objective 1:

1. Determine whether each relation represented by a graph or an equation is a function.

a.

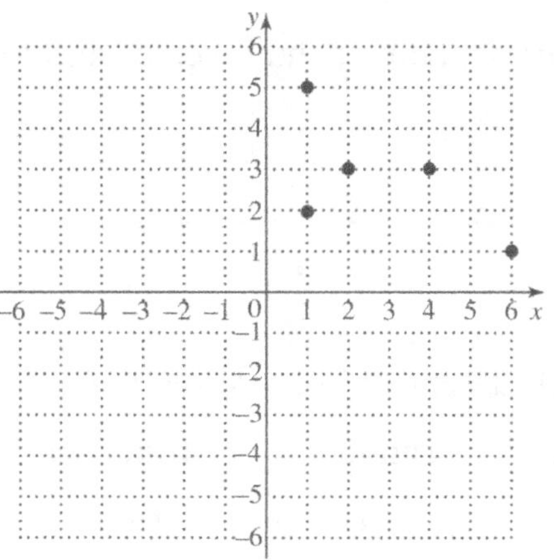

Because there are two ordered pairs with first component 1, this is not the graph of a function.

b.

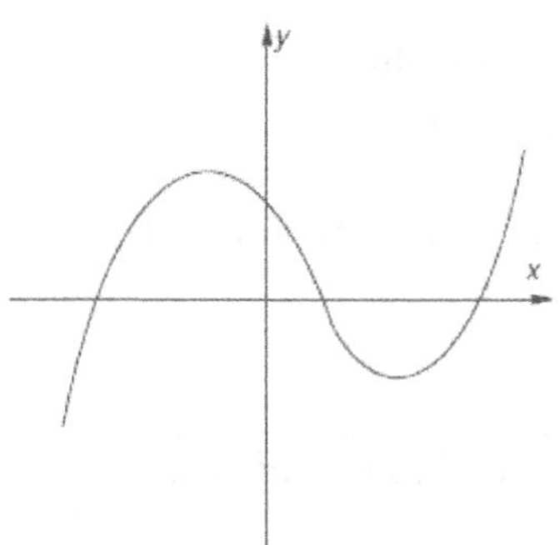

Use the vertical line test. Any vertical line intersects the graph just once, so this is the graph of a function.

c. $y = 4x - 2$

This linear equation is in the form $y = mx + b$. Since the graph of this equation is a line that is not vertical, the equation defines a function.

Now Try:

1. Determine whether each relation represented by a graph or an equation is a function.

a.

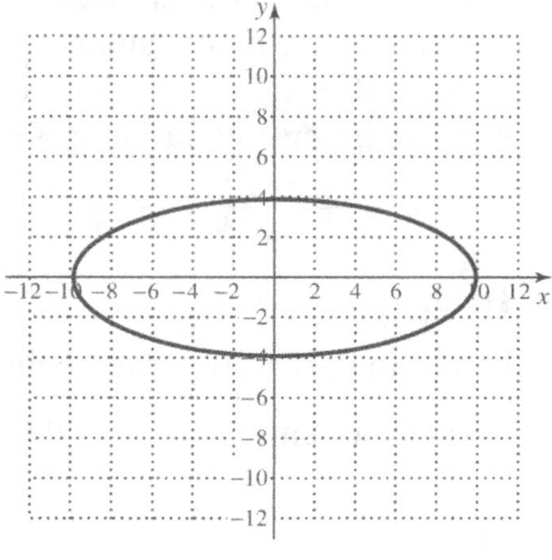

____

b.

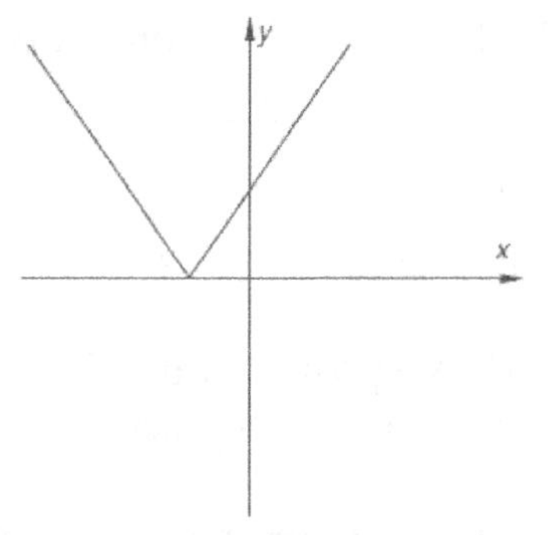

____

c. $y = -3x + 5$

____

Objective 1 Practice Exercises

Use the vertical line test to determine whether each relation graphed is a function.

1.

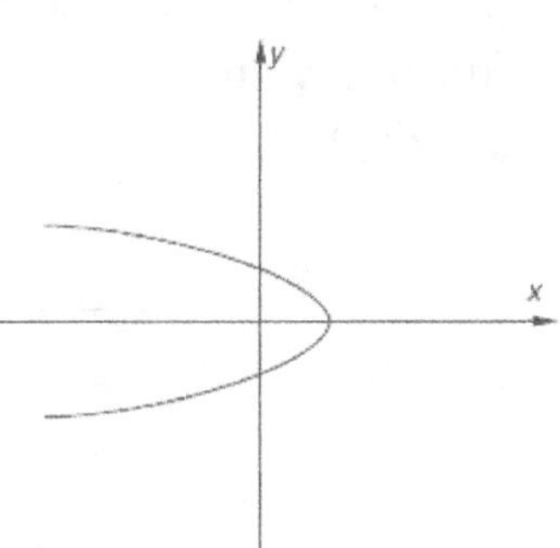

1. _________________

2.

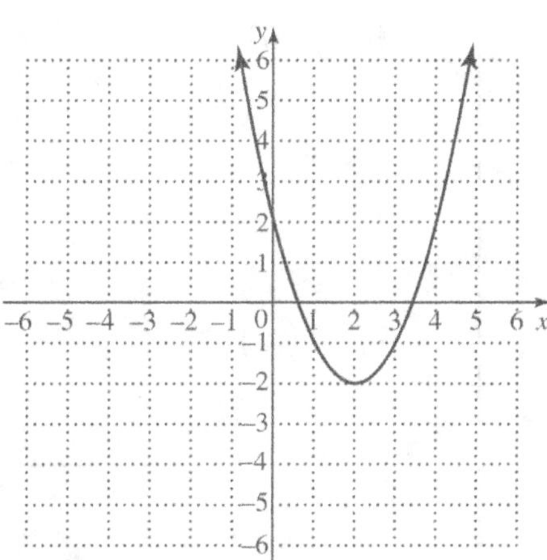

2. _________________

Decide whether the equation defines y as a function of x.

3. $y = 7$

3. _________________

Objective 2   Use function notation to evaluate a function at a specified value.

Review these examples for Objective 2:	**Now Try:**

2. For the function $f(x) = x^2 - 5$, find each function value.

2. For the function $f(x) = 5x - 10$, find each function value.

a. $f(3)$

Substitute 3 for x.

$$f(x) = x^2 - 5$$
$$f(3) = 3^2 - 5$$
$$f(3) = 9 - 5$$
$$f(3) = 4$$

a. $f(3)$

____

b. $f(0)$

$$f(0) = 0^2 - 5$$
$$f(0) = 0 - 5$$
$$f(0) = -5$$

b. $f(0)$

____

c. $f(-4)$

$$f(-4) = (-4)^2 - 5$$
$$f(-4) = 16 - 5$$
$$f(-4) = 11$$

c. $f(-1)$

____

Objective 2 Practice Exercises

For each function f, find (a) $f(-2)$, (b) $f(0)$, *and* (c) $f(4)$.

4. $f(x) = 3x - 7$

4. a. _____________

b. _____________

c. _____________

5. $f(x) = x^2 + 2$

5. a. _____________

b. _____________

c. _____________

6. $f(x) = 9$

6.a.

b. _______________

c. _______________

Objective 3   Find domains and ranges.

Review this example for Objective 3:

3. Find the domain and range of the function.

$$y = 4x - 5$$

Any number may be input for x, so the domain is the set of all real numbers, or $(-\infty, \infty)$. Any number may be the output for y, so the range is also the set of all real numbers, or $(-\infty, \infty)$.

Now Try:

3. Find the domain and range of the function.
$$y = -5x + 6$$

____

____

Objective 3 Practice Exercises

Find the domain and range of each function.

7. $y = -2x + 3$

7. _____________

____

8. $y = -x - 1$

8. _____________

____

9. $y = 2x^2$

9. _____________

____

Objective 4   Determine whether situations and equations represent direct and inverse variations.

Review these examples for Objective 4:
Determine whether the following situation suggests direct or inverse variation.

4.   **a.** The distance that a runner runs and the times it takes to complete the run.

   As the distance increases the time it takes to complete the run increases. So, this is an example of direct variation.

   b. The number of friends cleaning a house and the time it takes to complete the task.

   The more people there are to do a job, the less time it takes. So, this is an example of indirect variation.

Now Try:
Determine whether the following situation suggests direct or inverse variation.

4.   **a.** The time to empty a tank and the rate of pumping.

   ____

   b. The time of an investment and the amount of interest earned.

   ____

Objective 4 Practice Exercises

Determine whether the following situation suggests direct or inverse variation.

10.   The number of calories you eat and the amount of weight you gain.

10.   _______________

11.   The rate at which you drive and how far you travel.

11.   _______________

12.   The number of painters to paint a house and how long it takes to paint the house.

12.   _______________

Objective 5   Solve direct variation problems.

Review these examples for Objective 5:

5. If w varies directly as v, and $w = 24$ when $v = 20$, find w when $v = 25$.

Step 1 Since w varies directly as v, there is a constant k such that $w = kv$.

Step 2 We let $w = 24$, $v = 20$, and solve for k.
$$w = kv$$
$$24 = k \cdot 20$$
$$\frac{6}{5} = k$$

Step 3 Since $w = kv$ and $k = \frac{6}{5}$, we have
$$w = \frac{6}{5}v.$$

Step 4 Now we can find the value of w when $v = 25$.
$$w = \frac{6}{5} \cdot 25 = 30$$

Thus, $w = 30$ when $v = 25$.

6. The force required to compress a spring varies directly as the change in the length of the spring. If a force of 25 pounds is required to compress a spring 4 inches, how much force is required to compress the spring 8 inches?

Step 1 If F represents the force and l represents the length of the spring, then there is a constant k such that $F = kl$.

Step 2 We let $F = 25$, $l = 4$, and solve for k.
$$F = kl$$
$$25 = k \cdot 4$$
$$\frac{25}{4} = k$$

Step 3 Since $F = kl$ and $k = \frac{25}{4}$, we have
$$F = \frac{25}{4}l.$$

Step 4 Now we can find the value of F when $l = 8$.

Now Try:

5. If a varies directly as b, and $a = 61.5$ when $b = 82$, find a when $b = 224$.

____

6. For a given height, the area of a triangle varies directly as its base. Find the area of a triangle with a base of 4 centimeters, if the area is 9.6 square centimeters when the base is 3 centimeters.

____

$$F = \frac{25}{4} \cdot 8 = 50$$

Thus, $F = 50$ lbs when $l = 8$.

Objective 5 Practice Exercises

Solve each problem involving direct variation.

13. If y varies directly as x, and $x = 14$ when $y = 42$, find y when $x = 4$.

13. _______________

14. If c varies directly as d, and $c = 100$ when $d = 5$, find c when $d = 3$.

14. _______________

15. For a given rate, the distance that an object travels varies directly with time. Find the distance an object travels in 5 hours if the object travels 165 miles in 3 hours.

15. _______________

Objective 6   Solve inverse variation problems.

Review these examples for Objective 6:	**Now Try:**

Review these examples for Objective 6:

7. If g varies inversely as f, and $g = 6$ when $f = 12$, find g when $f = 18$.

Since g varies inversely as f, there is a

constant k such that $g = \dfrac{k}{f}$. We know that

$g = 6$ when $f = 12$, so we can find k.

$$g = \frac{k}{f}$$

$$6 = \frac{k}{12}$$

$$72 = k$$

Since $g = \dfrac{72}{f}$, we let $f = 18$ and solve for g.

$$g = \frac{72}{f} = \frac{72}{18} = 4$$

Therefore, when $f = 18$, $g = 4$.

8. For a specified distance, time varies inversely with speed. If Ramona walks a certain distance on a treadmill in 40 minutes at 4.2 miles per hour, how long will it take her to walk the same distance at 3.5 miles per hour?

   Let $t = $ time and $s = $ speed.
Since t varies inversely as s, there is a constant k

such that $t = \dfrac{k}{s}$. Recall that $40 \text{ min} = \dfrac{40}{60} \text{ hr}$.

$$t = \frac{k}{s}$$

$$\frac{40}{60} = \frac{k}{4.2}$$

$$2.8 = k$$

Now use $t = \dfrac{k}{s}$ to find the value of t

when $s = 3.5$.

$$t = \frac{2.8}{3.5} = \frac{4}{5}$$

It takes $\dfrac{4}{5}$ hr, or 48 min to walk the same

distance.

Now Try:

7. If y varies inversely as x, and $y = 10$ when $x = 2$, find y when $x = 4$.

____

8. If the temperature is constant, the pressure of a gas in a container varies inversely as the volume of the container. If the pressure is 9 pounds per square foot in a container of 6 cubic feet, what is the pressure in a container of 7.5 cubic feet?

____

Objective 6 Practice Exercises

Solve each problem involving indirect variation.

16. If y varies inversely as x, and $y = 20$ when $x = 4$,   16. _______________
    find y when $x = 10$.

17. If n varies inversely as m, and $n = 10.5$ when   17. _______________
    $m = 1.2$, find n when $m = 5.6$.

Solve the problem.

18. The length of a violin string varies inversely with the   18. _______________
    frequency of its vibrations. A 10-inch violin string
    vibrates at a frequency of 512 cycles per second.
    Find the frequency of an 8-inch string.

Chapter 10 Roots, Radicals, and Root Functions

Learning Objectives
  Find square roots.
  Find the cube, fourth, and other roots.
  Simplify square root radicals using the product rule.
  Simplify radical sums and differences.
  Rationalize denominators with square roots.

Key Terms

Use the vocabulary terms listed below to complete each statement in exercises 1−12.

square root	**principal square root**	**radicand**
radical	**radical expression**	**perfect square**
irrational number	**cube root**	
like radicals	**index**	**unlike radicals**

1. The number or expression inside a radical sign is called the _________________.

2. A number with a rational square root is called a _________________________.

3. In a radical of the form $\sqrt[n]{a}$, the number n is the _________________________.

4. The number b is a _____________________ of a if $b^2 = a$.

5. The expression $\sqrt[n]{a}$ is called a _________________________________.

6. The expressions $2\sqrt{2}$ and $6\sqrt[3]{2}$ are _______________________.

7. The expressions $2\sqrt{2}$ and $7\sqrt{2}$ are _______________________.

8. The positive square root of a number is its _________________________.

9. A real number that is not rational is called an _____________________.

10. A _________________ is a radical sign and the number or expression in it.

11. The number b is a _____________________ of a if $b^3 = a$.

Objective 1   Find square roots.

Review these examples for Objective 1:	**Now Try:**

Review these examples for Objective 1:

1. Find the square roots of 64.

   What number multiplied by itself equals 64?
   $8^2 = 64$ and $(-8)^2 = 64$.
   Thus, 64 has two square roots: 8 and –8.

2. Find each square root.

   a. $\sqrt{121}$

   $11^2 = 121$, so $\sqrt{121} = 11$.

   b. $-\sqrt{\dfrac{16}{25}}$

   $-\sqrt{\dfrac{16}{25}} = -\dfrac{4}{5}$

3. Find the square of each radical expression.

   a. $\sqrt{17}$

   The square of $\sqrt{17}$ is $\left(\sqrt{17}\right)^2 = 17$.

   b. $\sqrt{w^2 + 3}$

   $\left(\sqrt{w^2 + 3}\right)^2 = w^2 + 3$

Now Try:

1. Find the square roots of 81.

   ____

2. Find each square root.

   a. $\sqrt{169}$

   ____

   b. $-\sqrt{\dfrac{9}{49}}$

   ____

3. Find the square of each radical expression.

   a. $\sqrt{19}$

   ____

   b. $\sqrt{n^2 + 5}$

   ____

Objective 1 Practice Exercises

Find the square root.

1. 625

   1. ______________

2. $\dfrac{121}{196}$

   2. ______________

Find the square of the radical expression.

3. $\sqrt{x^2 - 5}$

   3. ______________

Objective 2   Find the cube, fourth, and other roots.

Review these examples for Objective 2:

4.   Find each cube root.

   a. $\sqrt[3]{729}$

   $\sqrt[3]{729} = 9,$ because $9^3 = 729.$

   b. $\sqrt[3]{-64}$

   $\sqrt[3]{-64} = -4,$ because $(-4)^3 = -64.$

   c. $\sqrt[3]{\dfrac{125}{8}}$

   $\sqrt[3]{\dfrac{125}{8}} = \dfrac{5}{2},$ because $\left(\dfrac{5}{2}\right)^3 = \dfrac{125}{8}.$

5.   Find each root.

   a. $\sqrt[4]{81}$

   $\sqrt[4]{81} = 3,$ because 3 is positive and $3^4 = 81.$

   b. $\sqrt[5]{-1024}$

   $\sqrt[5]{-1024} = -4,$ because $(-4)^5 = -1024.$

Now Try:

4.   Find each cube root.

   a. $\sqrt[3]{343}$

   ____

   b. $\sqrt[3]{-125}$

   ____

   c. $\sqrt[3]{\dfrac{64}{27}}$

   ____

5.   Find each root.

   a. $\sqrt[4]{1296}$

   ____

   b. $\sqrt[5]{-3125}$

   ____

Objective 2 Practice Exercises

Find each root.

4. $\sqrt[3]{-64}$

5. $\sqrt[4]{256}$

6. $\sqrt[7]{-1}$

4. _____________

5. _____________

6. _____________

Objective 3   Simplify square root radicals by using the product rule.

Review these examples for Objective 3:	**Now Try:**

Review these examples for Objective 3:

6. Simplify each radical.

   a. $\sqrt{40}$

$$\sqrt{40} = \sqrt{4 \cdot 10}$$
$$= \sqrt{4} \cdot \sqrt{10}$$
$$= 2\sqrt{10}$$

   b. $\sqrt{75}$

$$\sqrt{75} = \sqrt{25 \cdot 2}$$
$$= \sqrt{25} \cdot \sqrt{2}$$
$$= 5\sqrt{2}$$

   c. $\sqrt{48}$

$$\sqrt{48} = \sqrt{16 \cdot 3} = \sqrt{16} \cdot \sqrt{3} = 4\sqrt{3}$$

7. Find each product and simplify.

   a. $\sqrt{16} \cdot \sqrt{20}$

$$\sqrt{16} \cdot \sqrt{20} = 4\sqrt{20}$$
$$= 4\sqrt{4 \cdot 5}$$
$$= 4\sqrt{4} \cdot \sqrt{5}$$
$$= 4 \cdot 2 \cdot \sqrt{5}$$
$$= 8\sqrt{5}$$

   b. $\sqrt{27} \cdot \sqrt{50}$

$$\sqrt{27} \cdot \sqrt{50} = \sqrt{27 \cdot 50}$$
$$= \sqrt{9 \cdot 3 \cdot 25 \cdot 2}$$
$$= \sqrt{9} \cdot \sqrt{25} \cdot \sqrt{3 \cdot 2}$$
$$= 3 \cdot 5 \cdot \sqrt{6}$$
$$= 15\sqrt{6}$$

Now Try:

6. Simplify each radical.

   a. $\sqrt{12}$

   ____

   b. $\sqrt{98}$

   ____

   c. $\sqrt{80}$

   ____

7. Find each product and simplify.

   a. $\sqrt{25} \cdot \sqrt{18}$

   ____

   b. $\sqrt{6} \cdot \sqrt{12}$

   ____

Objective 3 Practice Exercises

Simplify the radical.

7. $\sqrt{405}$

7. _________________

Find each product and simplify.

8. $\sqrt{11} \cdot \sqrt{33}$

8. _________________

9. $\sqrt{18} \cdot \sqrt{24}$

9. _________________

Objective 4   Simplify radical sums and differences.

Review these examples for Objective 4:

8. Add or subtract, as indicated.

  a. $4\sqrt{5} + \sqrt{20}$

$$4\sqrt{5} + \sqrt{20} = 4\sqrt{5} + \sqrt{4 \cdot 5}$$
$$= 4\sqrt{5} + \sqrt{4} \cdot \sqrt{5}$$
$$= 4\sqrt{5} + 2\sqrt{5}$$
$$= 6\sqrt{5}$$

  b. $2\sqrt{28} + 8\sqrt{63}$

$$2\sqrt{28} + 8\sqrt{63} = 2\left(\sqrt{4} \cdot \sqrt{7}\right) + 8\left(\sqrt{9} \cdot \sqrt{7}\right)$$
$$= 2\left(2\sqrt{7}\right) + 8\left(3\sqrt{7}\right)$$
$$= 4\sqrt{7} + 24\sqrt{7}$$
$$= 28\sqrt{7}$$

  c. $6\sqrt[3]{54} + 2\sqrt[3]{3}$

$$6\sqrt[3]{54} + 2\sqrt[3]{3} = 6\left(\sqrt[3]{27} \cdot \sqrt[3]{3}\right) + 2\sqrt[3]{3}$$
$$= 6\left(3\sqrt[3]{3}\right) + 2\sqrt[3]{3}$$
$$= 18\sqrt[3]{3} + 2\sqrt[3]{3}$$
$$= 20\sqrt[3]{3}$$

Now Try:

8. Add or subtract, as indicated.

  a. $3\sqrt{6} + \sqrt{150}$

  ____

  b. $3\sqrt{24} + 7\sqrt{54}$

  ____

  c. $4\sqrt[3]{128} + 7\sqrt[3]{2}$

  ____

Objective 4 Practice Exercises

Simplify and add or subtract wherever possible.

10. $4\sqrt{128} + 2\sqrt{32}$

10. ________________

11. $2\sqrt[3]{16} - 5\sqrt[3]{2}$

11. ________________

12. $5\sqrt{32} - 8\sqrt{18} + 2\sqrt{20}$

12. ________________

Objective 5   Rationalize denominators with square roots.

Review these examples for Objective 5:

9.   Rationalize each denominator.

   a. $\dfrac{10}{\sqrt{5}}$

$$\frac{10}{\sqrt{5}} = \frac{10 \cdot \sqrt{5}}{\sqrt{5} \cdot \sqrt{5}} \quad \text{Multiply by } \frac{\sqrt{5}}{\sqrt{5}} = 1.$$

$$= \frac{10\sqrt{5}}{5}$$

$$= 2\sqrt{5} \qquad \text{Write in lowest terms.}$$

   b. $\dfrac{18}{\sqrt{27}}$

$$\frac{18}{\sqrt{27}} = \frac{18}{3\sqrt{3}}$$

$$= \frac{18 \cdot \sqrt{3}}{3\sqrt{3} \cdot \sqrt{3}} \quad \text{Multiply by } \frac{\sqrt{3}}{\sqrt{3}} = 1.$$

$$= \frac{18\sqrt{3}}{3 \cdot 3}$$

$$= 2\sqrt{3}$$

Now Try:

9.   Rationalize each denominator.

   a. $\dfrac{14}{\sqrt{7}}$

____

   b. $\dfrac{5}{\sqrt{75}}$

____

Objective 5 Practice Exercises

Rationalize each denominator.

13. $\dfrac{15}{\sqrt{10}}$

13. _______________

14. $\dfrac{6}{\sqrt{28}}$

14. _______________

15. $\dfrac{3\sqrt{5}}{\sqrt{125}}$

15. _______________

Chapter 11 Quadratic Equations, Inequalities, and Functions

Learning Objectives
Evaluate exponential expressions.
Use the four steps for solving a linear equation to solve equations.
Factor trinomials with a coefficient of 1 for the second-degree term.
Factor trinomials using the FOIL method.
Solve quadratic equations using the zero-factor property.

Key Terms

Use the vocabulary terms listed below to complete each statement in exercises 1−11.

linear equation	solution set	equivalent equations
prime polynomial	factoring	greatest common factor
quadratic equation	standard form	double solution
coefficient	trinomial	

1. Equations that have exactly the same solutions sets are called

   ______________________________.

2. An equation that can be written in the form $Ax + B = C$, where A, B, and C are real numbers and $A \neq 0$, is called a ______________________________.

3. In the term $6x^2 y$, 6 is the ______________________________.

4. An equation that can written in the form $ax^2 + bx + c = 0$, with $a \neq 0$, is a

   ______________________.

5. The ______________________________ of a polynomial is the greatest term that is a factor of all the terms in the polynomial.

6. A ______________________________ is a polynomial that cannot be factored using only integers.

7. A polynomial with three terms is a ______________________________.

8. An equation written in the form $ax^2 + bx + c = 0$ is written in the ______________________________ of a quadratic equation.

9. Two factors are identical and both lead to the same solution, called a

   ______________________.

10. The set of all numbers that satisfy an equation is called its ______________________.

11. ______________________________ is the process of writing a polynomial as a product.

Objective 1   Evaluate exponential expressions.

Review this example for Objective 1:

1.   Find the value of the exponential expression.

$$6^2$$

6^2 means $6 \cdot 6$, which equals 36.

Now Try:

1.   Find the value of the exponential expression.

7^2

____

Objective 1 Practice Exercises

Find the value of each exponential expression.

1. 3^3

1.   _____________

2. $\left(\dfrac{2}{3}\right)^4$

2.   _____________

3. $(0.4)^2$

3.   _____________

Objective 2   Use the four steps for solving a linear equation.

Review these examples for Objective 2: | **Now Try:**

2. Solve $-5x+8=23$.

2. Solve $-8x+11=59$.

____

Step 1 There are no parentheses, fractions, or decimals in this equation, so this step is not necessary.

$$-5x+8=23$$

Step 2
$$-5x+8-8=23-8$$
$$-5x=15$$

Step 3
$$\frac{-5x}{-5}=\frac{15}{-5}$$
$$x=-3$$

Step 4 Check by substituting –3 for x in the original equation.
$$-5x+8=23$$
$$-5(-3)+8\overset{?}{=}23$$
$$15+8\overset{?}{=}23$$
$$23=23 \quad \text{True}$$

The solution, –3, checks, so the solution set is $\{-3\}$.

3. Solve $4x+3=6x-11$.

3. Solve $5x+4=8x-20$.

____

Step 1 There are no parentheses, fractions, or decimals in this equation, so begin with Step 2.
$$4x+3=6x-11$$

Step 2
$$4x+3-4x=6x-11-4x$$
$$3=2x-11$$
$$3+11=2x-11+11$$
$$14=2x$$

Step 3
$$\frac{14}{2}=\frac{2x}{2}$$
$$7=x$$

Step 4 Check by substituting 7 for x in the original equation.
$$4x+3=6x-11$$
$$4(7)+3\overset{?}{=}6(7)-11$$
$$28+3\overset{?}{=}42-11$$
$$31=31 \quad \text{True}$$

The solution, 7, checks, so the solution set is $\{7\}$.

4. Solve $9a - (4 + 3a) = 2a + 5$.

$$9a - (4 + 3a) = 2a + 5$$

Step 1 $\quad 9a - 4 - 3a = 2a + 5$

$$6a - 4 = 2a + 5$$

Step 2 $\quad 6a - 4 - 2a = 2a + 5 - 2a$

$$4a - 4 = 5$$

$$4a - 4 + 4 = 5 + 4$$

$$4a = 9$$

Step 3 $\quad \dfrac{4a}{4} = \dfrac{9}{4}$

$$a = \dfrac{9}{4}$$

Step 4 Check that the solution set is $\left\{\dfrac{9}{4}\right\}$.

4. Solve $10a - (11 + 3a) = 5a + 4$.

____

Objective 2 Practice Exercises

Solve each equation and check your solution.

4. $\quad 7t + 6 = 11t - 4$

4. _______________

5. $\quad 3a - 6a + 4(a - 4) = -2(a + 2)$

5. _______________

6. $\quad 3(t + 5) = 6 - 2(t - 4)$

6. _______________

Objective 3   Factor trinomials with coefficient 1 for the second-degree term.

Review these examples for Objective 3: | **Now Try:**

5.   Factor $m^2 + 8m + 15$.

Look for integers whose product is 15 and whose sum is 8. Only positive signs are needed.

Factors of 15	Sums of Factors
15, 1	$15 + 1 = 16$
5, 3	$5 + 3 = 8$

From the table, 5 and 3 are the required integers.

$m^2 + 8m + 15$ factors as $(m+5)(m+3)$

Check   Use the FOIL method.

$$(m+5)(m+3) = m^2 + 3m + 5m + 15$$
$$= m^2 + 8m + 15$$

5.   Factor $x^2 + 11x + 24$.

____

6.   Factor the trinomial.

$x^2 - 7x + 18$

Look for integers whose product is 18 and whose sum is -7. Since the numbers have a positive product and a negative sum, we consider only pairs of negative integers.

Factors of 18	Sums of Factors
$-18, -1$	$-18 + (-1) = -19$
$-9, -2$	$-9 + (-2) = -11$
$-6, -3$	$-6 + (-3) = -9$

None of the pairs of integers has a sum of -7.

$x^2 - 7x + 18$ cannot be factored. It is a prime polynomial.

6.   Factor the trinomial.

$m^2 - 7m + 5$

____

7.   Factor $x^2 - 6xy - 7y^2$.

Here, the coefficient of x in the middle term is $-6y$, so we need to find two expressions whose product is $-7y^2$ and whose sum is $-6y$.

Factors of $-7y^2$	Sums of Factors
$7y, -y$	$7y + (-y) = 6y$
$-7y, y$	$-7y + y = -6y$

$x^2 - 6xy - 7y^2$ factors as $(x-7y)(x+y)$

7.   Factor $p^2 - 5pq - 14q^2$.

____

Check   Use the FOIL method.
$$(x-7y)(x+y)=x^2+xy-7xy-7y^2$$
$$=x^2-6xy-7y^2$$

Objective 3 Practice Exercises

Factor completely. If a polynomial cannot be factored, write prime.

7. r^2+r+3

7. _______________

8. $x^2-11x+28$

8. _______________

9. $x^2-8x-33$

9. _______________

Objective 4   Factor trinomials using the FOIL method.

| **Review these examples for Objective 4:** | **Now Try:** |

Review these examples for Objective 4: **Now Try:**

8. Factor $6x^2 + 13x + 7$.

8. Factor $15x^2 + 26x + 7$.

The number 6 has several possible pairs of factors, but 7 has only 1 and 7, or -1 and -7. We choose positive factors since all coefficients in the trinomial are positive.

$$(\underline{\quad} + 7)(\underline{\quad} + 1)$$

The possible pairs of $6x^2$ are $6x$ and x, or $3x$ and $2x$.

$$(3x + 7)(2x + 1)$$
gives middle term $3x + 14x = 17x$. Incorrect.
$$(2x + 7)(3x + 1)$$
gives middle term $2x + 21x = 23x$. Incorrect.
$$(6x + 7)(x + 1)$$
gives middle term $6x + 7x = 13x$. Correct.

$6x^2 + 13x + 7$ factors as $(6x + 7)(x + 1)$.

Check. Multiply $(6x + 7)(x + 1)$ to obtain $6x^2 + 13x + 7$.

9. Factor $6x^2 - x - 15$.

9. Factor $8x^2 + 2x - 21$.

The integer 6 has several possible pairs of factors, as does -15. Since the constant term is negative, one positive factor and one negative factor of -15 are needed. Since the coefficient of the middle term is relatively small, it is wise to avoid large factors. We try $3x$ and $2x$ as factors of $6x^2$ and 5 and -3 as factors of -15.

$$(3x + 5)(2x - 3)$$
has middle term $-9x + 10x = x$. Incorrect.
$$(3x - 5)(2x + 3)$$
has middle term $9x - 10x = -x$. Correct.

$6x^2 - x - 15$ factors as $(3x - 5)(2x + 3)$.

10. Factor $18x^2 - 3xy - 28y^2$.

There are several factors of $18x^2$, including
18x and x, 9x and 2x, and 6x and 3x.
There are many possible pairs of factors of
$-28y^2$, including
28y and $-y$, $-28y$ and y, 14y and $-2y$,
$-14y$ and 2y, 7y and $-4y$, $-7y$ and 4y.

Once again, since the coefficient of the middle
term is relatively small, avoid the larger factors.
Try the factors of 6x and 3x, and 4y and $-7y$.

$(6x + 4y)(3x - 7y)$ Incorrect

The first binomial has a common factor of 2.

$(6x - 7y)(3x + 4y)$

has middle term $24xy - 21xy = 3xy$. Incorrect.
Interchange the signs of the last two terms.

$(6x + 7y)(3x - 4y)$

has middle term $-24xy + 21xy = -3xy$. Correct.

Thus, $18x^2 - 3xy - 28y^2$ factors as
$(6x + 7y)(3x - 4y)$

11. Factor the trinomial.

$$-105a^3 + 65a^2 - 10a$$

The common factor is $-5a$. Then use trial
and error.

$$-105a^3 + 65a^2 - 10a = -5a(21a^2 - 13a + 2)$$
$$= -5a(3a - 1)(7a - 2)$$

Check.

$$-5a(3a - 1)(7a - 2) = -5a(21a^2 - 13a + 2)$$
$$= -105a^3 + 65a^2 - 10a$$

10. Factor $24x^2 - 2xy - 15y^2$.

____

11. Factor the trinomial.

$$-18a^3 + 66a^2 - 60a$$

____

Objective 4 Practice Exercises

Factor each trinomial completely.

10. $8q^2 + 10q + 3$

10. __________________

11. $3a^2 + 8ab + 4b^2$ **11.** ________________

12. $4c^2 + 14cd - 8d^2$ **12.** ________________

Objective 5   Solve quadratic equations using the zero-factor property.

Review these examples for Objective 5:	**Now Try:**

Review these examples for Objective 5:

12. Solve each equation.

   a. $(x+9)(5x-6)=0$

By the zero-factor property, either $x+9=0$ or $5x-6=0$.

$$x+9=0 \quad \text{or} \quad 5x-6=0$$
$$x=-9 \quad \text{or} \quad 5x=6$$
$$x=\frac{6}{5}$$

Check:
Let $x=-9$.
$$(x+9)(5x-6)=0$$
$$(-9+9)[5(-9)-6]\overset{?}{=}0$$
$$0(-51)\overset{?}{=}0$$
$$0=0 \text{ True}$$

Let $x=\frac{6}{5}$.
$$(x+9)(5x-6)=0$$
$$\left(\frac{6}{5}+9\right)\left[5\left(\frac{6}{5}\right)-6\right]\overset{?}{=}0$$
$$\left(\frac{51}{5}\right)(6-6)\overset{?}{=}0$$
$$0=0 \text{ True}$$
Both values check, so the solution set is
$$\left\{-9,\ \frac{6}{5}\right\}.$$

   b. $x(8x-11)=0$

Use the zero-factor property.
$$x=0 \quad \text{or} \quad 8x-11=0$$
$$8x=11$$
$$x=\frac{11}{8}$$

Check these solutions by substituting each in the original equation. The solution set is $\left\{0,\ \frac{11}{8}\right\}.$

Now Try:

12. Solve each equation.

   a. $(x+12)(4x-7)=0$

____

   b. $x(6x-11)=0$

____

13. Solve $8p^2 + 30 = 46p$.

$$8p^2 + 30 = 46p$$

$$8p^2 - 46p + 30 = 0 \quad \text{Standard form}$$

$$2(4p^2 - 23p + 15) = 0 \quad \text{Factor out 2.}$$

$$4p^2 - 23p + 15 = 0 \quad \text{Divide each side by 2}$$

$$(4p - 3)(p - 5) = 0 \quad \text{Factor.}$$

$$4p - 3 = 0 \quad \text{or} \quad p - 5 = 0 \quad \text{Zero-factor}$$

$$p = \frac{3}{4} \quad \text{or} \quad p = 5 \quad \text{property}$$

The solution set is $\left\{\frac{3}{4}, 5\right\}$.

13. Solve $15p^2 + 36 = 57p$.

____

Objective 5 Practice Exercises

Solve each equation and check your solutions.

13. $2x^2 - 3x - 20 = 0$

13. _______________

14. $25x^2 = 20x$

14. _______________

15. $c(5c + 17) = 12$

15. _______________

Chapter 12 Inverse, Exponential and Logarithmic Functions

Learning Objectives
Solve equations of the form $x^2 = k$, where $k > 0$.
Solve radical equations having square root radicals.
Determine whether a graph or an equation represents a function.
Find domain and range of given functions.

Key Terms

Use the vocabulary terms listed below to complete each statement in exercises 1–9.

quadratic equation	**zero-factor property**	**range**
radical equation	**extraneous solution**	**domain**
components	**relation**	**function**

1. In an ordered pair (x, y), x and y are the _________________________.

2. An equation that can be written in the form $ax^2 + bx + c = 0$ is a
    _____________________.

3. An _____________________ is a potential solution to an equation that does
    not satisfy the equation.

4. An equation with a variable in the radicand is a _____________________.

5. The _____________________ states that if a product equals 0, then at least
    one of the factors of the product also equals zero.

6. Any set of ordered pairs is called a _____________________.

7. The set of all second components in the ordered pairs of a relation is the
    _____________________ of the relation.

8. A _____________________ is a set of ordered pairs in which each first
    component corresponds to exactly one second component.

9. The set of all first components in the ordered pairs of a relation is the
    _____________________ of the relation.

Objective 1   Solve equations of the form $x^2 = k$, where $k > 0$.

Review these examples for Objective 1:	**Now Try:**

Review these examples for Objective 1:

1. Solve each equation. Write radicals in simplified form.

a. $x^2 = 36$

By the square root property, if $x^2 = 36$, then
$$x = \sqrt{36} = 6 \text{ or } x = -\sqrt{36} = -6$$
The solution set is $\{-6, 6\}$, or $\{\pm 6\}$.

b. $x^2 = 13$

The solutions are $x = \sqrt{13}$ or $x = -\sqrt{13}$
The solution set is $\{-\sqrt{13}, \sqrt{13}\}$, or $\{\pm\sqrt{13}\}$.

c. $5p^2 - 100 = 0$

$$5p^2 - 100 = 0$$
$$5p^2 = 100$$
$$p^2 = 20$$
$$p = \sqrt{20} \quad \text{or} \quad p = -\sqrt{20}$$
$$p = 2\sqrt{5} \quad \text{or} \quad p = -2\sqrt{5}$$

Check

$$\begin{array}{c|c} 5p^2 - 100 = 0 & 5p^2 - 100 = 0 \\ 5(2\sqrt{5})^2 - 100 \overset{?}{=} 0 & 5(-2\sqrt{5})^2 - 100 \overset{?}{=} 0 \\ 5(20) - 100 \overset{?}{=} 0 & 5(20) - 100 \overset{?}{=} 0 \\ 0 = 0 \text{ True} & \text{True } 0 = 0 \end{array}$$

The solution set is $\{2\sqrt{5}, -2\sqrt{5}\}$, or $\{\pm 2\sqrt{5}\}$.

d. $5x^2 + 16 = 31$

$$5x^2 + 16 = 31$$
$$5x^2 = 15$$
$$x^2 = 3$$
$$x = \sqrt{3} \quad \text{or} \quad x = -\sqrt{3}$$

Now Try:

1. Solve each equation. Write radicals in simplified form.

a. $x^2 = 81$

____

b. $x^2 = 23$

____

c. $3x^2 - 54 = 0$

____

d. $4x^2 + 14 = 22$

____

2. Solve $p^2 = -64$.

Because –64 is a negative number and because the square of a real number cannot be negative, there is no real number solution of this equation. The solution set is $\varnothing$.

2. Solve $n^2 = -25$.

____

Objective 1 Practice Exercises

Solve each equation by using the square root property. Express all radicals in simplest form.

1. $r^2 = 900$

1. _______________

2. $3x^2 - 5 = 22$

2 _______________

3. $p^2 = -144$

3. _______________

Integrated Review Worksheets for *Beginning and Intermediate Algebra*, 8th edition

Objective 2   Solve radical equations having square root radicals.

Review these examples for Objective 2: | **Now Try:**

3.   Solve $\sqrt{p+2}=5$.

$$\left(\sqrt{p+2}\right)^2 = 5^2$$
$$p+2 = 25$$
$$p = 23$$

Check $\sqrt{p+2} = 5$
$$\sqrt{23+2} \overset{?}{=} 5$$
$$\sqrt{25} \overset{?}{=} 5$$
$$5 = 5 \quad \text{True}$$

The solution set is $\{23\}$.

3.   Solve $\sqrt{p+5}=4$.

____

4.   Solve $4\sqrt{x}=\sqrt{x+30}$.

$$\left(4\sqrt{x}\right)^2 = \left(\sqrt{x+30}\right)^2$$
$$4^2\left(\sqrt{x}\right)^2 = \left(\sqrt{x+30}\right)^2$$
$$16x = x+30$$
$$15x = 30$$
$$x = 2$$

Check $4\sqrt{x} = \sqrt{x+30}$
$$4\sqrt{2} \overset{?}{=} \sqrt{2+30}$$
$$4\sqrt{2} \overset{?}{=} \sqrt{32}$$
$$4\sqrt{2} = 4\sqrt{2} \quad \text{True}$$

The solution set is $\{2\}$.

4.   Solve $5\sqrt{x}=\sqrt{x+72}$.

____

Objective 2 Practice Exercises

Solve each equation.

4. $\sqrt{3x+1}=3$

4.   _________________

    Copyright © 2025 Pearson Education, Inc.

5. $\sqrt{2+4k} = 3\sqrt{k}$

5. ___________________

6. $\sqrt{4x+3} = \sqrt{3x+5}$

6. ___________________

Objective 3   Determine whether a graph or an equation represents a function.

Review these examples for Objective 3:

5. Determine whether each relation represented by a graph or an equation is a function.

a.

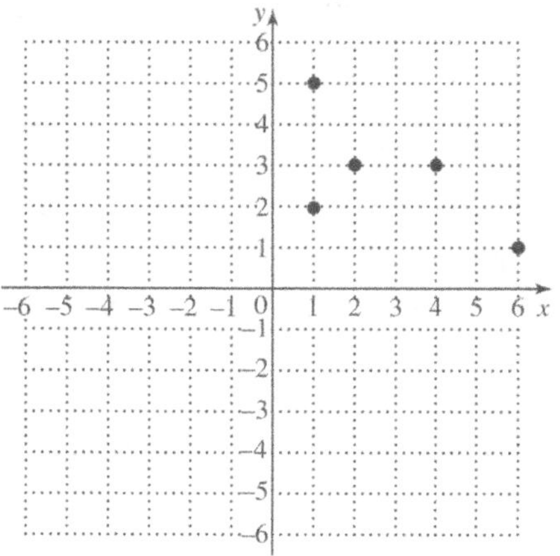

Because there are two ordered pairs with first component 1, this is not the graph of a function.

b.

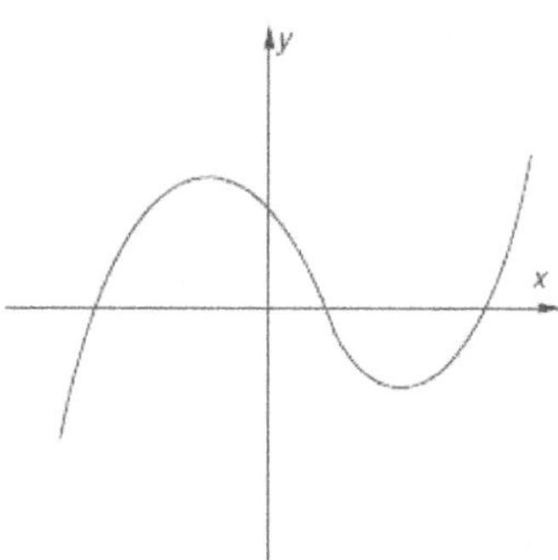

Use the vertical line test. Any vertical line intersects the graph just once, so this is the graph of a function.

c. $y = 4x - 2$

This linear equation is in the form $y = mx + b$. Since the graph of this equation is a line that is not vertical, the equation defines a function.

Now Try:

5. Determine whether each relation represented by a graph or an equation is a function.

a.

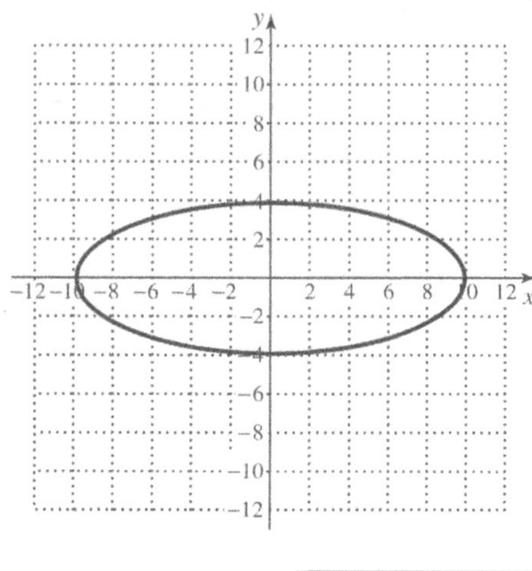

____

b.

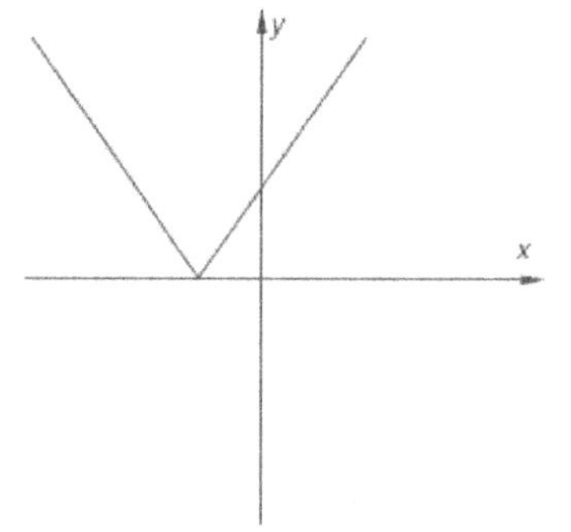

____

c. $y = -3x + 5$

____

Objective 3 Practice Exercises

Use the vertical line test to determine whether each relation graphed is a function.

7.

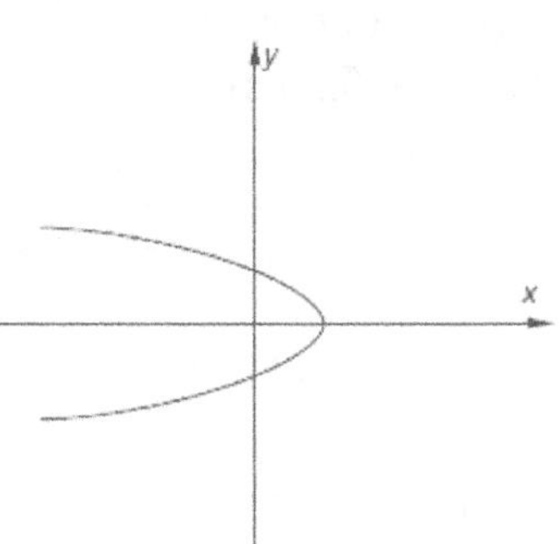

7. _______________

8.

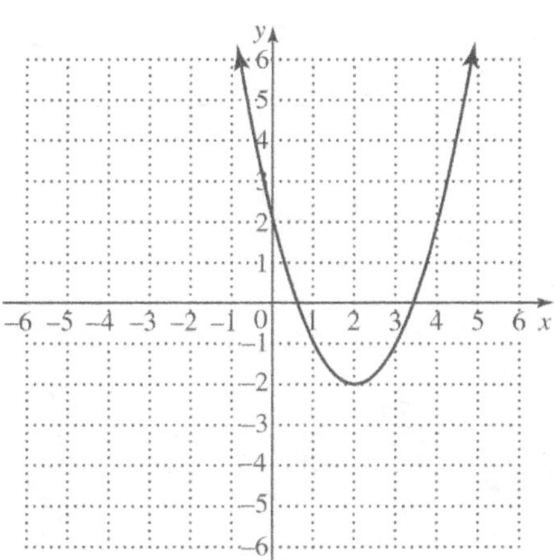

8. _______________

Decide whether the equation defines y as a function of x.

9. $y = 7$

9. _______________

Objective 4   Find domains and ranges.

Review this example for Objective 4:

6.   Find the domain and range of the function.

$$y = 4x - 5$$

Any number may be input for x, so the domain is the set of all real numbers, or $(-\infty, \infty)$.
Any number may be the output for y, so the range is also the set of all real numbers, or $(-\infty, \infty)$.

Now Try:

6.   Find the domain and range of the function.
$$y = -5x + 6$$

____

____

Objective 4 Practice Exercises

Find the domain and range of each function.

10. $y = -2x + 3$

10. _____________

____

11. $y = -x - 1$

11. _____________

____

12. $y = 2x^2$

12. _____________

____

Chapter 13 Nonlinear Functions, Conic Sections, and Nonlinear Systems

---

Learning Objectives

Midpoint and Distance in the Coordinate Plane.

Solve quadratic equations by completing the square when the leading coefficient is 1.

Graph a line using its slope and a point on the line.

Graph quadratic equations.

Solve systems of linear inequalities by graphing.

Key Terms

Use the vocabulary terms listed below to complete each statement in exercises 1–12.

completing the square   perfect square trinomial   square root property

slope-intercept form   point-slope form   standard form

parabola   vertex   axis of symmetry   line of symmetry

system of linear inequalities   solution set of a system of linear inequalities

1. A ______________________________ can be written in the form $x^2 + 2kx + k^2$ or $x^2 - 2kx + k^2$

2. The ______________________________ says that, if k is positive and $a^2 = k$, then $a = \pm\sqrt{k}$.

3. Use the process called ______________________________ in order to rewrite an equation so it can be solved using the square root property.

4. A linear equation in the form $y - y_1 = m(x - x_1)$ is written in

______________________________.

5. A linear equation in the form $Ax + By = C$ is written in

______________________________.

6. A linear equation in the form $y = mx + b$ is written in

______________________________.

7. If a graph is folded on its______________________________, the two sides coincide.

8. The _______________________________ of a parabola that opens upward or downward is the lowest or highest point on the graph.

9. The _______________________________ of a parabola that opens upward or downward is a vertical line through the vertex.

10. The graph of the quadratic equation $y = ax^2 + bx + c$ is called a _______________________.

11. All ordered pairs that make all inequalities of the system true at the same time is called the _______________________________.

12. A _______________________________ contains two or more linear inequalities (and no other kinds of inequalities).

Objective 1   Midpoint and Distance in the Coordinate Plane.

Review these examples for Objective 1:	**Now Try:**

Review these examples for Objective 1:

1. Find the midpoint of the line segment with endpoints of coordinates: $(3, 4)$ and $(-1, 2)$

   To find the midpoint use the midpoint formula

   $$\left(\frac{x_1 + x_2}{2}, \frac{y_1 + y_2}{2}\right)$$

   $$\left(\frac{3 + -1}{2}, \frac{4 + 2}{2}\right) = \left(\frac{2}{2}, \frac{6}{2}\right) = (1, 3)$$

   So, the midpoint of the line segment is $(1, 3)$.

2. Find the distance between the coordinates: $(3, 4)$ and $(-1, 2)$

   To find the distance use the distance formula:

   $$d = \sqrt{(x_2 - x_1)^2 + (y_2 - y_1)^2}.$$

   $$d = \sqrt{(-1 - 3)^2 + (2 - 4)^2}$$

   $$= \sqrt{(-4)^2 + (-2)^2}$$

   $$= \sqrt{16 + 4} = \sqrt{20} = 2\sqrt{5}.$$

Now Try:

1. Find the midpoint of the line segment with endpoints of coordinates: $(3, -4)$ and $(-2, 8)$

   ____

2. Find the distance between the coordinates: $(3, -4)$ and $(-2, 8)$

   ____

Objective 1 Practice Exercises

Given the following coordinates a) find the midpoint of the line segment b) find the distance between the given coordinates.

1. $(0, 3)$ and $(-4, 7)$

   1. Midpoint: _________________
      Distance: _________________

2. $(-2, 3)$ and $(1, 7)$

   2. Midpoint: _________________
      Distance: _________________

3. $(1, 1)$ and $(2, -4)$

3. Midpoint: _________________
Distance: _________________

Objective 2   Solve quadratic equations by completing the square when the coefficient of the second-degree term is 1.

Review these examples for Objective 1:

3. Complete each trinomial so that it is a perfect square. Then factor the trinomial.

   a. $x^2 + 24x +$ _______

   The perfect square trinomial will have the form $x^2 + 2kx + k^2$. Thus, the middle term $24x$, must equal $2kx$.

   $$24x = 2kx$$
   $$12 = k$$

   Therefore, $k = 12$ and $k^2 = 12^2 = 144$. The perfect square trinomial is $x^2 + 24x + 144$, which factors as $(x + 12)^2$.

   b. $x^2 - 36x +$ _______

   The perfect square trinomial will have the form $x^2 - 2kx + k^2$. Thus, the middle term $-36x$, must equal $-2kx$.

   $$-36x = -2kx$$
   $$18 = k$$

   Therefore, $k = 18$ and $k^2 = 18^2 = 324$. The required perfect square trinomial is $x^2 - 36x + 324$, which factors as $(x - 18)^2$.

4. Solve $x^2 + 8x + 3 = 0$.

   $$x^2 + 8x + 3 = 0$$
   $$x^2 + 8x = -3$$

   The expression on the left must be written as a perfect square trinomial in the form $x^2 + 2kx + k^2$.

   $$x^2 + 8x + \underline{\quad}$$

   Here, $2kx = 8x$, so $k = 4$ and $k^2 = 16$. The required perfect square trinomial is $x^2 + 8x + 16$ which factors as $(x + 4)^2$.

   Therefore, if we add 16 to each side of

Now Try:

3. Complete each trinomial so that it is a perfect square. Then factor the trinomial.

   a. $x^2 + 10x +$ _______

   ____

   b. $x^2 - 22x +$ _______

   ____

4. Solve $x^2 + 10x - 7 = 0$.

   ____

$x^2 + 8x = -3$, the equation will have a perfect square trinomial on the left side, as needed.

$$x^2 + 8x + 16 = -3 + 16$$

$$(x + 4)^2 = 13$$

Use the square root property.

$$x + 4 = \sqrt{13} \qquad \text{or} \quad x + 4 = -\sqrt{13}$$

$$x = -4 + \sqrt{13} \quad \text{or} \qquad x = -4 - \sqrt{13}$$

Check by substituting $-4 + \sqrt{13}$ and then $-4 - \sqrt{13}$ for x in the original equation. The solution set is $\left\{ -4 + \sqrt{13},\ -4 - \sqrt{13} \right\}$.

Objective 2 Practice Exercises

Solve each equation by completing the square.

4. $r^2 + 8r = -4$

4. _______________

5. $x^2 - 4x = 2$

5. _______________

6. $x^2 + 2x = 63$

6. _______________

Objective 3   Graph a line by using its slope and a point on the line.

Review this example for Objective 3:

5. Graph the equation by using the slope and y-intercept.

$$2x - 3y = 6$$

Step 1 Solve for y to write the equation in slope-intercept form.

$$2x - 3y = 6$$
$$-3y = -2x + 6$$
$$y = \frac{2}{3}x - 2$$

Step 2 The y-intercept is $(0, -2)$. Graph this point.

Step 3 The slope is $\frac{2}{3}$. By definition,

$$\text{slope } m = \frac{\text{change in } y \text{ (rise)}}{\text{change in } x \text{ (run)}} = \frac{2}{3}$$

From the y-intercept, count up 2 units and to the right 3 units to obtain the point $(3, 0)$.

Step 4 Draw the line through the points $(0, -2)$ and $(3, 0)$ to obtain the graph.

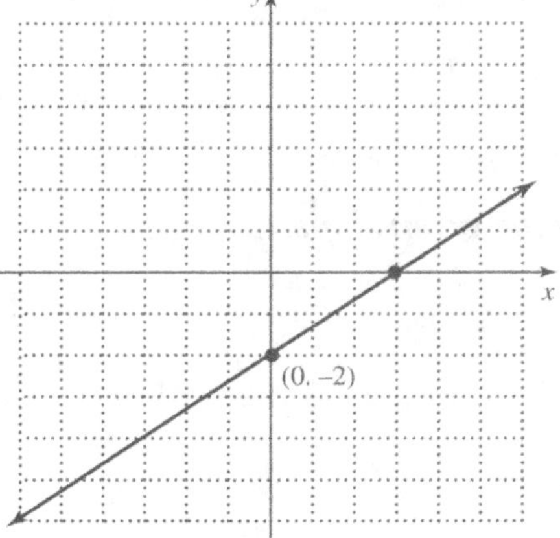

Now Try:

5. Graph the equation by using the slope and y-intercept.

$$y = \frac{2}{3}x$$

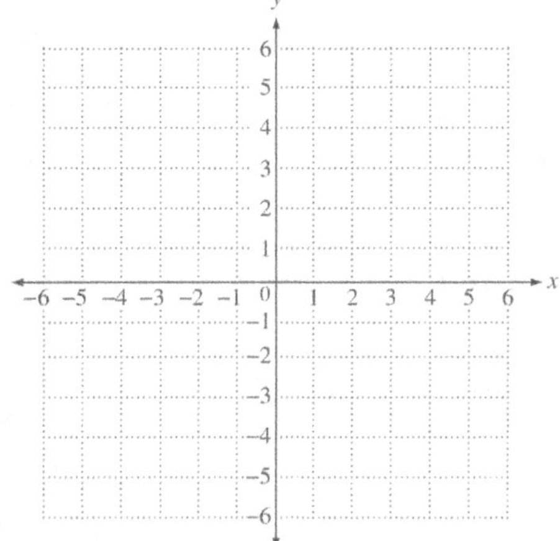

Objective 3 Practice Exercises

Graph each equation by using the slope and y-intercept.

7. $4x - y = 4$

7.

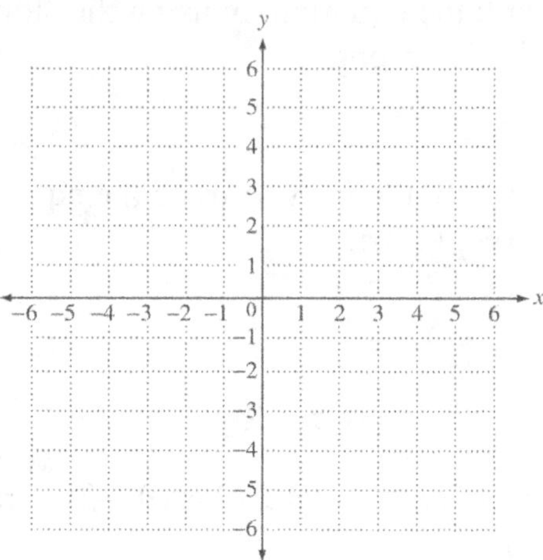

8. $y = -3x + 6$

8.

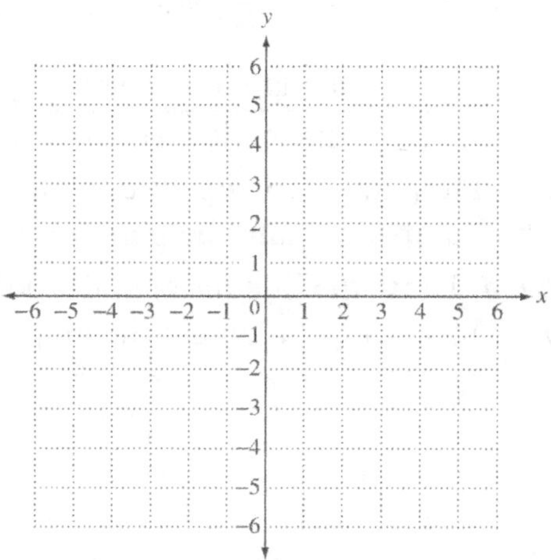

Graph the line passing through the given point and having the given slope.

9. $(4, -2)$; $m = -1$

9.

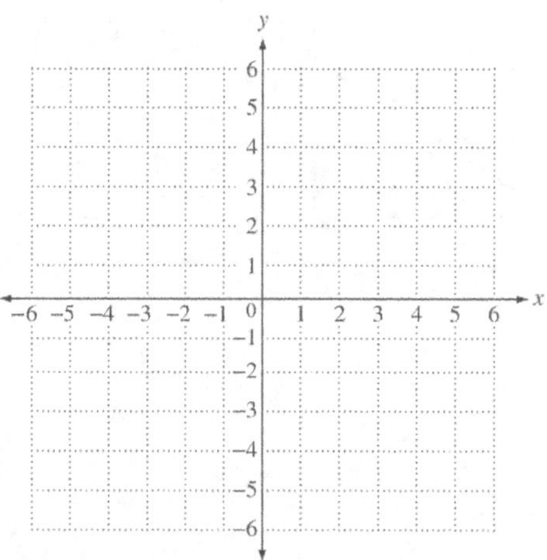

Copyright © 2025 Pearson Education, Inc.

Objective 4   Graph quadratic equations.

Review this example for Objective 4:

6. Graph $y = 9 - x^2$. Identify the vertex.

Let $x = 3$.

$$y = 9 - 3^2$$
$$y = 9 - 9$$
$$y = 0$$

Ordered pair: $(3, 0)$

Let $x = 0$.

$$y = 9 - 0^2$$
$$y = 9 - 0$$
$$y = 9$$

Ordered pair: $(0, 9)$

Let $x = -3$.

$$y = 9 - (-3)^2$$
$$y = 9 - 9$$
$$y = 0$$

Ordered pair: $(-3, 0)$

Select several values for x and find the corresponding values for y. Plot the ordered pairs and join them with a smooth curve.

x	y
-4	-7
-3	0
-2	5
0	9
3	0
4	-7

The vertex is the highest point on the graph, $(0, 9)$.

Now Try:

6. Graph $y = -x^2 - 1$. Identify the vertex.

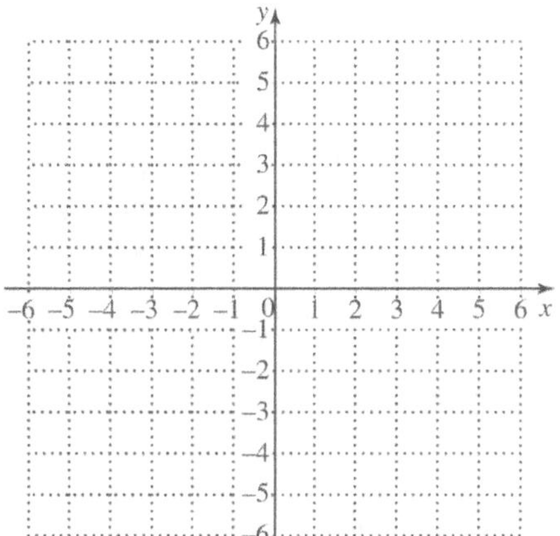

Objective 4 Practice Exercises

Graph each equation. Give the coordinates of the vertex in each case.

10. $y = -x^2$

10. vertex: _______________________

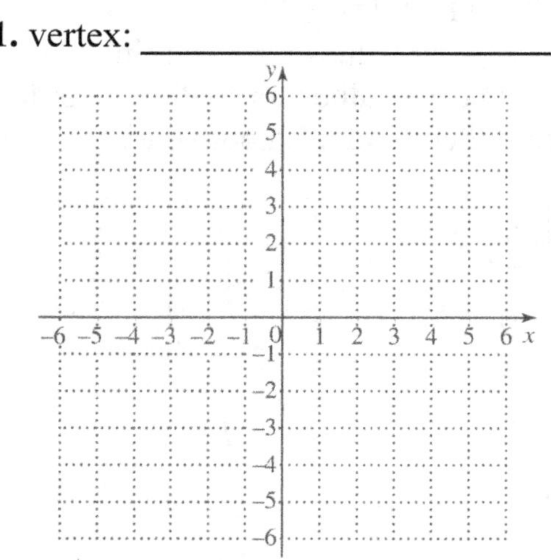

11. $y = x^2 - 1$

11. vertex: _______________________

Copyright © 2025 Pearson Education, Inc.

Objective 5   Solve systems of linear inequalities by graphing.

Review these examples for Objective 5:	**Now Try:**

Review these examples for Objective 5:

7. Graph the solution set of the system.

$$x + y \le 3$$

$$5x - y \ge 5$$

To graph $x + y \le 3$, graph the solid boundary line $x + y = 3$ using the intercepts $(0, 3)$ and $(3, 0)$. Determine the region to shade using $(0, 0)$ as a test point.

$$x + y \le 3$$

$$0 + 0 \overset{?}{\le} 3$$

$$0 \le 3 \quad \text{True}$$

Shade the region containing $(0, 0)$.

To graph $5x - y \ge 5$, graph the solid boundary line $5x - y = 5$ using the intercepts $(0, -5)$ and $(1, 0)$. Determine the region to shade using $(0, 0)$ as a test point.

$$5x - y \ge 5$$

$$5(0) - 0 \overset{?}{\ge} 5$$

$$0 \ge 5 \quad \text{False}$$

Shade the region that does not contain $(0, 0)$.

The solution set of this system includes all points in the intersection (overlap) of the graph of the two inequalities. This intersection is the gray shaded region and portions of the two boundary lines that surround it.

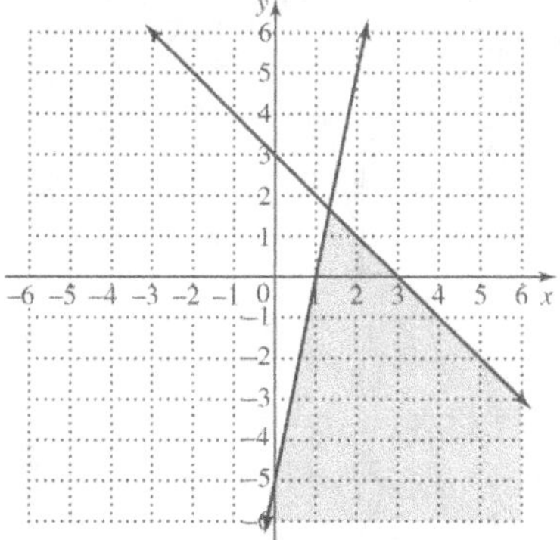

Now Try:

7. Graph the solution set of the system.

$$3x - y \le 3$$

$$x + y \le 0$$

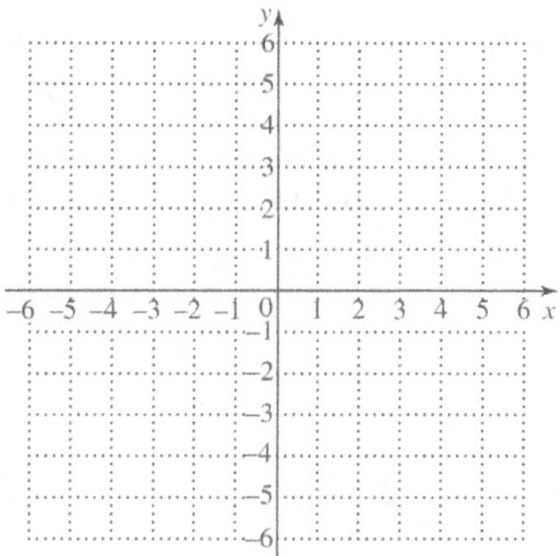

8. Graph the solution set of the system.

$$4x - y > 2$$

$$x + y > -2$$

8. Graph the solution set of the system.

$$6x - y > 6$$

$$2x + 5y < 10$$

Graph $4x - y > 2$ using a dashed line for

$4x - y = 2$ and intercepts $(0, -2)$ and $\left(\frac{1}{2},\ 0\right)$.

Using the test point $(0, 0)$, we shade the region that does not contain the point $(0, 0)$.

Graph $x + y > -2$ using a dashed line for $x + y = -2$ and intercepts $(-2, 0)$ and $(0, -2)$.

Using the test point $(0, 0)$, we shade the region that does contain the point $(0, 0)$.

The solution set is marked in gray. The solution set does not include either boundary line.

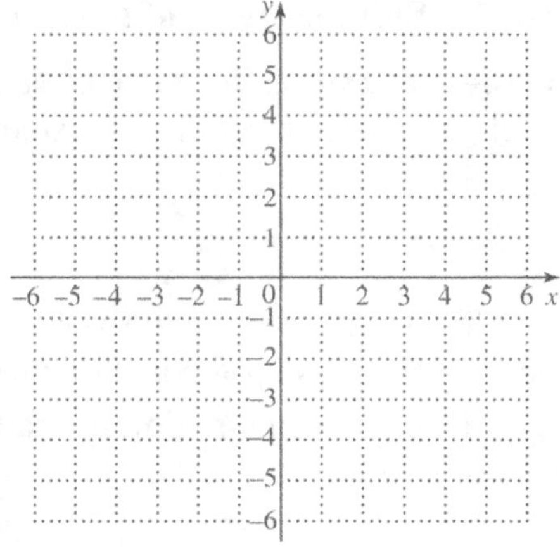

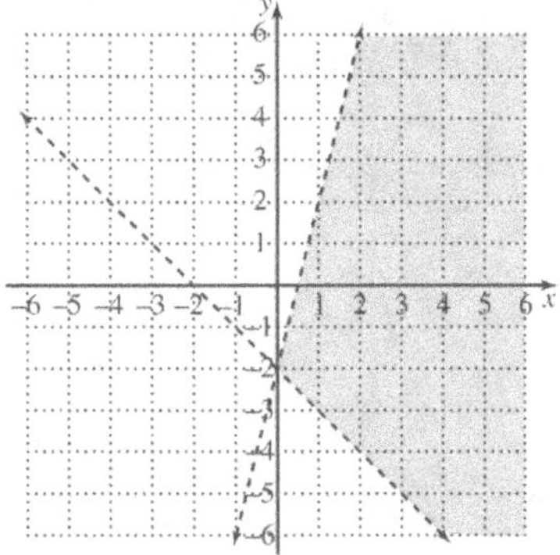

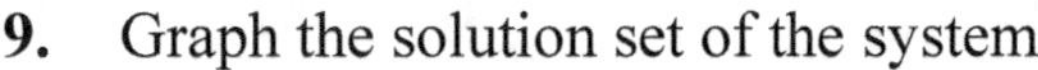

9. Graph the solution set of the system.
$$y \geq -1$$
$$2x - y > -1$$

Recall that $y = -1$ is a horizontal line through the point $(0, -1)$. Use a solid line, and shade the region with the point $(0, 0)$.

The graph of $2x - y > -1$ is created using a dashed line through the intercepts $(0, 1)$ and $\left(-\frac{1}{2},\ 0\right)$. Shade the region with the point $(0, 0)$.

9. Graph the solution set of the system.
$$x - 3y \leq -7$$
$$x < 2$$

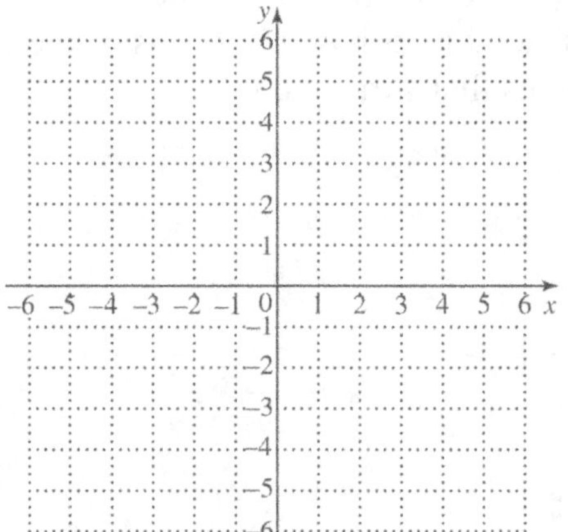

The solution set is marked in gray. The solution set includes the boundary line $y \geq -1$.

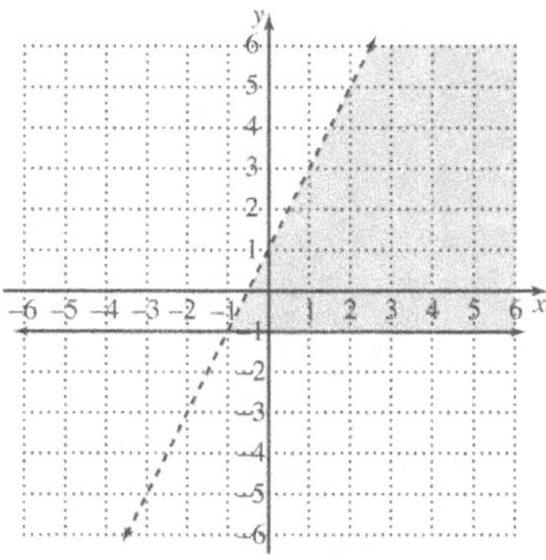

Objective 5 Practice Exercises

Graph the solution of each system of linear inequalities.

12. $4x + 5y \leq 20$

$\quad\quad y \leq x + 3$

12.

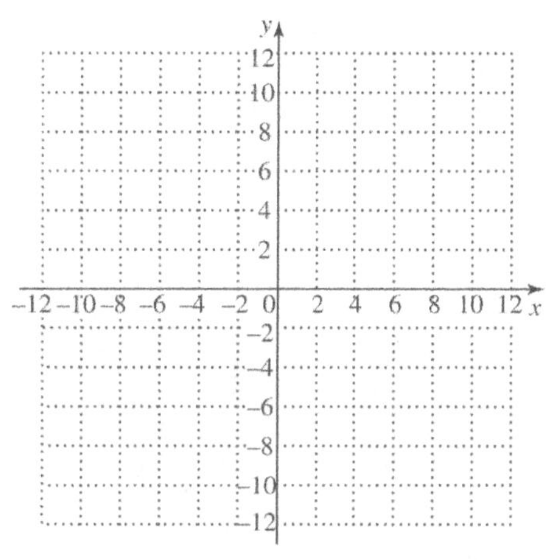

13. $x < 2y + 3$

$\quad\quad 0 < x + y$

13.

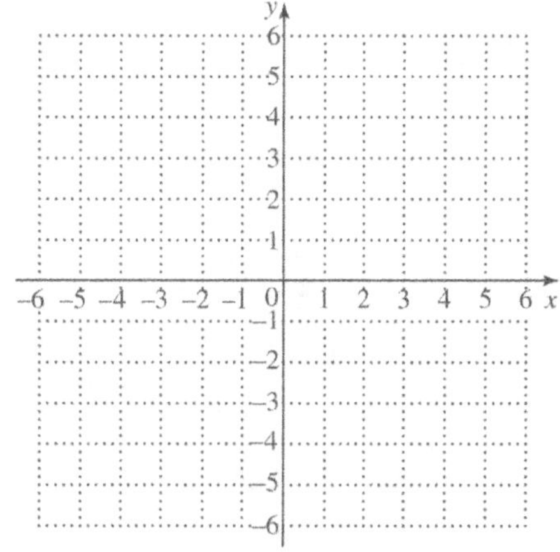

14. $y < 4$

$x \geq -3$

14. 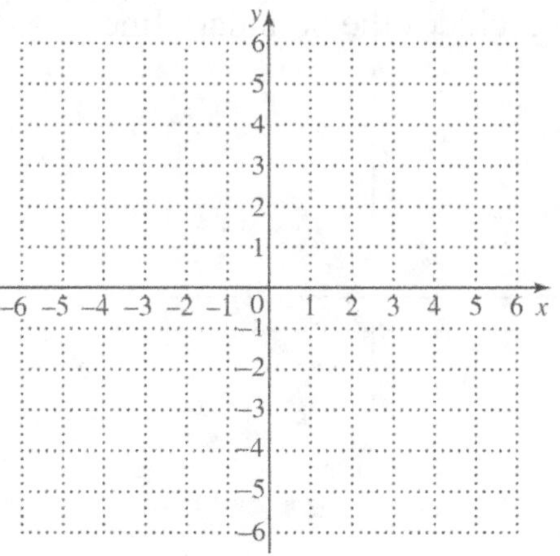

$x \geq -3$

Chapter 14 Further Topics in Algebra

Learning Objectives
Evaluate algebraic expressions, given values for the variables.
Use point-slope form to write an equation of a line.
Find greater powers of binomials.

Key Terms

Use the vocabulary terms listed below to complete each statement in exercises 1−7.

variable	constant	algebraic expression
equation	solution	set elements

1. A(n) ________________________ is a statement that says two expressions are equal.

2. The objects that belong to a set are its ________________________ .

3. A ________________________ is a symbol, usually a letter, used to represent an unknown number.

4. A collection of numbers, variables, operation symbols, and grouping symbols is an________________________.

5. A ________________________ is collection of objects.

6. Any value of a variable that makes an equation true is a(n) ________________ of the equation.

7. A ________________________ is a fixed, unchanging number.

Objective 1   Evaluate algebraic expressions, given values for the variables.

Review these examples for Objective 2:

1. Evaluate each expression for $x = 4$.

   a. $3x - 5$

   $$3x - 5 = 3(4) - 5 \quad \text{Let } x = 4.$$
   $$= 12 - 5 \quad \text{Multiply.}$$
   $$= 7 \quad \text{Subtract.}$$

   b. $5x^2$

   $$5x^2 = 5 \cdot 4^2 \quad \text{Let } x = 4.$$
   $$= 5 \cdot 16 \quad \text{Square 4.}$$
   $$= 80 \quad \text{Multiply.}$$

3. Find the value of each expression for $x = 7$ and $y = 6$.

   $3x + 4y + 2$

   Replace x with 7 and y with 6.
   $$3x + 4y + 2 = 3 \cdot 7 + 4 \cdot 6 + 2$$
   $$= 21 + 24 + 2 \quad \text{Multiply.}$$
   $$= 47 \quad \text{Add.}$$

Now Try:

1. Evaluate each expression for $x = 6$.

   a. $4x - 7$

   ____

   b. $7x^2$

   ____

2. Find the value of each expression for $x = 8$ and $y = 4$.

   $5x + 6y + 1$

   ____

Objective 1 Practice Exercises

Find the value of each expression if $x = 2$ and $y = 4$.

1. $9x - 3y + 2$

1. _______________

2. $\dfrac{2x + 3y}{3x - y + 2}$

2. _______________

3. $\dfrac{3y^2 + 2x^2}{5x + y^2}$

3. _______________

Objective 2   Use point-slope form to write an equation of a line.

Review this example for Objective 3:	**Now Try:**

Review this example for Objective 3:

3. Write an equation of each line. Give the final answer in slope-intercept form.

The line passing through $(5, -3)$ with slope $\frac{5}{4}$.

$$y - y_1 = m(x - x_1)$$
$$y - (-3) = \frac{5}{4}(x - 5)$$
$$y + 3 = \frac{5}{4}x - \frac{25}{4}$$
$$y = \frac{5}{4}x - \frac{37}{4}$$

Now Try:

3. Write an equation of each line. Give the final answer in slope-intercept form.
The line passing through $(6, 11)$ with slope $-\frac{2}{3}$.

____

Objective 2 Practice Exercises

Write an equation for the line passing through the given point and having the given slope. Write the equations in slope-intercept form, if possible.

4. $(-3, 4)$; $m = -\frac{3}{5}$

4. ______________

5. $(-4, -3)$; $m = -2$

5. ______________

6. $(2, 2)$; $m = -\frac{3}{2}$

6. ______________

Objective 3   Find greater powers of binomials.
Review these examples for Objective 5:

	Now Try:
4. Find each product.	**4.** Find each product.

a. $(x+4)^3$

$(x+4)^3$

$= (x+4)^2 (x+4)$

$= (x^2 + 8x + 16)(x+4)$

$= x^3 + 8x^2 + 16x + 4x^2 + 32x + 64$

$= x^3 + 12x^2 + 48x + 64$

a. $(x+6)^3$

____

b. $(5y-4)^4$

$(5y-4)^4$

$= (5y-4)^2 (5y-4)^2$

$= (25y^2 - 40y + 16)(25y^2 - 40y + 16)$

$= 625y^4 - 1000y^3 + 400y^2$

$\qquad - 1000y^3 + 1600y^2 - 640y$

$\qquad + 400y^2 - 640y + 256$

$= 625y^4 - 2000y^3 + 2400y^2 - 1280y + 256$

b. $(3x-5)^4$

____

Objective 3 Practice Exercises

Find each product.

7. $(a-3)^3$

7. _______________

8. $(j+3)^4$

8. _______________

9. $(4s+3t)^4$

9. _______________

   Copyright © 2025 Pearson Education, Inc.

Chapter R PREALGEBRA REVIEW

R.1   Fractions

Key Terms

1. equivalent fractions
2. improper fraction
3. numerator
4. proper fraction
5. denominator
6. composite number
7. prime factorization
8. prime number
9. lowest terms

Objective 1

Now Try

1a. prime

1c. composite; $2 \cdot 3 \cdot 5 \cdot 7$

Practice Exercises

1. composite; $2 \cdot 7 \cdot 7$
2. prime
3. composite; $2 \cdot 3 \cdot 7 \cdot 13$

Objective 2

Now Try

2. $\dfrac{3}{5}$

Practice Exercises

5. $\dfrac{5}{6}$

Objective 3

Now Try

3. $14\dfrac{4}{5}$

4. $\dfrac{86}{7}$

Practice Exercises

7. $21\dfrac{2}{5}$

9. $\dfrac{244}{11}$

Objective 4

Now Try

5. $\dfrac{1}{10}$

6. $\dfrac{16}{21}$

Practice Exercises

11. $\dfrac{7}{5}$ or $1\dfrac{2}{5}$

Objective 5

Now Try

7. $\dfrac{3}{4}$

8. $\dfrac{23}{24}$

9. $\dfrac{2}{9}$

Practice Exercises

13. $\dfrac{256}{225}$ or $1\dfrac{31}{225}$

15. $\dfrac{119}{24}$ or $4\dfrac{23}{24}$

Objective 6
Now Try

10. $8\dfrac{3}{4}$ cups

Practice Exercises

17. $11\dfrac{3}{4}$ yards

Objective 7
Now Try

11a. Home; $\dfrac{1}{50}$

11b. 104 workers

Practice Exercises

19. Lunch Room

21. 754 workers

R.2 Decimals and Percents

Key Terms

1. decimals

2. place value

3. percent

Objective 1
Now Try

1a. $\dfrac{72}{100}$

1b. $\dfrac{53}{1000}$

1c. $\dfrac{37,058}{10,000}$

Practice Exercises

1. $\dfrac{7}{1000}$

3. $\dfrac{300,005}{10,000}$

Objective 2
Now Try

2a. 37.871

2b. 27.282

Practice Exercises

5. 755.098

Objective 3
Now Try

3a. 251.116

3b. 0.028

4a. 32.4

4b. 1.43

Practice Exercises
7. 2.3424   9. 0.4292

Objective 4
  Now Try
6a. 0.35   6b. $3.\overline{5}$ or 3.555…

  Practice Exercises
11. $0.\overline{4}$ or 0.444...

Objective 5
  Now Try
9a. 0.91   9b. 0.06   9c. 43%

9d. 520%

  Practice Exercises
13. 3.62   15. 8.4%

Objective 6
  Now Try
10a. $\dfrac{3}{10}$ 10b. $1\dfrac{1}{4}$ 11a. 15%

11b. $5\dfrac{5}{9}\%$ exact, or 5.6% rounded

  Practice Exercises
17. $\dfrac{139}{250}$

Objective 7
  Now Try
12. discount: $10.64, sale price: $8.36

  Practice Exercises
19. $14; sale price: $56   21. $42; sale price: $14

Chapter 1 THE REAL NUMBER SYSTEM

1.1   Exponents, Order of Operations, and Inequality

Key Terms
1. exponential expression   2. inequality

3. base   4. exponent

Objective 1
  Now Try
  1. 49

  Practice Exercises
  1. 27   3. 0.16

Objective 2
  Now Try
  2. 23

  Practice Exercises
  5. 45

Objective 3
  Now Try
  3. 215

  Practice Exercises
  7. −8   9. 48

Objective 4
  Now Try
  4. true

  Practice Exercises
11. false

Objective 5
  Now Try
  5. $19 \leq 11 + 8$

  Practice Exercises
13. $7 = 13 - 6$ 15. $20 \geq 2 \cdot 7$

Objective 6
  Now Try

  6. $11 < 15$

  Practice Exercises
17. $8 \leq 12$

1.2 Variables, Expressions, and Equations

Key Terms

1. equation
2. elements
3. variable
4. algebraic expression
5. set
6. Solution
7. constant

Objective 1
Now Try

1b. 17   1c 252   2. 65

Practice Exercises

1. 8   3. $\dfrac{28}{13}$

Objective 2
Now Try

3. $20x$

Practice Exercises
5. $8x - 11$

Objective 3
Now Try

4. no

Practice Exercises
7. no   9. yes

Objective 4
Now Try

5. $x + 7 = 11$; 4

Practice Exercises
11. $5 + x = 14$; 9

Objective 5
Now Try

6. expression

Practice Exercises
13. expression   15. equation

1.3 Real Numbers and the Number Line

Key Terms

1. whole numbers
2. additive inverse
3. integers
4. natural numbers
5. absolute value
6. number line
7. irrational number
8. coordinate
9. negative number
10. positive number
11. real numbers
12. graph
13. set-builder notation
14. rational number
15. signed numbers

Objective 1

Now Try

1. –282

2. 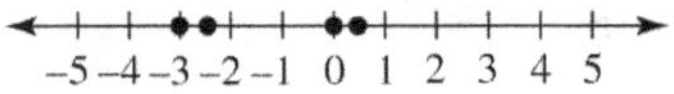

3a. 0, 7

3b. –10, 0, 7

3c. $-10, -\frac{5}{8}, 0, 0.\overline{4}, 5\frac{1}{2}, 7, 9.9$

3d. $\sqrt{5}$

Practice Exercises

1. –75 pounds

3. 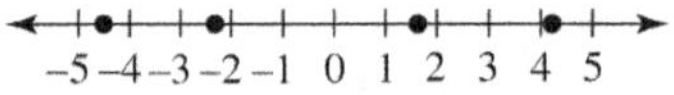

Objective 2

Now Try

4. true

Practice Exercises

5. false

Objective 3

Now Try

5. –8, 4, –0.26, 0

Practice Exercises

7. 25

9. –4.5

Objective 4

Now Try

6a. 10

6b. –10

6c. 3

Practice Exercises

11. 1.22

Objective 5

Now Try

7. Milk and Electricity

Practice Exercises

13. Gasoline

15. Eggs in 2012 to 2013

1.4   Adding and Subtracting Real Numbers

Key Terms

1. minuend
2. subtrahend
3. difference
4. sum
5. addends

Objective 1
Now Try
1. -5
2. -12

Practice Exercises
1. -18
3. $-5\frac{5}{8}$

Objective 2
Now Try
3. 3
4. $-\frac{7}{10}$

Practice Exercises
5. $-\frac{1}{6}$

Objective 3
Now Try
6a. -3
6b. 2

Practice Exercises
7. -25
9. 0

Objective 4
Now Try
7. -4

Practice Exercises
11. -6

Objective 5
Now Try
8. $-10 + 11 + 2$; 3
9. $-17 - 9$; -26
10. 5464 ft

Practice Exercises
13. $-4 - 4$; -8
15. $-51.2°C$

Objective 6
Now Try
11. $-\$0.006$

Practice Exercises
17. $-\$0.061$

1.5   Multiplying and Dividing Real Numbers

Key Terms

1. quotient   2. reciprocals   3. product

Objective 1

Now Try

1. −56

Practice Exercises

1. −28   3. −13.12

Objective 2

Now Try

2a. 30

Practice Exercises

5. $\dfrac{4}{5}$

Objective 3

Now Try

3. −12, −6, −4, −3, −2, −1, 1, 2, 3, 4, 6, and 12

Practice Exercises

7. −8, −4, −2, −1, 1, 2, 4, 8

9. −42, −21, −14, −7, −6, −3, −2, −1, 1, 2, 3, 6, 7, 14, 21, 42

Objective 4

Now Try

4a. −2   4b. 3

Practice Exercises

11. 0

Objective 5

Now Try

5a. 28   5d. $-\dfrac{5}{4}$

Practice Exercises

13. 64   15. $\dfrac{16}{21}$

Objective 6

Now Try

6. −432

Practice Exercises

17. 25

Objective 7

Now Try

7. $\dfrac{5}{6}[-8+(-4)]$; -10

Practice Exercises

19. $(-7)(3)+(-7)$; -28

21. $-12+\dfrac{49}{-7}$; -19

Objective 8

Now Try

9. $\dfrac{36}{x}=-4$

Practice Exercises

23. $-8x=72$

1.6 Properties of Real Numbers

Key Terms

1. identity element for addition

2. identity element for multiplication

Objective 1

Now Try

1a. -12

1b. 2

Practice Exercises

1. 4

3. $(4+z)$

Objective 2

Now Try

2a. 4

2b. $[(-3)\cdot 4]$

3a. associative

3b. commutative

3c. both

4. $39x+20$

Practice Exercises

5. $[(-4+3y)]$

Objective 3

Now Try

5a. 0

5b. 1

6a. $\dfrac{7}{9}$

6b. $\dfrac{2}{3}$

Practice Exercises

7. 4

9. $\dfrac{6}{7}$

Objective 4
 Now Try

7a. 11   7b. -8 7d. $\dfrac{5}{8}$

7e. -10 8. 6

 Practice Exercises
11. 0; identity

Objective 5
 Now Try

9a. 54   9b. $12y+72+12x$ 9e. $17x-102$

9f. $-8x+20$ 10a. $-3x-4$ 10b. $8k+9$

10c. $4x+5y-z$

 Practice Exercises
13. $2an-4bn+6cn$ 15. $2k-7$

1.7   Simplifying Expressions

Key Terms
 1. numerical coefficient   2. term   3. like terms

Objective 1
 Now Try

1a. $35x-21y$ 1d. $11-7x$

 Practice Exercises
 1. $8x+27$ 3. $10x+3$

Objective 2
 Practice Exercises

5. $\dfrac{7}{9}$

Objective 3
 Practice Exercises
 7. like   9. unlike

Objective 4
 Now Try

2b. $23r$ 2c. $19x$ 2d. $8x^2$

3b. $37k-24$ 3f. $-\dfrac{5}{4}x+6$

 Practice Exercises
11. $2x-14$

                Copyright © 2025 Pearson Education, Inc.

Objective 5
Now Try
4. $11 + 10x + 8x + 4x;\ 11 + 22x$

Practice Exercises
13. $6x + 12 + 4x = 10x + 12$

15. $4(2x - 6x) + 6(x + 9) = -10x + 54$

Chapter 2   LINEAR EQUATIONS AND INEQUALITIES IN ONE VARIABLE

2.1   The Addition Property of Equality

Key Terms
1. equivalent equations
2. linear equation
3. solution set

Objective 1
Practice Exercises
1. no
3. yes

Objective 2
Now Try
1. $\{21\}$
3. $\left\{\dfrac{4}{15}\right\}$
4. $\{-19\}$
6. $\{13\}$
7. $\{-3\}$

Practice Exercises
5. $\dfrac{1}{2}$

Objective 3
Now Try
8. $\{29\}$
9. $\{8\}$

Practice Exercises
7. 7
9. 7.2

2.2 The Multiplication Property of Equality

Key Terms
1. multiplication property of equality
2. addition property of equality

Objective 1
Now Try

1. $\{14\}$ 3. $\{5.7\}$ 4. $\{24\}$

5. $\{36\}$ 6. $\{-3\}$

Practice Exercises
1. $\{-17\}$ 3. $\{6.4\}$

Objective 2
Now Try

7. $\{4\}$ 8. $\{-4\}$

Practice Exercises
5. $\{-5\}$

2.3 Solving Linear Equations Using Both Properties of Equality

Key Terms
1. identity 2. conditional equation 3. contradiction; empty set

Objective 1
Now Try

1. $\{-6\}$ 2. $\{8\}$ 4. $\left\{\dfrac{15}{2}\right\}$

Practice Exercises

1. $\left\{\dfrac{5}{2}\right\}$ 3. $\left\{-\dfrac{1}{5}\right\}$

Objective 2
Now Try

6. $\{$all real numbers$\}$ 7. $\varnothing$

Practice Exercises
5. infinitely many

Objective 3
Now Try

8. $67 - t$

Practice Exercises
7. $36 - m$ 9. $6x + 4y$

2.4 Clearing Fractions and Decimals When Solving Linear Equations

Objective 1
Now Try

1. $\{6\}$ 2. $\{-18\}$ 3. $\{2\}$

Practice Exercises

1. $\{6\}$ 3. $\{-14\}$

Objective 2
Now Try

4. $\{-10\}$ 5. $\{2\}$

Practice Exercises

5. $\{10\}$

2.5 Applications of Linear Equations

Key Terms

1. supplementary angles   2. complementary angles

3. right angle   4. straight angle   5. consecutive integers

Objective 1
Practice Exercises

1. Read the problem; assign a variable to represent the unknown; write an equation; solve the equation; state the answer; check the answer.

Objective 2
Now Try

2. 16

Practice Exercises

3. $-2(4 - x) = 24$; 16

Objective 3
Now Try

3. 52   6. 8 feet

Practice Exercises

5. $x + (x + 5910) = 34,730$; Mt. Rainier: 14,410 ft; Denali: 20,320 feet

7. $x + (5 + 3x) + 4x = 29$; Mark: 3 laps; Pablo: 14 laps; Faustino: 12 laps

Objective 4
Now Try

8. $-2, 0, 2, 4$

Practice Exercises

9. 27, 28

Objective 5
 Now Try
10. 66°

 Practice Exercises
11. 133°   13. 27°

2.6   Formulas and Additional Applications from Geometry

Key Terms
 1. vertical angles   2. formula   3. perimeter
 4. area

Objective 1
 Now Try
 1. $W = 5.5$

 Practice Exercises
 1. $a = 36$ 3. $h = 12$

Objective 2
 Now Try
 2. 12 ft   3. 15 ft, 20 ft, 30 ft

 Practice Exercises
 5. 1.5 years

Objective 3
 Now Try
 5. 54°, 126°

 Practice Exercises
 7. 35°, 35°   9. 129°, 51°

Objective 4
 Now Try
 6. $r = \dfrac{d}{t}$ 7. $a = P - b - c$ 8. $h = \dfrac{2A}{b + B}$

 Practice Exercises
 11. $n = \dfrac{S}{180} + 2 \text{ or } n = \dfrac{S + 360}{180}$

2.7   Ratio, Proportion, and Percent

Key Terms
1. proportion   2. ratio   3. terms; extremes; means
4. cross products

Objective 1
Now Try

1a. $\dfrac{11}{17}$ 1b $\dfrac{3}{8}$ 1c. $\dfrac{4}{15}$

2. 24-ounce jar, $0.054 per oz

Practice Exercises

1. $\dfrac{8}{3}$ 3. 45-count box

Objective 2
Now Try

3. True   5. $\left\{-\dfrac{3}{4}\right\}$

Practice Exercises

5. $\left\{\dfrac{10}{3}\right\}$

Objective 3
Now Try
6. $651.20   7a. 16 minutes   7b. No, you will be 1 min late

Practice Exercises
7. 15 inches   9. $135

Objective 4
Now Try
8a. 140   8b. 950   8c. 85%

9. 87 out-of-state students   10. $10

Practice Exercises
11. 2%

2.8   Further Applications of Linear Equations

Objective 1
Now Try
1. $117

Practice Exercises
1. 1100 students   3. 15,000 students

          A-15

Objective 2
Now Try
2. 8 oz

Practice Exercises
5. 60 lb

Objective 3
Now Try
4. 5%: $1600; 7%: $3500

Practice Exercises
7. 7%: $800; 9%: $300

Objective 4
Now Try
5. 32 quarters; 18 nickels

Practice Exercises
9. 25 large jars; 55 small jars

11. 1500 general admission; 750 reserved

Objective 5
Now Try
6. 270 miles   7. 5 hours

Practice Exercises
13. 450 miles

2.9   Solving Linear Inequalities

Key Terms
1. three-part inequality   2. interval

3. linear inequality   4. inequalities   5. interval notation

Objective 1
Now Try

1. 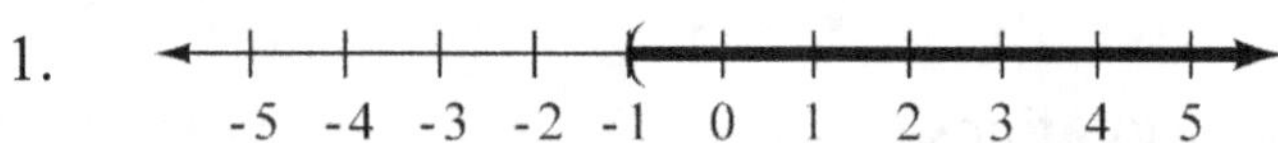

Practice Exercises

1. $(3, \infty)$;

3. $(-\infty, -4)$;

Objective 2
  Now Try
  2. $[-3, \infty)$

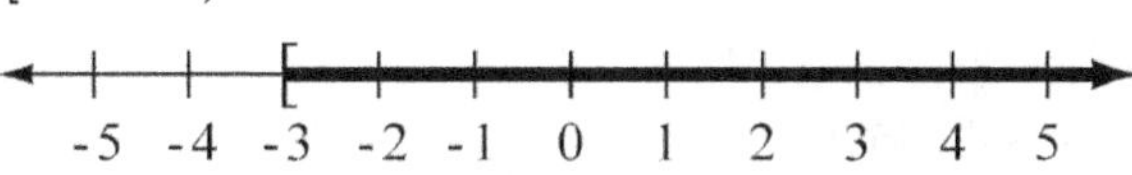

Practice Exercises

  5. $(2, \infty)$;

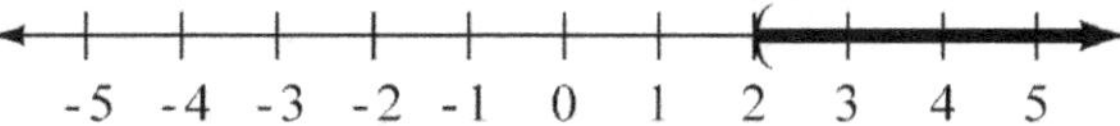

Objective 3
  Now Try
3a. $(-\infty, -5]$

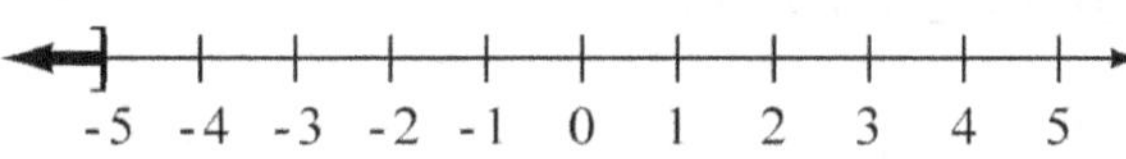

3b. $(-\infty, -4)$

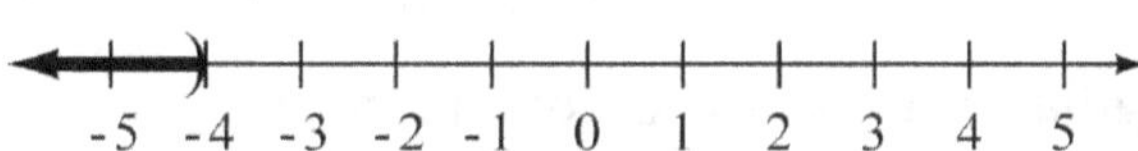

Practice Exercises

  7. $(-2, \infty)$;

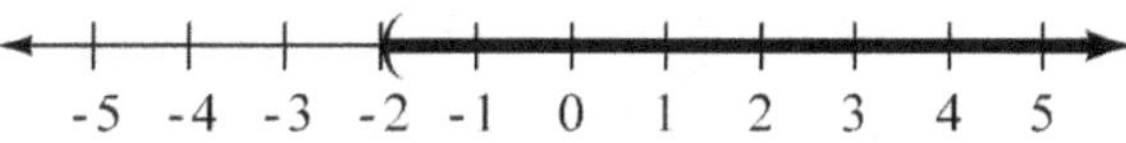

  9. $(-\infty, 4]$

Objective 4
  Now Try
  4. $[2, \infty)$

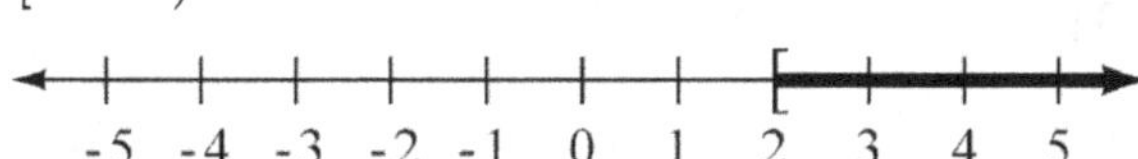

Practice Exercises

  11. $(-\infty, 5]$;

Objective 5
  Now Try
  7. 10 feet

  Practice Exercises
13. 89   15. all numbers greater than 5

Objective 6
  Now Try
  9. $[2, 4)$

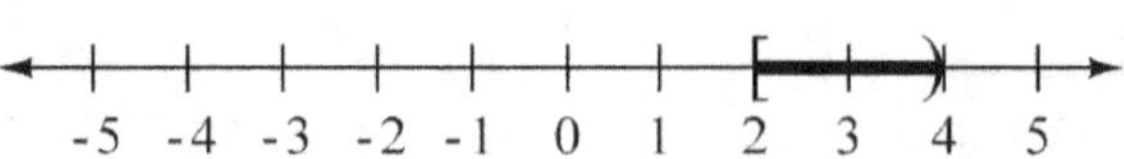

  Practice Exercises

  17. $[-5, -3)$;

Chapter 3 LINEAR EQUATIONS IN TWO VARIABLES

3.1   Linear Equations and Rectangular Coordinates

Key Terms
   1. line graph   2. linear equation in two variables

   3. coordinates   4. x-axis   5. y-axis

   6. ordered pair   7. rectangular (Cartesian) coordinate system

   8. quadrants   9. origin   10. Plot

   11. scatter diagram   12. table of values   13. plane

Objective 1
  Now Try
  1a. 2020-2021, 2022-2023, 2024-2025   1b. 300 degrees

  Practice Exercises
  1. 2021-2022, 2023-2024

Objective 2
  Practice Exercises
  3. $(4, 7)$ 5. $(0.2, 0.3)$

Objective 3
  Now Try
  2a. yes   2b. no

  Practice Exercises
  7. not a solution

Objective 4
  Now Try
  3. $(5, 13)$

  Practice Exercises
  9. (a) $(2, -1)$; (b) $(0, -5)$; (c) $(4, 3)$; (d) $(-1, -7)$; (e) $(7, 9)$

Objective 5
Now Try

4.

x	y
1	−4
5	12
2	0
3	4

$(1, -4), (5, 12), (2, 0), (3, 4)$

Practice Exercises

11. $(1, -3), (1, 0), (1, 5)$

13. $(0, 4), (3, 0), \left(\dfrac{15}{4}, -1\right)$

Objective 6
Now Try

5.

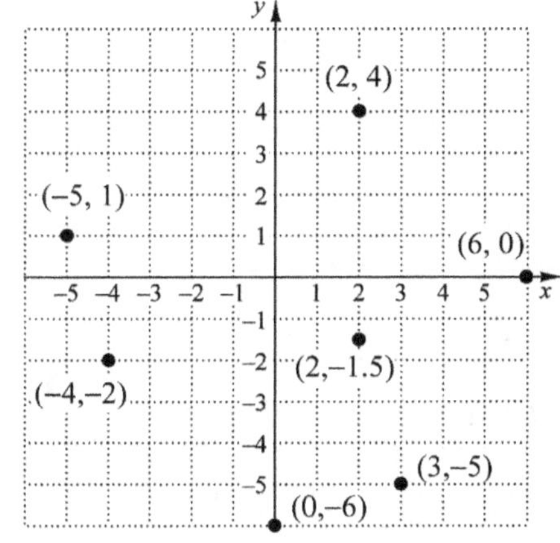

Practice Exercises

15.

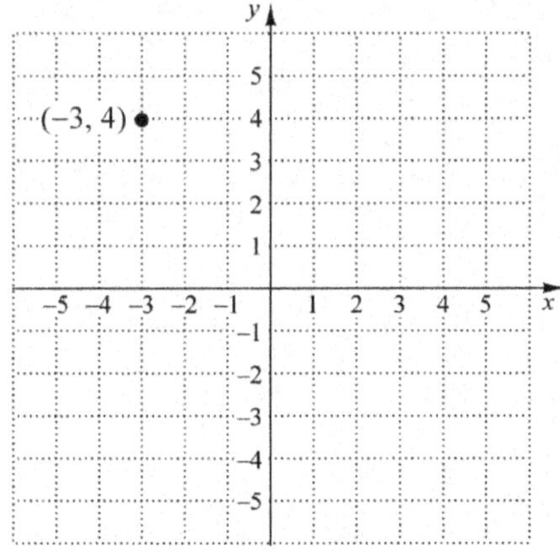

3.2   Graphing Linear Equations in Two Variables

Key Terms

1. *y*-intercept   2. *x*-intercept   3. graphing

4. graph

Objective 1

Now Try

2.

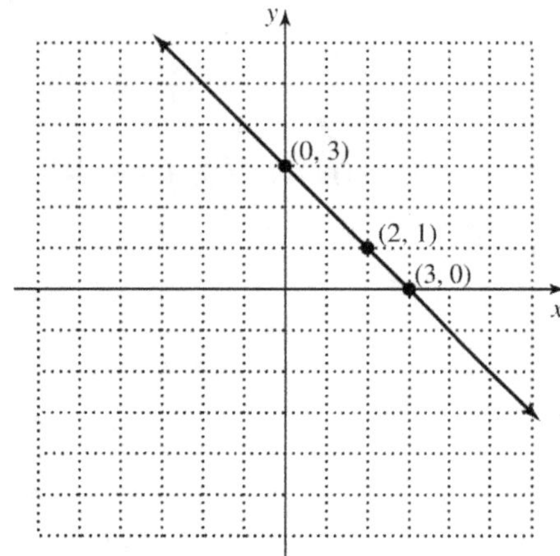

Practice Exercises

1. $\left(0,-2\right),\left(\dfrac{2}{3},0\right),\left(2,4\right)$

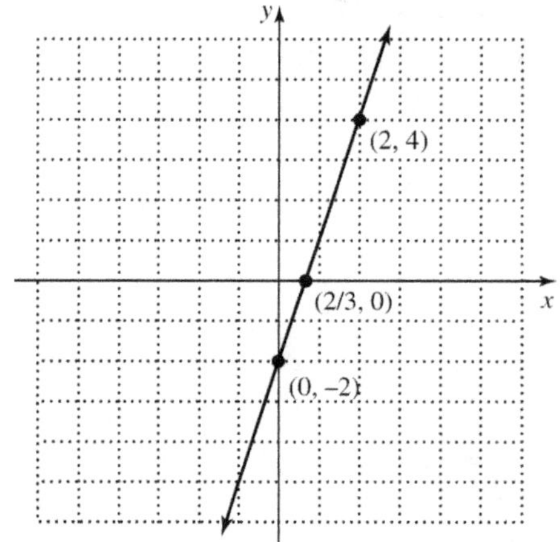

3. $\left(0,-\dfrac{1}{2}\right),\left(1,0\right),\left(-3,-2\right)$

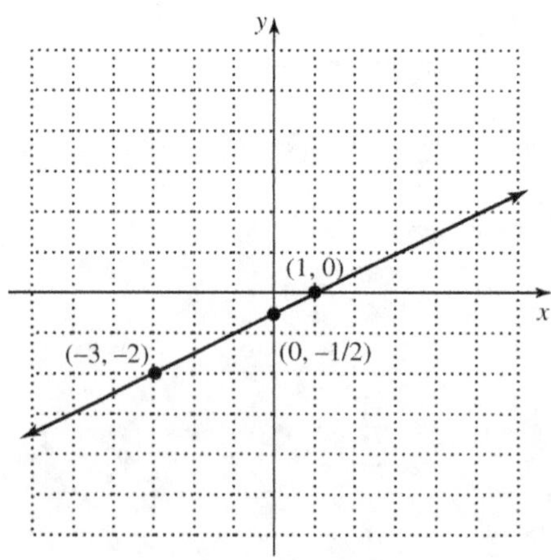

        Copyright © 2025 Pearson Education, Inc.

Objective 2
Now Try

3.

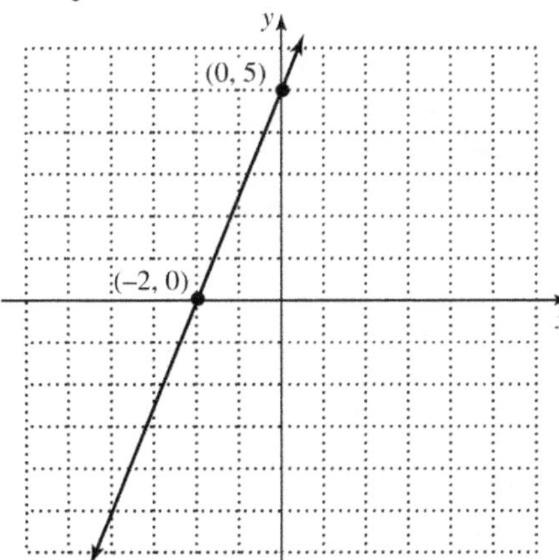

4.

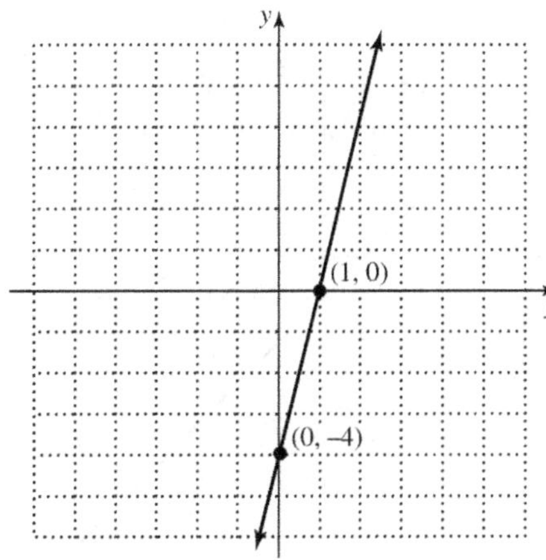

Practice Exercises

5.

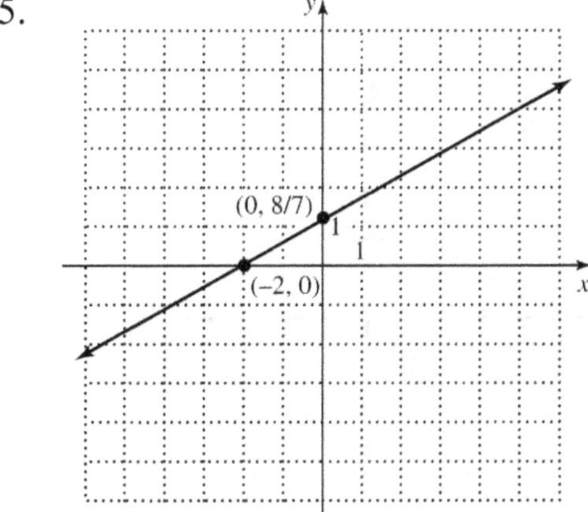

Objective 3
Now Try

5.

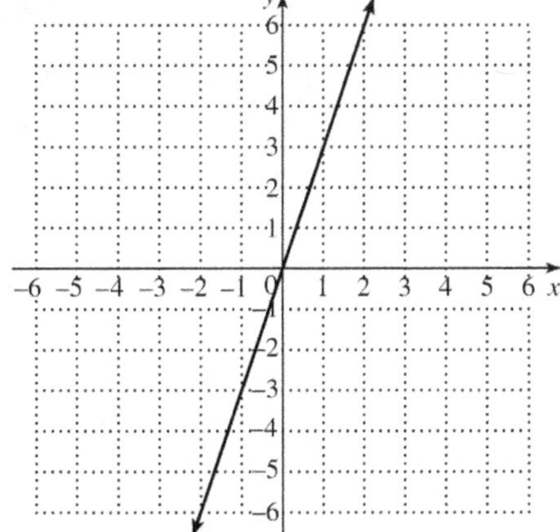

Practice Exercises

7. $x + y = 0$

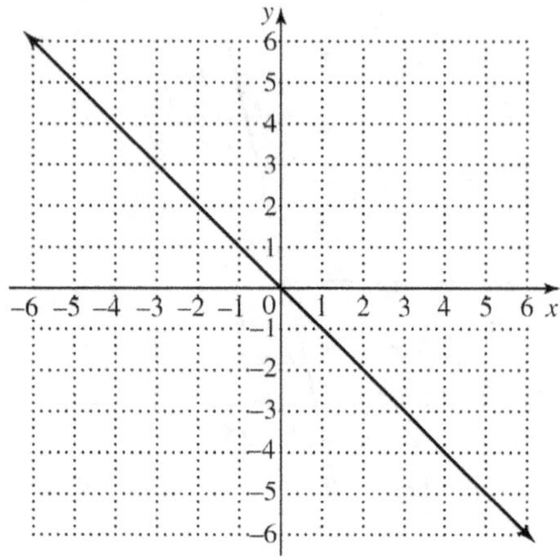

Objective 4
Now Try

6.

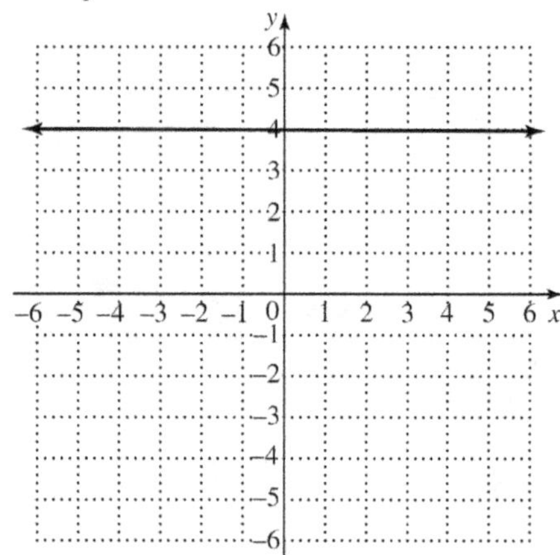

7.

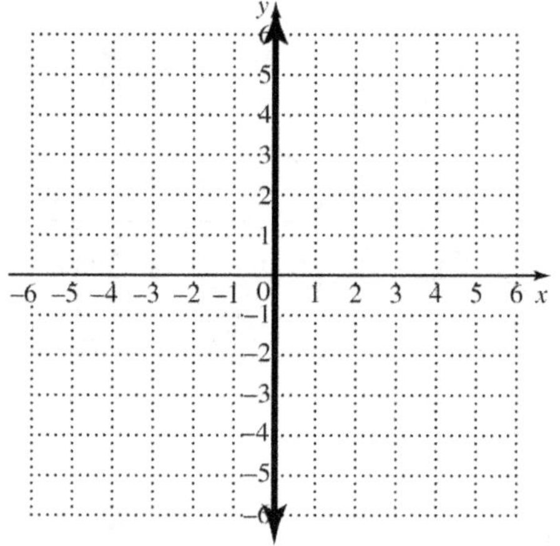

Practice Exercises

9. $x - 1 = 0$

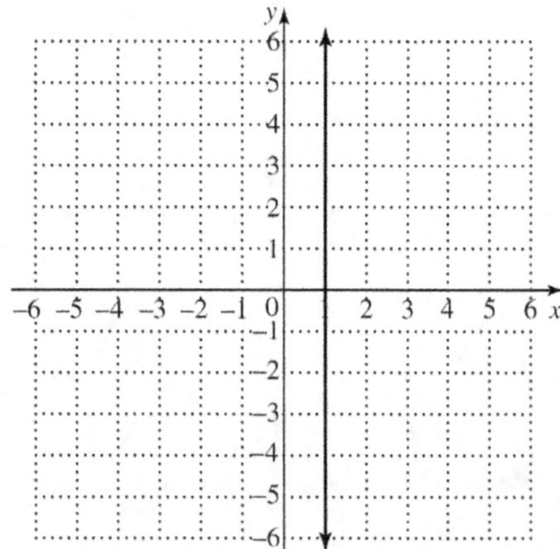

Objective 5

Now Try

8a. 0 calculators, $45
5000 calculators, $42
20,000 calculators, $33
45,000 calculators, $18

8b. (0, $45), (5, $42), (20, $33), (45, $18)

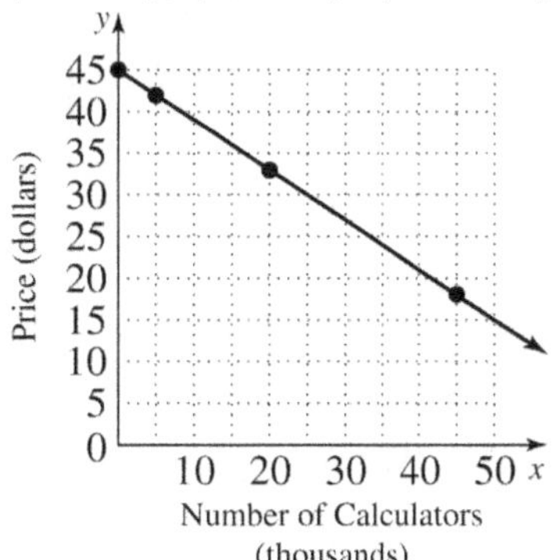

8c. 30,000 calculators, $27

Practice Exercises

11. 2020, 4.9 million;
2021, 5.53 million;
2022, 6.16 million;
2023, 6.79 million

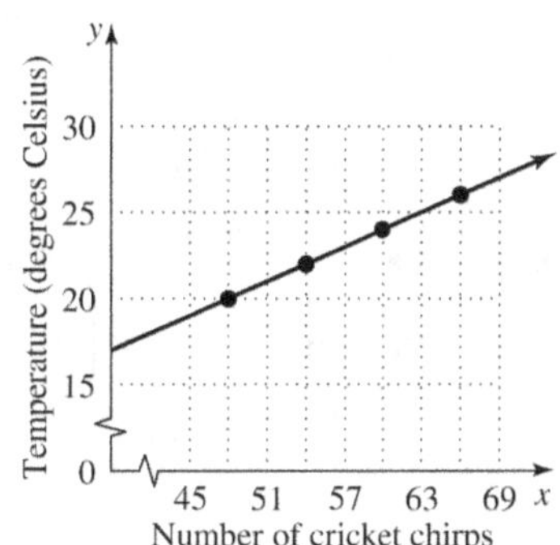

13. 48, 20°C;
54, 22°C;
60, 24°C;
66, 26°C

3.3   The Slope of a Line

Key Terms

1. perpendicular lines   2. slope   3. rise
4. parallel lines   5. run

Objective 1

Now Try

1. $\dfrac{4}{1}$, or 4   2. $-\dfrac{16}{9}$ 3. 0

4. undefined slope

Practice Exercises

1. -2 3. 0

Objective 2

Now Try

5. $\dfrac{7}{4}$

Practice Exercises

5. $\dfrac{2}{3}$

Objective 3

Now Try

6a. parallel   6b. perpendicular

Practice Exercises

7. $1; 1;$ parallel   9. $-3; \dfrac{1}{3};$ perpendicular

3.4   Slope-Intercept Form of a Linear Equation

Key Terms

1. point-slope form   2. standard form   3. slope-intercept form

Objective 1

Now Try

1a. slope: -12; y-intercept: $(0, 6)$ 1b. slope: $-\dfrac{1}{7}$; y-intercept: $\left(0, -\dfrac{7}{5}\right)$

Practice Exercises

1. slope: $\dfrac{3}{2}$; y-intercept: $\left(0, -\dfrac{2}{3}\right)$

Objective 2

Now Try

2.

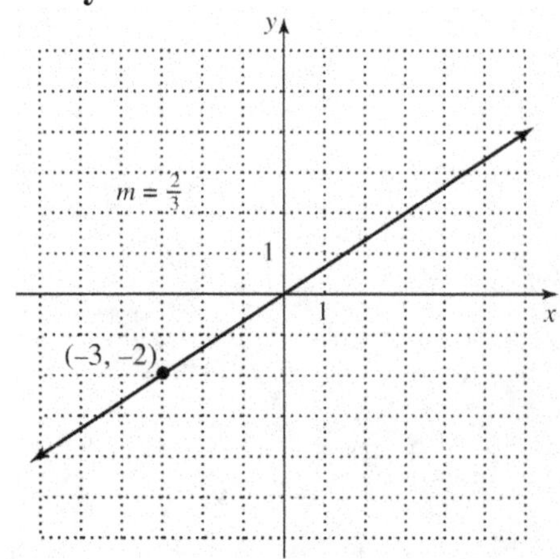

           Copyright © 2025 Pearson Education, Inc.

Practice Exercises

3. 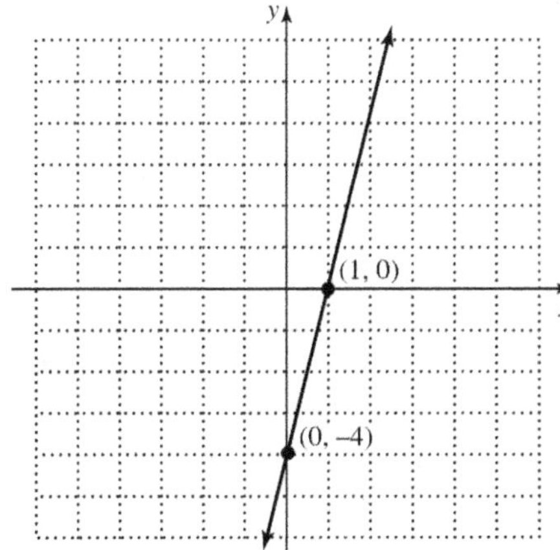

5. 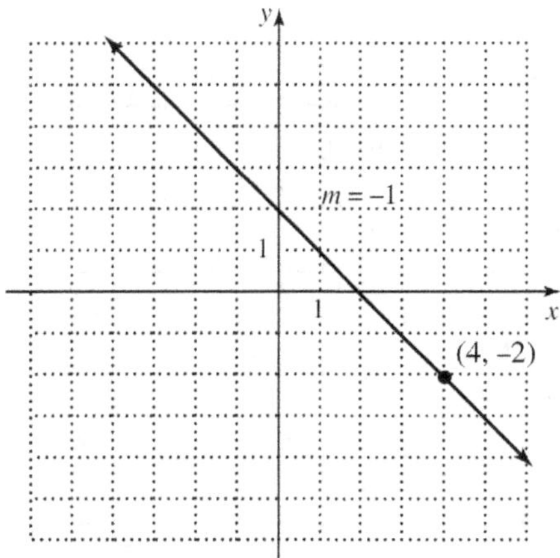

Objective 3
  Now Try

4a. $y = \dfrac{3}{4}x - 5$

4b. $y = 6x + 10$

5. $y = 4x - 4$

  Practice Exercises
  7. $y = -2x + 12$

Objective 4
  Now Try

6a. 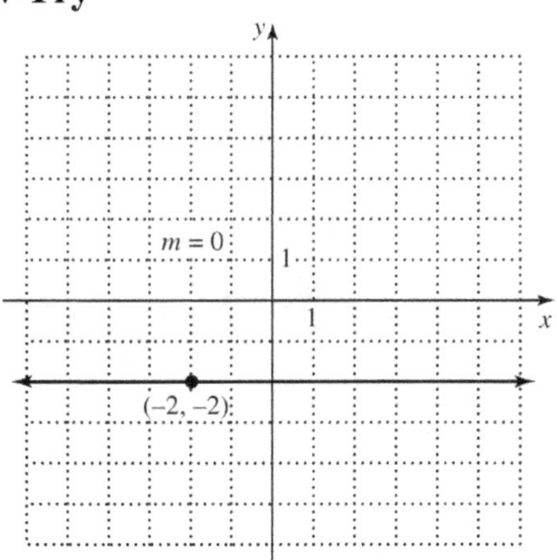

6b. 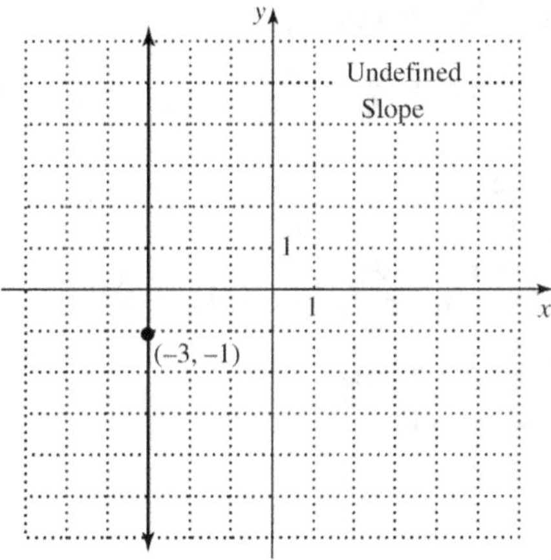

7a. $y = 5$

7b. $x = -5$

Practice Exercises

9. 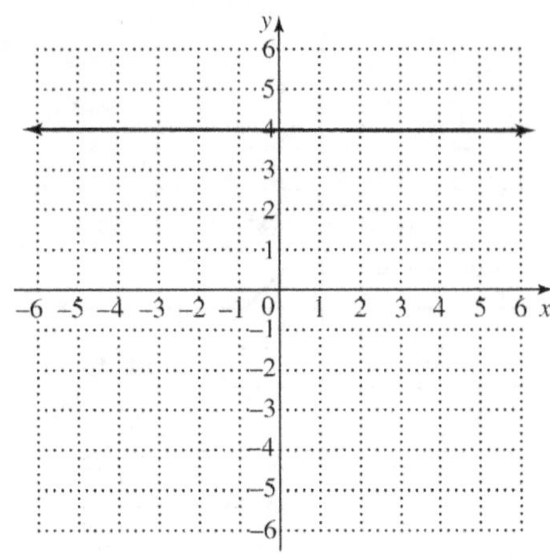

11. $x = -3$

3.5   Point-Slope Form of a Linear Equation and Modeling

Key Terms

1. standard form   2. slope-intercept form   3. point-slope form

Objective 1

Now Try

1. $y = -\dfrac{2}{3}x + 15$

Practice Exercises

1. $y = -\dfrac{3}{5}x + \dfrac{11}{5}$ 3. $y = -\dfrac{3}{2}x + 5$

Objective 2

Now Try

2. $y = -\dfrac{3}{4}x + \dfrac{81}{4}$; $3x + 4y = 81$ 3. $y = \dfrac{5}{2}x + 5$; $5x - 2y = -10$

Practice Exercises

5. $3x - 2y = 0$

Objective 3

Now Try

4. 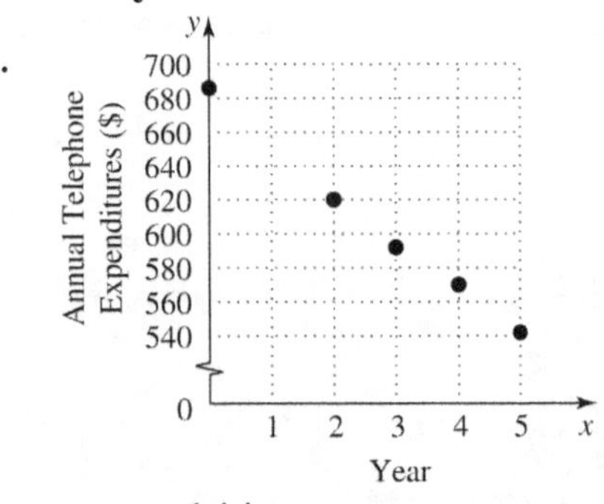

$$y = -\dfrac{144}{5}x + 686$$

Practice Exercises

7.

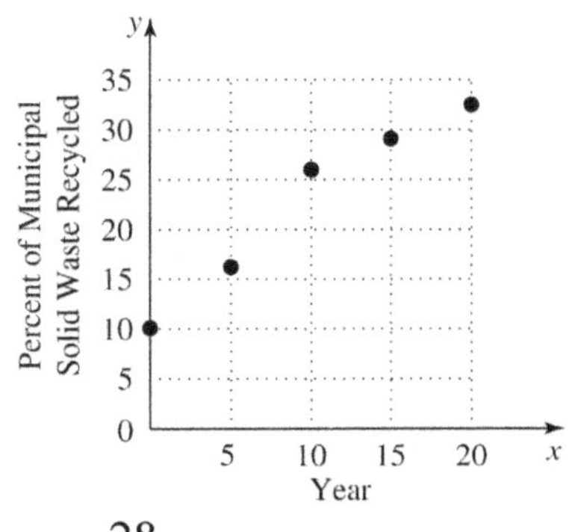

$$y = \frac{28}{25}x + 10.1$$

Chapter 4 EXPONENTS AND POLYNOMIALS

4.1   The Product Rule and Power Rules for Exponents

Key Terms

1. power   2. exponential expression   3. base

Objective 1

  Now Try

1. 4^5 2a. base: 2; exponent: 6; value: 64

2b. base: −2; exponent: 6; value 64

  Practice Exercises

1. $\dfrac{1}{243}$ 3. −6561; base: 3; exponent: 8

Objective 2

  Now Try

3a. 9^{13} 3c. m^{27} 3e. 108

3f. $-18x^{11}$

  Practice Exercises

5. $8c^{15}$

Objective 3

  Now Try

4a. 7^8 4c. x^{30}

  Practice Exercises

7. 7^{12} 9. $(-3)^{21}$

      A-27

Objective 4
  Now Try
5a. $64a^3b^3$ 5d. -7^{10}

  Practice Exercises
11. $-0.008a^{12}b^3$

Objective 5
  Now Try
6. $\dfrac{1}{1024}$

  Practice Exercises
13. $-\dfrac{8x^3}{125}$ 15. $-\dfrac{128a^7}{b^{14}}$

Objective 6
  Now Try
7a. $\dfrac{5^5}{2^3}$, or $\dfrac{3125}{8}$ 7d. $-x^{27}y^{13}$

  Practice Exercises
17. $32a^9b^{14}c^5$

Objective 7
  Now Try
8. $28x^5$

  Practice Exercises
19. $36x^5$ 21. $28q^{11}$

4.2   Integer Exponents and the Quotient Rule

Key Terms
  1. power rule for exponents   2. base; exponent

  3. product rule for exponents

Objective 1
  Now Try
1a. 1   1c. −1   1f. 1   1g. 0

  Practice Exercises
  1. −1   3. 0

Objective 2
  Now Try

2a. $\dfrac{1}{27}$ 2c. 25   2d. $\dfrac{8}{27}$ 2f. $\dfrac{1}{8}$ 2i. $\dfrac{1}{p^5}$

3a. $\dfrac{125}{36}$ 3b. $\dfrac{y^2}{x^7}$ 3c. $\dfrac{qr^5}{4p^3}$

  Practice Exercises

5. $\dfrac{1}{m^{18}n^9}$

Objective 3
  Now Try

4a. 9   4d. z^{10} 4f. $(a-b)^2$ 4h. $\dfrac{36b^7}{a^8}$

  Practice Exercises

7. $\dfrac{k^4 m^5}{2}$ 9. $\dfrac{p^8}{3^5 m^3}$ or $\dfrac{p^8}{243m^3}$

Objective 4
  Now Try

5a. 6   5c. $3125b^5$ 5d. $\dfrac{243}{32p^{20}}$ 5e. $\dfrac{x^{13}y^2}{343z}$

  Practice Exercises
11. $a^{16}b^{22}$

4.3   Scientific Notation

Key Terms
  1. scientific notation   2. power rule   3. quotient rule

Objective 1
  Now Try
1b. 4.771×10^{10} 1c. 4.63×10^{-2}

  Practice Exercises
  1. 2.3651×10^4 3. -2.208×10^{-4}

Objective 2
  Now Try
2a. 27,960,000   2b. 0.000164

  Practice Exercises
  5. 0.0064

Objective 3
Now Try
3a. 2.7×10^8, or 270,000,000

3b. 3×10^{-8}, or 0.00000003

4. 9×10^{23} grains of sand

Practice Exercises
7. 2.53×10^2

9. 4.86×10^{19} atoms

4.4   Adding, Subtracting, and Graphing Polynomials

Key Terms
1. degree of a term
2. descending powers
3. term
4. trinomial
5. polynomial
6. monomial
7. degree of a polynomial
8. binomial
9. like terms
10. line of symmetry
11. vertex
12. axis
13. parabola
14. coefficient; leading term

Objective 1
Now Try
1. 1, 7, –2; three terms

Practice Exercises
1. three terms; 3, –2, 1   3. three terms; 8, –1, –1

Objective 2
Now Try
2. $15m^3 + 29m^2$

Practice Exercises
5. $2.6z^8 - 0.9z^7$

Objective 3
Now Try
3a. $3x^3 - 7x^2 + 4x$; degree 3; trinomial

3b. $4w^5 - 3w^2$; degree 5; binomial

Practice Exercises
7. $n^8 - n^2$; degree 8; binomial

9. $5c^5 + 3c^4 - 10c^2$; degree 5; trinomial

Objective 4
Now Try
4. 1285

Practice Exercises

11. a. 71; b. −19

Objective 5

Now Try

6a. $4x^3 + x + 12$ 6b. $7x^3 + 8x^2 - 14x - 1$ 7. $-11x^3 - 7x + 1$

9. $7 + 8x - 4x^2$ 10. $x^2y + 4xy$

Practice Exercises

13. $3r^3 + 7r^2 - 5r - 2$ 15. $-7x^2y + 3xy + 5xy^2$

Objective 6

Now Try

11.

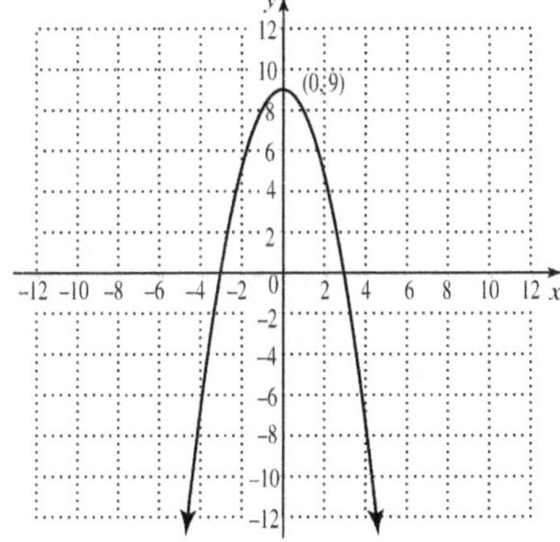

Practice Exercises

17.

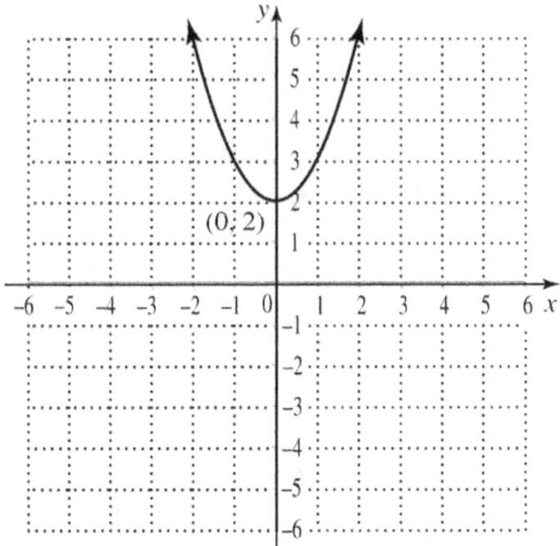

Vertex: (0, 2)

4.5   Multiplying Polynomials

Key Terms

1. inner product   2. FOIL   3. outer product

Objective 1

Now Try

1a. $-18a^5$ 1b. $15p^7q^4$

Practice Exercises

1. $36p^5$ 3. $-12x^7$

Objective 2

Now Try

2. $32x^4 + 64x^3$

Practice Exercises

5. $6m + 14m^3 + 6m^4$

Objective 3

Now Try

3. $4x^7 - 2x^5 + 37x^4 - 18x^2 + 9x$

4. $28x^4 - 33x^3 + 51x^2 + 17x - 15$

5. $2x^5 - 6x^4 + 10x^3 - 29x^2 + 5$

Practice Exercises

7. $x^3 + 27$ 9. $6x^4 + 11x^3 - 9x^2 - 4x$

Objective 4

Now Try

6. $x^2 + 3x - 54$ 7. $16xy + 72y - 14x - 63$

8. $15k^2 + 44kn + 32n^2$

Practice Exercises

11. $3 + 10a + 8a^2$

4.6 Special Products

Key Terms
1. binomial
2. conjugate
3. difference of two squares

Objective 1
Now Try

2a. $c^2 - 8c + 16$

2c. $4a^2 + 36ak + 81k^2$

2e. $9p^2 + p + \dfrac{1}{36}$

Practice Exercises
1. $49 + 14x + x^2$

3. $16y^2 - 5.6y + 0.49$

Objective 2
Now Try

3a. $x^2 - 81$

3b. $\dfrac{25}{36} - a^2$

4a. $121x^2 - y^2$

4b. $4p^5 - 144p$

Practice Exercises
5. $64k^2 - 25p^2$

Objective 3
Now Try

5a. $x^3 + 18x^2 + 108x + 216$

5b. $81x^4 - 540x^3 + 1350x^2 - 1500x + 625$

Practice Exercises
7. $a^3 - 9a^2 + 27a - 27$ 9. $256s^4 + 768s^3t + 864s^2t^2 + 432st^3 + 81t^4$

4.7 Dividing Polynomials

Key Terms
1. dividend
2. quotient
3. divisor

Objective 1
Now Try

1. $10x^3 - 5x$

2. $3n^2 - 4n - \dfrac{2}{n}$

3. $\dfrac{7z^4}{2} - 4z^3 - \dfrac{5}{z} - \dfrac{3}{z^2}$

4. $2a^3b^2 + 4a^2b - 3$

Practice Exercises
1. $2a^3 - 3a$

3. $-13m^2 + 4m - \dfrac{5}{m^2}$

Objective 2
Now Try

5. $4x + 3$

6. $2x^2 - 2x - 2 + \dfrac{-5}{5x - 1}$

7. $x^2 + 10x + 100$

8. $3x^2 + 5x + 5 + \dfrac{8x + 29}{x^2 - 4}$

Practice Exercises

5. $3x^2 - 6x + 2 + \dfrac{13x - 7}{2x^2 + 3}$

Objective 3
Now Try

10. $L = 2r^2 - r + 5$ units

Practice Exercises

7. $4y^2 + 24y + 100$ units

Chapter 5 FACTORING AND APPLICATIONS

5.1   Greatest Common Factors; Factoring by Grouping

Key Terms

1. factoring   2. factored form   3. greatest common factor

4. common factor   5. factor

Objective 1
 Now Try
1a. 6   1b. 8   1c. 1

 Practice Exercises
1. 28   3. 1

Objective 2
 Now Try
2. $6x^4$

 Practice Exercises
5. $k^2 m^4 n^4$

Objective 3
 Now Try
3. $4y^2\left(5y^2 - 3y + 1\right)$ 5a. $(y+8)(y+4)$ 5b. $(z+5)\left(z^2 - 11\right)$

 Practice Exercises
7. $10x\left(2x + 4xy - 7y^2\right)$ 9. $-13x^8\left(-2 + x^4 - 4x^2\right)$

Objective 4
 Now Try
6a. $(9+t)(4x+1)$ 6b. $(x-7)(4x+5y)$ 6c. $(x+7)\left(x^2 - 2\right)$
7. $(8x - 3y)(7x + 4)$

 Practice Exercises
11. $(2x + y)(x - 7y)$

5.2   Factoring Trinomials

Key Terms
1. factoring
2. greatest common factor
3. prime polynomial

Objective 1
Now Try
1. $(x+3)(x+8)$
2. $(y-7)(y-5)$
3. $(p+9)(p-3)$
5. prime
6. $(p-7q)(p+2q)$

Practice Exercises
1. prime
3. $(x-11)(x+3)$

Objective 2
Now Try
7a. $7x^4(x-5)(x-2)$
7b. $-3y^4(y+5)(y-4)$

Practice Exercises
5. $2ab(a-3b)(a-2b)$

5.3   More on Factoring Trinomials

Key Terms
1. coefficient
2. trinomial
3. inner product
4. FOIL
5. outer product

Objective 1
Now Try
1. $(5x+2)(x+3)$
2a. $(7x-5)(2x+1)$
2b. $(3m-7)(m+2)$
2c. $(5x+3y)(2x-y)$
3. $3x^3(5x-3)(2x+7)$

Practice Exercises
1. $(4b+3)(2b+3)$
3. $(5c-7t)(2c-3t)$

Objective 2
Now Try
5. $(3x+1)(5x+7)$
6. $(4x-1)(5x-2)$
7. $(4x+7)(2x-3)$
8. $(6x-5y)(4x+3y)$
9. $-6a(3a-5)(a-2)$

Practice Exercises
5. $(a+2b)(3a+2b)$

5.4   Special Factoring Techniques

Key Terms
1. difference
2. sum of cubes
3. difference of cubes
4. perfect square trinomial

Objective 1
Now Try
1. $(z+6)(z-6)$
2a. $(2x+9)(2x-9)$
2b. $(5t+7)(5t-7)$
3a. $10(3x+7)(3x-7)$
3b. $\left(p^2+16\right)(p+4)(p-4)$

Practice Exercises
1. $(x-7)(x+7)$
3. prime

Objective 2
Now Try
4. $(p+8)^2$
5a. prime
5b. $5x(2x+5)^2$
5c. $(8m+3)^2$

Practice Exercises
5. $(3j+2)^2$

Objective 3
Now Try
6a. $(t-6)(t^2+6t+36)$
6b. $(3k-y)(9k^2+3ky+y^2)$
6c. $(3x+7y^2)(9x^2-21xy^2+49y^4)$

Practice Exercises
7. $(2a-5b)(4a^2+10ab+25b^2)$
9. $2n(3m^2+n^2)$

Objective 4
Now Try
7a. $(6x+1)(36x^2-6x+1)$
7b. $6(x+2y)(x^2-2xy+4y^2)$

Practice Exercises
11. $8(a+2b)(a^2-2ab+4b^2)$

5.5   Solving Quadratic Equations Using the Zero-Factor Property

Key Terms

1. standard form   2. double solution   3. quadratic equation

Objective 1

Now Try

1a. $\left\{-12, \frac{7}{4}\right\}$ 1b. $\left\{0, \frac{11}{6}\right\}$ 3. $\left\{\frac{4}{5}, 3\right\}$

4. $\left\{-\frac{3}{10}, \frac{3}{10}\right\}$

Practice Exercises

1. $\left\{-\frac{5}{2}, 4\right\}$ 3. $\left\{-4, \frac{3}{5}\right\}$

Objective 2

Now Try

6. $\{-5, 0, 5\}$ 7. $\left\{\frac{2}{5}, 2, 9\right\}$ 8. $\{-4, 3\}$

Practice Exercises

5. $\{-9, 0, 1\}$

5.6   Applications of Quadratic Equations

Key Terms

1. legs   2. hypotenuse   3. consecutive

Objective 1

Now Try

1. width: 3 m, length: 5 m

Practice Exercises

1. width: 8 in., length: 24 in.   3. height: 4 ft, width: 6 ft

Objective 2

Now Try

2. 0, 1, 2, or 5, 6, 7

Practice Exercises

5. 6, 8

Objective 3

Now Try

3. 16 ft

Practice Exercises

7. 45 m, 60 m, 75 m   9. 20 mi

Objective 4
Now Try
4. 1 sec

Practice Exercises
11. 40 items or 110 items

Chapter 6 RATIONAL EXPRESSIONS AND APPLICATIONS

6.1   The Fundamental Property of Rational Expressions

Key Terms
1. rational expression   2. lowest terms

Objective 1
Now Try
1. 14

Practice Exercises
1. a. $-\dfrac{11}{9}$; b. -4 3. a. $-\dfrac{1}{6}$; b. $-\dfrac{7}{2}$

Objective 2
Now Try

2a. $y \neq \dfrac{1}{7}$ 2b. never undefined   2c. $m \neq 5,\ m \neq -4$

Practice Exercises
5. none

Objective 3
Now Try

3. $\dfrac{3}{k^3}$ 4a. $\dfrac{7}{9}$ 4c. $\dfrac{m+6}{2m+3}$

5. -1

Practice Exercises

7. $\dfrac{-5b}{8c}$ 9. $\dfrac{9(x+3)}{2}$

Objective 4
Now Try

7. $\dfrac{-(10x-7)}{4x-3},\ \dfrac{-10x+7}{4x-3},\ \dfrac{10x-7}{-(4x-3)},\ \dfrac{10x-7}{-4x+3}$

Practice Exercises

11. $\dfrac{-(2p-1)}{-(1-4p)};\ \dfrac{1-2p}{4p-1};\ \dfrac{-(2p-1)}{4p-1};\ \dfrac{1-2p}{-(1-4p)}$

6.2   Multiplying and Dividing Rational Expressions

Key Terms
1. reciprocal   2. rational expression   3. lowest terms

Objective 1
  Now Try

1a. $\dfrac{4}{15}$ 1b. $\dfrac{4}{3x}$ 2. $\dfrac{s^2}{6(r-s)}$

3. $\dfrac{35}{x}$

  Practice Exercises

1. $\dfrac{10m^3n}{3}$ 3. $\dfrac{x+4}{2x-8}$

Objective 2
  Now Try

4a. $\dfrac{7y}{2x^3}$ 4b. $\dfrac{4}{3x}$

  Practice Exercises

5. $\dfrac{6z^2-2z+1}{13}$

Objective 3
  Now Try

5a. $\dfrac{15}{2}$ 5b. $\dfrac{y-2}{6(y+2)}$ 7. $\dfrac{8x}{(x-3)^2}$

8. $\dfrac{-(m-8)}{5m(m+9)}$

  Practice Exercises

7. $-\dfrac{1}{2}$ 9. $\dfrac{-3(6k-1)}{3k-1}$

6.3   Least Common Denominators

Key Terms
1. equivalent expressions   2. least common denominator

Objective 1
  Now Try

1. 72   2. $120a^4$ 3a. $9w(w-2)$

3b. $(b+4)(b+1)(b-4)^2$ 3c. $p-14$ or $14-p$

Practice Exercises

1. $108b^4$

3. $w(w+3)(w-3)(w-2)$

Objective 2
  Now Try

4. $\dfrac{65}{30}$

5a. $\dfrac{76}{24c-20}$

5b. $\dfrac{3(z+2)}{z(z-7)(z+2)}$ or $\dfrac{3z+6}{z^3-5z-14z}$

Practice Exercises

5. $30r$

6.4   Adding and Subtracting Rational Expressions

Key Terms

1. greatest common factor

2. least common multiple

Objective 1
  Now Try

1a. $\dfrac{4}{5}$

1b. $2x$

Practice Exercises

1. $\dfrac{4}{w^2}$

3. $\dfrac{1}{x-2}$

Objective 2
  Now Try

2a. $\dfrac{137}{315}$

2b. $\dfrac{46}{63y}$

4. $\dfrac{7x^2+11x+8}{(x+2)(x+1)(x-4)}$

Practice Exercises

5. $\dfrac{7z^2-z-6}{(z+2)(z-2)^2}$

Objective 3
  Now Try

8. 7

9. $\dfrac{8x^2+37x+15}{(x+5)(x-5)^2}$

Practice Exercises

7. $\dfrac{8z}{(z-2)(z+2)}$ or $\dfrac{8z}{z^2-4}$

9. $\dfrac{2m^2-m+2}{(m-2)(m+2)^2}$

6.5   Complex Fractions

Key Terms
1. complex fraction   2. LCD

Objective 1, Objective 2
  Now Try

1. $\dfrac{90}{x}$ 2. $\dfrac{bc^2}{a^2}$ 3. $\dfrac{-9x+65}{2x-5}$

  Practice Exercises

1. $\dfrac{7m^2}{2n^3}$ 3. $\dfrac{9s+12}{6s^2+2s}$ or $\dfrac{3(3s+4)}{2s(3s+1)}$

Objective 3
  Now Try

4. $\dfrac{12}{x}$ 5. $\dfrac{5n-9}{3(6n+5)}$

  Practice Exercises

5. $\dfrac{(x-2)^2}{x(x+2)}$

Objective 4
  Now Try

7. $\dfrac{(x+y)(x+1)}{x}$

  Practice Exercises

7. $\dfrac{2z^3+xy^2z^3}{x}$ 9. $\dfrac{2y^3}{x^2y^3+3x^2}$

6.6   Solving Equations with Rational Expressions

Key Terms
1. extraneous solution   2. proposed solution

Objective 1
  Now Try

1a. expression; $\dfrac{1}{2}x$ 1b. equation; $\{14\}$

  Practice Exercises

1. equation; $\{-2\}$ 3. expression; $\dfrac{41x}{15}$

Objective 2
  Now Try
  4. $\varnothing$ 7. $\{-4\}$

  Practice Exercises
  5. $\{-4, 16\}$

Objective 3
  Now Try
  9. $b = aq + c$ 10. $y = \dfrac{x - wz}{w}$, or $\dfrac{x}{w} - z$ 11. $x = \dfrac{yz}{y + z}$

  Practice Exercises
  7. $f = \dfrac{d_0 d_1}{d_0 + d_1}$ 9. $q = \dfrac{2pf - Ab}{Ab}$ or $\dfrac{2pf}{Ab} - 1$

6.7   Applications of Rational Expressions

Key Terms
  1. numerator   2. denominator   3. reciprocal

Objective 1
  Now Try
  1. 3

  Practice Exercises
  1. $-\dfrac{2}{3}$ or 1   3. $\dfrac{3}{5}$

Objective 2
  Now Try
  2. 3 miles per hour

  Practice Exercises
  5. 24 miles per hour

Objective 3
  Now Try
  3. $\dfrac{2}{5}$ hour

  Practice Exercises
  7. $2\dfrac{2}{5}$ hr   8. 3 hr

Chapter 7 LINEAR EQUATIONS, GRAPHS, AND SYSTEMS

7.1   Review of Graphs and Slopes of Lines

Key Terms

1. y-intercept   2. linear equation in two variables

3. coordinate   4. origin   5. x-intercept

6. x-axis   7. quadrant   8. ordered pair

9. y-axis   10. components   11. plot

12. rectangular (Cartesian) coordinate system   13. slope

14. graph of an equation   15. linear equation in two variables

16. rise   17. run

Objective 2
Now Try

1. 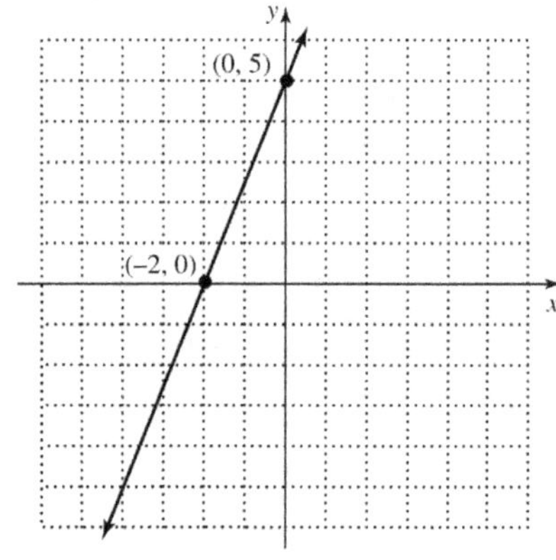   2. 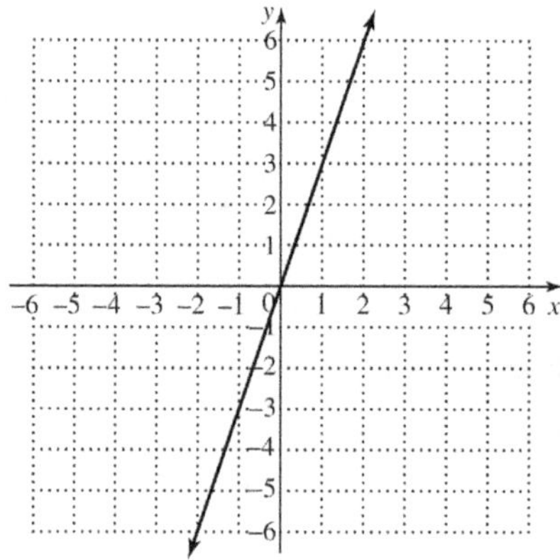

Practice Exercises

1. 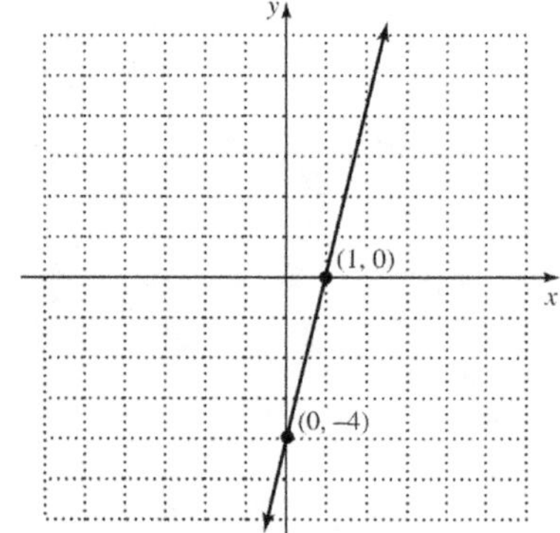

Objective 3
Now Try

3a.

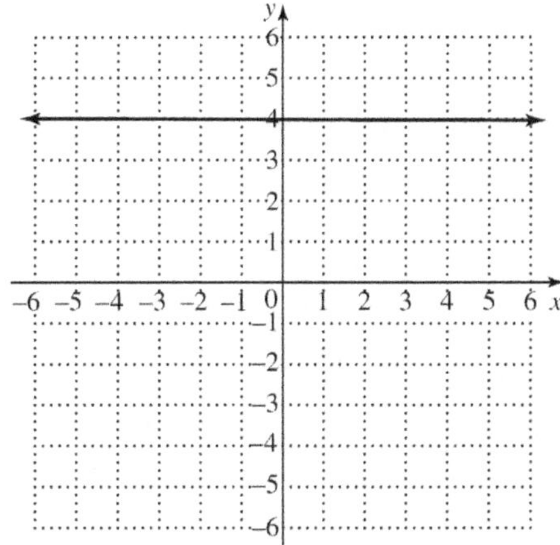

3b.

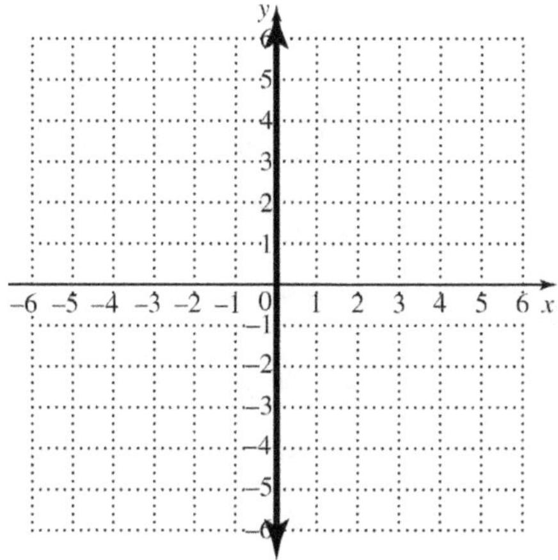

Practice Exercises

3. $x - 1 = 0$

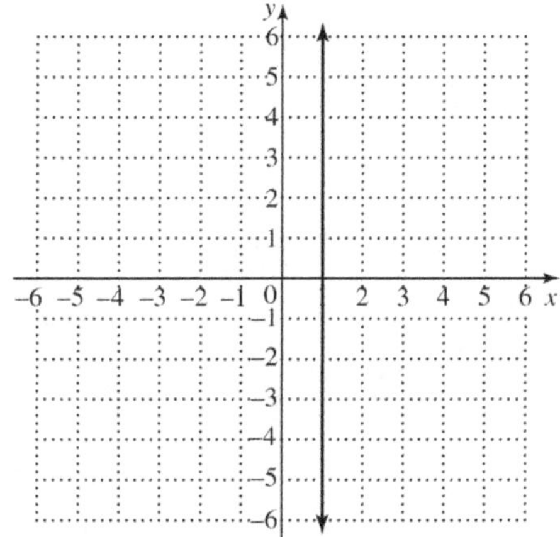

Objective 4
Now Try

4. $(5, -2)$

Practice Exercises

5. $(2, 2)$ 7. $(-5.3, -1.8)$

Objective 5
Now Try

5. $-\dfrac{16}{9}$

Practice Exercises

9. -3

Objective 6
Now Try
8.

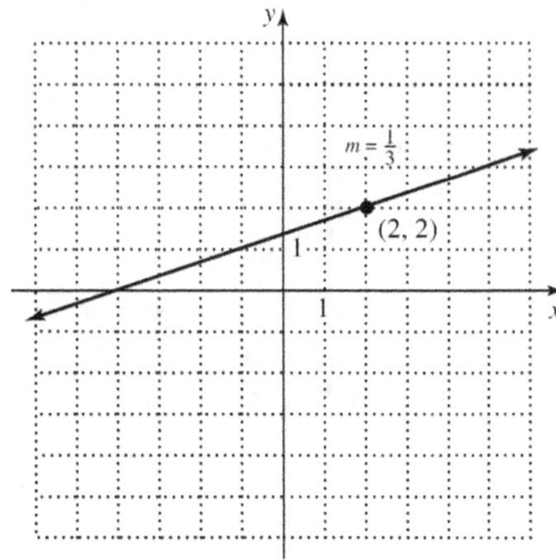

Practice Exercises
11.

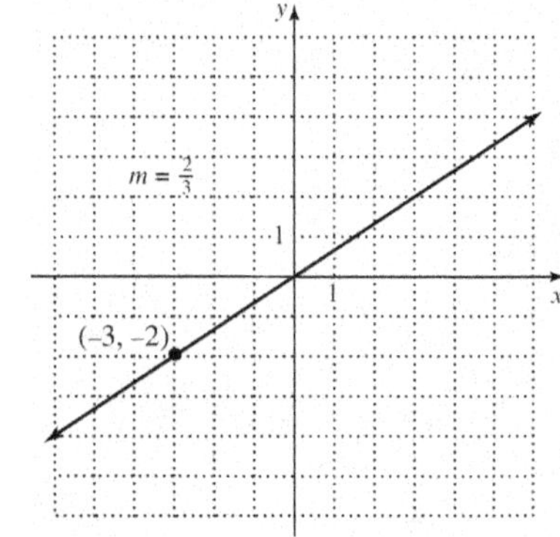

Objective 7
Now Try
9. parallel

10a. perpendicular

Practice Exercises
13. neither

15. perpendicular

Objective 8
Now Try
11. 113.5 ft/min

12. an increase of 5 employees/yr

Practice Exercises
17. 4500 people/yr

7.2   Review of Equations of Lines; Linear Models

Key Terms

1. point-slope form   2. standard form   3. slope-intercept form

Objective 1

Now Try

1. $y = \dfrac{7}{9}x + 8$

Practice Exercises

1. $y = \dfrac{3}{2}x - \dfrac{2}{3}$ 3. $y = -\dfrac{6}{5}x + \dfrac{2}{5}$

Objective 2

Now Try

2.

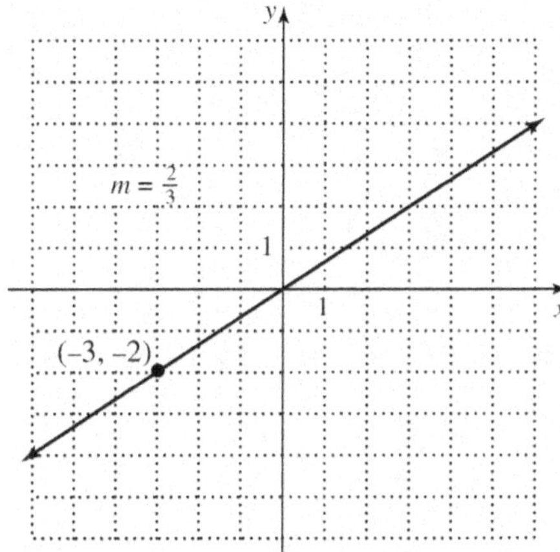

Practice Exercises

5.

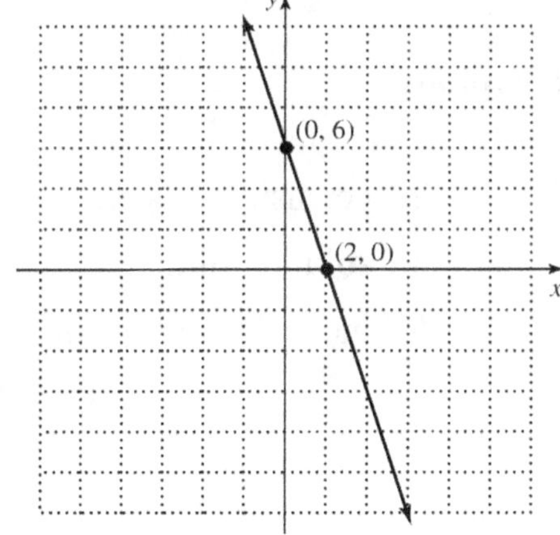

Objective 3

Now Try

3. $y = \dfrac{4}{5}x - \dfrac{44}{5}$

Practice Exercises

7. $3x + 5y = 11$ 9. $2x - 3y = -8$

Objective 4
  Now Try

4. $y = -\dfrac{3}{4}x + \dfrac{81}{4};\ \ 3x + 4y = 81$

  Practice Exercises

11. $x + 3y = -1$

Objective 5
  Now Try

5a. $y = 5$ 5b. $x = -5$

  Practice Exercises

13. $x = -1$ 15. $y = -5$

Objective 6
  Now Try

6a. $y = -\dfrac{3}{4}x + \dfrac{7}{2}$ 6b. $y = \dfrac{4}{3}x + 16$

  Practice Exercises

17. $4x - 3y = -17$

Objective 7
  Now Try

7a. $y = -28.8x + 686$ 7b. $(1, 657.2)$

  Practice Exercises

19. (a) $y = 1.25x + 25$; (b) $(0, 25), (5, 31.25), (10, 37.50)$

7.3   Solving Systems of Linear Equations by Graphing

Key Terms

1. independent equations   2. consistent system

3. solution set of the system   4. solution of the system

5. dependent equations   6. inconsistent system

7. system of linear equations

Objective 1
  Now Try

1. no

  Practice Exercises

1. no   3. no

Objective 2
  Now Try
2.

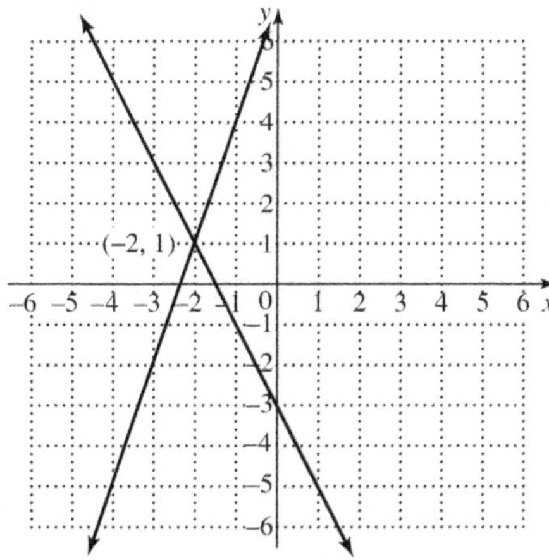

Practice Exercises
  5.

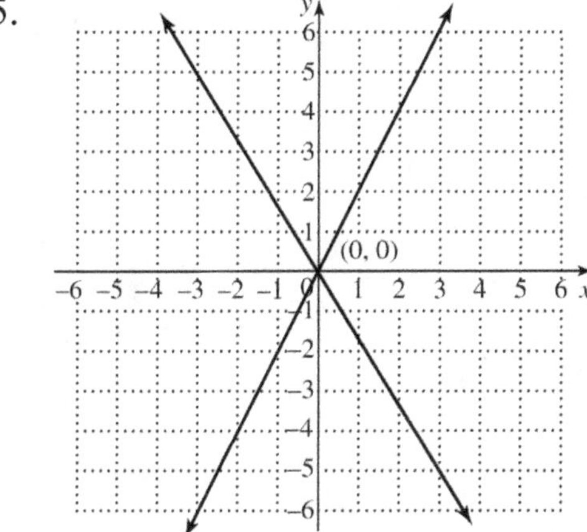

Objective 3
  Now Try
  3. $\varnothing$ 4. $\{(x, y)\mid 4x - 2y = 8\}$

  Practice Exercises
  7. no solution

Objective 4
  Now Try
  5a. neither   5b. intersecting lines   5c. exactly one solution

  Practice Exercises
  9. (a) neither   (b) intersecting lines   (c) one solution

  11. (a) dependent   (b) one line   (c) infinitely many solutions

7.4 Solving Systems of Linear Equations by Substitution

Key Terms
1. ordered pair
2. substitution
3. dependent system
4. inconsistent system

Objective 1
Now Try
1. $\{(1, 6)\}$
2. $\{(9, -4)\}$
3. $\{(8, -2)\}$

Practice Exercises
1. $(2, 4)$
3. $(4, -9)$

Objective 2
Now Try
4. $\varnothing$
5. $\{(x, y)\mid 5x + 4y = 20\}$

Practice Exercises
5. $\{(x, y)\mid x - 2y = -6\}$

Objective 3
Now Try
6. $\{(-2, 5)\}$
7. $\{(3, -5)\}$

Practice Exercises
7. $(-9, -11)$
9. $\{(x, y)\mid 0.3x + 0.4y = 0.5\}$

7.5 Solving Systems of Linear Equations by Elimination

Key Terms
1. elimination method
2. addition property of equality
3. substitution

Objective 1
Now Try
1. $\{(8, 3)\}$

Practice Exercises
1. $(8, 3)$
3. $(5, 0)$

Objective 2
Now Try
4. $\{(-4, 9)\}$

Practice Exercises
5. $\left(\dfrac{1}{2}, 1\right)$

Objective 3
Now Try

5. $\left\{\left(\dfrac{9}{17}, \dfrac{11}{17}\right)\right\}$

Practice Exercises
7. $(2, -4)$　　　　　　9. $(3, -2)$

Objective 4
Now Try

6a. $\left\{(x, y)\,|\,9x - 7y = 5\right\}$　6b. $\varnothing$

Practice Exercises
11. $\left\{(x, y)\,|\,-x - 2y = 3\right\}$

7.6　Systems of Linear Equations in Three Variables

Key Terms
1. ordered triple　　　2. dependent system　　　3. inconsistent system
4. focus variable　　　5. working equation

Objective 1
Practice Exercises
1. The planes intersect in one point.

Objective 2
Now Try
1. $\{(-3, 1, 2)\}$

Practice Exercises
3. $\{(-1, 2, 1)\}$　　　　　5. $\{(4, -4, 1)\}$

Objective 3
Now Try
2. $\{(2, -5, 3)\}$

Practice Exercises
7. $\{(4, 2, -1)\}$

Objective 4
Now Try
3. $\varnothing$　　　　　　4. $\left\{(x, y, z)\,|\,x - 5y + 2z = 0\right\}$　　5. $\varnothing$

Practice Exercises
9. $\varnothing$; inconsistent system
11. $\left\{(x, y, z)\,|\,-x + 5y - 2z = 3\right\}$; dependent equations

　　　　　A-51

7.7   Applications of Systems of Linear Equations

Key Terms
1. elimination method   2. substitution

Objective 1
Now Try
1. length: 17 ft; width: 10 ft

Practice Exercises
1. square: 8 cm; triangle: 13 cm   3. 21 cm

Objective 2
Now Try
2. marigold: $12.29; carnation: $17.60

Practice Exercises
5. large: $6; small: $3

Objective 3
Now Try
3. water: 6 liters; 25% solution: 24 liters

Practice Exercises
7. $6 coffee: 100 lbs; $12 coffee: 50 lb

9. water: 9 oz; 80% solution: 3 oz

Objective 4
Now Try
4. Ashley: 5 mph; Taylor: 3 mph

Practice Exercises
11. distance from school: $1\frac{1}{8}$ mi; jogging time: $\frac{1}{8}$ hr or $7\frac{1}{2}$ min

Objective 5
Now Try
7. 20 lb of $4 candy; 30 lb of $6 candy; 50 lb of $10 candy

Practice Exercises
13. $8 coffee: 15 lb; $10 coffee: 12 lb; $15 coffee: 23 lb

Chapter 8 INEQUALITIES AND ABSOLUTE VALUE

8.1 Review of Linear Inequalities in One Variable

Key Terms

1. interval notation
2. linear inequality in one variable
3. inequality
4. interval

Objective 2
  Now Try

1. $(-\infty, -4)$

Practice Exercises

1. $[3, \infty);$

3. $(-\infty, 0];$

Objective 3
  Now Try

2. $(-\infty, -5]$

3. $(-\infty, -8]$

4. $(7, \infty)$

Practice Exercises

5. $[-4, \infty);$

Objective 4
  Now Try

5. $(-4, -2]$

Practice Exercises

7. $(2, 5];$

9. $(-1, 5)$

8.2   Set Operations and Compound Inequalities

Key Terms
1. union   2. compound inequality   3. intersection

Objective 2
Now Try
1. $\{30, 50\}$

Practice Exercises
1. $\{0, 2, 4\}$ 3. $\{1, 3, 5\}$

Objective 3
Now Try

2. $[12, 16]$ 

3. $(-\infty, -3)$ 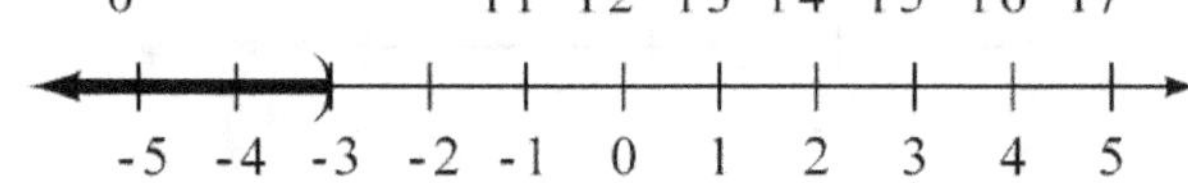

Practice Exercises

5. $[-4, 3)$

Objective 4
Now Try
5. $\{20, 30, 40, 50, 70\}$

Practice Exercises
7. $\{0, 1, 2, 3, 4, 5\}$ 9. $\{0, 1, 2, 3, 4, 5\}$

Objective 5
Now Try

6. $(-\infty, 2) \cup [5, \infty)$ 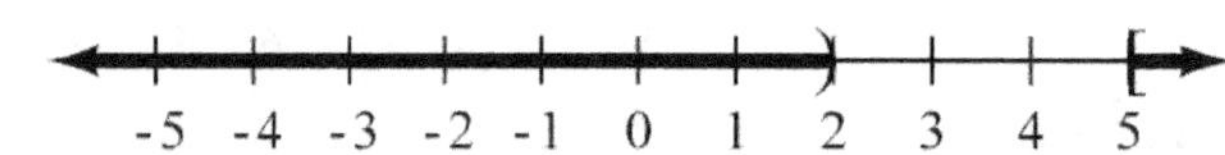

8. $(-\infty, \infty)$ 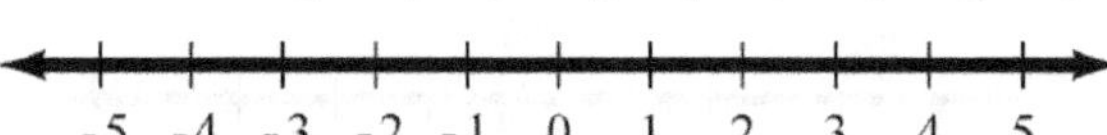

Practice Exercises

11. $(-\infty, \infty)$

8.3   Absolute Value Equations and Inequalities

Key Terms
1.   absolute value equation   2.   absolute value inequality

Objective 2
Now Try

1. $\left\{-3, \dfrac{7}{5}\right\}$ 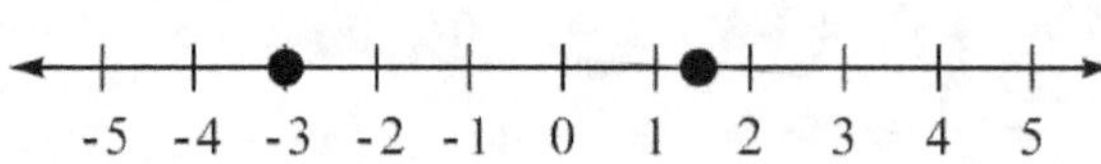

Practice Exercises

1. $\left\{-\dfrac{13}{2}, \dfrac{7}{2}\right\}$

3. $\{-2, 14\}$

Objective 3
Now Try

2. $(-\infty, -2) \cup (1, \infty)$

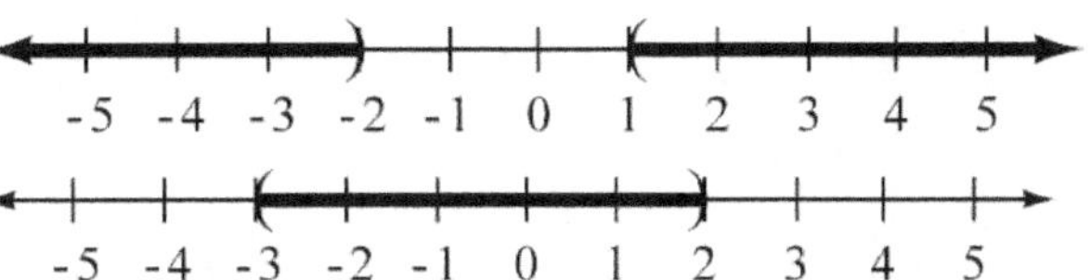

3. $(-3, 2)$

Practice Exercises

5. $(-\infty, -7] \cup [16, \infty)$

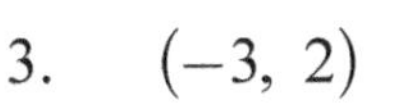

Objective 4
Now Try

5. $\{-4, 10\}$

6a. $(-\infty, -22] \cup [4, \infty)$

6b. $[-22, 4]$

Practice Exercises

7. $\{-2, 3\}$

9. $\left\{-\dfrac{5}{4}, -\dfrac{1}{4}\right\}$

Objective 5
Now Try

7. $\{-5, -1\}$

Practice Exercises

11. $\left\{-3, \dfrac{11}{3}\right\}$

Objective 6
Now Try

8a. $\varnothing$

8b. $\{-5\}$

9a. $(-\infty, \infty)$

9b. $\varnothing$

9c. $\{4\}$

Practice Exercises

13. $\{-14\}$

15. $\varnothing$

Objective 7
Now Try

10. between 16.055 and 17.745 oz, inclusive

Practice Exercises

17. between 16.6465 and 17.1535 oz, inclusive

8.4 Linear Inequalities and Systems in Two Variables

Key Terms

1. boundary line
2. solution set of a system of linear inequalities
3. system of linear inequalities
4. linear inequality in two variables

Objective 1

Now Try

1. 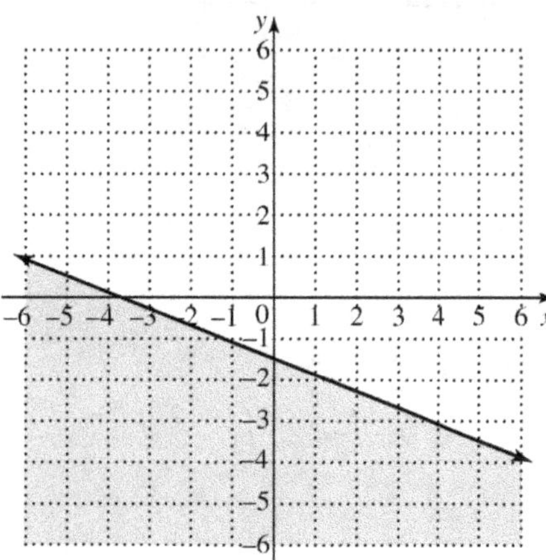

2. 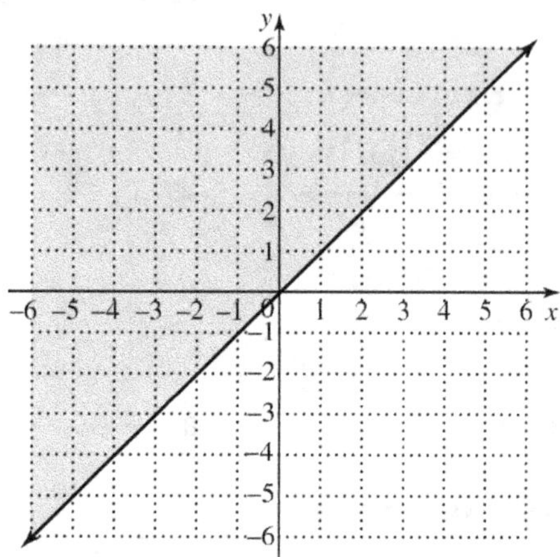

3. 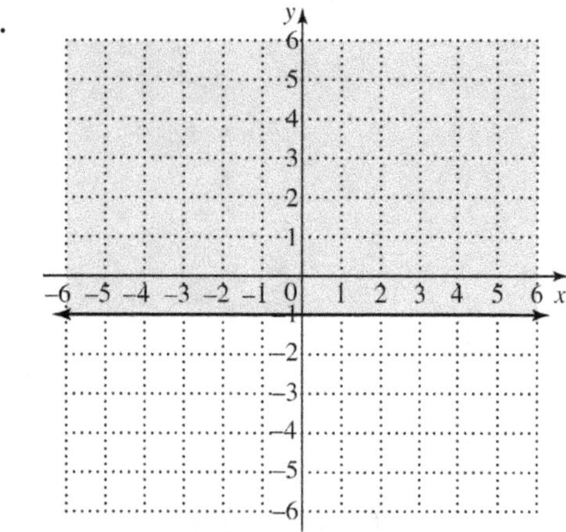

Practice Exercises

1. 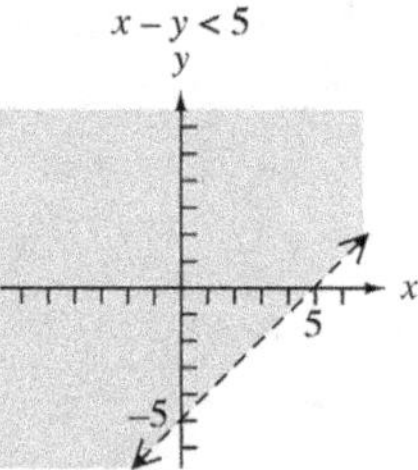

3. 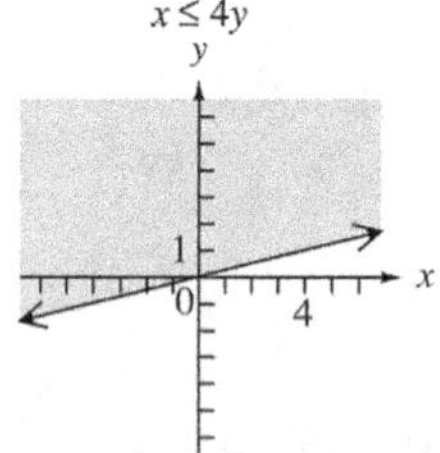

Copyright © 2025 Pearson Education, Inc.

Objective 2
Now Try
4.

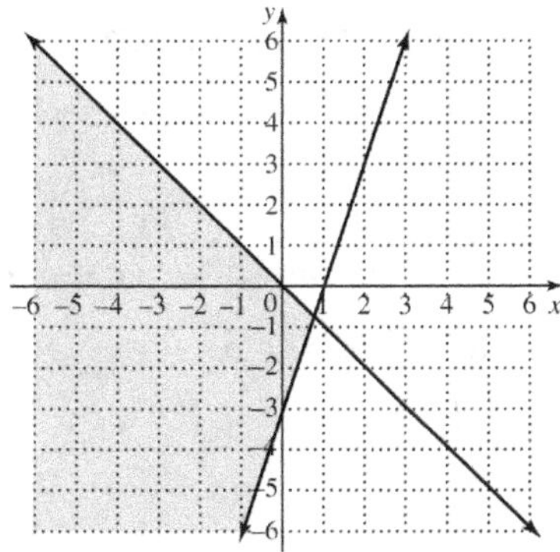

Practice Exercises
5.

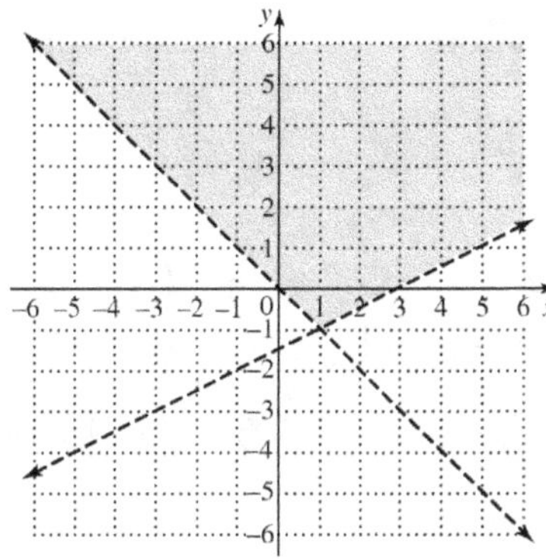

Chapter 9 RELATIONS AND FUNCTIONS

9.1   Introduction to Relations and Functions

Key Terms
1. range

2. relation

3. dependent variable

4. domain

5. function

6. independent variable

Objective 1
Now Try
1. $\{(1,\ 3),\ (2,\ 4),\ (3,\ 6),\ (5,\ 7)\}$

2a. function

2b. not a function

Practice Exercises
1. $\{(1,\ 3),\ (1,\ 4),\ (2,-1),\ (3,\ 7)\}$

3. not a function

Objective 2
Now Try
3. domain: $\{13\}$; range: $\{-2, -1, 4\}$; not a function

4a. domain: $(-\infty,\ \infty)$; range: $(-\infty,\ \infty)$

4b. domain: $[-4,\ \infty)$; range: $(-\infty,\ \infty)$

Practice Exercises
5. function; domain: $\{A, B, C, D, E\}$; range: $\{V, W, X, Z\}$

Objective 3
Now Try
5. function

6. function; domain: $(-\infty,\ \infty)$

Practice Exercises
7. function

9. function; $(-\infty, -6) \cup (-6,\ \infty)$

9.2   Function Notation and Linear Functions

Key Terms

1. linear function   2. function notation   3. constant function

Objective 1

Now Try

1. $f(3) = 17$ 2a. $f(-6) = 16$ 2b. $f(c) = -c^2 - 7c + 10$

3. $g(a-1) = 4a - 11$ 4. $f(-6) = 21$ 5a. $f(-2) = 3$

5b. $f(-3) = 2$ 6. $f(x) = -\dfrac{2}{3}x + \dfrac{7}{3}, \; f(-3) = \dfrac{13}{3}$

Practice Exercises

1. (a) -13; (b) -7; (c) $-3x - 7$ 3. (a) 9; (b) 9; (c) 9

Objective 2

Now Try

7. domain: $(-\infty, \infty)$;

   range: $(-\infty, \infty)$

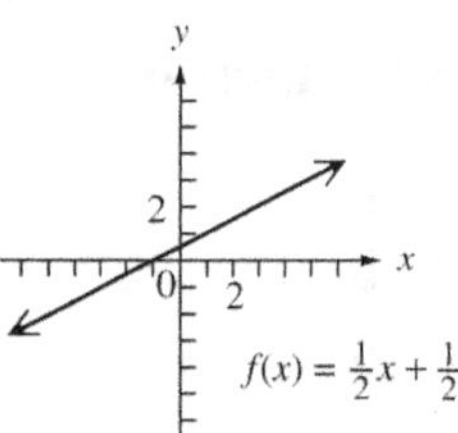

Practice Exercises

5. domain: $(-\infty, \infty)$

   range: $(-\infty, \infty)$

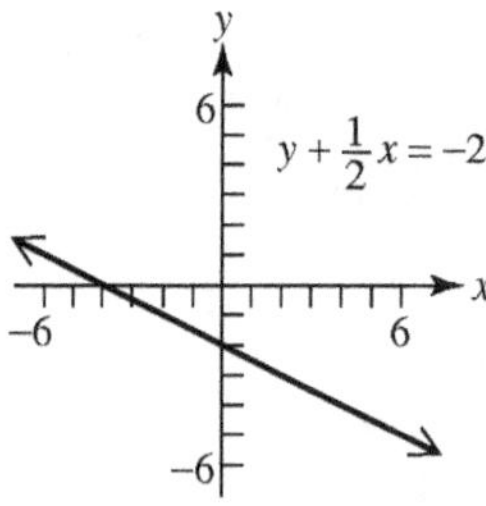

9.3   Polynomial Functions, Graphs, Operations, and Composition

Key Terms

1. cubing function   2. composite function

3. polynomial function of degree n 4. squaring function

5. identity function   6. composition

Objective 1

Now Try

1. 1

Practice Exercises

1. (a) -7; (b) -17 3. (a) 28; (b) 198

Objective 2
Now Try
2a. 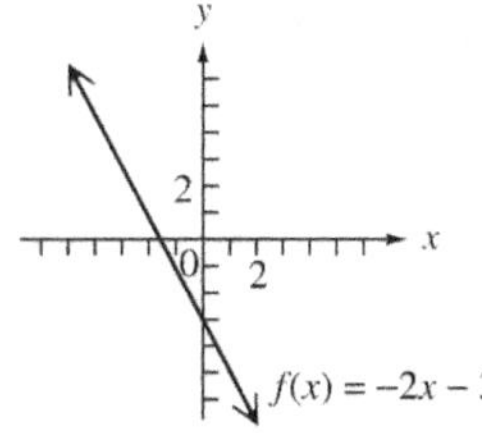

domain: $(-\infty, \infty)$;
range: $(-\infty, \infty)$

2b. 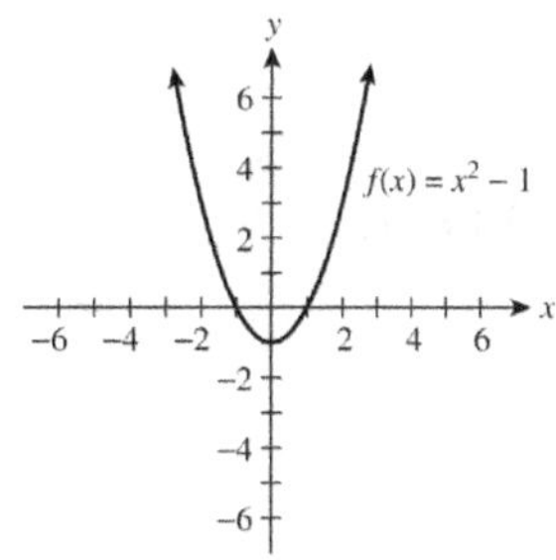

domain: $(-\infty, \infty)$;
range: $[-1, \infty)$

2c. 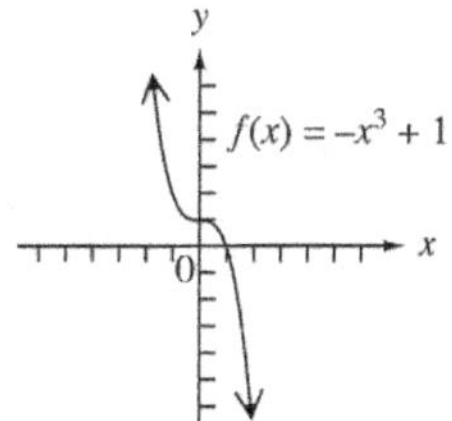

domain: $(-\infty, \infty)$;
range: $(-\infty, \infty)$

Practice Exercises
5. 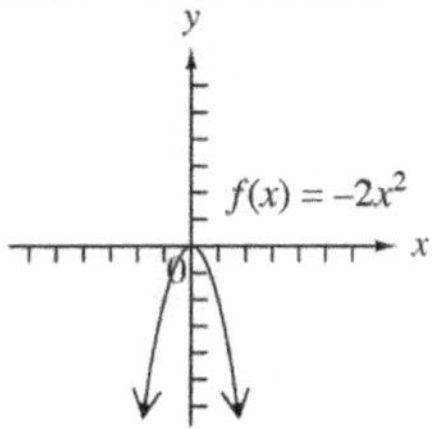

domain: $(-\infty, \infty)$;
range: $(-\infty, 0]$

Objective 3
Now Try
3a. $3x^2 - 6x + 21$　　　3b. $9x^2 - 8x + 3$　　　5. $42x^3 + 47x^2 + 10x;\ -741$

6. $2x + 5\ \ x \neq 4;\ 1$

Practice Exercises
7. (a) $x^2 + 7x - 13$; (b) $3x^2 + x + 3$　　　9. $4x + 9,\ x \neq 5$

Objective 4
Now Try
7. -15　　　9. 76　　　10a $g(x) = 0.40x$

10b. $f(x) = 0.30x$　　　10c. $(f \circ g)(x) = 0.12x$　　　10d. $48

Practice Exercises
11. a. $\dfrac{1}{5}$; b. 16; c. $\dfrac{3}{x^2} - \dfrac{4}{x} + 1$

9.4　Variation

Key Terms
1. constant of variation　　　2. varies inversely

3. varies directly　　　4. varies jointly

　　　

Objective 1; Objective 2

Now Try

1. $25; $P = 25x$ 2. 125 psi   3. 153.86 sq cm

Practice Exercises

1. $k = 0.25; y = 0.25x$ 3. 100 newtons

Objective 3

Now Try

4. 640 cycles per second 5. $\dfrac{9}{8}$

Practice Exercises

5. 40 lb

Objective 4

Now Try

6. 1280 psi

Practice Exercises

7. 96   9. 750°

Objective 5

Now Try

7. About 63.5 cm^3

Practice Exercises

11. 9 hr

Chapter 10 ROOTS, RADICALS, AND ROOT FUNCTIONS

10.1 Radical Expressions and Graphs

Key Terms
1. fourth root
2. radical
3. principal square root
4. radicand
5. cube root function
6. perfect square
7. negative square root
8. index (order)
9. cube root
10. square root function
11. square root
12. radical expression

Objective 1
Now Try

2a. 13

2b. 41

2c. $\dfrac{3}{7}$

3a. 19

3b. 37

3c. $n^2 + 5$

Practice Exercises

1. 25, −25

3. $\dfrac{30}{7}$

Objective 2
Now Try

4a. irrational

4b. rational

4c. not a real number

Practice Exercises

5. not a real number

Objective 3
Now Try

5a. 5

5b. −5

5c. 7

6a. 5

6b. −5

6c. none

6d. −5

6e. −5

Practice Exercises

7. −4

9. −1

Objective 4
Now Try

7a. 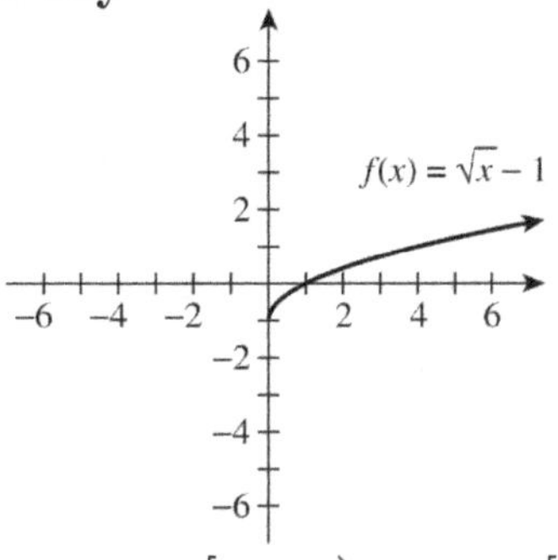
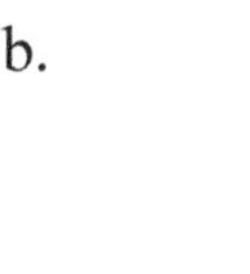

7b. 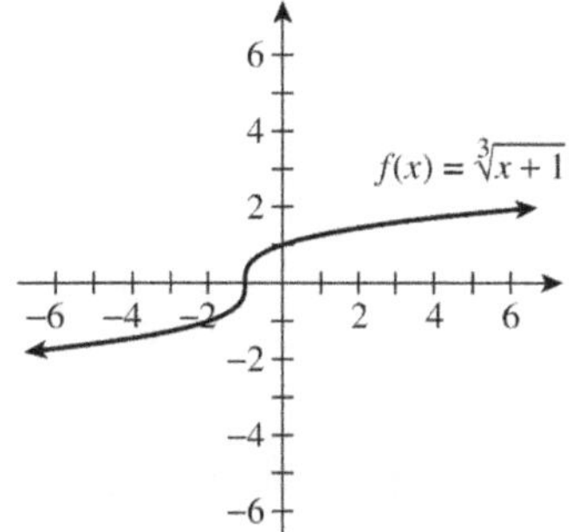

domain: $[0, \infty)$ range : $[-1, \infty)$ domain: $(-\infty, \infty)$ range: $(-\infty, \infty)$

Practice Exercises
11.

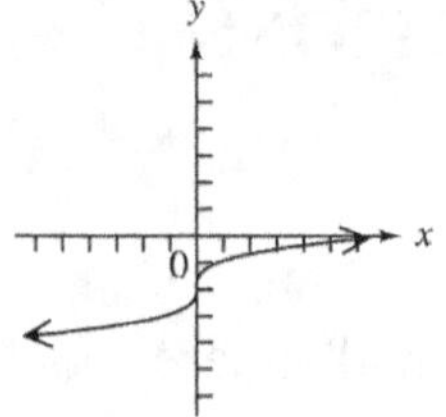

       domain: $(-\infty, \infty)$

       range: $(-\infty, \infty)$

Objective 5
Now Try

8a. 73	8d. $\lvert n \rvert$	9a. 4
9b. −5	9c. −2	9d. $-x^4$
9e. w^{10}	9f. $\lvert x^5 \rvert$	

Practice Exercises

13. 9	15. $-x^4$

Objective 6
Now Try

10a. 3.742	10b. −26.038	10c. 7.554
10d. 9.499	11. 23.2 mph	

Practice Exercises
17. −31.464

10.2   Rational Exponents

Key Terms
1. power rule for exponents
2. product rule for exponents
3. quotient rule for exponents

Objective 1
Now Try

1a. 6

1b. 11

1c. –2

1d. not a real number

1e. –3

1f. $\dfrac{1}{2}$

Practice Exercises

1. –4

3. –15

Objective 2
Now Try

2a. 9

2b. 8

2c. –216

2d. 9

2e. not a real number

3a. $\dfrac{1}{4}$

3b. $\dfrac{1}{625}$

3c. $\dfrac{16}{9}$

Practice Exercises

5. 7776

Objective 3
Now Try

4a. $\sqrt[3]{23}$

4b. $\left(\sqrt[4]{21}\right)^3$

4c. $5\left(\sqrt[4]{x}\right)^5$

4d. $\left(\sqrt[3]{2x}\right)^4 - 3\left(\sqrt[5]{x}\right)^2$

4e. $\dfrac{1}{\left(\sqrt{x}\right)^3}$

4f. $\sqrt[4]{x^2 - y^2}$

5a. $22^{1/2}$

5b. $10^2 = 100$

5c. $3^2 = 9$

5d. m

Practice Exercises

7. $4\left(\sqrt[5]{y}\right)^2 + \sqrt[5]{5x}$

9. $a^{1/4}$, or $\sqrt[4]{a}$

Objective 4
Now Try

6a. $5^{5/2}$

6b. $a^{2/15}$

6c. $x^2 y^{7/2}$

6d. $\dfrac{1}{c^{1/3} d^{2/3}}$

6e. $\dfrac{x^2}{y^{1/4}}$

6f. $r^{7/6} - r^{19/6}$

7a. $x^{7/6}$

7b. $y^{7/12}$

7c. $c^{3/4} d^{1/2}$

7d. $x^{1/4}$

Practice Exercises

11. $a^{5/6}$

10.3 Simplifying Radicals, the Distance Formula, and Circles

Key Terms

1. hypotenuse
2. index; radicand
3. legs
4. circle
5. center
6. radius

Objective 1

Now Try

1a. $\sqrt{14}$

1b. $\sqrt{35x}$

1c. $\sqrt{33mn}$

2a. $\sqrt[3]{21}$

2b. $\sqrt[3]{35xy}$

2c. $\sqrt[5]{8w^4}$

2d. cannot be simplified using the product rule

Practice Exercises

1. $\sqrt{42tx}$

3. cannot be simplified using the product rule

Objective 2

Now Try

3a. $\dfrac{6}{7}$

3b. $\dfrac{\sqrt{13}}{9}$

3c. $-\dfrac{7}{5}$

3d. $-\dfrac{a^2}{5}$

3e. $\dfrac{\sqrt[4]{m}}{3}$

Practice Exercises

5. $\dfrac{\sqrt[5]{7x}}{2}$

Objective 3

Now Try

4a. $2\sqrt{21}$

4b. $9\sqrt{2}$

4c. cannot be simplified

4d. $4\sqrt[3]{4}$

4e. $-2\sqrt[5]{16}$

5a. $10y\sqrt{y}$

5b. $4m^2r^4\sqrt{3mr}$

5c. $-2n^2t\sqrt[3]{4nt^2}$

5d. $-3y^2\sqrt[4]{5x^3y}$

6a. $\sqrt[4]{11^3}$, or $\sqrt[4]{1331}$

6b. $\sqrt[5]{z^4}$

6c. $w^3\sqrt{w}$

Practice Exercises

7. $\sqrt[3]{x^2}$

9. $5ab^2\sqrt[3]{10a^2b}$

Objective 4

Now Try

7. $\sqrt[6]{63}$

Practice Exercises

11. $\sqrt[8]{28}$

Objective 5
  Now Try
  8. $12\sqrt{2}$
  Practice Exercises
13. 26   15. $6\sqrt{2}$

Objective 6
  Now Try
  9. $\sqrt{34}$
  Practice Exercises
17. 5

Objective 7
  Now Try
  10. $x^2 + y^2 = 25$ 11. $(x+5)^2 + (y-4)^2 = 16$

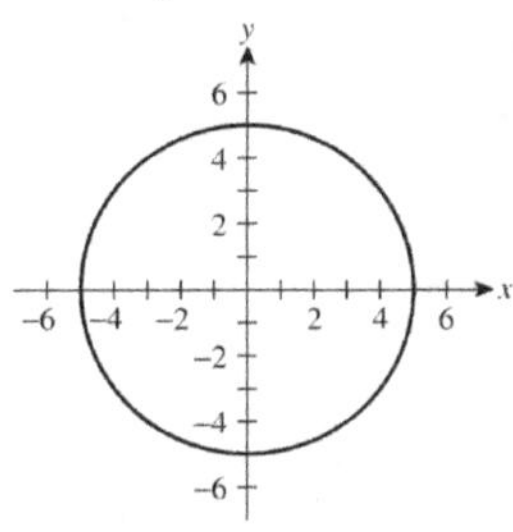 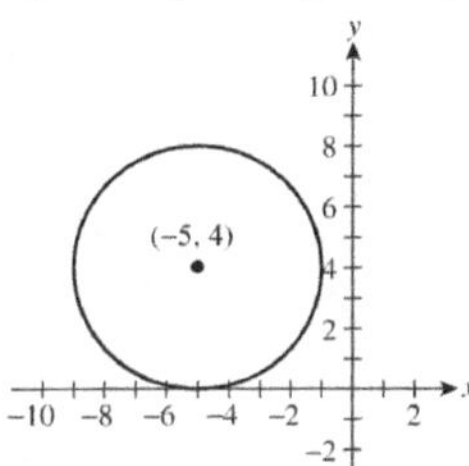

  12. $(x-3)^2 + (y+1)^2 = 6$
  Practice Exercises
  19. $(x-3)^2 + (y+4)^2 = 25$ 21. $x^2 + (y-3)^2 = 2$

10.4   Adding and Subtracting Radical Expressions

Key Terms
  1. unlike radicals   2. like radicals

Objective 1
  Now Try
1a. $8\sqrt{5}$ 1b. $-4\sqrt{3}$ 1c. $-\sqrt{6}$
1d. $13\sqrt{2z}$ 1e. cannot be simplified
2a. $-3\sqrt[3]{2}$ 2b. $5z\sqrt[4]{2y^2 z}$ 2c. $6x^2\sqrt[3]{2x} + 6x^3\sqrt{3x}$
3a. $\dfrac{2\sqrt{5}}{3}$ 3b. $\dfrac{17}{w^2}$

  Practice Exercises
  1. $12\sqrt{x}$ 3. $\dfrac{8y^2\sqrt[3]{y}}{15}$

10.5   Multiplying and Dividing Radical Expressions

Key Terms

1.   conjugate

2.   rationalizing the denominator

Objective 1

Now Try

2a. 12

2b. $6+3\sqrt{7}-2\sqrt{2}-\sqrt{14}$

2c. $11-6\sqrt{2}$

Practice Exercises

1. $\sqrt{10}-4\sqrt{5}+2\sqrt{3}-4\sqrt{6}$

3. $4-\sqrt[3]{25}$

Objective 2

Now Try

3. $\dfrac{2\sqrt{15}}{15}$

4. $-\dfrac{3\sqrt{10}}{8}$

5. $\dfrac{\sqrt[3]{10}}{5}$

Practice Exercises

5. $\dfrac{y\sqrt{21b}}{6b}$

Objective 3

Now Try

6. $-2\left(\sqrt{3}+2\right)$

Practice Exercises

7. $-4\sqrt{3}+8$

9. $\dfrac{\sqrt{3}+2\sqrt{6}+\sqrt{2}+4}{-7}$

Objective 4

Now Try

7. $\dfrac{3+2\sqrt{15}}{4}$

Practice Exercises

11. $\dfrac{2-9\sqrt{2}}{3}$

10.6   Solving Equations with Radicals

Key Terms

1. extraneous solution   2. radical equation

Objective 1

Now Try

1. $\{10\}$

2. $\varnothing$

Practice Exercises
1. {11}   3. {13}

Objective 2
Now Try
3. {5}   4. {3}   5. {4}

Practice Exercises
5. {5, 6}

Objective 3
Now Try
6. {2}

Practice Exercises
7. {−31}   9. {2}

Objective 4
Now Try
7. $h = \dfrac{3v}{\pi r^2}$

Practice Exercises
11. $C = \dfrac{1}{4\pi^2 f^2 L}$

10.7 Complex Numbers

Key Terms
1. complex number   2. complex conjugate   3. imaginary part

4. real part   5. standard form (of a complex number)

6. pure imaginary number   7. nonreal complex number

Objective 1
Now Try
1a. $4i$ 1b. $-12i$ 1c. $i\sqrt{11}$
1d. $8i\sqrt{2}$ 2. $-\sqrt{30}$ 3a. 5
3b. $4i$

Practice Exercises
1. $-9i\sqrt{2}$ 3. $6i$

Objective 2
Practice Exercises
5. imaginary

Objective 3
  Now Try
4. $10 - 9i$ 5. $24 + 4i$

  Practice Exercises
7. $-4 + i$ 9. $2 - 3i$

Objective 4
  Now Try
6a. $-14 + 8i$ 6b. $17 - 7i$ 6c. $17 + i$

  Practice Exercises
11. $-8 + 6i$

Objective 5
  Now Try
7. $\dfrac{18}{29} + \dfrac{13i}{29}$

  Practice Exercises
13. $\dfrac{4}{5} - \dfrac{7}{5}i$ 15. $\dfrac{15}{13} + \dfrac{16}{13}i$

Objective 6
  Now Try
8a. 1 8b. i 8c. $-i$
8d. $-i$

  Practice Exercises
17. i

Chapter 11 QUADRATIC EQUATIONS, INEQUALITIES, AND FUNCTIONS

11.1 Solving Quadratic Equations by the Square Root Property

Key Terms

1. quadratic equation   2. zero-factor property

Objective 1

Now Try

1a. $\{-7, -1\}$ 1b. $\{-10, 10\}$

Practice Exercises

1. $\{-4, -2\}$ 3. $\{-7, 5\}$

Objective 2

Now Try

2a. $\{13, -13\}$, or $\{\pm 13\}$ 2b. $\left\{\sqrt{13}, -\sqrt{13}\right\}$, or $\left\{\pm\sqrt{13}\right\}$

2c. $\left\{3\sqrt{2}, -3\sqrt{2}\right\}$, or $\left\{\pm 3\sqrt{2}\right\}$ 3. About 2.7 seconds

Practice Exercises

5. $\left\{-7\sqrt{2}, 7\sqrt{2}\right\}$

Objective 3

Now Try

5. $\left\{\dfrac{3 \pm 4\sqrt{2}}{7}\right\}$

Practice Exercises

7. $\{-6, 2\}$ 9. $\left\{\dfrac{1}{5}, \dfrac{4}{5}\right\}$

Objective 4

Now Try

6a. $\left\{-4i\sqrt{2}, 4i\sqrt{2}\right\}$ 6b. $\{-2 - 7i, -2 + 7i\}$

Practice Exercises

11. $\{-1 - 6i, -1 + 6i\}$

11.2 Solving Quadratic Equations by Completing the Square

Key Terms
 1. perfect square trinomial
 2. square root property
 3. completing the square

Objective 1
 Now Try
 1. $\left\{-\dfrac{1}{3}, \dfrac{2}{5}\right\}$
 3. $\left\{\dfrac{11-\sqrt{89}}{2}, \dfrac{11+\sqrt{89}}{2}\right\}$

 Practice Exercises
 1. $\left\{-4-2\sqrt{3}, \ -4+2\sqrt{3}\right\}$
 3. $\{-9, 7\}$

Objective 2
 Now Try
 4. $\left\{\dfrac{5}{4}, \dfrac{11}{4}\right\}$
 5. $\left\{\dfrac{-3-\sqrt{11}}{2}, \dfrac{-3+\sqrt{11}}{2}\right\}$

 Practice Exercises
 5. $\left\{\dfrac{-2-\sqrt{10}}{6}, \dfrac{-2+\sqrt{10}}{6}\right\}$

Objective 3
 Now Try
 8. $\left\{\dfrac{-3\pm\sqrt{33}}{2}\right\}$

 Practice Exercises
 7. $\left\{\dfrac{-1-\sqrt{13}}{2}, \dfrac{-1+\sqrt{13}}{2}\right\}$
 9. $\left\{-2-\sqrt{2}, \ -2+\sqrt{2}\right\}$

11.3 Solving Quadratic Equations by the Quadratic Formula

Key Terms
 1. discriminant
 2. quadratic formula

Objective 1
Objective 2
 Now Try
 1. $\left\{\dfrac{4}{3}, \dfrac{3}{2}\right\}$
 2. $\left\{-\dfrac{3}{5}\right\}$
 3. $\left\{\dfrac{1-\sqrt{7}}{2}, \dfrac{1+\sqrt{7}}{2}\right\}$
 4. $\{1-2i, \ 1+2i\}$

 Practice Exercises
 1. $\{2, 4\}$
 3. $\{5-3i, 5+3i\}$

Objective 3
  Now Try
5a. 81; two rational solutions; factoring

5b. 96; two irrational solutions; quadratic formula

5c. −48; two nonreal complex solutions; quadratic formula

5d. 0; one rational solution; factoring

Practice Exercises
5. C

11.4   Equations That Lead to Quadratic Methods

Key Terms
  1. standard form   2. quadratic in form

Objective 1
  Now Try
  1. $\left\{-7, \dfrac{5}{4}\right\}$

Practice Exercises
  1. $\left\{-\dfrac{35}{4}, -3\right\}$ 3. $\left\{-7, -\dfrac{7}{2}\right\}$

Objective 2
  Now Try
  2. 2 mph   3. Tom: 12 hours; Huck: 24 hours

Practice Exercises
5. 550 mph

Objective 3
  Now Try
4a. {2}   4b. {8}

Practice Exercises
7. {2, 5}   9. {9}

Objective 4
  Now Try
  6. {−2, −1, 1, 2}   7a. {−2, 10}   7b. {−1, 27}

Practice Exercises
  11. $\left\{-27, -3\sqrt{3},\ 3\sqrt{3},\ 27\right\}$

11.5 Formulas and Further Applications

Key Terms
1. quadratic function 2. Pythagorean theorem

Objective 1
Now Try

1a. $t = \pm \dfrac{\sqrt{mxF}}{F}$ 1b. $z = \dfrac{y^2}{6p^2}$ 2. $q = \dfrac{-k \pm k\sqrt{5}}{2p}$

Practice Exercises

1. $d = \dfrac{k^2 l^2}{F^2}$ 3. $a = \dfrac{-c \pm c\sqrt{2}}{b}$

Objective 2
Now Try
3. south: 72 mi; east 54 mi

Practice Exercises
5. 10 ft

Objective 3
Now Try
4. 2.5 in.

Practice Exercises
7. 2.5 ft 9. 4 ft

Objective 4
Now Try
5. 4.1 sec

Practice Exercises
11. 1.2 sec

11.6   Graphs of Quadratic Functions

Key Terms
1. axis
2. vertex
3. quadratic function
4. parabola

Objective 1; Objective 2
Now Try

1. vertex: $(0,-1)$, axis: $x = 0$;
   domain: $(-\infty, \infty)$; range: $[-1, \infty)$

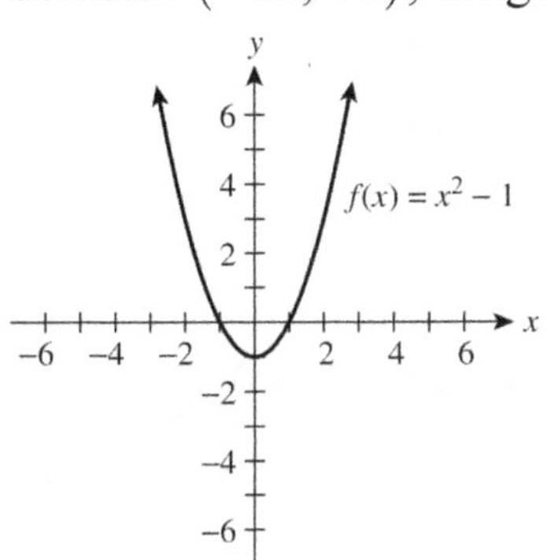

2. Vertex: $(-3, 0)$; axis: $x = -3$
   domain: $(-\infty, \infty)$; range: $[0, \infty)$

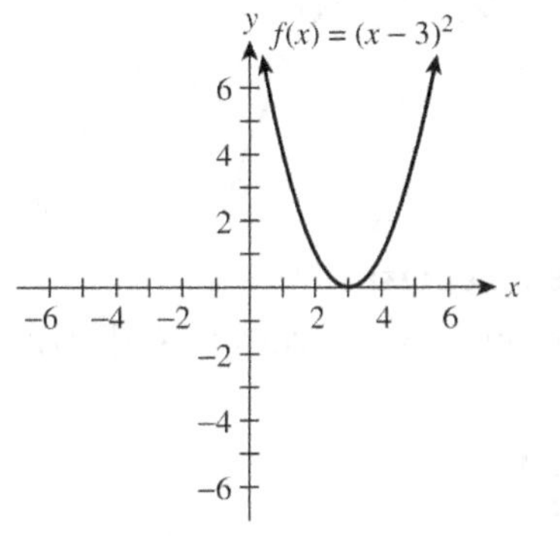

3. vertex: $(-2,-1)$; axis: $x = -2$
   domain: $(-\infty, \infty)$; range: $[-1, \infty)$

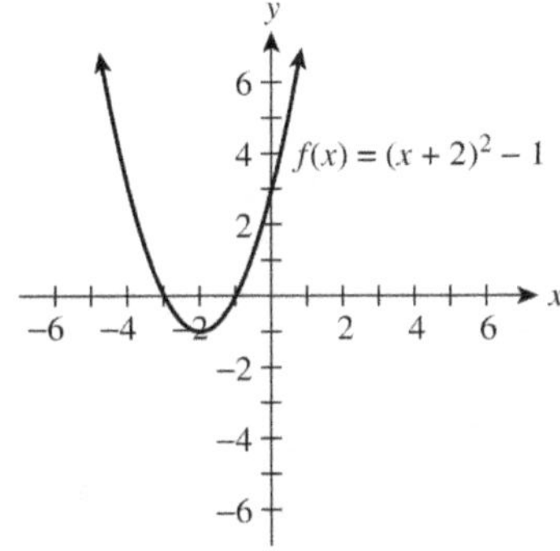

Practice Exercises

1. 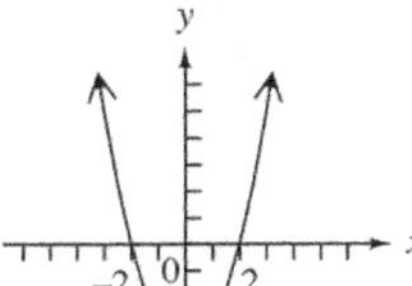

Vertex: $(0, -4)$
Axis: $x = 0$
Domain: $(-\infty, \infty)$
Range: $[-4, \infty)$

3. 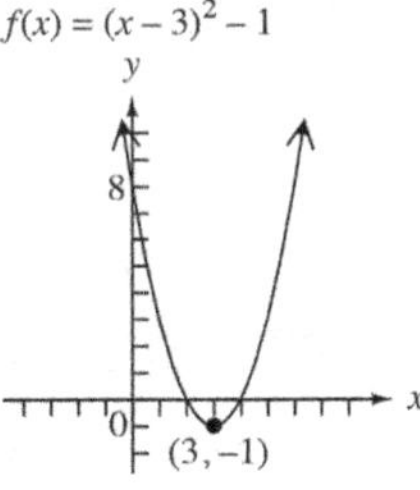

Vertex: $(3, -1)$
Axis: $x = 3$
Domain:
$(-\infty, \infty)$
Range: $[-1, \infty)$

Objective 3
Now Try

4. vertex: $(0, 0)$; axis: $x = 0$
   domain: $(-\infty, \infty)$; range: $(-\infty, 0]$

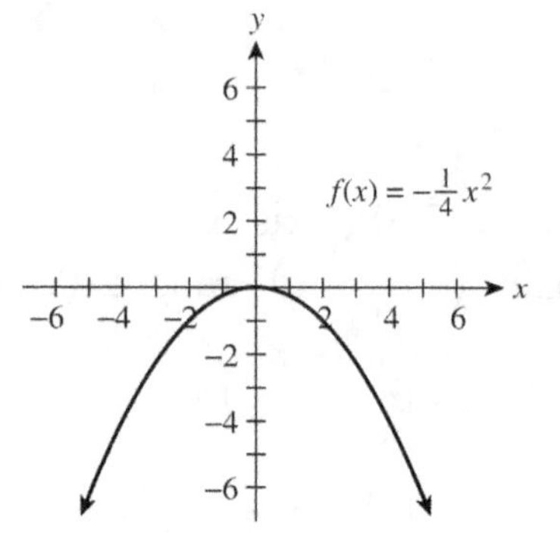

5. vertex: $(1, 1)$; axis: $x = 1$
   domain: $(-\infty, \infty)$; range: $[1, \infty)$

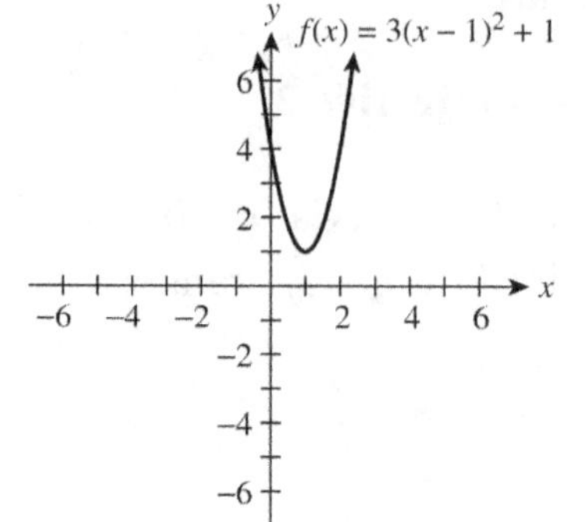

Practice Exercises

5. down; narrower; vertex: $(-1, 0)$; domain: $(-\infty, \infty)$; range: $[0, \infty)$

Objective 4
Now Try

6. $y = 0.01x^2 + 0.77x + 8.36$

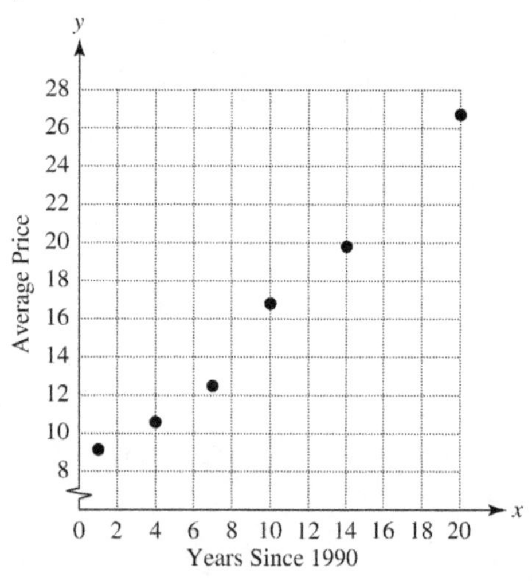

Practice Exercises

7. linear; positive

9.

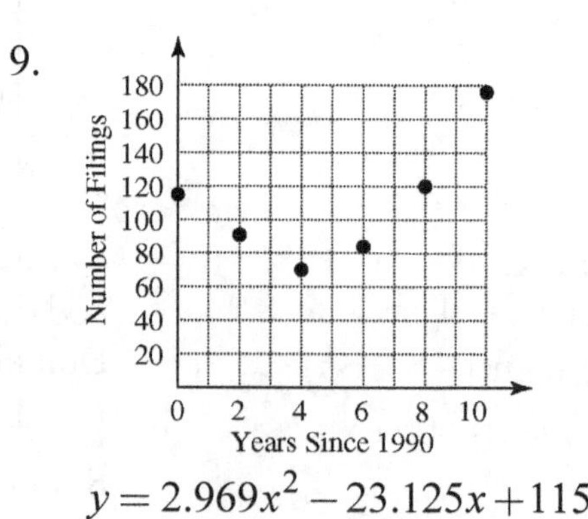

$y = 2.969x^2 - 23.125x + 115$

11.7   More about Parabolas and Their Applications

Key Terms

1. discriminant   2. vertex

Objective 1

Now Try

1. $(3, -5)$ 2. $(1, 1)$ 3. $\left(\dfrac{3}{2}, \dfrac{1}{2}\right)$

Practice Exercises

1. $(1, 3)$ 3. $(-6, 0)$

Objective 2

Now Try

4. vertex: $(-2, 1)$, axis: $x = -2$;
   domain: $(-\infty, \infty)$; range: $[1, \infty)$

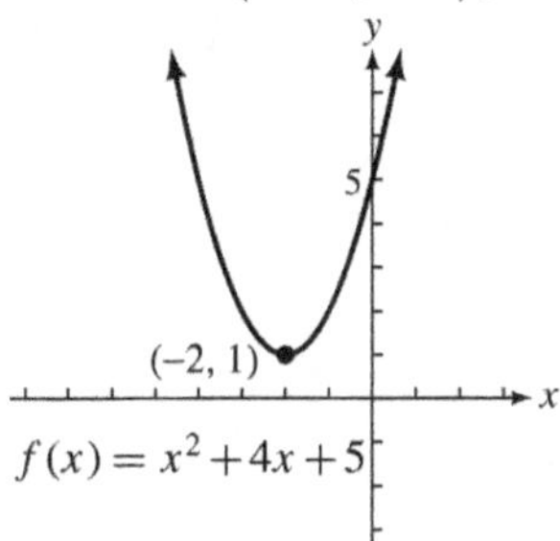

Practice Exercises

5. 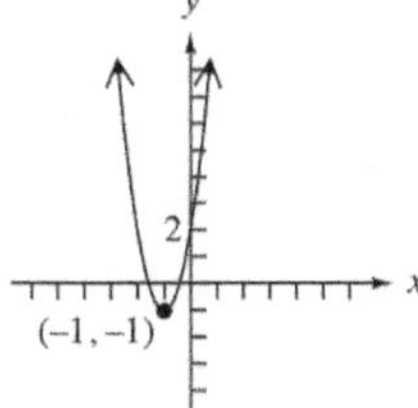

Vertex: $(-1, -1)$
Axis: $x = -1$
Domain: $(-\infty, \infty)$
Range: $[-1, \infty)$

Objective 3

Now Try

5. 1

Practice Exercises

7. 0   9. 1

Objective 4
Now Try
6. maximum area: 125,000 sq yd; length: 500 yd; width: 250 yd
7. maximum height: 286 feet after 1.5 seconds

Practice Exercises
11. 24 (a square)

Objective 5
Now Try
9. vertex: $(0, 2)$; axis: $y = 2$
   domain: $(-\infty, 0]$; range: $(-\infty, \infty)$

Practice Exercises
13. 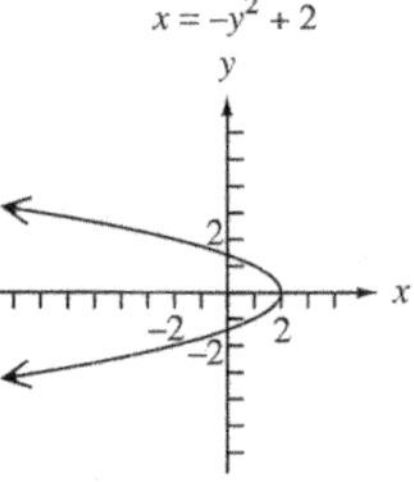

Vertex: $(2, 0)$
Axis: $y = 0$
Domain: $(-\infty, 2]$
Range: $(-\infty, \infty)$

15. 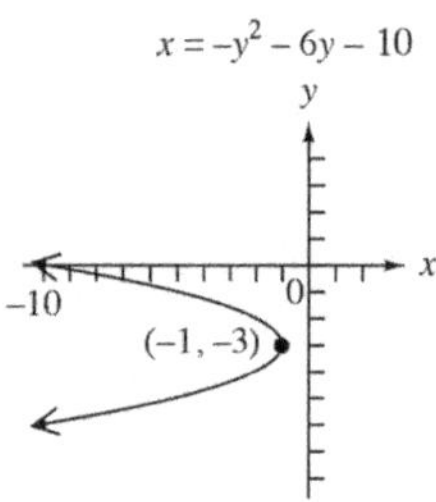

Vertex: $(-1, -3)$
Axis: $y = -3$
Domain: $(-\infty, -1]$
Range: $(-\infty, \infty)$

11.8 Polynomial and Rational Inequalities

Key Terms
1. rational inequality   2. quadratic inequality

Objective 1
Now Try
1a. $(-\infty, 2) \cup (6, \infty)$ 1b. $(2, 6)$ 2. $(-1, 2)$

4a. $(-\infty, \infty)$ 4b. $\varnothing$

    Copyright © 2025 Pearson Education, Inc.

Practice Exercises

1. $[-1, 2]$

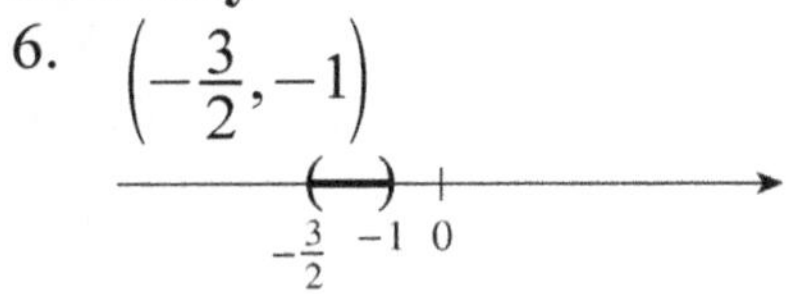

3. $\varnothing$

Objective 2

Now Try

5. $\left(-\infty, -\dfrac{3}{2}\right] \cup \left[-\dfrac{1}{3}, \dfrac{1}{2}\right]$

Practice Exercises

5. $(-\infty, -5] \cup [-3, 1]$

Objective 3

Now Try

6. $\left(-\dfrac{3}{2}, -1\right)$

7. $(-\infty, 3) \cup [8, \infty)$

Practice Exercises

7. $(-\infty, 1) \cup [8, \infty)$

9. $[-2, 3)$

Chapter 12 INVERSE, EXPONENTIAL, AND LOGARITHMIC FUNCTIONS

12.1 Inverse Functions

Key Terms
1. one-to-one function
2. inverse of a function

Objective 1
Now Try

1a. One-to-one; $G^{-1} = \{(2,\ 3),\ (-2,-3),\ (3,\ 2),\ (-3,-2)\}$

1b. Not one-to-one   1c. Not one-to-one

Practice Exercises
1. One-to-one; $\{(-1,-3), (2,-2), (3,-1), (4, 0)\}$

3. One-to-one; $\{(0, 0), (1, 1), (-1, -1), (2, 2), (-2, -2)\}$

Objective 2
Now Try
2a. One-to-one   2b. Not one-to-one

Practice Exercises
6. Not one-to-one

Objective 3
Now Try

3a. $f^{-1}(x) = \frac{1}{4}x + \frac{1}{4}$ 3b. Not one-to-one   3c. $f^{-1}(x) = \sqrt[3]{\dfrac{x+3}{2}}$

5. $f^{-1}(x) = x^2 - 7,\ x \geq 0$

Practice Exercises
7. $f^{-1}(x) = \dfrac{x+5}{2}$ 9. Not one-to-one

Objective 4
Now Try
5.

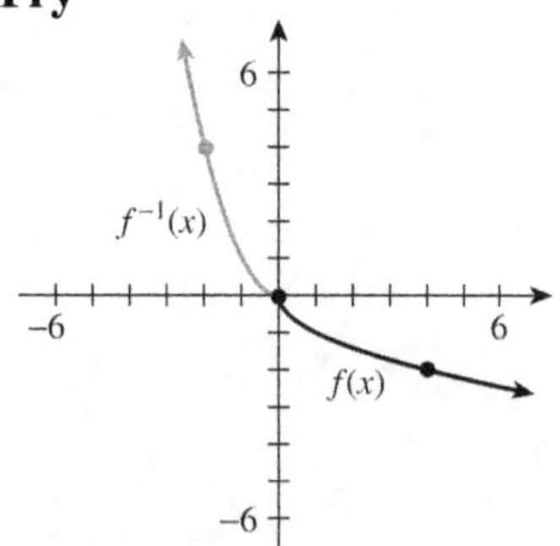

Practice Exercises
11. Not one-to-one

12.2 Exponential Functions

Key Terms
1. inverse   2. exponential equation   3. asymptote

Objective 1
Now Try
1a. 8.064   1b. 0.172   1c. 1.246

Practice Exercises
1. 3.737   3. 1.442

Objective 2
Now Try
2. 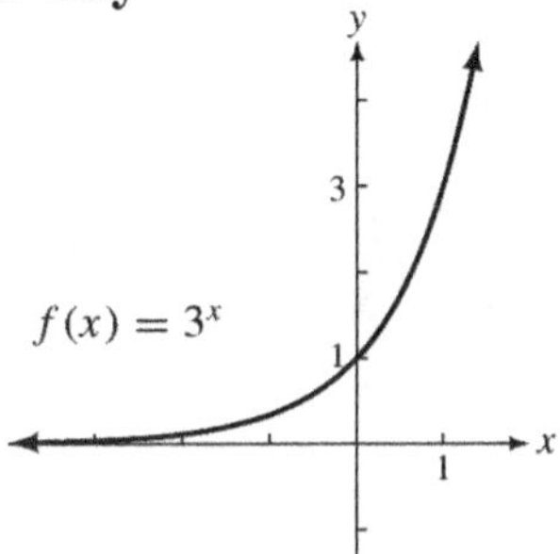   3. 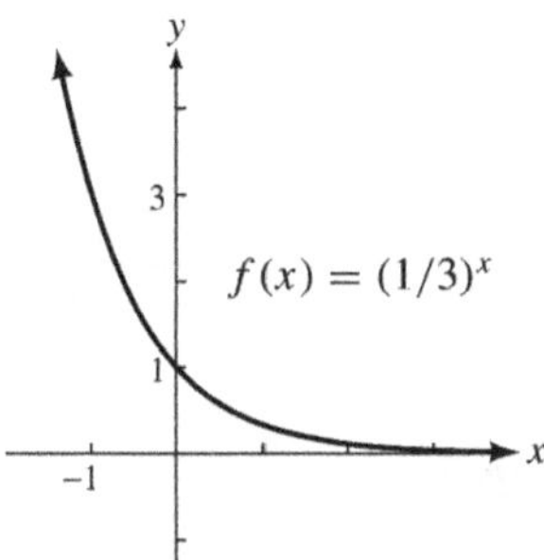

4. 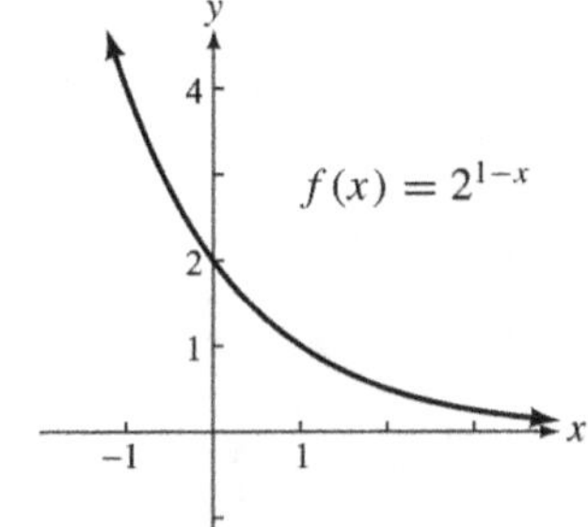

Practice Exercises
5. 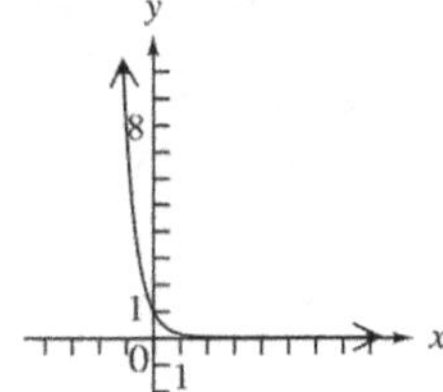

Objective 3
Now Try
5. $\left\{\dfrac{3}{2}\right\}$ 6a. $\{-2\}$ 6b. $\{-5\}$

6c. $\{-4\}$

Practice Exercises
7. $\left\{\dfrac{1}{2}\right\}$ 9. $\{-2\}$

Objective 4
  Now Try
  7. 256,000   8. about 32,656

  Practice Exercises
  11. 1 gram

12.3   Logarithmic Functions

Key Terms
  1. logarithm   2. logarithmic equation

Objective 1
Objective 2
  Now Try

  1a. $\log_8 64 = 2$ 1b. $16^{-1/2} = \dfrac{1}{4}$ 2a. 8

  2b. -5 2c. $\dfrac{1}{2}$ 2d. -4

  Practice Exercises
  1. $10^{-3} = 0.001$ 3. $\dfrac{1}{4}$

Objective 3
  Now Try
  3a. 1   3c. 0   3d. 0

  3e. 3   3g. 9

  Practice Exercises
  5. 3

Objective 4
  Now Try
  4a. $\left\{\dfrac{1}{64}\right\}$ 4b. $\{40\}$ 4c. $2\sqrt{3}$

  4d. $\left\{\dfrac{1}{6}\right\}$

  Practice Exercises
  7. $\left\{\dfrac{3}{5}\right\}$ 9. $\{16\}$

Objective 5
Now Try
5.

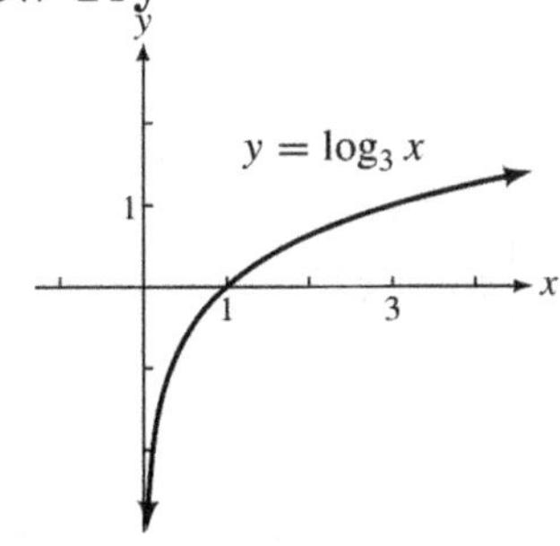

6.

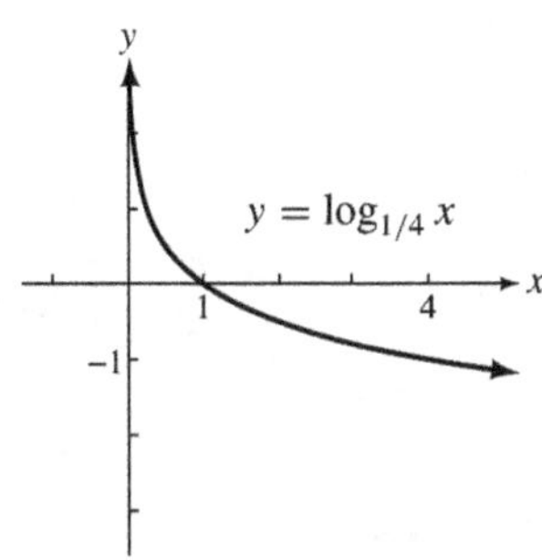

Practice Exercises
11.

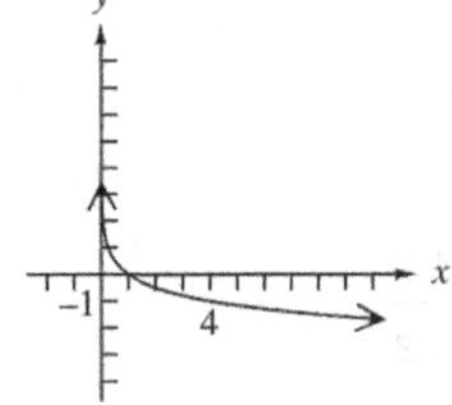

Objective 6
Now Try
7. 16 fish

Practice Exercises
13. 100 mites

12.4   Properties of Logarithms

Key Terms
1. special properties

2. quotient rule for logarithms

3. product rule for logarithms

4. power rule for logarithms

Objective 1
Now Try
1a. $\log_6 5 + \log_6 3$

1b. $\log_5 21$

1c. $1 + \log_4 x$

1d. $4 \log_5 x$

Practice Exercises
1. $\log_7 5 + \log_7 m$

3. $\log_4 21$

Objective 2
Now Try

2a. $\log_5 4 - \log_5 9$

2b. $\log_6 \dfrac{x}{3}$

2c. $2 - \log_4 11$

2d. $\left\{ \dfrac{1}{6} \right\}$

Practice Exercises
5. $\log_6 k - \log_6 3$

Objective 3
Now Try

3a. $3\log_6 4$

3c. $\frac{1}{2}\log_b 13$

3d. $\frac{3}{4}\log_3 x$

3e. $-7\log_3 x$

Practice Exercises
7. $7\log_m 2$

9. $\sqrt[3]{7}$

Objective 4
Now Try

4a. $2 + 3\log_6 x$

4b. $\frac{1}{2}(\log_b x - \log_b 3)$

4d. $\log_b \frac{xy^4}{z}$

4e. $\log_2 \frac{x^2(x-1)}{\sqrt[3]{x^2+1}}$

4f. cannot be rewritten

5a. 7.6293

5b. -3.4594

5c. 9.5097

6a. False

6b. False

Practice Exercises
11. 6.1700

12.5 Common and Natural Logarithms

Key Terms
1. natural logarithm

2. common logarithm

Objective 1
Now Try
1. 2.9928

Practice Exercises
1. 1.7576

3. 4.9401

Objective 2
Now Try
2. pH $= 7.2076$; rich fen

3. 2.5×10^{-4}

4. 85 dB

Practice Exercises
5. 134 dB

Objective 3
Now Try
5. 4.5850

Practice Exercises
7. 4.3347   9. 3.9120

Objective 4
  Now Try
6. 11.9 years

  Practice Exercises
11. 3240 years

Objective 5
  Now Try
7. 3.2266

  Practice Exercises
13. 1.1887   15. −2.3219

12.6   Exponential and Logarithmic Equations; Further Applications

Key Terms
  1. continuous compounding   2. compound interest

Objective 1
  Now Try
1. 2.465   2. 6.770

  Practice Exercises
1. {1}   3. {−4.324}

Objective 2
  Now Try
5. $-1+\sqrt[3]{36}$ 7. {1}

  Practice Exercises
5. {5}

Objective 3
  Now Try
8. $12,201.90   9. 13.89 years   10a. 5309.18
10b. 34.7 years

  Practice Exercises
7. $55,200.99   9. 7.7 years

Objective 4
  Now Try
11a. 11.3 mg   11b. 30.0 years

  Practice Exercises
11. 29 years

Chapter 13 NONLINEAR FUNCTIONS, CONIC SECTIONS, AND NONLINEAR SYSTEMS

13.1 Additional Graphs of Functions

Key Terms

1. step function
2. greatest integer function
3. square root function
4. reciprocal function
5. asymptotes
6. absolute value function

Objective 1

Now Try

1. 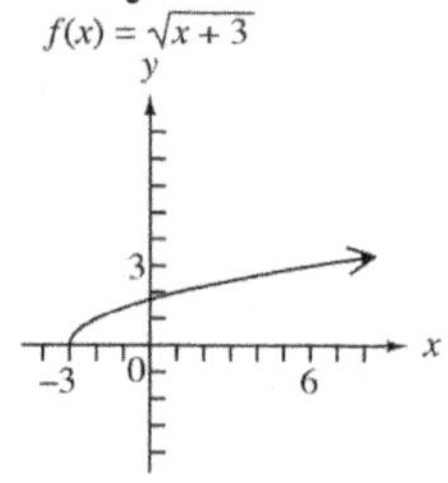

Domain: $[-3, \infty)$

Range: $[0, \infty)$

2. 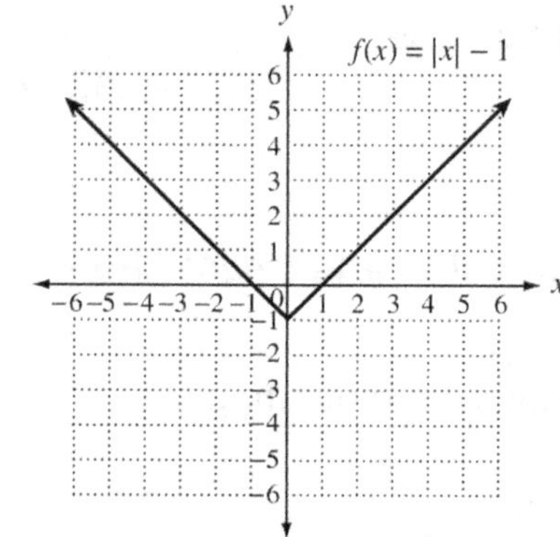

Domain: $(-\infty, \infty)$

Range: $[-1, \infty)$

3. 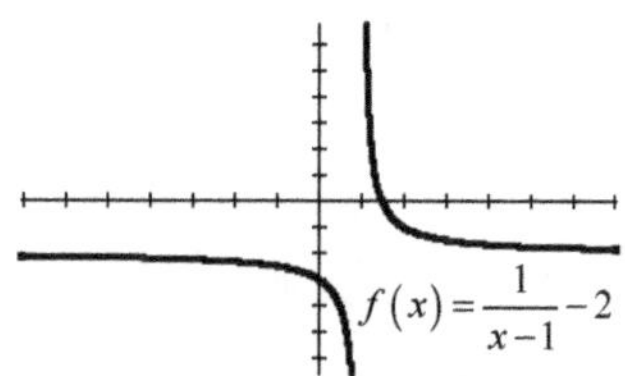

Domain: $(-\infty, 1) \cup (1, \infty)$

Range: $(-\infty, -2) \cup (-2, \infty)$

Practice Exercises

1. 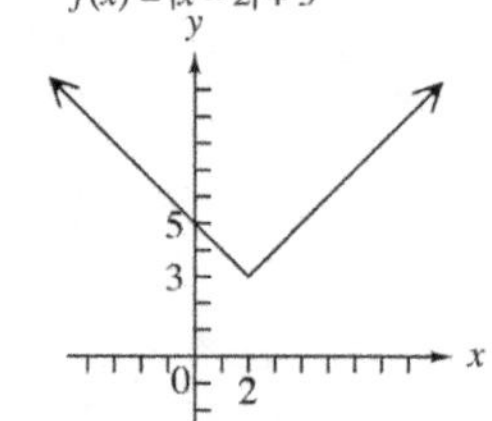

Domain: $(-\infty, \infty)$

Range: $[3, \infty)$

3. 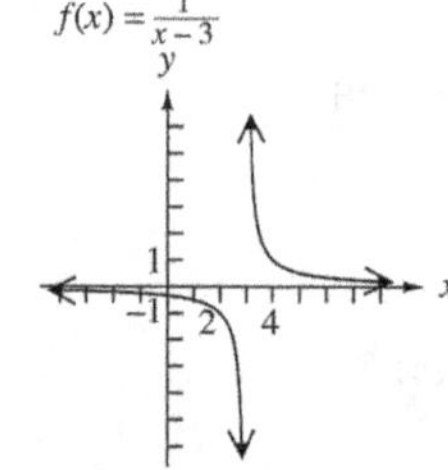

Domain: $(-\infty, 3) \cup (3, \infty)$

Range: $(-\infty, 0) \cup (0, \infty)$

Objective 2
Now Try

4d. 0   4e. 1   4f. −11

5.
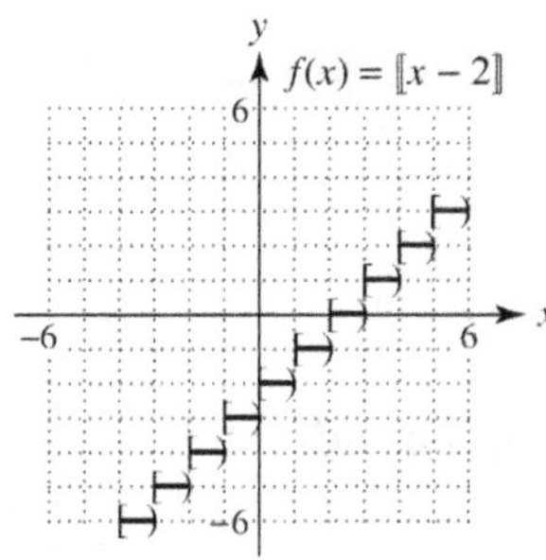

The domain is $(-\infty, \infty)$.

The range is $\{\ldots, -2, -1, 0, 1, 2, \ldots\}$ or the set of integers.

6.
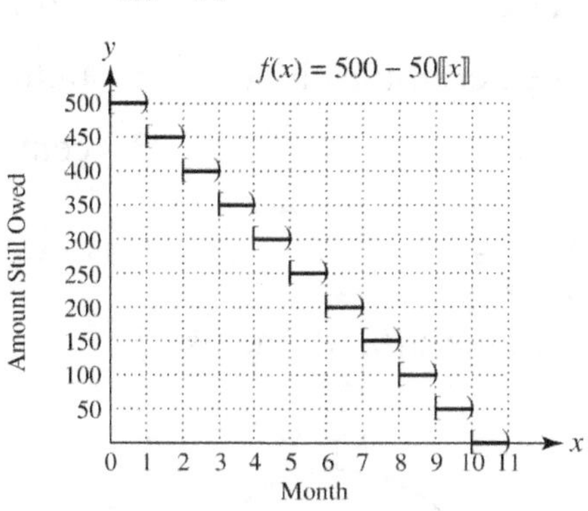

Practice Exercises

5.
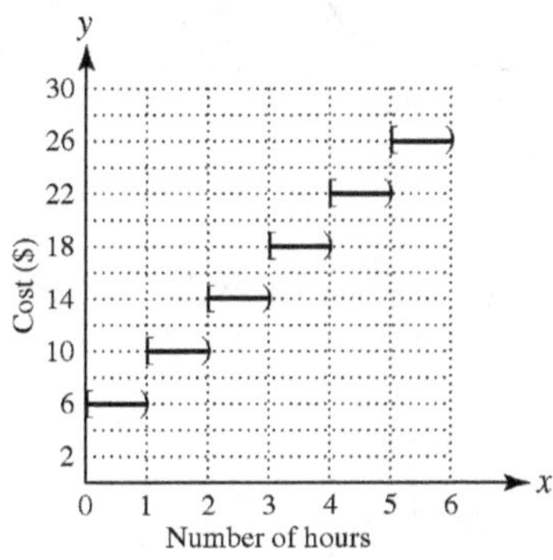

13.2 Circles Revisited and Ellipses

Key Terms

1. circle
2. ellipse
3. conic sections
4. center of the circle
5. radius
6. center-radius form
7. foci
8. center

Objective 1
Now Try

1a. Center: (0, 0); radius : 5

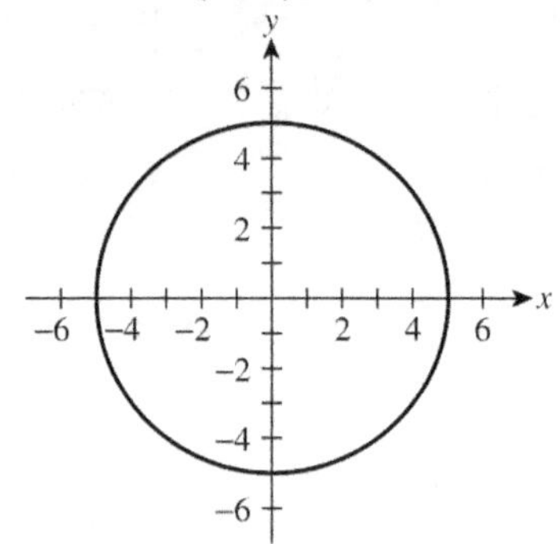

1b. Center: (−5, 4); radius: 4

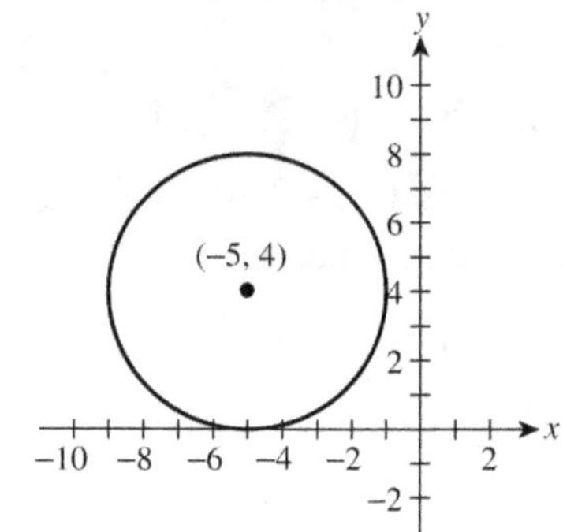

Practice Exercises

1. Center: (1, 4); radius 2

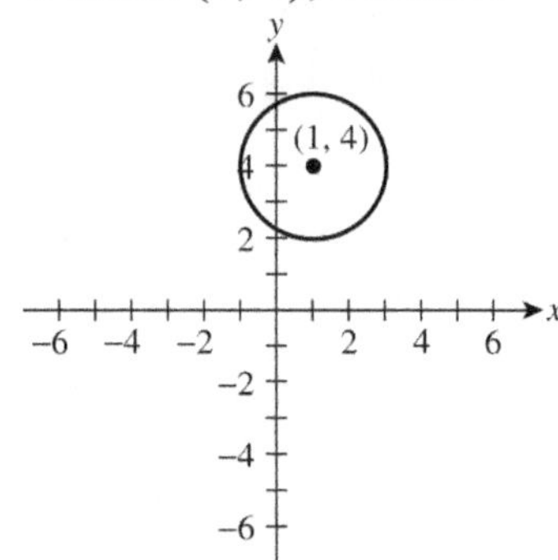

3. Center: (0, 5); radius: 3

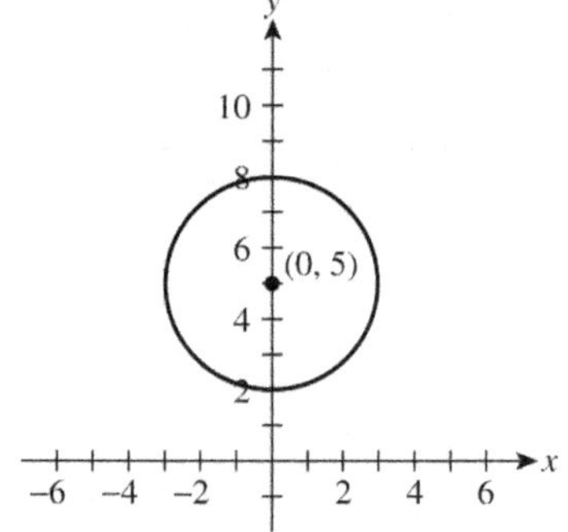

Objective 2
Now Try

2. $x^2 + (y-3)^2 = 2$

Practice Exercises

5. $(x-4)^2 + y^2 = 11$

Objective 3
Now Try

3. Center: (4, 1); radius: $\sqrt{2}$

Practice Exercises

7. Center: (2, −4); radius: 3

9. Center: (−5, −2); radius: 6

Objective 4

Objective 5

Now Try

4.

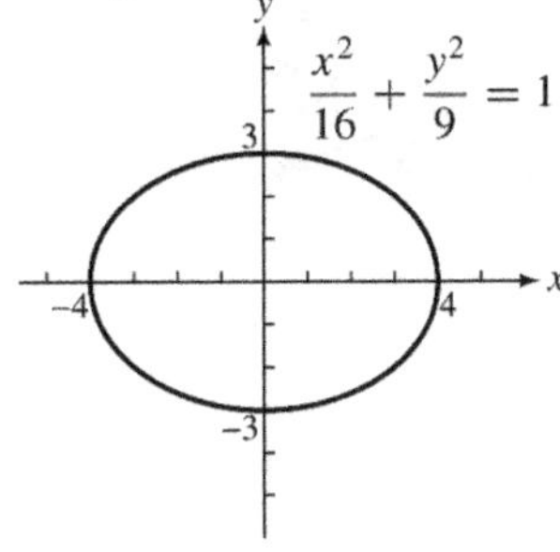

6.

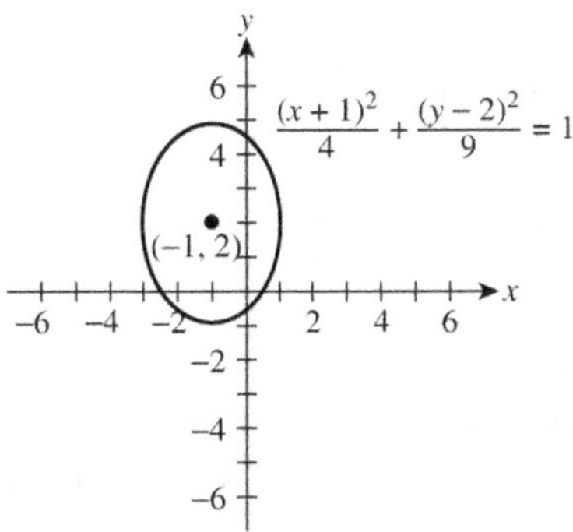

Practice Exercises

11.

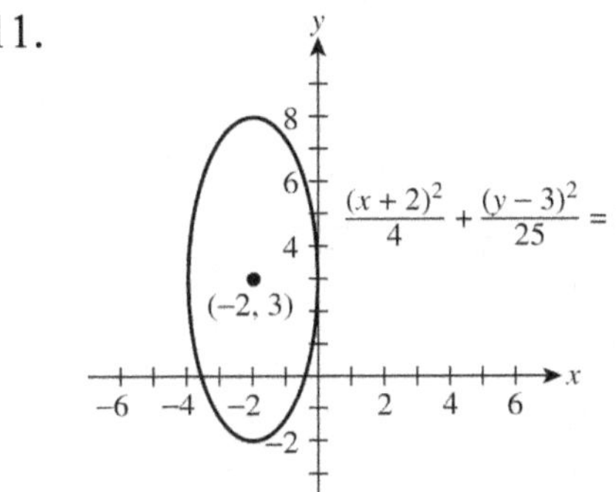

13.3   Hyperbolas and Functions Defined by Radicals

Key Terms

1. hyperbola
2. asymptotes
3. fundamental rectangle
4. transverse axis
5. generalized square root function

Objective 1

Objective 2

Now Try

1.

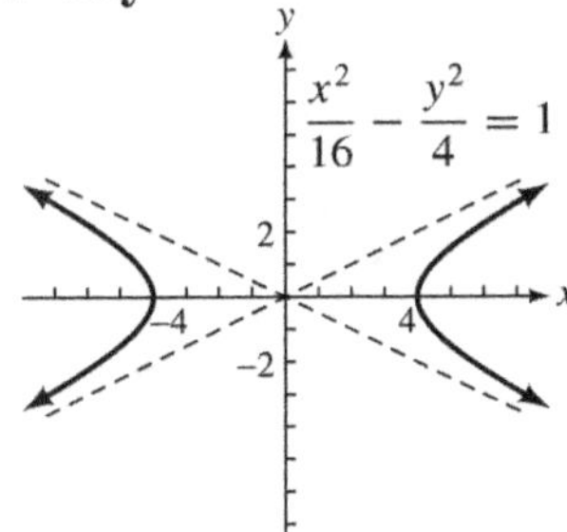

Practice Exercises

1.
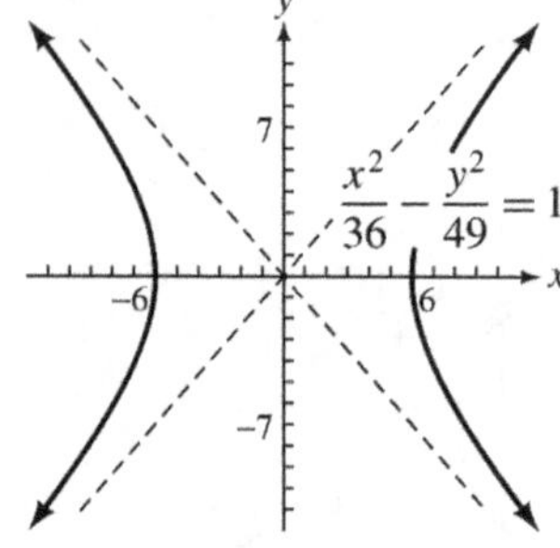

3.
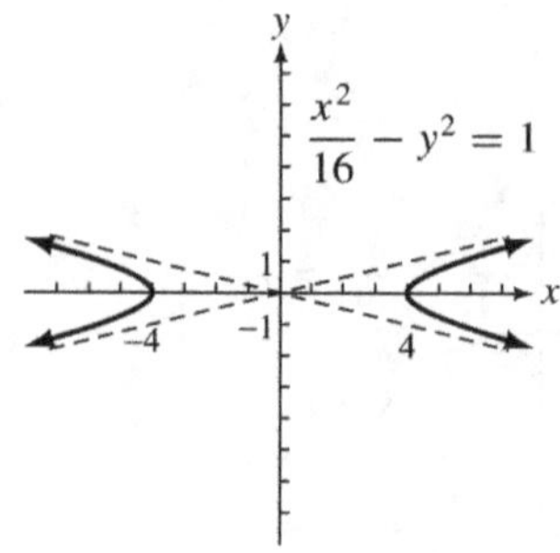

Objective 3
Now Try

3a. hyperbola   3b. ellipse   3c. parabola

Practice Exercises
5. hyperbola

Objective 4
Now Try

4.
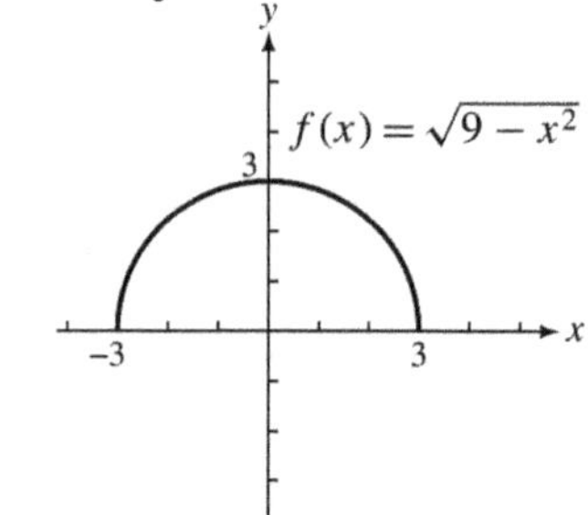

Domain: $[-3, 3]$
Range: $[0, 3]$

5.
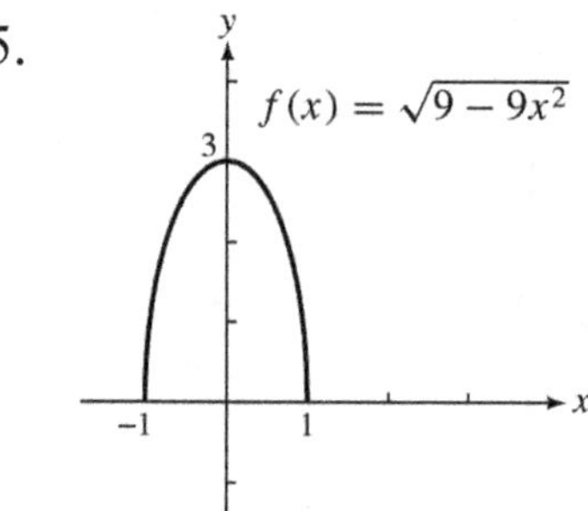

Domain: $[-1, 1]$
Range: $[0, 3]$

Practice Exercises

7.
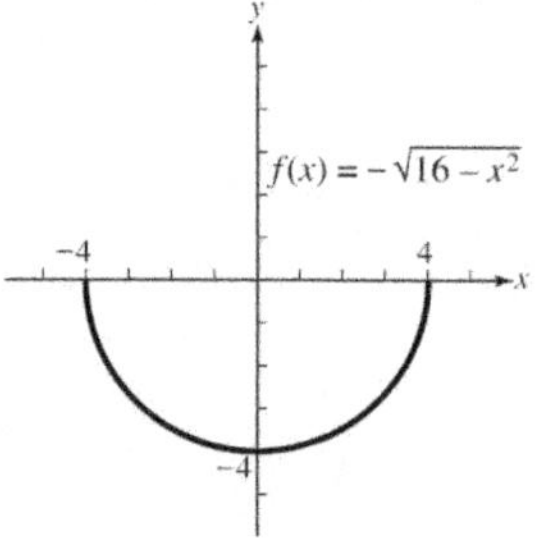

Domain: $[-4, 4]$
Range: $[-4, 0]$

9.
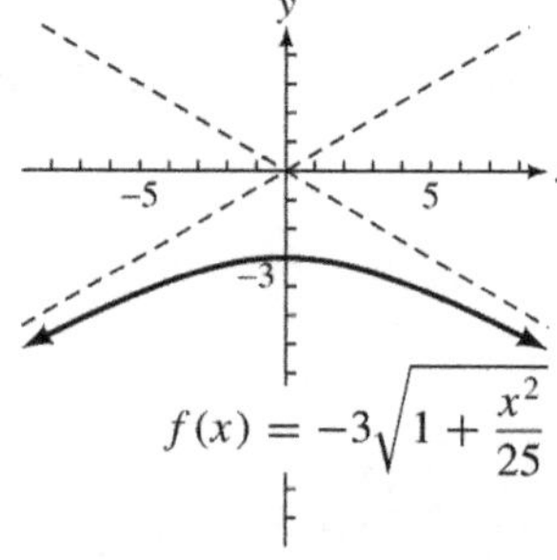

Domain: $(-\infty, \infty)$

Range: $(-\infty, -3]$

13.4   Nonlinear Systems of Equations

Key Terms

1. nonlinear equation   2. nonlinear system of equations

Objective 1

Now Try

1. $\{(-4, 1), (1,-4)\}$ 2. $\{(1, 1)\}$

Practice Exercises

1. $\{(12,-17), (2, 3)\}$ 3. $\{(3, 8), (-4,-6)\}$

Objective 2

Now Try

3. $\left\{(0,-4),\ (2\sqrt{3},\ 2),\ (-2\sqrt{3},\ 2)\right\}$

Practice Exercises

5. $\{(3, 1), (3,-1), (-3, 1), (-3,-1)\}$

Objective 3

Now Try

4. $\{(-4, 1), (-1, 4), (1,-4), (4,-1)\}$

Practice Exercises

7. $\{(2,-3), (-2, 3), (3,-2), (-3, 2)\}$

9. $\{(2, 2), (-2,-2), (2i,-2i), (-2i, 2i)\}$

13.5   Second-Degree Inequalities and Systems of Inequalities

Key Terms

1. system of inequalities   2. second-degree inequality

Objective 1

Now Try

1. 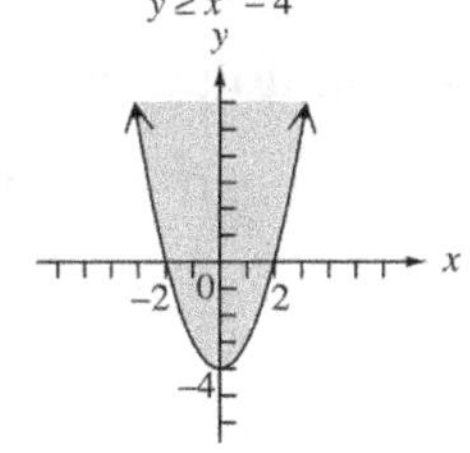

3. 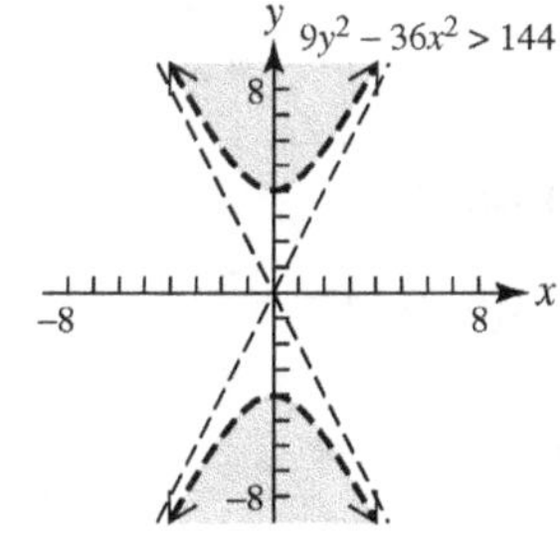

Practice Exercises

1. 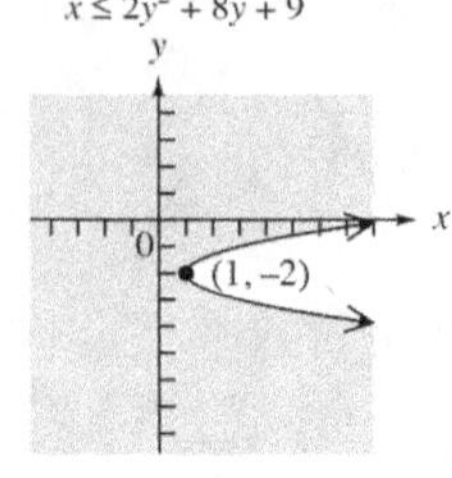

    $x \le 2y^2 + 8y + 9$

3. 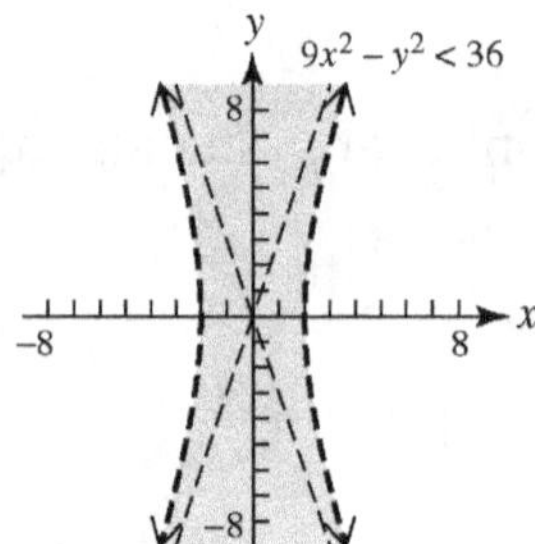

    $9x^2 - y^2 < 36$

Objective 2

Now Try

5. 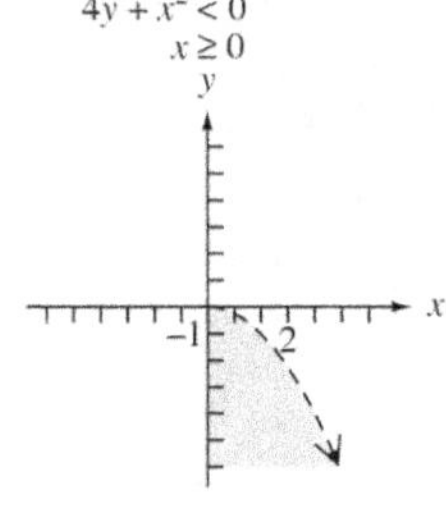

    $4y + x^2 < 0$
    $x \ge 0$

7. 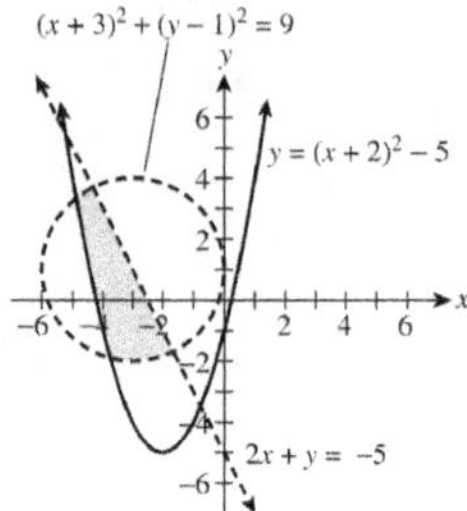

    $(x + 3)^2 + (y - 1)^2 = 9$
    $y = (x + 2)^2 - 5$
    $2x + y = -5$

Practice Exercises

5. $x^2 + y^2 \le 25$
    $3x - 5y > -15$

Chapter 14 FURTHER TOPICS IN ALGEBRA

14.1   Sequences and Series

Key Terms

1. infinite sequence   2. series   3. summation notation

4. index of summation   5. general term   6. arithmetic mean (average)

7. terms of a sequence   8. finite sequence

Objective 1

Objective 2

Now Try

1. $\dfrac{3}{5}, \dfrac{4}{10} = \dfrac{2}{5}, \dfrac{5}{15} = \dfrac{1}{5}, \dfrac{6}{20} = \dfrac{3}{10}, \dfrac{7}{25}$

Practice Exercises

1. $-1, 1, -1, 1, -1$ 3. $\dfrac{11}{21}$

Objective 3
Now Try

2. $a_n = (-1)^{n-1} 2^n$

Practice Exercises

5. $a_n = \left(\sqrt{3}\right)^n$

Objective 4
Now Try

3. Month 1: \$620; Month 2: \$607.60; Month 3: \$595.45; Month 4: 583.54
    Remaining balance: \$3593.41

Practice Exercises

7. 16 g

Objective 5
Now Try

4. $1 + 5 + 9 + 13 + 17 + 21 + 25 = 91$

Practice Exercises

9. $5 + 7 + 9 + 11 = 32$ 11. $2 + 5 + 10 + 17 + 26 + 37 = 97$

Objective 6
Now Try

5. $\displaystyle\sum_{i=1}^{6} 4i$

Practice Exercises

13. $\displaystyle\sum_{i=1}^{5} \dfrac{1}{3i+2}$

Objective 7
Now Try

6. About 3.38 in.

Practice Exercises

15. 8

14.2   Arithmetic Sequences

Key Terms
1. arithmetic sequence (arithmetic progression)   2. common difference

Objective 1
  Now Try
1. 2   2. $9, 4, -1, -6, -11$

  Practice Exercises
1. 4   3. $6, 3, 0, -3, -6$

Objective 2
  Now Try
3. $a_n = 31 + 4n$; $a_{15} = 91$

  Practice Exercises
5. $a_n = \dfrac{1}{3}n + \dfrac{1}{6}$

Objective 3
  Now Try
4. \$41,800

  Practice Exercises
7. \$10,900   9. \$16.75

Objective 4
  Now Try
5a. 93   5b. -44 6. 16

  Practice Exercises
11. $\dfrac{41}{2}$

Objective 5
  Now Try
7. 39   8. 806   9. 365

  Practice Exercises
14. 80

14.3   Geometric Sequences

Key Terms

1. common ratio   2. annuity

3. geometric sequence (geometric progression)   4. future value of an annuity

5. ordinary annuity   5. term of the annuity   6. payment period

Objective 1

Now Try

1. $\dfrac{1}{4}$

Practice Exercises

1. $\sqrt{2}$

Objective 2

Now Try

2. $a_n = \left(\sqrt{3}\right)^n$

Practice Exercises

3. $a_n = 6 \cdot 2^{n-1}$ 5. $a_n = -4\left(\dfrac{1}{5}\right)^n$

Objective 3

Now Try

3a. -729 3b. $\dfrac{1}{384}$ 4. $0.8, -4, 20, -100, 500$

Practice Exercises

7. -162

Objective 4

Now Try

5. $\dfrac{728}{3}$ 6. 682

Practice Exercises

9. 2186 11. -258

Objective 5

Now Try

7a. $328,988.05   7b. $132,877.70

Practice Exercises

13. $3464.34

Objective 6
  Now Try
  8. $\dfrac{25}{6}$ 9. $|r| > 1,$ so the sum does not exist.

  Practice Exercises
  13. $\dfrac{3}{2}$

14.4   The Binomial Theorem

Key Terms
  1. binomial theorem (general binomial expansion)   2. Pascal's triangle

Objective 1
  Now Try
  1. 3,628,800   2a. 1   2b. 36

2c. 9   2d. 126   3. 45

  4. $81r^4 - 216r^3 s + 216r^2 s^2 - 96rs^3 + 16s^4$

  5. $-\dfrac{x^5}{32} + \dfrac{5}{8}x^4 - 5x^3 + 20x^2 - 40x + 32$

  Practice Exercises
  1. 35

  3. $32y^5 + 240y^4 z + 720y^3 z^2 + 1080y^2 z^3 + 810yz^4 + 243z^5$

Objective 2
  Now Try
  6. $-5103m^6 r^5$

  Practice Exercises
  5. $-4320x^2 y^3$

Beginning and Intermediate Algebra Integrated Review Answers

Chapter 1 The Real Number System

Key Terms

1. number line   2. sum   3. addends

4. factors   5. product   6. order of operations

7. exponent

Objective 1
Now Try
1. a. >   1b. >   1c. >

Practice Exercises
1. >   3. <

Objective 2
Now Try
2. −5   3a. −36   3b. 40

4a. −16   4b. 21

Practice Exercises
5. −15

Objective 3
Now Try
5. −38

Practice Exercises
7. 0   9. −22

Objective 4
Now Try
6a. −42   6b. 60   6c. −64   6d. −54

Practice Exercises
11. −62

Objective 5
Now Try
7a. −17   7b. 16   7c. −53   7d. −24

Practice Exercises
13. 17   15. −16

Objective 6
Now Try

8. The coefficient of the variable h is 1 and the constant term is -6.

Practice Exercises

17. The coefficient of the variable w is -2.

Objective 7

Now Try

9a and b.   10a and b.

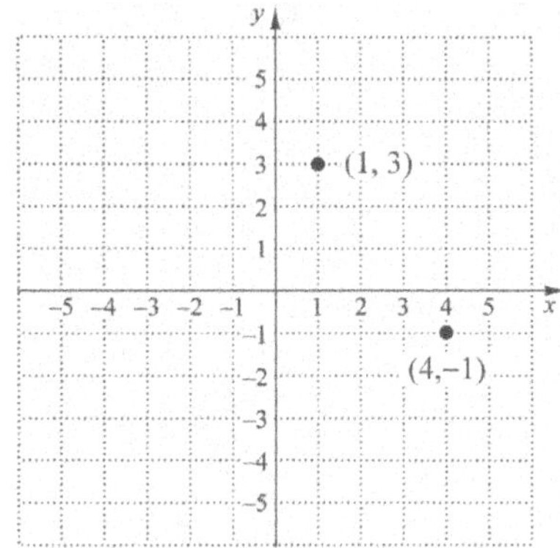 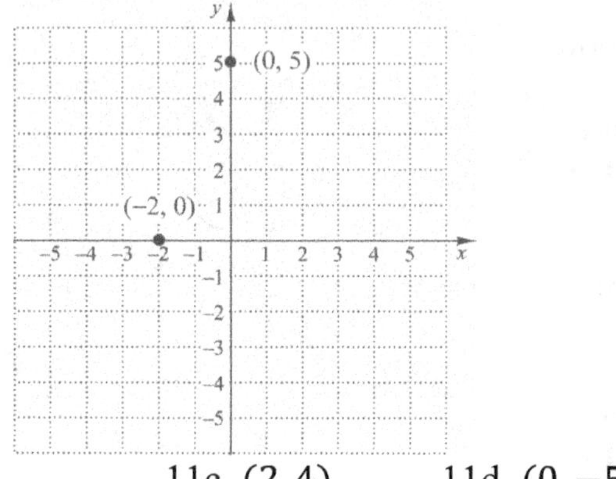

11a. $(-4, -4)$ 11b. $(1, -3)$ 11c. $(2, 4)$ 11d. $(0, -5)$

Practice Exercises

19a -c.

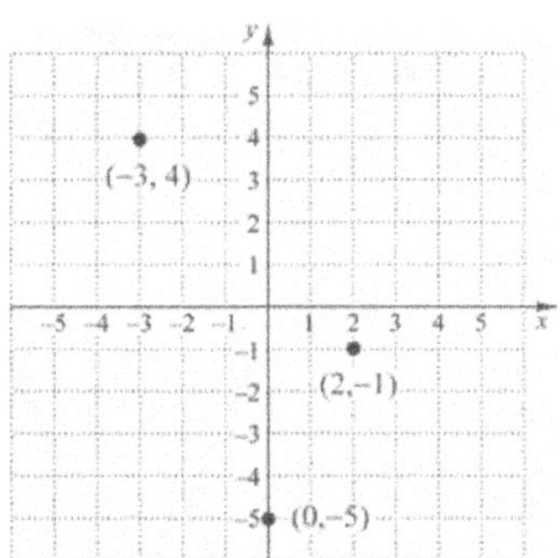

Chapter 2 Linear Equations and Inequalities in One Variable

Key Terms

1. like terms   2. term   3. simplify an expression

4. variable term   5. constant term   6. evaluate an expression

7. increase by, sum   8. product, double   9. quotient, per

10. less than, difference

Objective 1

Now Try

1a. 29   1b. 50   2. 32 yards   3. 96

Practice Exercises
1a. 140 Newtons   1b. 920 Newtons   3a. $6000   3b. $7484

Objective 2
  Now Try
  4. $20x$

  Practice Exercises
  5. $7d$

Objective 3
  Now Try
  5a. 43%   5b. 230%   5c. 75.1%

  Practice Exercises
  7. 20%   9. 493%

Objective 4
  Now Try
  6a. $28x - 112$ 6b. $30x + 12$ 6c. $-54x - 36$ 7. $6x - 6$

  Practice Exercises
  11. $-24s + 32$

Objective 5
  Now Try
  8. $93\frac{3}{4}\%$

  Practice Exercises
  13. 94%   15. $85\frac{1}{3}\%$

Objective 6
  Now Try
  9. $-8x - 2$

  Practice Exercises
  17. $-42m + 43$

Objective 7
  Now Try
  10a. $x + (-20)$ 10b. $48 - x$ 11. $6x - 81$

  Practice Exercises
  19. $-6x$ 21. $1 + 3x$

Chapter 3 Linear Equations in Two Variables

Key Terms
  1. variable   2. constant   3. algebraic expression   4. number line

  5. coordinate   6. linear inequality   7. Inequalities   8. interval

Objective 1
  Now Try
  1. 29

  Practice Exercises
  1. $r = 12$ 3. $y = 7$

Objective 2
  Now Try
  2. $x = -6$ 3. $x = 8$ 4. $a = 15/2$

  Practice Exercises
  5. $a = 4$

Objective 3
  Now Try
  5. 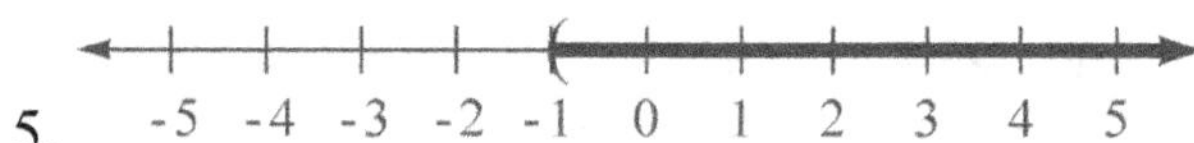

  Practice Exercises

  7. $(3, \infty)$; 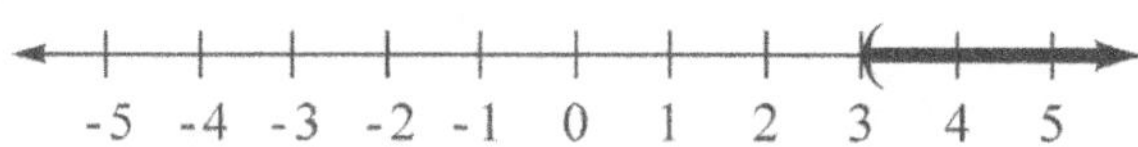

  9. $(-\infty, -4)$; 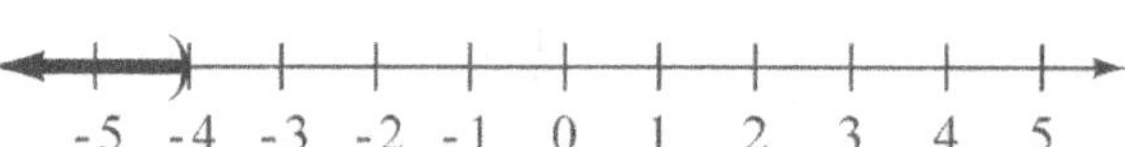

Objective 4
  Now Try
  6. $[2, \infty)$ 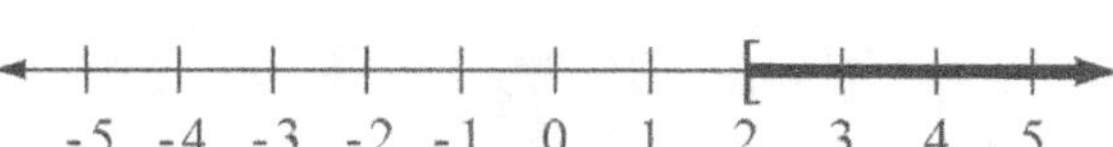

  Practice Exercises

  11. $(-\infty, 5]$; 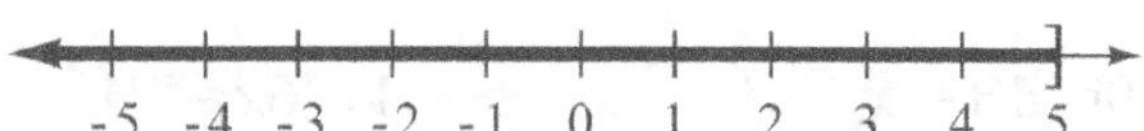

Chapter 4 Exponents and Polynomials

Key Terms
  1. base   2. Exponential expression   3. like terms   4. power

  5. exponential expression   6. base

Objective 1
Now Try
1a. base: 2, exponent: 6, value: 6   1b. base: 2, exponent: 6, value: -64

1c. base: -2, exponent: 6, value: 64

Practice Exercise
1. $\left(\frac{1}{3}\right)^5 = \frac{1}{243}$ **3.** base: 3, exponent: 8, value: -6561

Objective 2
Now Try
2. $18x^{11}$

Practice Exercises
5. $8c^{15}$

Objective 3
Now Try
3. n^{30}

Practice Exercises
7. 7^{12} 9. 2^{21}

Objective 4
Now Try
4a. $5x^4$ 4b. $-8x^7$ 4c. $44m^2$ 4d. $5p^2q$

Practice Exercises
11. $-0.9z^7 + 2.6z^8$

Chapter 5 Factoring and Applications

Key Terms
1. factorizations   2. composite number   3. prime factorization

4. prime number   5. factors

Objective 1
Now Try
1. $2 \cdot 2 \cdot 2 \cdot 3$ 2. $2 \cdot 2 \cdot 2 \cdot 7$ 3. $2 \cdot 3 \cdot 5 \cdot 5$

Practice Exercises
1. $2 \cdot 2 \cdot 7$ 3. $2 \cdot 3 \cdot 3 \cdot 5 \cdot 5$

Objective 2
Now Try
4. $32x^4 + 64x^3$

Practice Exercises
5. $6m + 14m^3 + 6m^4$

Objective 3

Now Try

5. $4x^7 - 2x^5 + 10x^4 - 18x^2 + 9x$　　　　　　6. $28x^4 - 33x^3 + 33x^2 + 53x + 15$

7. $2x^5 - 6x^4 + 10x^3 - 29x^2 + 5$

Practice Exercises

7. $x^3 + 27$　　　　　9. $6x^4 + 11x^3 - 9x^2 - 4x$

Chapter 6 Rational Expressions and Applications

Key Terms

1. lowest terms　　　　　2. common factor　　　　　3. equivalent fractions

4. mixed number　　　　　5. improper fraction　　　　　6. unlike fractions

7. like fractions　　　　　8. least common denominator

Objective 1

Now Try

1a. $\dfrac{2}{3}$　　　　　1b. $\dfrac{3}{5}$

Practice Exercises

1. $\dfrac{2}{7}$　　　　　3. $\dfrac{5}{7}$

Objective 2

Now Try

2a. $\dfrac{5}{6x}$　　　　　2b. $\dfrac{1}{3}$　　　　　2c. $\dfrac{b^2}{4a}$

Practice Exercises

5. $\dfrac{2c}{5}$

Objective 3

Now Try

3a. $\dfrac{-40}{143}$　　　　　3b. $\dfrac{-2}{3}$　　　　　3c. $\dfrac{3}{20}$

Practice Exercises

7. $\dfrac{-1}{7}$　　　　　9. 45

Objective 4
  Now Try

4a. $\dfrac{14}{15}$

4b. $\dfrac{4c}{5}$

  Practice Exercises

11. $\dfrac{5}{2m}$

Objective 5
  Now Try

5a. $b = 21$

5b. $x = -30$

5c. $n = \dfrac{2}{5}$

  Practice Exercises

13. $b = -36$

15. $h = \dfrac{3}{16}$

Objective 6
  Now Try

6a. $\dfrac{17}{18}$

6b. $-\dfrac{1}{6}$

6c. $\dfrac{41}{72}$

Objective 7
  Now Try

7a. $\dfrac{10+3x}{15}$

7b. $\dfrac{5x-64}{8x}$

Practice Exercises

19. $\dfrac{15+3n}{5n}$

21. $\dfrac{12-ab}{4a}$

Chapter 7 Linear Equations, Graphs, and Systems

Key Terms
  1. y-intercept

2. x-intercept

3. parallel lines

  4. graphing

5. graph

6. perpendicular lines

Objective 1
Now Try
1.

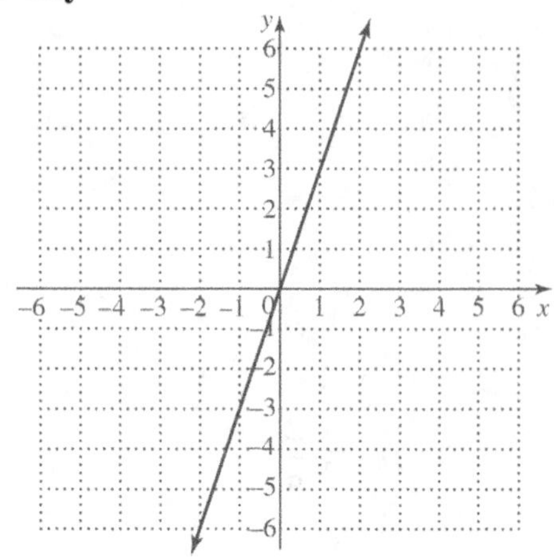

Practice Exercises
1. $-3x - 2y = 0$ 3. $y = 2x$

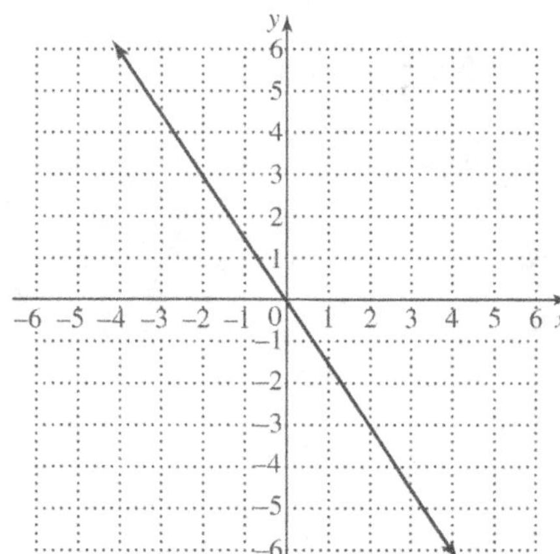 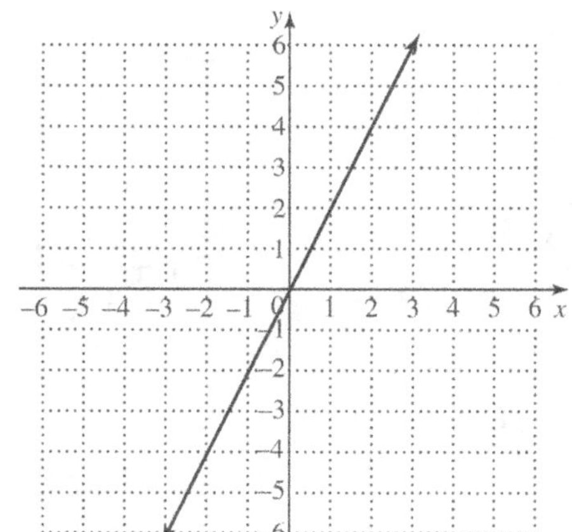

Objective 2
Now Try
1. positive

Practice Exercises
5. negative

Objective 3
Now Try
3a. parallel   3b. perpendicular

Practice Exercises
7. parallel   9. perpendicular

Objective 4
Now Try
4a. $y = \frac{3}{4}x - 5$ 4b. $y = 6x + 10$

Practice Exercises
11. $y = 2x$

Objective 5
Now Try
5. not a solution

Practice Exercises
13. not a solution   15. not a solution

Objective 6
Now Try
6. $(1, 6)$ 7. $(9, 4)$

Practice Exercises
17. $(2, 7)$

Chapter 8 Inequalities and Absolute Value

Key Terms
1. equivalent equations   2. linear equation   3. solution set

Objective 1
Now Try
1. $t = 29$ 2. $x = 8$

Practice Exercises
1. $t = -7$ 3. $p = 7.2$

Objective 2
Now Try
3. $m = 4$ 4. $x = -4$

Practice Exercises
5. $w = -5$

Objective 3
Now Try
5. $[2, \infty)$

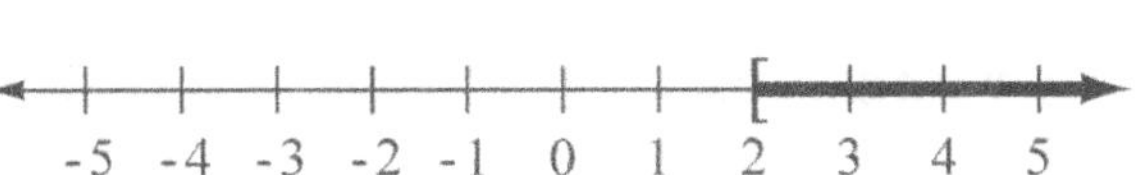

Practice Exercises

7. $(4, \infty)$;

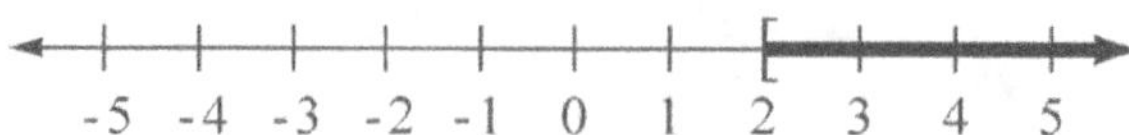

9. $[2, \infty)$;

Objective 4
Now Try

6.

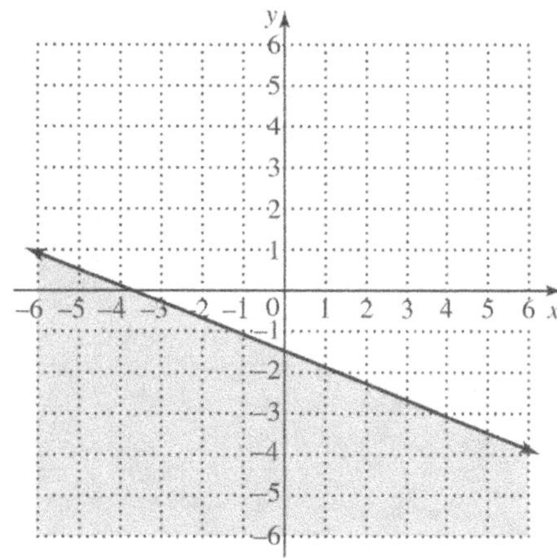

7.

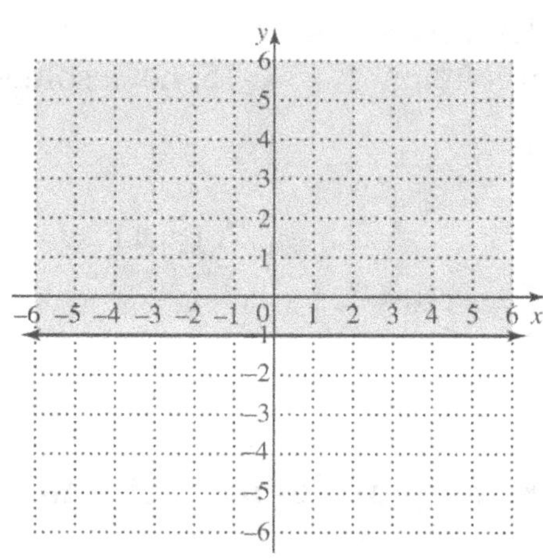

Practice Exercises

2.

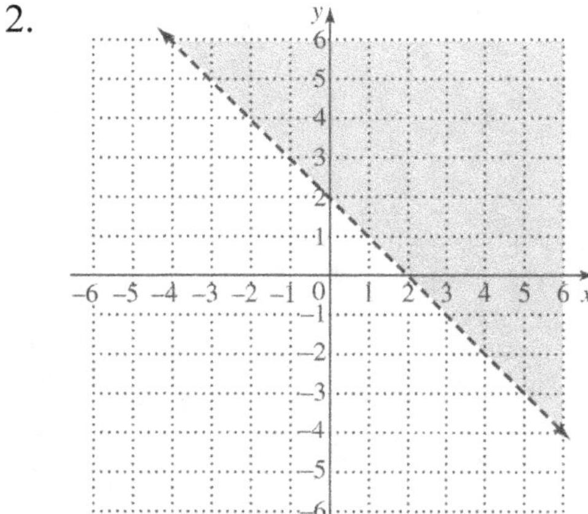

Chapter 9 Relations and Functions

Key Terms

1. constant of variation   2. relation   3. range   4. function

5. direct variation   6. components   7. inverse variation   8. domain

Objective 1
Now Try
1a. not a function   1b. function   1c. function

Practice Exercises
1. not a function   3. yes

Objective 2
Now Try
2a. 5   2b. -10 2c. -15

Practice Exercises
5. (a) 6; (b) 2 ; (c) 18

Objective 3
Now Try
3. Domain: $(-\infty, \infty)$, Range: $(-\infty, \infty)$

Practice Exercises
7. Domain: $(-\infty, \infty)$, Range: $(-\infty, \infty)$ 9. Domain: $(-\infty, \infty)$, Range: $(0, \infty)$

Objective 4
Now Try
4a. inverse variation   4b. direct variation.

Practice Exercises
11. direct variation

Objective 5
Now Try
5. 168   6. 12.8 cm

Practice Exercises
13. 12   15. 275 mi

Objective 6
Now Try
7. 5   8. 7.2 pounds per square foot

Practice Exercises
17. 2.25

Chapter 10 Roots, Radicals, and Root Functions

Key Terms

1. radicand   2. perfect square   3. index

4. square root   5. radical expression   6. unlike radicals

7. like radicals   8. Principal square root   9. irrational number

10. radical   11. cube root

Objective 1
  Now Try

1. 9 and -9   2a. 13   2b. $-\dfrac{3}{7}$

3a. 19   3b. $n^2 + 5$

  Practice Exercises

1. 25   3. $x^2 - 5$

Objective 2
  Now Try

4a. 7   4b. -5 4c. $\dfrac{4}{3}$ 5a. 6   5b. -5

  Practice Exercises

5. 4

Objective 3
  Now Try

6a. $2\sqrt{3}$ 6b. $2\sqrt{7}$ 6c. $4\sqrt{5}$ 7a. $15\sqrt{2}$ 7b. $6\sqrt{2}$

  Practice Exercises

7. $9\sqrt{5}$ 9. $12\sqrt{3}$

Objective 4
  Now Try

8a. $8\sqrt{6}$ 8b. $33\sqrt{6}$ 8c. $23\sqrt[3]{2}$

  Practice Exercises

11. $-\sqrt[3]{2}$

Objective 5
  Now Try

9a. $2\sqrt{7}$ 9b. $\dfrac{\sqrt{3}}{3}$

  Practice Exercises

13. $\dfrac{3\sqrt{10}}{2}$ 15. $\dfrac{3}{5}$

Chapter 11 Quadratic Equations, Inequalities, and Functions

Key Terms

1. equivalent equations

2. linear equation

3. coefficient

4. quadratic equation

5. greatest common factor

6. prime polynomial

7. trinomial

8. standard form

9. double solution

10. solution set

11. factoring

Objective 1
Now Try
1. 49

Practice Exercises
1. 27   3. 0.16

Objective 2
Now Try
2. $x = -6$ 3. $x = 8$ 4. $a = \frac{15}{2}$

Practice Exercises
5. $a = 4$

Objective 3
Now Try
5. $(x + 8)(x + 3)$ 6. prime   7. $(p - 7q)(p + 2q)$

Practice Exercises
7. prime   9. $(x - 11)(x + 3)$

Objective 4
Now Try
8. $(3x + 1)(5x + 7)$ 9. $(4x + 7)(2x - 3)$ 10. $(6x - 5y)(4x + 3y)$

11. $-6a(3a - 5)(a - 2)$

Practice Exercises
11. $(3a + 2b)(a + 2b)$

Objective 5
Now Try
12a. $x = -12, \frac{7}{4}$ 12b. $x = 0, \frac{11}{6}$ 13. $p = 3, \frac{4}{5}$

Practice Exercises
13. $x = 4, \frac{-5}{2}$ 15. $c = -4, \frac{3}{5}$

Chapter 12 Inverse, Exponential and Logarithmic Functions

Key Terms

1. components   2. quadratic equation   3. extraneous solution

4. radical equation   5. zero-factor property   6. relation

7. range   8. function   9. domain

Objective 1

Now Try

1a. $x = 9 \text{ or } -9$ 1b. $x = \pm\sqrt{23}$ 1c. $x = \pm 3\sqrt{3}$

1d. $x = \pm\sqrt{2}$ 2. no real solution

Practice Exercises

1. $r = 300 \text{ or } -300$ 3. no real solution

Objective 2

Now Try

1. $p = 11$ 4. $x = 3$

Practice Exercises

5. $k = \dfrac{2}{5}$

Objective 3

Now Try

5a. no   5b. yes   5c. yes

Practice Exercises

7. no   9. yes

Objective 4

Now Try

6. $D: (-\infty, \infty), R: (-\infty, \infty)$

Practice Exercises

11. $D: (-\infty, \infty), R: (-\infty, \infty)$

Chapter 13 Nonlinear Functions, Conic Sections, and Nonlinear Systems

Key Terms

1. perfect square trinomial   2. square root property   3. completing the square

4. point-slope form   5. standard form   6. slope-intercept form

7. line of symmetry 8. vertex 9. axis of symmetry

10. parabola 11. solution set of a system of linear inequalities

12. system of linear inequalities

Objective 1
Now Try
1. $\left(\frac{1}{2}, 2\right)$ 2. 13

Practice Exercises
1. Midpoint: $(1, 3)$, Distance: $4\sqrt{2}$ 3. Midpoint: $\left(\frac{3}{2}, -\frac{3}{2}\right)$, Distance: $\sqrt{26}$

Objective 2
Now Try
3a. $+ 25, (x + 5)^2$ 3b. $+ 121, (x - 11)^2$ 4. $x = -5 \pm 4\sqrt{2}$

Practice Exercises
5. $x = 2 \pm 2\sqrt{6}$

Objective 3
Now Try
5.

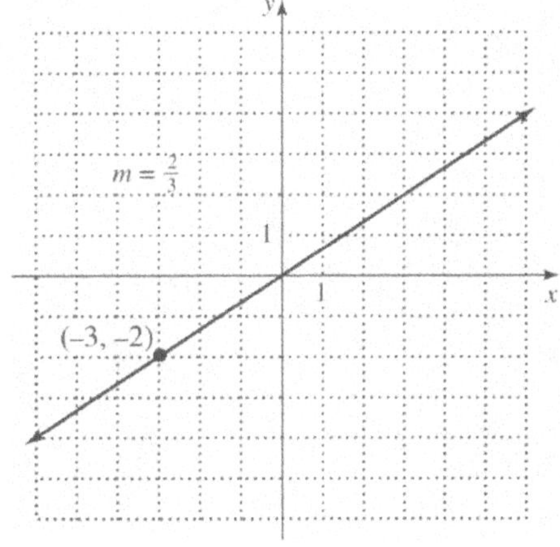

Practice Exercises

7.

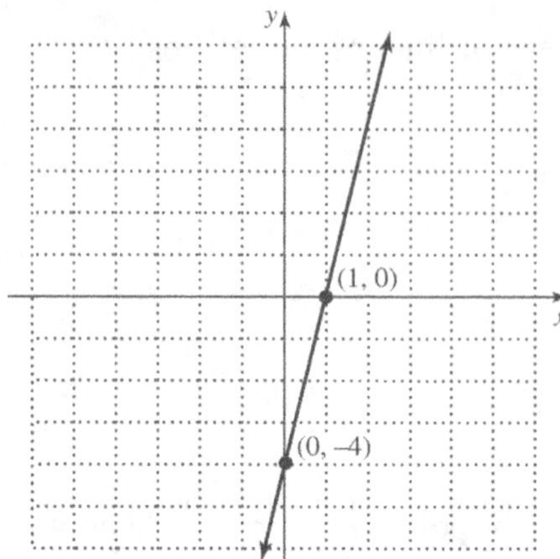

9.

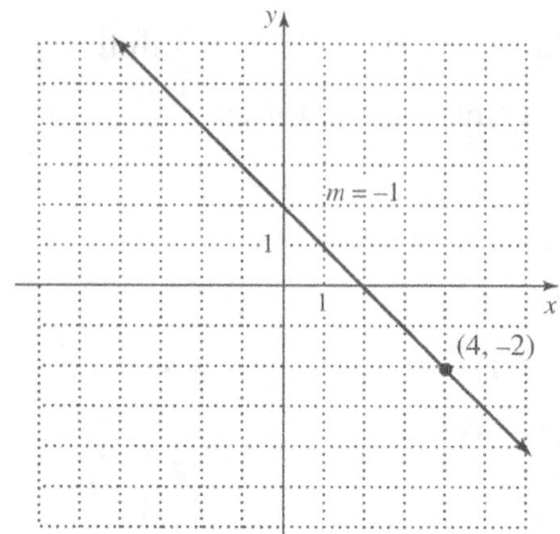

Objective 4

Now Try

6.

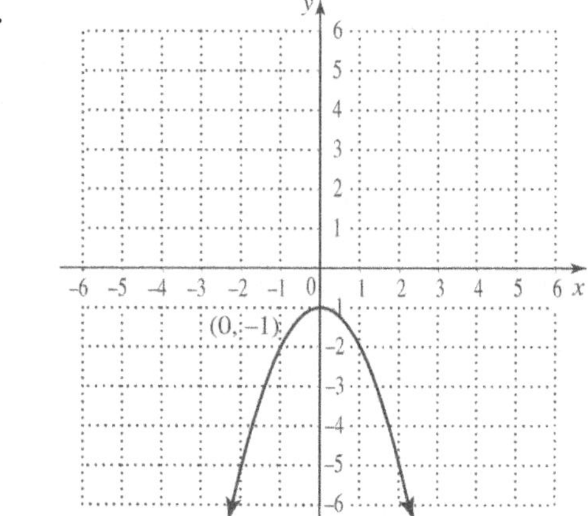

Vertex: $(0, -1)$

Practice Exercises

11.

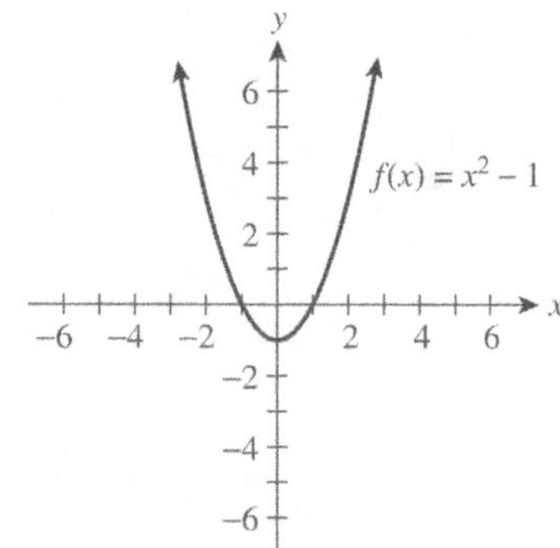

Vertex: $(0, -1)$

Objective 5
Now Try

7.
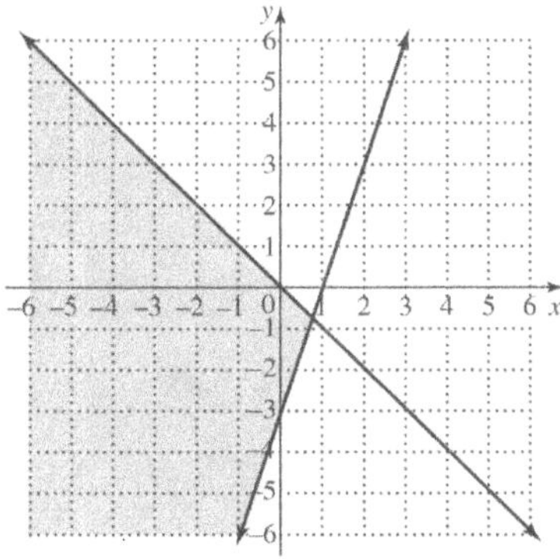

8.
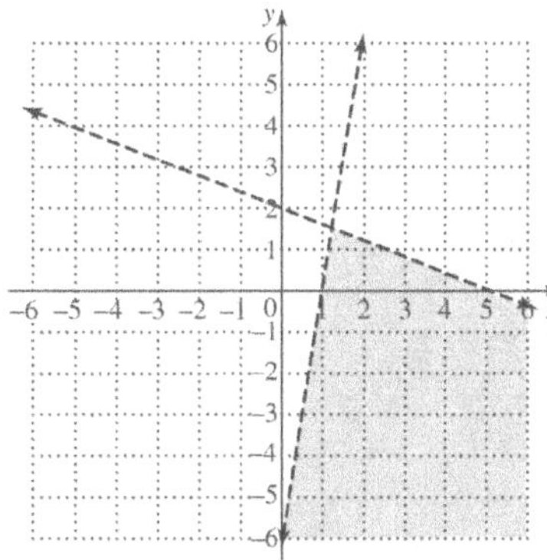

9.
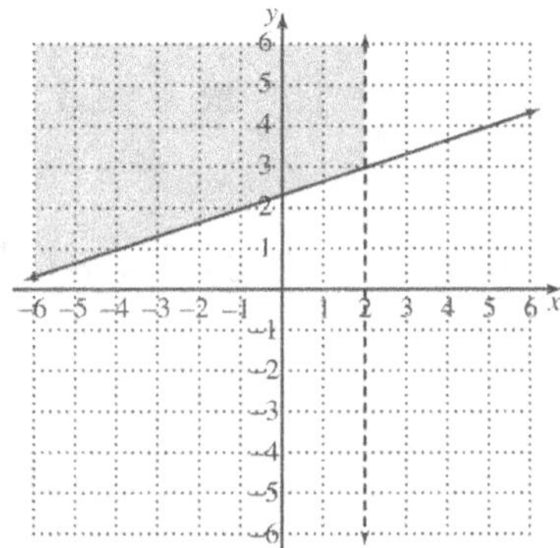

Practice Exercises

13.
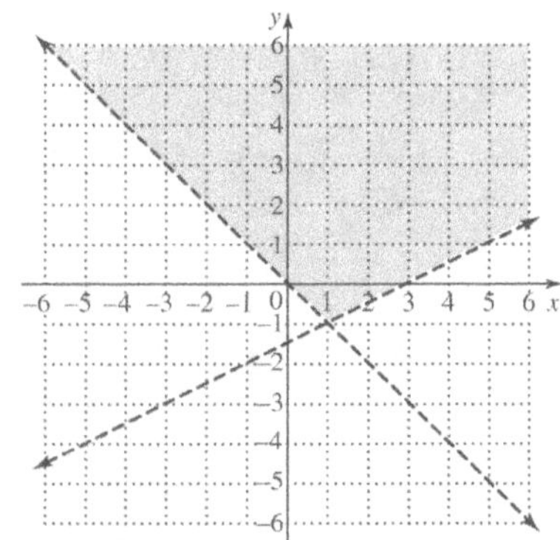

Chapter 14 Further Topics in Algebra

Key Terms

1. equation   2. elements   3. variable   4. algebraic expression

5. set   6. solution   7. constant

Objective 1
Now Try

1a. **17**   1b. **252**   2. **65**

Practice Exercises

1. 8   3. $\dfrac{28}{13}$

Objective 2

Now Try

3. $y = -\dfrac{2}{3}x + 15$

Practice Exercises

5. $y = -2x - 11$

Objective 3

Now Try

4a. $x^3 + 18x^2 + 108x + 216$ 4b. $81x^4 - 540x^3 + 1050x^2 + 1500x + 625$

Practice Exercises

7. $a^3 - 9a^2 + 27a - 27$ 9. $256s^4 + 768s^3t + 864s^2t^2 + 432st^3 + 81t^4$

